The Ecological Importance of Mixed-Severity Fires

Nature's Phoenix

Dedication

A special dedication is in order: To Ariela Fay DellaSala, Dr. DellaSala's ten-year-old daughter, who on hikes to the Grizzly Peak high-severity burn in southwest Oregon was especially excited to see the woodpeckers, butterflies, and kaleidoscope of flowering plants thriving in the snag forest. We are encouraged by the thought that if a ten-year-old can see beauty in a post-fire landscape, then maybe someday others will look twice before declaring these areas a wasteland. Also to Rachel Fazio, Dr. Hanson's wife, for her perspective, support, and patience throughout the writing of this book.

The Ecological Importance of Mixed-Severity Fires

Nature's Phoenix

Edited by

Dominick A. DellaSala
Chief Scientist, Geos Institute, Ashland, Oregon

Chad T. Hanson
Ecologist, the John Muir Project of Earth Island Institute, Berkeley, California

ELSEVIER

AMSTERDAM • BOSTON • HEIDELBERG • LONDON • NEW YORK • OXFORD
PARIS • SAN DIEGO • SAN FRANCISCO • SINGAPORE • SYDNEY • TOKYO

Elsevier
Radarweg 29, PO Box 211, 1000 AE Amsterdam, Netherlands
The Boulevard, Langford Lane, Kidlington, Oxford OX5 1GB, UK
225 Wyman Street, Waltham, MA 02451, USA

Library of Congress Cataloging-in-Publication Data
A catalog record for this book is available from the Library of Congress

British Library Cataloguing in Publication Data
A catalogue record for this book is available from the British Library

For information on all Elsevier publications
visit our website at http://store.elsevier.com/

ISBN: 978-0-12-802749-3

Contents

Section 2
Global and Regional Perspectives on Mixed- and High-Severity Fires

List of Contributors

Numbers in Parenthesis indicate the pages on which the author's contributions begin.

Ronald W. Abrams (223, 372), Dru Associates, Inc., Glen Cove, New York, USA

André Arsenault (247), Natural Resources Canada, Canadian Forest Service–Atlantic Forestry Centre, Corner Brook, NL, Canada; Thompson Rivers University, Kamloops, BC, Canada; Memorial University at Grenfell, Corner Brook, NL, Canada

William L. Baker (3, 373), Program in Ecology and Department of Geography, University of Wyoming, Laramie, WY, USA

Colden V. Baxter (118), Stream Ecology Center, Department of Biological Sciences, Idaho State University, Pocatello, ID, USA

Laurence E. Berry (210, 372), Conservation and Landscape Ecology Group, Fenner School of Environment and Society, The Australian National University, Canberra, ACT, Australia

Monica L. Bond (55, 89), Wild Nature Institute, Hanover, NH, USA

Dominick A. DellaSala (3, 23, 55, 265, 313, 372), Geos Institute, Ashland, OR, USA

Jim Furnish (313), Consulting Forester, Rockville, MD, USA

Richard W. Halsey (177, 372), Chaparral Institute, Escondido, CA, USA

Chad T. Hanson (3, 23, 265, 313, 372), John Muir Project of Earth Island Institute, Berkeley, CA, USA

Petr Heneberg (240, 372), Charles University in Prague, Prague, Czech Republic

Richard L. Hutto (3, 55, 372), Division of Biological Sciences, University of Montana, Missoula, MT, USA

Timothy Ingalsbee (348), Firefighters United for Safety, Ethics, and Ecology, Eugene, OR, USA

Breeanne K. Jackson (118), School of Environment and Natural Resources, The Ohio State University, Columbus, OH, USA

Dominik Kulakowski (149), Graduate School of Geography, Clark University, Worcester, MA, USA

David B. Lindenmayer (313), Fenner School of Environment and Society, The Australian National University, Canberra, Australia

Rachel L. Malison (118), Norwegian Institute for Nature Research, Trondheim, Norway

Stephen Mitchell (290), Duke University, Durham, NC, USA

Dennis C. Odion (372), Department of Environmental Studies, Southern Oregon University, Ashland, OR, USA

Urooj Raja (348), Department of Sociology, University of Colorado, Boulder, CO, USA

Rosemary L. Sherriff (3), Department of Geography, Humboldt State University, Arcata, CA

Holly Sitters (210, 372), Fire Ecology and Biodiversity Group, School of Ecosystem and Forest Sciences, University of Melbourne, Creswick, VIC, Australia

S. Mažeika P. Sullivan (118), School of Environment and Natural Resources, The Ohio State University, Columbus, OH, USA

Alexandra D. Syphard (177), Senior Research Ecologist, Conservation Biology Institute, Corvallis, OR, USA

Thomas T. Veblen (3, 149), Department of Geography, University of Colorado-Boulder, Boulder, CO, USA

Cathy Whitlock (265), Department of Earth Sciences and Montana Institute on Ecosystems, Montana State University, Bozeman, MT, USA

Shaye Wolf (265), Center for Biological Diversity, San Francisco, CA, USA

Biographies

EDITORS

Dominick A. DellaSala, PhD, is President and Chief Scientist of the Geos Institute (www.geosinstitute.org) in Ashland, Oregon; has served two terms as president of the Society for Conservation Biology, North America Section; and is Courtesy Professor at Oregon State University. He is an internationally renowned author of over 200 technical papers on forest and fire ecology, conservation biology, endangered species, and landscape ecology. He has received conservation leadership awards from the World Wildlife Fund (2000, 2004) and Wilburforce Foundation (2006), *Choice Publisher*'s "academic excellence" for "Temperate and Boreal Rainforests of the World: Ecology and Conservation (Island Press)," and is on the Fulbright Specialist roster as an ecologist. He has appeared on numerous nature documentaries (PBS) and national and international news syndicates, testified in support of endangered species many times in the US Congress, and has given keynote addresses at numerous conferences and international meetings, including the United Nations Earth Summit. He is motivated by his passion to leave a living planet for his daughter and all those who follow.

Chad T. Hanson, PhD, is Director and Staff Ecologist of the John Muir Project of Earth Island Institute. His PhD is in ecology from the University of California, Davis, and his research focus is on fire ecology in conifer forest ecosystems. Studies published by Dr. Hanson cover topics such as habitat selection of rare wildlife species associated with habitat created by high-severity fire; postfire conifer responses and adaptations; fire history, especially historical versus current rates of high-severity fire occurrence; and current fire patterns. Dr. Hanson lives in the San Bernardino Mountains of southern California and conducts research in conifer forests of the western United States, primarily in forests of California.

CHAPTER AUTHORS

Ronald W. Abrams, PhD, CEP, is President and Principal Ecologist of Dru Associates, Inc., in Glen Cove, New York (www.dru-associates.com), an independent ecological research firm. He has served as finance officer for the Africa Section of the Society for Conservation Biology and chair of the Ecological Footprint Committee for the Society and has been an adjunct at Long Island University. He is on the Fulbright Specialist Roster in Ecological Restoration, consulting internationally in wetland management, conservation biology, and endangered species impact assessment. Dr. Abrams' three decades of conservation consultation has been a bridge between ecological science and land use planning and development. Early in his career he served as a senior research officer at the University of Cape Town for the Southern Ocean Program, and he continues to conduct research in South Africa in the field of ecosystem services.

André Arsenault, PhD, is a forest ecologist with the Canadian Forest Service in Corner Brook, Newfoundland, and an adjunct professor in biology at Thompson Rivers University and in Environmental Science at Grenfell campus, Memorial University. His research program is focused on disturbance ecology and conservation biology and how to apply this information into planning, operations, and policy. He had the great privilege to study a wide range of forest ecosystems in British Columbia, including coastal and interior cedar-hemlock rainforests, dry forests, and high-elevation forests. André more recently expanded his program to the boreal forest of Newfoundland and Labrador. Recent accomplishments include a book on inland rainforests, a special issue of *Forest Ecology and Management* on forest biodiversity, and a number of manuscripts that challenge conventional wisdom on fire ecology and management of dry forests in western North America.

William L. Baker, PhD, is Emeritus Professor in the Program in Ecology and Department of Geography at the University of Wyoming, where he taught Fire Ecology and Landscape Ecology for 22 years and benefited from interactions with more than 30 energetic graduate students. His research interests are fire history and ecology in forests, shrublands, and grasslands of the western United States. He is the author of *Fire Ecology in Rocky Mountain Landscapes*. He has recently been studying historical vegetation using the General Land Office surveyors.

Colden V. Baxter, PhD, is Associate Professor and Director of the Stream Ecology Center, Idaho State University, Pocatello, Idaho. He grew up farming and ranching, principally in northwest Montana, and received his academic training in biology and geology (BA, University of Oregon), ecology (MS, University of Montana), fisheries science and philosophy of science (PhD, Oregon State University), and ecosystem studies (postdoctoral work, Colorado State University and Hokkaido University, Japan). His research program focuses on rivers and streams and more generally on the ecological linkages between water and land. Reciprocal connections such as those between streams, floodplains, and riparian forests couple land and water in their vulnerability to the agents of global environmental change, including changing fire regimes. Dr. Baxter's work is aimed at improving our understanding of the basic nature of such connections and contributing to better-informed conservation and stewardship.

Laurence E. Berry, Bsc(Hons), is a PhD scholar of Conservation and Landscape Ecology in the Fenner School of Environment and Society at the

Australian National University. He is part of the Australian Research Council Centre of Excellence for Environmental Decisions and the National Environmental Research Program Environmental Decision Hub. Mr. Berry is a member of the International Association of Wildland Fire and the Australian Ecological Society. His work uses a landscape approach to ecosystem disturbances, and he has published several articles on the ecology of fire refuges in the tall mountain ash forests of the Central Highlands of Victoria, Australia. He also has worked on bird responses to fire-induced fragmentation in the semiarid Mallee woodland ecosystems of South Australia and on assessing the health of River Red Gum communities in the floodplain ecosystems of the lower Murray-Darling Basin.

Monica L. Bond, MS, is a wildlife biologist and biodiversity advocate with the Wild Nature Institute (www.wildnatureinstitute.org) who has published 15 scientific papers on the ecology of wildlife in fire-affected forests. She is a graduate of Green Corps, the field school for environmental organizing, and has worked for the National Wildlife Federation and the Center for Biological Diversity, where she focused on improving management practices to conserve threatened and endangered plants and animals. Ms. Bond received the "Keeper of the Forests" award from Environment Now Foundation in 2002 in recognition of her efforts to safeguard California's national forests. She has conducted field research on a diverse range of taxa including gray-tailed voles, Western burrowing owls, spotted owls, black-backed woodpeckers, arboreal salamanders, northern elephant seals, Hawaiian monk seals, and Masai giraffe. She resides in California but travels around the world researching and advocating for imperiled wildlife and their habitats.

Jim Furnish is currently a consulting forester in the Washington, DC, area following a 34-year career with the US Department of Agriculture Forest Service. He last served as deputy chief for national forests—9% of all lands in the United States. He also served as Siuslaw National Forest Supervisor in Corvallis, Oregon, during the spotted owl crisis, reforming management from timber production to restoration principles. Mr. Furnish was also a principle Forest Service leader in creating protections for over 23 million ha of roadless areas in 2001. He has served on the board of directors for Wildlands CPR, Evangelical Environmental Network, and Geos Institute. His memoir *Toward a Natural Forest* (Oregon State University Press) speaks to forest management that works in concert with nature.

Richard W. Halsey, MA, is the director of the California Chaparral Institute, a nonprofit research and educational organization focusing on the ecology of California's chaparral ecosystem, dynamics of wildland fire, and promotion of nature education in a way that encourages communities to better connect with their surrounding natural environments. Halsey teaches natural history throughout California, works as a consultant on wildfire issues, and has published papers on chaparral ecology, fire behavior, and natural history. He taught biology for over 20 years in both public and private schools and was honored as the teacher of the year for San Diego City Schools in 1991. The second edition of his book, *Fire, Chaparral, and Survival in Southern California*," was published in 2008. Halsey also has been trained as a type II wildland firefighter with the US Forest Service.

Petr Heneberg, PhD, is a biologist at Charles University in Prague, Czech Republic. His research interests include the impact of human-induced disturbances and other changes of a cultural landscape on species and ecosystems. Dr. Heneberg's research has focused on industrial and post-industrial sites affected by mining of aggregates or other minerals and by mining processes such as spoil heaps formation, fly ash depositing, recultivation and reclamation. Among his recent projects was an analysis of the response of bees and wasps (Hymenoptera: Aculeata) to Central European pine forest fires in the České Švýcarsko National Park, Czech Republic, Central Europe.

Richard L. Hutto, PhD, is Professor Emeritus in the Division of Biological Sciences at the University of Montana (http://cas.umt.edu/dbs/labs/hutto/). Over the past 38 years, Dr. Hutto has published more than 100 papers dealing with his research on migratory landbirds in Mexico in winter, the southwest during spring and autumn, and in the Northern Rockies in summer. In 1990 he developed the US Forest Service Northern Region Landbird Monitoring Program, and since the Yellowstone fires of 1988, he has focused his research on the ecological effects of fire, as evidenced by bird distribution patterns. Dr. Hutto was host of "Birdwatch," a nationally televised PBS series that ran from 1998-2001. In 2004 Dr. Hutto established the Avian Science Center on the University of Montana campus to promote ecological awareness and informed decision making by listening to what western birds tell us about the ecological effects of human land use practices. Dr. Hutto maintains a Fire Ecology Lab Facebook page (https://www.facebook.com/FireEcologyLab) devoted to exposing the ecological importance of severe fire.

Timothy Ingalsbee, PhD, is cofounder and Executive Director of Firefighters United for Safety, Ethics, and Ecology (FUSEE) and is codirector of the Association for Fire Ecology. He is also an adjunct professor and research associate at the University of Oregon. Dr. Ingalsbee is a frequent speaker and writer on fire ecology and management issues, and he began his career as a wildland firefighter for the US Forest Service and National Park Service. In 1993 he received the Oregon Conservationist of the Year Award from the Oregon Natural Resources Council. He cofounded and directed the Cascadia Fire Ecology Education Project from 1994-2004, created and directed the Western Fire Ecology Center from 1997-2004, and is a senior wildland fire ecologist certified by the Association for Fire Ecology.

Breeanne K. Jackson, MS, is a PhD candidate at The Ohio State University in the School of Environment and Natural Resources. She has studied wildfires in the Frank Church "River of No Return" Wilderness in central Idaho and in the Sierra Nevada. Ms. Jackson is recipient of a National Science Foundation Dissertation Improvement Grant and a Joint Fire Sciences Graduate Research and Innovation Award for her original research on antecedent wildfire effects on stream-riparian food web dynamics. She has worked as a wildland firefighter and fire lookout in the central Cascades of Washington and now serves as a resource advisor during fire suppression efforts in Yosemite National Park.

Dominik Kulakowski, PhD, is Associate Professor of Geography at Clark University. His research and teaching focus on the effects of climate, disturbances, and land use on mountain forest ecosystems in the US Rocky Mountains and in Europe. His recent research themes include effects of climatic variability on forest disturbances such as fires and insect outbreaks; effects of climate on tree mortality and forest dieback; interactions and feedback among forest disturbances under climatic variability, especially how previous disturbances affect the spread and severity of subsequent disturbances and how compounding disturbances affect ecological dynamics. Dr. Kulakowski has authored over 40 peer-reviewed articles on various aspects of ecology and ecosystem management, has been widely interviewed in the popular press, and has testified before the US Congress on several occasions about integrating forest ecology and federal policy.

David Lindenmayer, PhD, is Research Professor at The Australian National University. He currently runs five large-scale, long-term research programs in southeastern Australia, primarily associated with developing ways to conserve biodiversity in reserves, national parks, wood production forests, plantations, and on farm land. His greatest achievement has been helping more than 50 students complete their postgraduate PhD or MSc degrees. He has written more than 920 scientific articles (including 535 papers in peer-reviewed international journals) and 38 books on forest ecology and management, forest and woodland biodiversity, conservation in agricultural landscapes, the ecology and management of fire, and conservation science and natural resource management. He is a member of the Australian Academy of Science and the New York Academy of Science, winner of the Eureka Prize (twice), Whitely Award (six times), the Australian Natural History Medal, and the Serventy Medal for Ornithology. Dr. Lindenmayer was awarded a prestigious 5-year Australian Research Council Laureate Fellowship in 2013 and an Order of Australia in mid-2014.

Rachel L. Malison, PhD, is a Research Ecologist at the Norwegian Institute for Nature Research in Trondheim, Norway. She received a Marie Curie International Incoming Fellowship to study the impact of Eurasian beavers on Atlantic salmon and sea trout in the Trøndelag region of Norway. Dr. Malison received her PhD from the University of Montana, studying the influence of North American beavers on Pacific salmon in an Alaskan floodplain river and her MS from Idaho State University where she studied the effects of wildfire on aquatic-terrestrial linkages in the Frank Church 'River of No Return' Wilderness in central Idaho. Her research focuses on rivers and streams and she is interested in the different physical and biological drivers that influence ecological linkages between different habitats.

Stephen R. Mitchell, PhD, is Research Scientist at the Nicholas School of the Environment at Duke University in Durham, North Carolina. Dr. Mitchell combines field-based ecological research and landscape-level ecosystem simulation modeling approaches to examine the effects of landscape-level disturbances on terrestrial ecosystems. Of particular interest to him are the impacts of ongoing climatic change on the frequency and severity of landscape-level disturbance regimes and how changing disturbance regimes may impact the structure and function of forest ecosystems. Additional areas of prior and ongoing research include an examination of the impacts of forest bioenergy production on forest carbon storage and the effectiveness of fire and fire surrogate treatments as a means to restore the flora and fauna of fire-suppressed ecosystems. Dr. Mitchell lives in Chapel Hill, North Carolina, with his wife and two children.

Dennis Odion, PhD, received his doctorate in geography from the University of California, Santa Barbara, and is a researcher at both University of California, Santa Barbara, and Southern Oregon University. His research interests include spatial patterns of fire and the ensuing effects on vegetation. He has published numerous papers on this general topic, as well as topics ranging from disturbance and species diversity, invasive species biology, ecological monitoring, and vegetation mapping. Dr. Odion has helped develop ecological monitoring programs for the National Park Service and mapped the vegetation in national parks.

Urooj Raja is a PhD candidate in Environmental Sociology at the University of Colorado, Boulder, focusing on institutional responses to natural disasters. Previously she was a Princeton University Pace Center High Meadows Fellow, serving as a research analyst and multimedia fellow at Climate Central. She also worked as a Saha fellow in Accra, Ghana, implementing clean-water treatment systems in rural villages. She worked as an adviser to the Permanent Mission from Pakistan to the United Nations and the Observer Mission from the Organization of Islamic Conference, where she advised delegations on humanitarian and environmental issues. She has previously worked as an instructor in Columbia University's Community Impact initiative, the Harlem Children's Zone, and as a community liaison for New York State Assembly Member Deborah Glick.

Rosemary L. Sherriff, PhD, is Associate Professor and Chair of Geography and faculty in the Forestry and Wildland Sciences Graduate Program at Humboldt State University. Before 2009 she was Assistant Professor at University of Kentucky (2007-2009) and University of Hawaii at Hilo (2005-2007). Her primary research has focused on forest disturbance ecology, climate-vegetation interactions, and forest stand dynamics in the western United States and in southwest Alaska. Recent research themes include stand dynamics and tree-growth response to climate, disturbance, and forest management practices in redwood, oak, and mixed-conifer forests of northwest California; climate and spruce beetle effects across white spruce ecosystems near the North American boreal-tundra margin in southwest Alaska; and mixed-severity fire regimes in montane forests of the Colorado wildland-urban interface and the broader western United States.

Holly Sitters, PhD, recently completed her doctorate investigating the effects of fire on bird diversity across a range of scales and vegetation types in southeast Australia. She is currently working as a postdoctoral research fellow with the University of Melbourne's Fire Ecology and Biodiversity Group. Her research focuses on how attributes of fire regimes influence animal diversity, and she is particularly interested in using patterns in genetic diversity to reveal the influence of past disturbances on animal movement.

Mažeika Sullivan, PhD, is an associate professor in the School of Environment and Natural Resources (senr.osu.edu) at The Ohio State University in Columbus, Ohio, where his research interests lie in aquatic and riparian ecology and conservation, especially in the ecological and biogeochemical connections between land and water. Dr. Sullivan is also the director of The Ohio State University's Wilma H. Shiermeier Olentangy River Wetland Research Park (swamp.osu.edu). He is a distinguished university teacher and has recently been selected as a Fulbright Distinguished Chair in Biodiversity and Sustainable Development. Dr. Sullivan has a deep appreciation for the complexity and beauty of the natural world, which he shares with his wife and three children.

Alexandra D. Syphard, PhD, is a senior research scientist at the Conservation Biology Institute. Her research focuses broadly on global change and its implications for biodiversity conservation. She uses a variety of spatial analytical and modeling methods to investigate how and why landscape change occurred in the past, how it is likely to occur in the future, and what types of ecological and social effects are likely to result. She also envisions how alternate management scenarios may differentially impact the biological and social integrity of different landscapes. Dr. Syphard has published on vegetation dynamics and wildfire in Mediterranean ecosystems; fire science and ecology in California; effects of multiple threats to native plant populations; biogeography and climate change; land use/land cover change; and influence of humans on fire regimes.

Thomas T. Veblen, PhD, is Professor of Geography at University of Colorado, Boulder and was Professor of Plant Ecology at Universidad Austral de Chile (1975-1979) and a research fellow with the Forest Research Institute in New Zealand (1979-1981). Dr. Veblen's research has been on how stand-scale and landscape-scale forest patterns result from interactions among natural disturbances, human activities, and recent climatic variation in Guatemala, Chile, New Zealand, Argentina, and Colorado. He has published more than 250 peer-reviewed publications, and the National Science Foundation has supported his research continuously since the mid-1980s. At the University of Colorado he has been the primary supervisor of more than 50 PhD and masters students. His honors include a Guggenheim Fellowship (1985), honorary membership in the Royal Society of New Zealand (1991), Professor of Distinction by the University of Colorado College of Arts and Sciences (2006), and Fellow of the American Association for the Advancement of Science (2008).

Cathy Whitlock, PhD, is Professor of Earth Sciences and co-director of the Montana Institute on Ecosystems. She is recognized internationally for her scholarly contributions and leadership activities in the area of long-term environmental and climate change. Dr. Whitlock studies vegetation, fire, and climate history of the western United States and other temperate regions, and she has published over 150 peer-reviewed publications on this topic. Following the 1988 Yellowstone fires, she helped develop analytical tools and modeling approaches to reconstruct past fires from lake-sediment charcoal; these methods are now the standard for fire history research around the world. Continuous grant funding from the National Science Foundation, Joint Fire Sciences Program, National Park Service, US Department of Agriculture Forest Service, Department of Energy, and other agencies has supported her research. Dr. Whitlock was elected as Fellow of the American Association for the Advancement of Science in 2012 and received the international EO Wilson Biodiversity Technology Pioneer Award in 2014.

Shaye Wolf, PhD, is the climate science director for the Center for Biological Diversity (www.biologicaldiversity.org) in San Francisco, California. Her research interests focus on examining the effects of current and projected climate change on species and ecosystems with an emphasis on identifying climate-vulnerable species. She aims to improve conservation action and management for species and ecosystems threatened by climate change through science-based mitigation and adaptation tools. Before working with the center, Dr. Wolf conducted conservation-focused field research across a variety of species in the United States and internationally. She worked as part of an interdisciplinary, cross-boundary partnership among conservation nongovernmental organizations, academic institutions, and government agencies in Mexico, the United States, and Canada to understand climate change effects on population dynamics using tools in population modeling, remote sensing, and climate modeling.

Preface

Dominick A. DellaSala and Chad T. Hanson

1 MYTHOLOGY MEETS FIRE ECOLOGY

If you are a curious reader with a knack for the analytical, you may be asking yourself, Why start a book about fire ecology with a mythological figure? And if you are a tried-and-true scientist, like we are, you may also be asking, Isn't it a bit risky to mix myth with science, fact with fiction, observation with mystique, nature with reincarnation?

But the mythological phoenix is exactly the right place to begin an ecological story of mixed- and high-severity fires. We open with ancient phoenix mythology stories as told by Greeks, Egyptians, Romans, Christians, Native Americans, Chinese, and, yes, even in *Harry Potter and the Order of the Phoenix*, where fire metaphorically engulfs the old in order to give birth to the new. To anyone that has marveled at the remarkable restorative powers of a postfire landscape, this is exactly what unfolds when a forest, woodland, or shrubland is burned by intense fire. So why not mix in a little myth before we give you a heavy dose of reality?

Greek mythology states that the phoenix (or phenix; φοῖνιξ) lives for 500 years, builds a nest when it barely has any life force left, and calls upon the Sun to ignite the nest it is perched on, which then consumes the bird in flames, only for the phoenix to be reborn from its own ashes in brilliance. In fire's essence, the phoenix symbolizes a desire for immortality, resurrection, life everlasting—ah, to be young again—along with the qualities of inspiration, beauty, and self-awareness; rise up and be reborn! (see http://www.phoenixarises.com/phoenix/legends/legends.htm; accessed January 23, 2015).

In severely burned areas there is also an ecological resurrection of sorts; the newborn fire-dependent biota emerges from the ancestral "corpses" (ashes, snags) of its fire-killed parents. But this is not a story about death or loss; rather, in the aftermath of fire-kill is the essence of nature and its fire-mediated, life-giving force. Hence, severe fire, much like the phoenix, is the ultimate change agent. Whether in mythology or in ecology, death is but a new beginning! Life and death are joined in the relentless march of time spanning ecosystems and human cultures.

The Phoenix (Greek and Roman ancient mythology)

"He knows his time is out! And doth provide
New principles of life; herbs he brings dried
From the hot hills, and with rich species frames
A Pile shall burn, and Hatch him with his flames
On this the weakling sits; salutes the Sun
With the pleasant noise, and prays and begs for some
Of his own fire, that quickly may restore
The youth and vigor, which he had before
Whom soon as Phoebus spies, stopping his rays
He makes a stand, and thus allays his pains
He shakes his locks, and from his golden head
Shoots on bright beam, which smites with vital fire
The willing bird; to burn is his desire
That we may live again; he's proud in death
And goes in haste to gain a better breath
The spice heap fired with celestial rays
Doth burn the aged Phoenix, when straight stays
The Chariot of the amazed Moon; the pole
Resists the wheeling, swift Orbs, and the whole
Fabric of Nature at a stand remains
Till the old bird anew, young begins again."

http://www.phoenixarises.com/phoenix/legends/greek.htm; accessed January 23, 2015.

The phoenix myth is similar to the poetic writings of naturalist John Muir, who also was a fire enthusiast:

By forces seemingly antagonistic and destructive Nature accomplishes her beneficent designs–now a flood of fire, now a flood of ice, now a flood of water; and again in the fullness of time an outburst of organic life…

"Mt. Shasta" in *Picturesque California* (1888-1890)

One is constantly reminded of the infinite lavishness and fertility of Nature—inexhaustible abundance amid what seems enormous waste. And yet when we look into any of her operations that lie within reach of our minds, we learn that no particle of her material is wasted or worn out. It is eternally flowing from use to use, beauty to yet higher beauty; and we soon cease to lament waste and death, and rather rejoice and exult in the imperishable, unspendable wealth of the universe, and faithfully watch and wait the reappearance of everything that melts and fades and dies about us, feeling sure that its next appearance will be better and more beautiful than the last.

"My First Summer in the Sierra" (1911)

Thus we begin this book with fire as nature's phoenix by calling attention to the ecological importance of mixed- and high-severity fires precisely because most researchers have minimized the breadth of these fires, land managers would rather suppress them, and the public fears them. While this book is about mixed- and high-severity fires, which are the most misunderstood components of mixed-severity fire regimes, for simplicity we use the term "higher-severity fire" as short-hand for both severities.

2 FLIGHT OF THE PHOENIX

As a global change agent, fire has been around since the dawn of terrestrial plants some 400 million years ago. It acts in concert with climate by triggering dramatic changes in ecosystems that reverberate across landscapes over long time lines. Tree fire scars, charcoal sediments long buried in lake deposits, and the fossil record provide clues of fire's indelible imprint.

Nearly every terrestrial biome and continent (except for polar regions) is influenced by fire to some degree; in some years, fires have been active across a significant proportion of the Earth's terrestrial surface (Figure 1; also see http://earthobservatory.nasa.gov/GlobalMaps/view.php?d1=MOD14A1_M_FIRE).

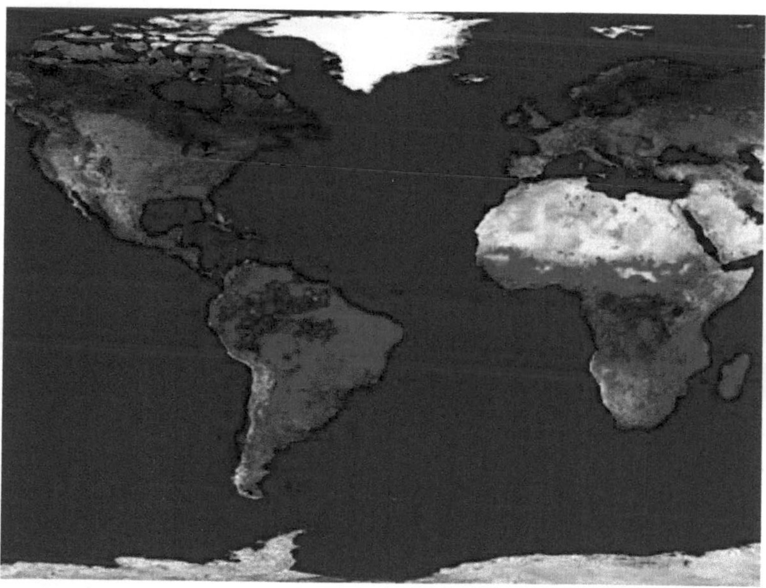

FIGURE 1 A MODIS Rapid Response System Global Fire Map showing active fire season over a 10-day period in November 2008. Locations of fires were detected by MODIS on board the Terra and Aqua satellites. Each colored dot is a location where at least one fire was detected during the 10-day period. Red dots are where the fire count was low, yellow are where the number of fires was large (http://modis.gsfc.nasa.gov/gallery/individual.php?db_date=2008-11-29; accessed January 12, 2015)

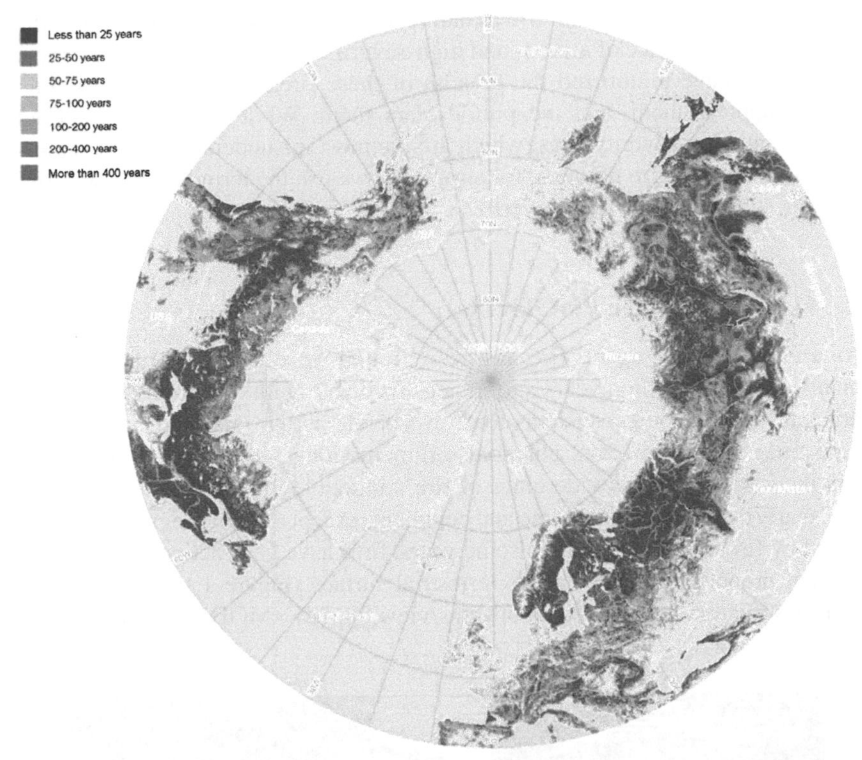

FIGURE 2 Fire-return intervals at the global scale. *(Map provided by Alexey Yaroshenko, Greenpeace International.)*

But fire is hardly uniform in its occurrence, intensity, or spatial patterns (Figure 2), as discussed throughout this book.

Interestingly, fire can be considered a self-directed force of nature, but humans have a questionable record in dealing with it: We try to subjugate (suppression) or dampen and homogenize it (most prescribed wildland fires), reduce its intensity via manipulation or removal of vegetation (fuels management), unknowingly build structures in its path, and willingly log the biologically rich postfire landscape, which results in many undesirable and unintended consequences. We are also changing the global climate in a way that could result in too much fire in some ecosystems and too little in others. For the large fires burning in fire-climate years (e.g., droughts), attempting to stop them over vast areas has proven risky to firefighters, costly, largely ineffective, and ecologically misinformed.

3 COMMAND AND CONTROL OF NATURE'S PHOENIX

Over thousands of years, aboriginal peoples developed a deep understanding of fire as a management tool, even though the extent to which human influence on natural fire events remains largely contested in ecology and anthropology circles. Nonetheless, at times, Native peoples purposefully set fire to prepare conditions for culturally important fire-loving plants and big game. Unfortunately, with the onset of the industrial revolution and domination of most of the Earth's biomes by humanity, fire phobia became engrained in our reptilian brain stem, underscored by shock-and-awe news accounts of flames racing toward towns each fire season. Consequently, we spend billions of dollars every year attempting to suppress fires all over the world, sending thousands of firefighters into hinterlands with military strike force precision. This seldom tames any of the large fires. In short, the primordial links between humans and fire were broken in the increasingly mechanized nineteenth and early twentieth centuries, as respect for fire shifted to fear and misunderstanding, leading to industrialized fire suppression.

Command-and-control attitudes toward fire are now pervasive worldwide, and such thinking has harmed fire-dependent plants and wildlife over vast landscapes. The popular press and even some fire ecologists sensationalize fire as a "destructive" and "catastrophic" event that "damages" verdant forest, terraforming living landscapes into "moonscapes." Others argue that forests cannot recover on their own without logging and artificial tree planting. Some well-intentioned conservationists who value green forests over blackened (instead of both) would much rather see an intensively thinned and "fire-proofed" landscape manicured to resist higher-severity fire rather than a mosaic of biologically rich postfire landscapes produced by higher-severity fire.

In the United States the fire-phobic media (with some notable exceptions) has shaped public opinion so much that when managers need to allow fires to take their course in wildlands for ecological reasons there is too often little public support (Kauffman 2004). For instance, attitudes about fires in the United States have certainly been molded to some extent by Smokey Bear, a fictitious character created in 1944 to symbolize fire suppression and whose mantra has been "Remember, ONLY YOU can prevent forest fires." Smokey was immensely popular with the public and led to important awareness about proper campfire etiquette, but it also unfortunately became the rallying call of the US Forest Service and its quest to put out every fire ignition by 10:00 AM the next day. In the era of Smokey Bear, the Forest Service would transform itself into the de facto "Fire Service" as the agency's fire suppression budgets skyrocketed while large fires marched on in years with a favorable climate, regardless of suppression forces. Firefighting became a political necessity rather than an ecological need, with angry members of Congress and the public breathing down the necks of a Forest Service caught in the middle of those calling for even more fire prevention versus those calling for use of fire as a management objective.

4 JUST WHAT ARE MIXED- AND HIGH-SEVERITY FIRES AND WHAT GOVERNS THEM?

The authors of chapters in this book prepared their work for readers who are either generally familiar with fire ecology or are coming up to speed. Some of the terms they use include "fire regime," the average spatial pattern, frequency, and intensity of a long series of fire events; "fire occurrence," defined in terms of a fire interval (period between fires) or fire rotation (amount of time required for an area equal to an area of interest to burn); and "fire intensity," the energy released by a fire (Keely, 2009).

Fire severity is generally defined as the change in above- and below-ground organic matter that elicits an ecosystem response, usually evaluated in terms of vegetation mortality. Severity is generally classified into three categories: (1) low severity, in which most plants survive the fire; (2) moderate severity, in which about one-third to three-fourths of overstory trees or dominant vegetation (if shrublands) is killed by fire; and (3) high severity, in which most (>75%) of the overstory trees or dominant vegetation is killed by fire (also known as "stand replacement" or "crown fire") (Keely 2009).

A mixed-severity fire is unique in that it creates a landscape mosaic of varying patch severities, including small to intermediate-size patches of lightly or unburned areas, intermediate-size patches of moderately burned areas, and small and large but often less abundant patches of intensely burned areas (Figure 3). Notably, fire scientists tend to quantify burn areas by first mapping the fire perimeter, defined as the outer edge or boundary of a particular burn, which often includes unburned patches as part of the fire boundary. Thus, given the heterogeneous nature of varying severities and patch sizes, mixed-severity fires are the most biologically complex and ecologically interesting of the fire regimes and are generally associated with high levels of biodiversity (DellaSala et al. 2014; see also Chapters 2 through 6). As fire guru Jim Agee (2005) aptly notes, mixed-severity fires are not simply intermediate states between low- and high-severity fires; rather, they are a unique disturbance dynamic that warrants careful study by ecologists.

Notably, fire severity is influenced by many factors, including slope (steep vs. gentle), aspect (cardinal direction of the slope), topographic position (low vs. higher elevations), fuels (vegetation), and weather (especially drought and high winds). In a mixed-severity fire, burn severity patterns are loosely influenced by the unequal effect of fuels and topography and are more directly governed by weather. For instance, in some regions low-severity patches within mixed-severity burns often occur along low-slope positions and on the northeast aspects (generally wetter or cooler areas), whereas high-severity patches tend to occur relatively more frequently on southwest and upper slopes and along ridgelines (generally hotter and more exposed areas; Agee 2005). This variability in fire severity patch patterns makes reconstructing fire histories for a particular area difficult; the patch mosaic shifts across the landscape from one fire to the next (dynamic) because of differences in weather and local fuel conditions (topographical influences remain constant). Notably, high-severity (crown) fires are frequently weather-driven events, with fuel being of secondary importance.

FIGURE 3 Mixed-severity fire patches from three areas: (a) the Biscuit fire of 2002, in southwest Oregon (photo by D. DellaSala, July 2012); (b) the Rim fire of 2013, in Stanislaus National Forest, California (photo by C. Hanson, May 2014); (c) and the King fire of 2014, in Eldorado National Forest, California (photo by C. Hanson, October 2014). Photos show unburned (U), light-severity (L), moderate-severity (M), and high-severity (H) patches based on cursory photo interpretation of tree canopy mortality.

5 THE MISUNDERSTOOD PHOENIX

Contrary to what many think, decades of fire suppression (mostly small fires are suppressed; large ones are nearly impossible to extinguish) may have contributed to a substantial *deficit* of higher-severity fire in places such as the montane forests of the western United States (Stephens et al. 2007, Marlon et al. 2012, Odion et al. 2014). The big fires ("mega-fires"; Chapter 2), which occur infrequently but affect proportionately large areas when they do occur, have gotten almost all of the attention from the press and politicians who have responded with unprecedented postfire logging proposals.

Despite the fire deficit in places, nature's fire balance sheet (fire hectares) is about to begin closing (e.g., perhaps the boreal and some dry fire regions) as global climate change primes the fire pump for more active fire seasons (Littell et al. 2009, Pechony and Shindell 2010; see Chapter 9). However, as climate shifts, some montane forests could actually see reductions in fire extent as a result of increases in precipitation and vegetation changes driven by climate change (Krawchuk et al. 2009). With increasing fires in some dry regions, concerns about greenhouse gas emissions produced by fire have been raised (Chapter 10). However, fire is a natural and necessary ecological process. The best way to address greenhouse gas emissions is to reduce our dependency on fossil fuels, given that fire emissions in living, growing, fire-adapted ecosystems cannot be equated with emissions from industrial smokestacks and automobile tailpipes; ecosystems adapted to mixed-severity fire do not merely emit some carbon during the fire, they also rapidly begin to uptake and store carbon after fire (Chapters 10). Moreover, with new science some surprises may emerge, such as recent research indicating that large high-severity fires can actually have a global cooling effect by increasing reflectance of solar radiation back into space ("albedo") due to increased snow and ice cover (Rogers et al. 2015).

Notably, given all the attention about fires (perhaps thousands of stories each fire season in the global popular press) and the growing potential that more of them will occur as a result of a changing climate in some areas, understanding the ecological benefits of fire—and not just its destructive effects on the human-built environment—is imperative. Ecologically speaking, fire is an essential natural force greatly underappreciated for its role as one of nature's principal architects. In nature there are short-term winners and losers in any natural change event, and fire provides no exception. Some species thrive (pyrogenic or pyrophilous) in the immediate postfire environment, whereas others move on (fire avoiders) (Taylor et al. 2012), but almost all species benefit at some point in postfire succession (see the data on birds in Chapter 3). Thus fire is nature's way of perpetuating what has been described as "pyrodiversity" (Moritz et al. 2014). By pyrodiversity we mean the spatial and temporal variability in fire regimes that result in postfire vegetation mosaics (Figure 3) indicative of the kinds of fires discussed throughout this book.

This book also uniquely demonstrates how plants and wildlife have evolved not only to cope with fires but also to thrive in the rich postfire environment. The geographic scope is mainly the western United States and Canada, but we include other ecosystems as often as possible to show broad patterns in fire behavior, the ecological effects of fires, and the conservation importance of postfire landscapes over vast geographies, including sub-Saharan Africa, southeastern Australia, portions of Europe, and areas where very large fires have influenced vegetation and wildlife dynamics (Chapter 8). We do not restrict our treatise solely to fires in montane forests because shrublands with a Mediterranean-like climate, such as California chaparral and South African fynbos, also exhibit codependence on fire's beneficial effects when the natural fire regime has not been altered so that too much fire occurs (Chapter 7).

In sum, while big, intense fires may initially seem damaging to the untrained eye, they soon result in the fire phoenix, triggered initially by an explosion of germinating seed banks long dormant before the first flames but that soon colonize in fire's aftermath, and the natural and beneficial changes in soil chemistry that prepare the way for new growth. Fire's perpetual occurrence means plants and wildlife have had a lot of time to coexist with it as a persistent agent of change and are quite resilient after it; many species actually depend on intense fire to germinate successfully (e.g., Odion et al. 2010; see Chapter 7). Some tree species are even labeled as mixed-severity fire indicators because they require severely burned patches in the midst of moderately burned areas to recruit successfully (Marcoux et al. 2015). Other species, like certain conifers, have produced unique adaptations to resist fires of moderate to high intensity, such as thick, fire-resistant tree bark acquired over eons of natural selection. Witness giant sequoia: Even when young it is perhaps the most fire-resistant conifer in the world, capable of surviving intense fire that scorches almost the entire tree crown. Numerous birds, small mammals, big game, invertebrates, and aquatic species prosper in postfire landscapes because the renewed plant growth brings a pulse of biological activity (Chapters 3–6). The ensuing biological legacies include fire-killed trees that then provide nesting and foraging sites, hiding cover, and "nurse logs" for seedlings. Resprouting native flowering shrubs also provide habitat for birds and small mammals that nest and den in montane chaparral, and for flying insects that are attracted to the flowers of the native shrubs and that, in turn, provide food for bats and fly-catching birds. Woody debris also assists in the development of stream channel morphology and hiding cover for aquatic organisms, providing food for native fish. The ecology of these process-driven interactions is what this book is also about.

6 WHAT WE COVER IN THIS BOOK

No book can cover all of the topics related to fire ecology (there are thousands of articles and books about different aspects). This one is unique in providing the first comprehensive global fire ecology reference on the benefits of mixed- and high-severity fires produced by large fire events. It is organized in 3 sections and 13 chapters. By fire ecology we mean the interdependence between fire and fire-adapted plants and animals, along with the factors that govern fire behavior, the role of fire in ecological and evolutionary processes (much like that of large floods, windstorms, and other large natural change events), and the effects of fire on past, present, and future ecosystem dynamics. Both Sections 1 and 2 describe the postfire environment, where nature resets the successional clock, bridging past with future events in a pulse of postfire productivity that is long lasting. The global Section 2 demonstrates the prevalence and broad ecological importance of large fire events.

The last, Section 3, includes chapters on the ecological use of fires over large landscapes. Fire-phobic and ecologically inappropriate land management policies have disrupted ancestral bonds with fire by suppressing fire before (via inappropriate landscape-scale thinning in mixed-severity systems), during (aerial retardants and ground-based suppression forces), and after (postfire logging, artificial planting and seeding, and eradication of native shrubs through mechanical means and herbicides) a fire event. Notably, postfire logging and related land uses are especially disruptive to the fire phoenix compared with the ecologically beneficial role that fire itself plays. It is our hope that, with a better understanding of the ecology of large mixed- and high-severity fires, a deeper appreciation for fire's restorative powers will emerge as a foundation for new policies that recognize fire's management importance.

Authors of chapters in this book have contributed their unique perspectives on the ecological importance of mixed- and high-severity fire based on an extensive research pedigree acquired on the ground from many places around the world. When putting the chapters together, we recognized that this book would be controversial with some, given that the information presented cuts against the grain of prevailing fire management policies and public attitudes, including policies that have resulted in billions of dollars of fire suppression and many pre- and postfire management actions that can do more ecological harm than good in these fire-dependent systems. During largely the past decade, however, the scientific literature has been making an increasingly powerful case for the importance of mixed- and high-severity fires. Barriers to ecologically based fire management are mostly a result of economic drivers associated with a rush to log postfire landscapes by devaluing them in an "all-bets-are-off" suspension of environmental laws and policies in order to facilitate efforts of land management agencies, such as the US Forest Service, to "get the cut out" (Chapter 11). Thus, some may view this book as fulfilling a timely need for spotlighting ecological shortcomings of the dominant postfire management responses, in particular our views on the ecological importance of megafires as a necessary natural process. Either way, we are okay with the criticism because scientific controversy is what often transforms status quo understandings about nature.

This book is also bound to get the juices flowing whether you are inquisitive about the ecological importance of large mixed- and high-severity fires or disagree with us about their ecological importance. We refute the dominant fire paradigm that these fires are ecologically destructive. We agree with fellow author Richard Hutto, who stated that "recovery" applies just as well to a green-tree forest that burns as it does to a charred forest that grows into a green forest, although we celebrate the importance of both as interconnected stages in the ongoing cycle of life and death and as agents of natural heterogeneity in fire-dynamic ecosystems.

7 NEVER JUDGE A POSTFIRE LANDSCAPE BY THE INITIATING DISTURBANCE EVENT

We also prepared this book because most people view a burned area as a single event in time, that is, right after a fire has "destroyed" their favorite hiking spot or forest. Few take the time to actually go back into the burn area 1, 3, or even 10 years later to witness the remarkable change and rejuvenation that has taken place when no postfire logging or other vegetation manipulation has occurred. Consequently, the story of fire as a restorative agent has yet to attract sufficient media coverage to present a compelling perspective that, over time, moves public attitudes. There is a need to infuse a bit of ecology into the discussion, especially where homes or lives are not at risk.

In 2003 editor Dominick DellaSala took a film crew from CNN into the Quartz Creek fire area of southwest Oregon, one of the few times the media were actually willing to explore a large burn area after the flames had subsided. This was repeated in 2012, 10 years after the historic Biscuit fire, also in southwest Oregon. In both cases members of the press were surprised by nature's resilience. Even in the most intensely burned patches life was flourishing shortly after the fire and continued to prosper a decade later, with the exception of areas that were logged after the fire. Similarly, Richard Hutto's field trip into two intensely burned forests in Montana, which was conducted in association with the Large Wildland Fire Conference of 2014, is available via a video documentary provided by the Northern Rockies Fire Science Network (http://nrfirescience.org/event/fire-effects-field-trip-dick-hutto).

The media's hesitancy to cover the ecology of postfire landscapes is nothing new. For instance, in 1988 news crews raced to cover the Yellowstone fires, reporting that fires had "destroyed" American's iconic park. There were calls to remove the park superintendent and even the Director of the National Park Service; clearly, someone was at fault! A year later scientists were busy counting the proliferation of lodgepole pine *(Pinus contorta)* seedlings carpeting the ground, while elk *(Cervus canadensis)* foraged on an explosion of flowering plants and previously rare bird species began to colonize the fire area. No film crews were present in Yellowstone that year, but the Park Service now gives routine fire ecology tours to photo-happy tourists because of the tapestry of wildflowers, naturally regenerating forests with abundant snags, and the prolific wildlife habitat created by the fire—especially in the areas that burned hottest in 1988. Fire ecology tutorials like these are vital in changing public attitudes by showing there is beauty and ecological value in forests that may seem lifeless to many shortly after fire, but which are actually representative of ecological renewal the moment the fire occurs and for decades thereafter. That beauty, like anything else, if properly appreciated, might lead to a sense of understanding, an intrinsic value of fire as a self-directed force, and an appreciation of postfire landscapes as ecologically valuable.

Unfortunately, we have a long way to go for a general ecological appreciation of postfire landscapes because postfire management most often sets up a destructive feedback loop; after a large fire burns beneficially through a landscape, including large patches of high-severity fire, it is often quickly logged. The most ecologically valuable burn areas—also economically valuable because of their large legacy trees—are then transformed into homogenized landscapes, with fine fuels left behind by logging (slash), converted to combustible tree farms that burn intensely and homogeneously in the next fire event (Ironically, such postfire management is typically promoted to the public as fire/fuel reduction). A portion of today's large fires are often the result of back-burns that, in turn, set the stage for high-severity fire patches that are then logged after the fire. This destructive management feedback loop has occurred many times, and its prevalence is increasing as fire begins to return to its natural range of variation in places influenced by climate change. For instance, in the western United States, the US Forest Service now routinely proposes mega-logging projects following very large fires only to scale logging proposals back somewhat in final planning to seem "reasonable." Land managers often tell the public that they are only logging a minor portion of an overall fire area but fail to candidly disclose that the logging would remove much or most of some of the rarest, most biodiverse, and most threatened habitat: complex early seral forest created by higher-severity fire and containing dense patches of medium-sized and large dead trees and abundant native shrubs (DellaSala et al. 2014). What was protected as a green forest in places is then put on the chopping block and devalued as "destroyed" by fire and in need of massive postfire logging and replanting as "restoration." This is generally the case in North American and Australian fire-dependent forests, but a deeper understanding of fire's beneficial role in shaping grassland and woodland systems of sub-Saharan Africa provides lessons learned about coexistence with fire (Chapter 8).

8 OUT WITH SMOKEY, IN WITH THE PHOENIX

In this book we take a long view of fires by examining them through the rearview mirror, using back-casting techniques that reconstruct the past fire environment and its ecosystem effects (e.g., Chapters 1, 2, and 9). By "long timelines" we mean 300-1000 years or more because fire cannot be fully understood by what happened last year or even in the past few decades or centuries. Examining long timelines is especially important for uncovering the cyclical nature of fire and its link to global climatic processes now poised to shift fire's behavior from fuel-limited to climate-limited fire systems (Willis et al. 2007, Littell et al. 2009). In ecology, knowledge of the past is important for understanding present conservation and restoration needs (Willis et al. 2007) and for determining whether a current or future event is characteristic. With multiple or continuous timelines, rather than a single point in time, scientists and managers can reconstruct proper historical envelopes, understand the natural fire regimes which native plants and animals evolved, properly gauge future projections, and then begin to purposely direct fire to perform its ecological functions while reducing the risks to human communities (Chapters 9, 12, 13).

We also provide the basis for a new and much needed communications framework (Chapter 13) so that we can eventually embrace fire's myriad ecosystem benefits while reducing risks to people through effective land- and fire-zoning approaches. That ecologists and conservation groups, as well as the media, develop more objective language when it comes to describing fire effects is especially crucial. We suggest that they abandon terms such as "catastrophic," "destroyed," "damaged," or "consumed" and replace them with "restored," "rejuvenated," and "recovered." In the United States, it is time that managers and ecologists replace Smokey Bear with nature's phoenix to begin educating the public about the benefits of fire when safely burning in the backcountry without suppression.

Importantly, there is no ecological justification for "salvage" logging, and the term itself needs to be replaced with what it truly is—postfire logging—as nothing ecologically is being "salvaged," or recovered, in the aftermath of fire. Rather, complex early seral forests (DellaSala et al. 2014), rich in ecological legacy structures and fire-dependent species, are damaged or eliminated for interests in "salvaging" economic value from postfire landscapes that are viewed as commodities. A change in terminology is the first step in translating fire science to the public, decision makers, and the press. Thus, we urge those in the profession of fire ecology, and especially conservationists, to lead by example and start the discussion around a less-loaded lexicon of coexistence and fire appreciation (Moritz et al. 2014).

There also needs to be substantial investments not only in the science of fire ecology but also in communications aimed at replacing Smokey Bear's fire-phobic messaging with ways to coexist in "firesheds," a term we use to describe the zone where homes are built in a watershed with similar fire risks (Chapter 13). As you will see throughout this book, we present compelling evidence of fire's ecological role that is likely to challenge many land managers, decision makers, and even some scientists and conservation groups. But if we are going to coexist with fire, an inevitable force of nature, then opening up a new dialogue that recognizes its rarity, beauty, and magic (in the words of Richard Hutto) as nature's phoenix is an important first step.

Wildfires will continue to shape both dry and moist regions through this millennium and the next, and we need to fight them when they are near homes. But instead of blaming environmental restrictions for the inability of land managers to "tame forest fires," as many politicians and land managers mistakenly believe, we should be working together, as many communities and conservation groups already are, to reduce fuels closest to where people live. We also need to encourage land managers to stop converting fire-resilient native forests to flammable tree plantations and, where plantations do exist, to responsibly treat slash piles to avoid homogeneously severe fires postfire logging that also contributes to future fire risks.

Finally, firefighters put their lives on the line every fire season to protect people and homes. To reduce the risks they encounter, we must all help create fire-safe communities and, through an appreciation of the ecological benefits of mixed- and high-severity fires, discourage backcountry firefighting that further threatens fire-dependent wildlife and puts firefighters at unnecessary risk. Thus, in this book we discuss ways for communities to live safely with fire by reducing homeowner risks while providing for ecological management of fire in wildlands. With fire safety properly addressed (Chapters 12 and 13), we encourage readers to take a hard look at what is actually happening to forests and shrublands after a fire. Go out and see the effects of megafires for yourself and come back to the area over time to witness the progression of fire's phoenix. See what these areas look like years after the fires are out; we hope you will agree that they are not wastelands by any means. In sum, fire is to fire-dependent ecosystems as rain is to rainforests—inseparable and necessary.

Below are listed some informative videos (accessed January 17, 2015) that complement this book and serve as an introduction to fire's ecological benefits for inquiring, open minds:

- International Wildlife Film Festival Award-winning video "Disturbance": (http://vimeo.com/8627070)
- Photos capturing the ecological magic of complex early seral forest created by intense fire: https://vimeo.com/75533376 and https://www.youtube.com/watch?v=4rXvP2r9W9c
- Fire video from PBS, "Exploring with Dick Hutto: A Burned Forest": http://www.youtube.com/watch?v=iTl-naywNyY&list=PL7F70F134E853F520&index=15
- "Forests Born of Fire" by Wild Nature Institute: http://www.youtube.com/watch?v=1BmTq8vGAVo&feature=youtu.be
- "Blacked-backed woodpeckers and fire": http://www.fs.usda.gov/detail/r5/news-events/audiovisual/?cid=stelprdb5431394

- A video of Chad Hanson's 2014 presentation on the restorative virtues of the Rim fire and the ecological value of complex early seral forest: https://vimeo.com/95535429.
- Richard Hutto's fire field trip: http://nrfirescience.org/event/fire-effects-field-trip-dick-hutto
- Contrast unlogged areas shown in the photos in this book with areas logged after fire, as shown in this slideshow: http://www.geosinstitute.org/images/stories/pdfs/Publications/Fire/SalvageLoggingBiscuitFire.pdf

REFERENCES

Agee, J.K., 2005. The complex nature of mixed severity fire regimes. In: Taylor, L., Zelnik, J., Cadwallader, S., Hughes, B. (Eds.), Mixed Severity Fire Regimes: Ecology and Management Symposium Proceedings. Washington State University, Pullman, Wash, Spokane, Wash., 17–19 November 2004. Association of Fire Ecology MISCO3.

DellaSala, D.A., Bond, M.L., Hanson, C.T., Hutto, R.L., Odion, D.C., 2014. Complex early seral forests of the Sierra Nevada: what are they and how can they be managed for ecological integrity? Nat. Areas J. 34, 310–324.

Kauffman, J.B., 2004. Death rides the forest: perceptions of fire, land use, and ecological restoration of western forests. Conserv. Biol. 18, 878–882.

Keely, J.E., 2009. Fire intensity, fire severity and burn severity: a brief review and suggested usage. Int. J. Wildland Fire 18, 116–126.

Krawchuk, M.A., Moritz, M.A., Parisien, M., Van Dorn, J., Hayhoe, K., 2009. Global pyrogeography: the current and future distribution of wildfire. PLoS One 4, e5102.

Littell, J.S., McKenzie, D., Peterson, D.L., Westerling, A.L., 2009. Climate and wildfire area burned in western U.S. ecoprovinces, 1916–2003. Ecol. Appl. 19, 1003–1021.

Marcoux, H.M., Daniels, L.D., Gergel, S.E., Da Silva, E., Gedalof, Z., Hessburg, P.F., 2015. Differentiating mixed- and high-severity fire regimes in mixed-conifer forests of the Canadian Cordillera. For. Ecol. Manage. http://dx.doi.org/10.1016/j.foreco.2014.12.027. (in press).

Marlon, J.R., Bartlein, P.J., Gavin, D.G., Long, C.J., Anderson, R.S., Brilese, C.E., Brown, K.J., Hallett, D.J., Power, M.J., Scharf, E.A., Walsh, M.K., 2012. Long-term perspective on wildfires in the western USA. PNAS www.pnas.org/lookup/suppl/.

Moritz, M.A., Batllori, E., Bradstock, R.A., Gill, A.M., Handmer, J., Hessburg, P.F., Leonard, J., McCaffrey, S., Odion, D.C., Schoennagel, T., Syphard, A.D., 2014. Learning to coexist with fire. Nature 515, 58–66. http://dx.doi.org/10.1038/nature13946.

Odion, D.C., Hanson, C.T., Arsenault, A., Baker, W.L., DellaSala, D.A., Hutto, R.L., Klenner, W., Moritz, M.A., Sherriff, R.L., Veblen, T.T., Williams, M.A., 2014. Examining historical and current mixed-severity fire regimes in ponderosa pine and mixed-conifer forests of western North America. PLoS One 9, 1–14.

Odion, D.C., Moritz, M.A., DellaSala, D.A., 2010. Alternative community states maintained by fire in the Klamath Mountains, USA. J. Ecol. 98, 96–105.

Pechony, O., Shindell, D.T., 2010. Driving forces of global wildfires over the past millennium and the forthcoming century. PNAS 107 (45), 19167–19170. http://dx.doi.org/10.1073/pnas.1003669107.

Rogers, B.M., Soja, A.J., Goulden, M.L., Randerson, J.T., 2015. Influence of tree species on continental differences in boreal fires and climate feedbacks. Nat. Geosci. (in press).

Stephens, S.L., Martin, R.E., Clinton, N.E., 2007. Prehistoric fire area and emissions from California's forests, woodlands, shrublands, and grasslands. For. Ecol. Manag. 251, 205–216.
Taylor, R.S., Watson, S.J., Nimmo, D.G., Kelly, L.T., Bennett, A.F., Clarke, M.F., 2012. Landscape-scale effects of fire on bird assemblages: does pyrodiversity beget biodiversity? Divers. Distrib. 18, 519–529.
Willis, K.J., Araújo, M.B., Bennett, K.D., Figueroa-Rangel, B., Froyd, C.A., Myers, N., 2007. How can knowledge of the past help conserve the future: biodiversity conservation and the relevance of long-term ecological studies. Phil. Trans. R. Soc. Biol. Sci. 362 (1478), 175–187.

Acknowledgments

We thank all of the chapter authors, photographers, and those people who provided us with data and figures for their detective-like exploration of the ecology of higher-severity post-fire landscapes. It is through their help that we were able to explain why post-fire landscapes are ecologically diverse but most often underappreciated. We especially thank Richard Hutto for his inspiring perspectives on fire and insightful reviews of many of the chapters of this book, and Doug Bevington and Amanda Stanley at the Environment Now and Wilburforce Foundations, respectively, for being leaders in funding fire ecology research and conservation science in western North America.

We also sincerely thank esteemed fire ecologist, Jon Keeley, of the University of California at Los Angeles and the U.S. Geological Survey, for his important input and contributions to this book. R. LeBlanc, C. Power, P. Baines, and A. Arsenault also provided helpful chapter edits and reviews.

Section 1

Biodiversity of Mixed- and High-Severity Fires

Chapter 1

Setting the Stage for Mixed- and High-Severity Fire

Chad T. Hanson[1], Rosemary L. Sherriff[2], Richard L. Hutto[3], Dominick A. DellaSala[4], Thomas T. Veblen[5] and William L. Baker[6]

[1]John Muir Project of Earth Island Institute, Berkeley, CA, USA, [2]Department of Geography, Humboldt State University, Arcata, CA, [3]Division of Biological Sciences, University of Montana, Missoula, MT, USA, [4]Geos Institute, Ashland, OR, USA, [5]Department of Geography, University of Colorado-Boulder, Boulder, CO, USA, [6]Program in Ecology and Department of Geography, University of Wyoming, Laramie, WY, USA

1.1 EARLIER HYPOTHESES AND CURRENT RESEARCH

In the late 19th century and early 20th century, fire—especially patches of high severity wherein most or all of the dominant vegetation is killed—was generally considered to be a categorically destructive force. Clements (1936) hypothesized that the mature/old state of vegetation would result in a stable "climax" condition and described natural disturbance forces such as fire as a threat to this state, characterizing mature forest that experienced high-severity fire as a "disclimax" state. One early report opined that there is no excuse or justification for allowing fires to continue to occur at all in chaparral and forest ecosystems (Kinney, 1900). After a series of large fires in North America in 1910, land managers established a policy goal of the complete elimination of fire from all North American forests (a "one size fits all" policy) through unsuccessful attempts to achieve 100% fire suppression (Pyne, 1982; Egan, 2010). Through the mid-20th century, and in recent decades, views have shifted to broadly acknowledge the importance of low- and low/moderate-severity fire. In this chapter we focus on drier montane forests of western North America as a case study of how diverse, competing, and rather complex sets of evidence are converging on a new story that embraces not just low-severity fire but also mixed- and high-severity fire in these ecosystems. Thus this chapter exemplifies how mixed- and high-severity fire is being better understood and appreciated as scientific evidence accumulates.

A commonly articulated hypothesis is that dry forests at low elevations in western North America were historically open and park-like, and heavily dominated by low-severity and low/moderate-severity fire (Weaver, 1943; Cooper,

1962; Covington, 2000; Agee and Skinner, 2005; Stephens and Ruth, 2005). Under this hypothesis, high-severity fire patches were rare, or at least were believed to be small to moderate in size, and larger patches (generally hundreds of hectares or larger) that burn today often are considered to be unnatural and ecologically harmful. While this model fits reasonably well in some low-elevation, xeric forest systems (Perry et al., 2011; Williams and Baker, 2012a, 2013), it has been extrapolated far beyond where it seems to apply best. That higher fire severities occurred historically, albeit at a wide variety of spatial and temporal scales, in most or all fire-dependent vegetation types of western North America is becoming increasingly clear (Veblen and Lorenz, 1986; Mast et al., 1998; Taylor and Skinner, 1998; Brown et al., 1999; Kaufmann et al., 2000; Heyerdahl et al., 2001, 2012; Wright and Agee, 2004; Sherriff and Veblen, 2006, 2007; Baker et al., 2007; Hessburg et al., 2007; Klenner et al., 2008; Amoroso et al., 2011; Perry et al., 2011; Schoennagel et al., 2011; Williams and Baker, 2012a; Marcoux et al., 2013; Odion et al., 2014; Hanson and Odion, 2015a).

A key extension of the concept of historical forests characterized by open structure coupled with a low- or low/moderate-severity fire regime is that current areas of dense forest structure—and larger, higher-severity fire patches in such areas—are the result of unnatural fuel accumulation from decades of fire suppression policies, leading to higher-severity fire effects outside the natural range of variability. The most fire-suppressed forests (i.e., those that have gone without fire for periods that exceed their "average" natural fire cycles) are, therefore, expected to experience unnaturally high proportions of higher-severity fire if they burn (Covington and Moore, 1994; Covington, 2000; Agee, 2002; Agee and Skinner, 2005; Stephens and Ruth, 2005; Roos and Swetnam, 2012; Williams, 2012; Stephens et al., 2013; Steel et al., 2015).

We recognize that the historical low-severity fire regime described above has not been applied to all forest types in western North America (e.g., Romme and Despain, 1989; Agee, 1993). The idea has, however, been widely applied in principle to most forest types, and widespread acceptance of the low- and low/moderate-severity fire regime has been the primary basis driving fire management policy in an overwhelmingly large proportion of montane forests in the western United States. Thus many management plans explicitly adopt a low-severity fire regime model without rigorously examining evidence of its applicability to the management of the ecosystem type under consideration. A key research need has been to determine the particular ecosystem types to which the low-severity fire regime applies. Scientists recently rigorously investigated the hypothesis that forests are burning in a largely unnatural fashion and found that historical forest structure and fire regimes were far more variable than previously believed, and that ecosystem responses to large, intense fires often differ from past assumptions (Figure 1.1; see also Chapters 2–5). We discuss these notions in greater depth throughout this book.

FIGURE 1.1 Natural regeneration of native vegetation—including conifers, deciduous trees, and shrubs—in large high-severity fire patches. Top: Star Fire of 2001 (photo by Chad Hanson, 2013); bottom: Storrie Fire of 2000 (photo by Chad Hanson, 2007) (see also Chapter 2).

Do Open and Park-Like Structures Provide an Accurate Historical Baseline for Dry Forest Types in Western US Forests?

Using spatially extensive tree ring field data, historical landscape photographs from the late 19th and early 20th centuries, early aerial photography from the 1930s through 1950s, and direct records from late 19th-century land surveyors, numerous recent studies have been able to reconstruct the historical structure of conifer forests in the western United States. A portion of the historical montane forest landscape in any given region undoubtedly comprised open forest dominated by low-severity fire (e.g., Brown et al., 1999, Fulé et al. 2009, Iniguez et al., 2009; Perry et al., 2011; Williams and Baker, 2012a; Hagmann et al., 2013; Baker, 2014), and some forest types (e.g., ponderosa pine [*Pinus ponderosa*]) often had a preponderance of low-severity fire in many low-elevation or xeric-type forest environments throughout western North America. Nevertheless, landscape-level evidence indicates that vast forested areas also comprised moderate to very dense forests characterized by a mixed-severity fire regime, wherein higher-severity fire patches of varying sizes occurred in a mosaic of low- and moderate-severity fire effects (Veblen and Lorenz, 1986, 1991; Baker et al., 2007; Sherriff and Veblen, 2007; Hessburg et al., 2007; Perry et al., 2011; Baker, 2012; Williams and Baker, 2012a,b; Baker, 2014; Baker and Williams, 2015; Hanson and Odion, 2015a). In general, in historical ponderosa pine and mixed-conifer forests of the western United States, local variability was substantial (Brown et al., 1999, Fulé et al. 2009, Iniguez et al., 2009; Hessburg et al., 2007; Perry et al., 2011; Baker, 2012; Williams and Baker, 2012a,b, 2013; Baker, 2014; Baker and Williams, 2015; Hanson and Odion, 2015a). In sum, these and other studies indicate that historically there was high variability in fire effects (low to high severity) and composition and structure at both small and large spatial scales, and these patterns varied greatly depending on the regional and biophysical setting.

Does Time Since Fire Influence Fire Severity?

The predominant view in North American fire science has been that as woody ecosystems age, they steadily increase in their potential for higher-severity fire. Thus in the fire exclusion/fuels buildup model applied to relatively dry conifer forests and woodlands (e.g., Covington and Moore, 1994), long fire-free intervals caused by effective fire suppression result in fuel accumulation and changes in fuel arrangements (e.g., vertical fuel continuity) that lead to increased fire severity. Likewise, even for forest ecosystems known to burn primarily in severe stand-replacing fires, many classical models of fire potential (in this case the instantaneous chance of fire occurrence) assume that fire severity increases with time since the last fire as a result of fuel load accumulation (Johnson and Gutsell, 1994); some research supports this (Steel et al., 2015). Nevertheless, empirical and modeling studies have demonstrated that in many ecosystem types, including temperate forests, flammability is still relatively stable with regard to time since fire (Kitzberger et al., 2012; Perry et al., 2012; Paritsis et al., 2014). We suggest that the predominance of the viewpoint in the western United States that flammability and potential fire severity inexorably increase with time since fire has been an important contributor to the expectation that 20th-century fire suppression—if assumed to have effectively reduced fire frequency—should result in increased, and unnaturally high, fire severity in the modern landscape. This relationship does not seem to hold in various ecosystem types and regions for a wide variety of reasons (Veblen, 2003; Odion et al., 2004; Baker, 2009; van Wagtendonk et al., 2012). Even but even if it held everywhere, this preoccupation with changes from historical

proportions of higher-severity fire skirts the key management questions of whether a change in the proportion of higher-severity patches renders a forest incapable of "recovery" after such fires (precious few papers deal with this key management question) or whether the overall spatiotemporal extent of higher-severity fire (i.e., rotation intervals) exceeds historical levels.

A second assumption about fire regimes in the western United States is implied by language commonly used to describe modern fire regimes in terms of "missed fire cycles." While fire cycle may be a useful descriptor of fire regimes, the assertion that a particular place or patch has missed one or more fire cycles implies a regularity to fire return intervals that is not supported by most studies of fire history; there is always variation around a mean. Even in dry forests characterized by relatively frequent fires, the historical fire frequency is typically characterized by such a high degree of variance that descriptors such as means or cycles are misleading. Using the term *missed fire cycle* in mixed-severity fire regimes, among which the frequency and severity of fires are inherently diverse, is particularly problematic. Usage of *missed fire cycles* connotes a consistency and degree of equilibrium in the historical fire regime that is not supported by actual fire history evidence, which shows large variations in fire intervals (e.g., Baker and Ehle, 2001; Baker, 2012). Though it seems to make intuitive sense that, with increasing time since fire, fuels would accumulate to create a higher probability of higher-severity fire effects, numerous countervailing factors modulate fire severity as stands mature since the previous fire.

Notably, many studies of this issue have found that, in some areas, the most long-unburned forests are burning mostly at low/moderate severity and are not experiencing higher levels of high-severity fire than forests that have experienced less fire exclusion (Odion et al., 2004, 2010; Odion and Hanson, 2006, 2008; Miller et al., 2012; van Wagtendonk et al., 2012). Further, forests with the largest amounts of surface fuels (based on prefire measurements) and small trees do not necessarily always experience more severe fire (Azuma et al., 2004). Debate about this issue remains, however. For example, Steel et al. (2015, Table 7 in particular) modeled time since fire and fire severity in California's forests and predicted that, in mixed-conifer forests, high-severity fire would range from 12% 10 years after fire to 20% 75 years after fire, though the modeling for mixed-conifer forests seems to have been based on what appears to be very limited data for forests that experienced fire less than 75 years earlier (Steel et al., Figure 4), weakening inferences about a time since fire/severity relationship. Regardless, the high-severity fire values reported by Steel et al.—even for forests that had not previously burned for 75-100 years—remain well within the range of natural variation of high-severity fire proportions in these forests found by most recent studies (Beaty and Taylor, 2001; Bekker and Taylor, 2001; Baker, 2014; Odion et al., 2014; Hanson and Odion, 2015a).

Although the notion that fire severity would not necessarily increase with time since fire is seemingly counterintuitive, a number of factors help explain it. For example, as forests mature with increasing time since the last fire, canopy cover increases, creating more cooling shade, facilitating moister surface conditions, and slowing wind speeds and thus rates of fire spread. Also, increasing shade in the forest understory can cause a reduction in sun-dependent shrubs and understory trees, making it more difficult to initiate or sustain crown fire (Odion et al., 2004, 2010; Odion and Hanson, 2006). Much more important, however, is that severe fire events are largely driven by weather (Finney et al., 2003) and often have relatively little to do with the amount of fuel available (Azuma et al., 2004).

An analogous assumption about the role of fuels was previously made regarding chaparral, one of the most fire-dependent plant communities in the world; that is, historically, there was less fuel and more moderate fire effects. This idea is also inconsistent with the scientific evidence (Keeley and Zedler, 2009); see also Chapter 7.

What is the Evidence for Mixed- and High-Severity Fire?

In recent decades a growing number of studies has investigated historical fire regimes using a variety of methods to determine the extent and frequency of mixed- and high-severity fire, particularly in the ponderosa pine and mixed-conifer forests of western North America (Table 1.1). Regardless of the method used, most landscape-level studies of dry forest types, for example, tend to find evidence for mixed-severity fire regimes that included low-, moderate-, and high-severity fire (both small and large patches) in most forest types and regional areas across the western United States, with few exceptions (Odion et al., 2014). Here we describe some of the more common methods that researchers have used to determine historical fire regimes, mostly in western North America.

Aerial Photos

Many researchers have used early aerial photos of montane forests to determine the historical occurrence of high-severity fire. Specifically, researchers have used such photos to determine (1) the number of emergent trees that survived previous high-severity fire (Beaty and Taylor, 2001, 2008; Bekker and Taylor, 2001, 2010); (2) broad stand-structure categories consistent with past low-, moderate-, and high-severity fire (Hessburg et al., 2007); and (3) levels of forest canopy mortality consistent with low-, moderate-, and high-severity fire. Such studies concluded that mixed- and high-severity fire effects were generally dominant in both lower- and middle-montane forests, including mixed-conifer forests, as well as upper montane forests. Comparisons of modern and historical aerial photographs have revealed important variability in forest changes along environmental gradients in areas experiencing similar land use and fire exclusion histories. For example, in the Colorado Front Range, comparison of aerial photographs showed no significant increase in tree densities in the upper montane zone of ponderosa pine and mixed-conifer forests from 1938 to 1999 (Platt and Schoennagel, 2009) in an area with mixed- and high-severity

TABLE 1.1 Summary of Historical Higher-Severity Fire Proportions Found in Various Reconstruction Study Areas at Least 1000 ha Within Mixed-Conifer and Ponderosa Pine Forests of Western North America

Region	Study	Study Area (ha)	Higher-Severity Fire (%)
Baja California	Minnich et al. (2000)	~75,000	16
Sierra Nevada	Baker (2014)	330,000	31-39
	Hanson and Odion, 2015a	65,296	26
Klamath	Taylor and Skinner (1998)	1570	12-31
Southern Cascades	Beaty and Taylor (2001)	1587	18-70
	Bekker and Taylor (2001)	2042	52-63
	Baker (2012)	400,000	26
Northern/Central Cascades	Hessburg et al. (2007)	303,156	37
Blue Mountains (Oregon)	Williams and Baker (2012a)	301,709	17
Front Range (Colorado)	Williams and Baker (2012a)	65,525	65
	Sherriff et al. (2014, Figure 6)	564,413	~72[a]
Southwestern United States (Arizona)	Williams and Baker (2012a)	556,294	15-55
British Columbia, Canada	Heyerdahl et al. (2012)	1105	10

[a]*Includes mixed- and high-severity fire.*

fire (Schoennagel et al., 2011), whereas low-elevation areas near the ecotone with the plains grasslands exhibited a moderate degree of invasion of grasslands by trees during the same period (Mast et al., 1997) in an area known to have had predominantly lower-severity fire (Veblen et al., 2000; Sherriff and Veblen, 2007; Sherriff et al., 2014).

Historical Reports

Scientists have reviewed evidence in early historical US government forest reports, finding widespread occurrence of small and large higher-severity fire patches in all forest types, including ponderosa pine and mixed-conifer forests

(Shinneman and Baker, 1997; Baker et al., 2007; Williams and Baker, 2012a [Appendix S1], 2014; Baker, 2014; Hanson and Odion, 2015a). Evidence in these reports includes detailed descriptions of low-, mixed-, and high-severity fires; maps of slightly to severely burned forests; estimates of total area burned at mixed and high severity; and photographs of the landscapes after these fires.

Direct Records and Reconstructions from Early Land Surveys

Field data from unlogged forests collected by the US General Land Office in the 19th century before fire suppression has been extensively analyzed across large landscapes, and historical stand structure has been correlated to fire severities that facilitated or stimulated those forest structures. Based on these analyses, substantial areas of ponderosa pine and mixed-conifer forests across the western United States were dominated by a mixed-severity fire regime that includes evidence of high-severity fire (Baker, 2012, 2014; Williams and Baker, 2012a,b, 2013) typically intermixed with areas of predominantly low/moderate-severity fire. Importantly, note that nearly all tree ring reconstructions that found open, park-like historical forests in some areas have been supported by these land survey reconstructions for those same areas, but the land surveys show definitively that these park-like forests grew only in portions of most dry forests in the western United States. Historical mixed- and high-severity fires shown by the land surveys led to diverse landscapes at scales of a few townships (e.g., 25,000 ha) within each region. These landscapes contained intermixed patches of open forests, dense forests, complex early seral forests, old-growth forests, dense shrub fields, and large patches of snag habitat important to wildlife. This landscape diversity was missed by tree ring reconstructions because using tree ring methods without abundant extant large, old trees is difficult, and thus more heavily burned historical forests were avoided or missed (Baker and Ehle, 2001; Williams and Baker, 2012a).

The land survey records also show (Baker and Williams, 2015) that historical dry forests were numerically dominated by small trees (e.g., <40 cm in diameter) but also included abundant large trees, which together provided "bet-hedging" resilience against a variety of forest disturbances that produce high levels of tree mortality (e.g., insect outbreaks, severe droughts, mixed- and high-severity fires). Large surviving trees are particularly important after severe fires, but smaller trees can differentially survive insect outbreaks and droughts (Baker and Williams, 2015).

Though some (e.g., Fule et al., 2014) have recently questioned some findings of Williams and Baker (2012a), an in-depth analysis of the critique found that most of its points were founded on mistakes, misunderstandings, and omission and misuse of evidence by critics (Williams and Baker, 2014). The accuracy of land survey methods has undergone extensive checking and cross-checking, and the findings are strongly corroborated by other published sources, including historical US government fire-severity mapping and reports (Williams and Baker, 2010, 2011, 2012a, 2014; Baker, 2012, 2014). These checks show that land survey reconstructions can achieve accuracies almost as high as those from tree ring reconstructions but can do so across very large land areas (e.g., ≥400,000 ha).

Tree Ring Reconstructions of Stand Densities and Fire History

Many scientists have used stand-age data from unlogged forests, often in combination with fire-scar dating of past fires, to reconstruct historical fire regimes and changes in the rate of new stand initiation from mixed- to high-severity fire. In mixed-conifer and ponderosa pine forests of western North America, researchers have found regional stand-age distributions consistent with a mixed-severity fire regime that maintained a mix of age classes and successional stages (e.g., Taylor and Skinner, 1998; Heyerdahl et al., 2012; Odion et al., 2014). Reconstructions of stand structures and fire history are most effective when supported by diverse evidence, gathered independently, that converges to the same overall interpretations. For example, in the Colorado Front Range, tree ring evidence, historical landscape photographs, and General Land Office surveys converge to the same conclusions demonstrating that the historical (i.e., before 1920) fire regime of ponderosa pine and mixed-conifer forests included low-severity fires (i.e., not lethal to large, fire-resistant trees) as well as high-severity fires (i.e., killing >70% of canopy trees) (Veblen and Lorenz, 1986, 1991; Mast et al., 1998; Schoennagel et al., 2011; Williams and Baker, 2012b). The conclusion that most of the montane zone forests dominated by ponderosa pine in the Front Range were characterized by a mixed-severity fire regime is further supported by independently conducted studies by noncollaborating researchers based on tree ring evidence of past fires and their ecological effects (Brown et al., 1999; Kaufmann et al., 2000; Huckaby et al., 2001).

Clear delineation of the spatial extent of past fire regimes is a major concern of ecosystem managers in the context of ecological restoration and management of wildfire. Fire histories and stand structures reconstructed using tree ring data are most useful for guiding management decisions where data sets are sufficiently robust to produce high-resolution spatial layers to compare historical and modern landscape conditions. As an example, in the Colorado Front Range a data set consisting of 7680 tree cores and 1262 fire-scarred tree samples collected at 232 field sites allowed for a spatially explicit comparison of historical fire severity (before fire exclusion in 1920) with observed modern fire severity and modeled potential fire behavior across 564,413 ha of montane forests (Sherriff et al., 2014). Forest structure and tree ring fire history were used to characterize fire severity at the 232 sites. Then, historical fire severity was spatially modeled across the entire study area using biophysical variables that had successfully predicted (retrodicted) fire severity at the 232 sampled sites.

Only 16% of the study area recorded a shift from historical low-severity fire to a higher potential for crown fire today. A historical fire regime of more frequent, low-severity fires at elevations below 2260 m is consistent with the view among land managers that these forests be thinned both to restore historical structure and to reduce fuels in this area of widespread exurban development. By contrast, at higher elevations in the upper montane zone (i.e., 2260-3000 m), mixed-severity fires were predominant historically and continue to be so today. Thus thinning treatments at higher elevations of the montane zone are inappropriate if the management goal is ecological restoration. Comparison of the severity of nine large fires that occurred between 2000 and 2012 with the severity of fires before the 20th century revealed no significant increase in fire severity from the historical to the modern period except for a few fires that occurred within the lowest elevations (16%) of the montane study area (Sherriff et al., 2014). This spatially extensive tree ring–based reconstruction is strongly corroborated by land survey records of higher-severity fire patches across the same area (Williams and Baker, 2012b).

Charcoal and Sediment Reconstructions

Paleoecologists have explored fire-induced sediment layers in alluvial fans (e.g., Pierce et al., 2004) and charcoal sediments (e.g., Whitlock et al., 2008; Colombaroli and Gavin, 2010; Jenkins et al., 2011; Marlon et al., 2012) to reconstruct historical fire occurrence. They found numerous periods of large and severe fire activity over the past several centuries and millennia in North American mixed-conifer and ponderosa pine forests (see Chapter 9 for many additional citations). Thus paleoecological methods and evidence further corroborate findings based upon other methods, discussed above, regarding historical mixed- and high-severity fire in these forests.

1.2 ECOSYSTEM RESILIENCE AND MIXED- AND HIGH-SEVERITY FIRE

Along with the surge in scientific investigation into historical fire regimes over the past 10-15 years has come enhanced understanding of the naturalness and ecological importance of mixed- and high-severity fire in many forest and shrub ecosystems. Contrary to the historical assumption that higher-severity fire is inherently unnatural and ecologically damaging, mounting evidence suggests otherwise. Ecologists now conclude that in vegetation types with mixed- and high-severity fire regimes, fire-mediated age-class diversity is essential to the full complement of native biodiversity and fosters ecological resilience and integrity in montane forests of North America (Hutto, 1995, 2008; Swanson et al., 2011; Bond et al., 2012; Williams and Baker, 2012a; DellaSala et al., 2014). Ecological resilience is essentially the opposite of "engineering resilience," which pertains to the suppression of natural disturbance to achieve stasis and control of resources

(Thompson et al., 2009). Ecological resilience is the ability to ultimately return to predisturbance vegetation types after a natural disturbance, including higher-severity fire. This sort of dynamic equilibrium, where a varied spectrum of succession stages is present across the larger landscape, tends to maintain the full complement of native biodiversity on the landscape (Thompson et al., 2009). Forests that are purported to be burning at unprecedented levels of high-severity fire are generally responding well in terms of the forest succession process and native biodiversity (see Chapters 2–5), so the widespread fear of too much severe fire seems to be unfounded in the vast majority of cases (see, e.g., Kotliar et al., 2002; Bond et al., 2009; Donato et al., 2009; Burnett et al., 2010; Malison and Baxter, 2010; Williams and Baker, 2012a, 2013; Buchalski et al., 2013; Baker, 2014; Odion et al., 2014; Sherriff et al., 2014; Hanson and Odion, 2015a). We acknowledge that more research is needed for some forest regions, such as some areas of the southwestern United States experiencing increasing fire severity (Dillon et al., 2011), to determine the effects of climate change on forest resilience.

As discussed above, in mixed-severity fire regimes, higher-severity fire occurs as patches in a mosaic of fire effects (Williams and Baker, 2012a; Baker, 2014). In conifer forests of North America, higher-severity fire patches create a habitat type, known as complex early seral forest (DellaSala et al., 2014), that supports levels of native biodiversity, species richness, and wildlife abundance that are generally comparable to, or even higher than, those in unburned old forest (Raphael et al., 1987; Hutto, 1995; Schieck and Song, 2006; Haney et al., 2008; Donato et al., 2009; Burnett et al., 2010; Malison and Baxter, 2010; Sestrich et al., 2011; Swanson et al., 2011; DellaSala et al., 2014). Many rare, imperiled, and declining wildlife species depend on this habitat (Hutto, 1995, 2008; Kotliar et al., 2002; Conway and Kirkpatrick, 2007; Hanson and North, 2008; Bond et al., 2009; Buchalski et al., 2013; Hanson, 2013, 2014; Rota, 2013; Siegel et al., 2013; DellaSala et al., 2014; Baker, 2015; see also Chapters 2–6). The scientific literature reveals the naturalness and ecological importance of multiple age classes and successional stages following higher-severity fire, as well as the common and typical occurrence of natural forest regeneration after such fire (Shatford et al., 2007; Donato et al., 2009; Crotteau et al., 2013; Cocking et al., 2014; Odion et al., 2014). These and other studies suggest that mixed-severity fire, including higher-severity fire patches, is part of the intrinsic ecology of these forests and has been shaping fire-dependent biodiversity and diverse landscapes for millennia (Figure 1.2).

1.3 MIXED- AND HIGH-SEVERITY FIRES HAVE NOT INCREASED IN FREQUENCY AS ASSUMED

Fire history studies show that for many montane forests, including mixed-conifer and ponderosa pine forests, fire frequencies in most forested regions were substantially less during the 20th century (and the early 21st century) compared with the previous few centuries (e.g., Odion et al., 2014). Nonetheless,

FIGURE 1.2 Though high-severity fire patches in montane forests may initially seem to be relatively lifeless landscapes, within the first weeks and months after fire, by the first spring after fire, and for many springs thereafter, native shrubs, conifers, and deciduous trees naturally regenerate, creating an ecologically rich habitat for numerous wildlife species. (a) Star Fire of 2001, Eldorado National Forest, Sierra Nevada *(photo by Chad Hanson, 2013)*. (b) McNally Fire of 2002, Sequoia National Forest, Sierra Nevada *(photo by Chad Hanson, 2014)*.

factors responsible for this decline in fire vary from region to region and include fire suppression, changes in forest structure as a result of timber harvesting, removal of fine fuels by livestock grazing, and climate change. The result is that all fire types, including high-severity fire, have been reduced substantially in broad regions since the early 20th century (Veblen et al., 2000, Odion and Hanson, 2013; Odion et al., 2014; Hanson and Odion, 2015a). Nevertheless, some forest types or local areas within regions may have more high-severity fire than they did historically (e.g., some low-elevation or other particular environments of xeric montane forests; Perry et al., 2011; Sherriff et al., 2014).

While some chaparral/shrub ecosystems (and some forests) are in close proximity to large human populations and associated unplanned human-caused ignitions resulting in an excess of fire relative to historical rates (see Chapter 7), these are the exception, not the rule—at least for conifer forests with mixed-severity fire regimes.

Recent climate-induced increases in fire frequency (Kasischke and Turetsky, 2006; Westerling et al., 2006; Dennison et al., 2014) have led to increases in total area burned in most regions of western North America, but most areas have not experienced trends in high-severity fire (for example, see Hanson et al., 2009; Dillon et al., 2011; Miller et al., 2012; Hanson and Odion, 2014; Hanson and Odion, 2015b; also see Chapter 9), though some have. For example, while the severity of fires (the proportion of high-severity fire effects) is not increasing in forests in the southwestern United States, the overall high-severity fire area has increased in recent decades (Dillon et al., 2011). In the southern Rocky Mountains, both high-severity fire area and proportion have increased in recent decades (Dillon et al., 2011).

Not only is the habitat created by higher-severity fire biodiverse and—in many forest regions—rare compared with historical conditions, it also is often severely threatened by the inertia of historical misconceptions about the effects of high-severity fire and the responses of ecosystems and biodiversity to such fire (Bond et al., 2012; DellaSala et al., 2014; Hanson, 2014; also see Chapter 11 and 13). This results in forest management policies that continue to focus on aggressive fire suppression, postfire logging, postfire shrub eradication and plantation establishment, homogenous low-severity prescribed burning designed to prevent mixed- and high-severity fire, and prefire mechanical thinning operations implemented across landscapes to further curb complex early seral forest habitat (Lindenmayer et al., 2004; Hutto, 2006; Hanson and North, 2008; Bond et al., 2009; DellaSala et al., 2014; Hanson, 2014).

1.4 CONCLUSIONS

Historical forest structure and fire regimes in mixed-conifer and ponderosa pine forests of western North America were far more variable than current management regimes assume, and mixed- and high-severity fires are a natural and ecologically beneficial part of many forests and shrublands. Yet the unique and ecologically rich habitat created by such fire remains demonized and, in nearly all places, is a habitat threatened by fire suppression, postfire logging (Chapter 11), and prefire management designed to reduce further the creation of postfire habitat. Ecologists are increasingly urging a shift in policies that would allow more mixed- and high-severity fire in the wildlands away from homes, while focusing on fuel reduction and fire suppression activities adjacent to homes to provide for public safety (Gibbons et al., 2012; Calkin et al., 2014; Moritz et al., 2014; see also Chapter 13). A paradigm shift in land management policies is needed to restore mixed-severity fire by allowing wildland fires to burn safely in the backcountry while protecting postfire habitat from the ecologically damaging practices of postfire logging, shrub removal, and artificial plantation establishment (Lindenmayer et al., 2004; Bond et al., 2012; DellaSala et al., 2014; Hanson, 2014).

REFERENCES

Agee, J.K., 1993. Fire ecology of Pacific Northwest forests. Island Press, Washington, DC.

Agee, J.K., 2002. The fallacy of passive management: managing for fire safe forest reserves. Conserv. Practice 3, 18–26.

Agee, J.K., Skinner, C.N., 2005. Basic principles of forest fuel reduction treatments. For. Ecol. Manage. 211, 83–96.

Amoroso, M.M., Daniels, L.D., Bataineh, M., Andison, D.W., 2011. Evidence of mixed-severity fires in the foothills of the Rocky Mountains of west-central Alberta, Canada. For. Ecol. Manage. 262, 2240–2249.

Azuma, D.L., Donnegan, J., Gedney, D., 2004. Southwest Oregon Biscuit fire: An analysis of forest resources and fire severity. U.S. Forest Service Research Paper PNW-RP-560, Pacific Northwest Research Station, Portland, OR, USA.

Baker, W.L., 2009. Fire Ecology in Rocky Mountain Landscapes. Island Press, Washington DC.

Baker, W.L., 2012. Implications of spatially extensive historical data from surveys for restoring dry forests of Oregon's eastern Cascades. Ecosphere 3, 23.

Baker, W.L., 2014. Historical forest structure and fire in Sierran mixed-conifer forests reconstructed from General Land Office survey data. Ecosphere 5, 79.

Baker, W.L., 2015. Historical Northern spotted owl habitat and old-growth dry forests maintained by mixed-severity wildfires. Landsc. Ecol. 30, 655–666.

Baker, W.L., Ehle, D., 2001. Uncertainty in surface-fire history: the case of ponderosa pine forests in the western United States. Can. J. Forest Res. 31, 1205–1226.

Baker, W.L., Veblen, T.T., Sherriff, R.L., 2007. Fire, fuels and restoration of ponderosa pine-Douglas fir forests in the Rocky Mountains, USA. J. Biogeogr. 34, 251–269.

Baker, W.L., Williams, M.A., 2015. Bet-hedging dry-forest resilience to climate-change threats in the western USA based on historical forest structure. Front. Ecol. Evol. (in press).

Beaty, R.M., Taylor, A.H., 2001. Spatial and temporal variation of fire regimes in a mixed conifer forest landscape, Southern Cascades, USA. J. Biogeogr. 28, 955–966.

Beaty, R.M., Taylor, A.H., 2008. Fire history and the structure and dynamics of a mixed-conifer forest landscape in the northern Sierra Nevada, Lake Tahoe Basin, California, USA. For. Ecol. Manage. 255, 707–719.

Bekker, M.F., Taylor, A.H., 2001. Gradient analysis of fire regimes in montane forests of the southern Cascade Range, Thousand Lakes Wilderness, California, USA. Plant Ecol. 155, 15–28.

Bekker, M.F., Taylor, A.H., 2010. Fire disturbance, forest structure, and stand dynamics in montane forest of the southern Cascades, Thousand Lakes Wilderness, California, USA. Ecoscience 17, 59–72.

Bond, M.L., Lee, D.E., Siegel, R.B., Ward Jr., J.P., 2009. Habitat use and selection by California Spotted Owls in a postfire landscape. J. Wildlife Manage 73, 1116–1124.

Bond, M.L., Siegel, R.B., Hutto, R.L., Saab, V.A., Shunk, S.A., 2012. A new forest fire paradigm: the need for high-severity fires. Wildlife Prof. 6, 46–49.

Brown, P., Kaufmann, M., Shepperd, W., 1999. Long-term, landscape patterns of past fire events in a montane ponderosa pine forest of central Colorado. Landsc. Ecol. 14, 513–532.

Buchalski, M.R., Fontaine, J.B., Heady III, P.A., Hayes, J.P., Frick, W.F., 2013. Bat response to differing fire severity in mixed-conifer forest, California, USA. PLoS One 8, e57884.

Burnett, R.D., Taillie, P., Seavy, N., 2010. Plumas Lassen Study 2009 Annual Report. U.S. Forest Service, Pacific Southwest Region, Vallejo, CA.

Calkin, D.E., Cohen, J.D., Finney, M.A., Thompson, M.P., 2014. How risk management can prevent future wildfire disasters in the wildland-urban interface. Proc. Natl. Acad. Sci. U. S. A. 111, 746–751.

Clements, F.E., 1936. Nature and structure of the climax. J. Ecol. 24, 252–284.

Cocking, M.I., Varner, J.M., Knapp, E.E., 2014. Long-term effects of fire severity on oak-conifer dynamics in the southern Cascades. Ecol. Appl. 24, 94–107.

Colombaroli, D., Gavin, D.G., 2010. Highly episodic fire and erosion regime over the past 2,000 y in the Siskiyou Mountains, Oregon. Proc. Natl. Acad. Sci. 107, 18909–18915.

Conway, C.J., Kirkpatrick, C., 2007. Effect of forest fire suppression on buff-breasted flycatchers. J. Wildlife Manage. 71, 445–457.

Cooper, C.F., 1962. Pattern in ponderosa pine forests. Ecology 42, 493–499.

Covington, W.W., 2000. Helping western forests heal: the prognosis is poor for U.S. forest ecosystems. Nature 408, 135–136.

Covington, W.W., Moore, M.M., 1994. Southwestern ponderosa forest structure: changes since Euro-American settlement. J. Forestry 92, 39–47.

Crotteau, J.S., Varner III, J.M., Ritchie, M.W., 2013. Post-fire regeneration across a fire severity gradient in the southern Cascades. For. Ecol. Manage. 287, 103–112.

DellaSala, D.A., Bond, M.L., Hanson, C.T., Hutto, R.L., Odion, D.C., 2014. Complex early seral forests of the Sierra Nevada: what are they and how can they be managed for ecological integrity? Nat. Areas J. 34, 310–324.

Dennison, P.E., Brewer, S.C., Arnold, J.D., Moritz, M.A., 2014. Large wildfire trends in the western United States, 1984–2011. Geophys. Res. Lett. 41, 2928–2933.

Dillon, G.K., Holden, Z.A., Morgan, P., Crimmins, M.A., Heyerdahl, E.K., Luce, C.H., 2011. Both topography and climate affected forest and woodland burn severity in two regions of the western US. 1984 to 2006. Ecosphere 2, 130.

Donato, D.C., Fontaine, J.B., Robinson, W.D., Kauffman, J.B., Law, B.E., 2009. Vegetation response to a short interval between high-severity wildfires in a mixed-evergreen forest. J. Ecol. 97, 142–154.

Egan, T., 2010. The Big Burn: Teddy Roosevelt and the Fire that Saved America. Houghton Mifflin Harcourt, New York.

Finney, M.A., Bartlette, R., Bradshaw, L., Close, K., Collins, B.M., Gleason, P., Min Hao, W., Langowski, P., McGinely, J., McHugh, C.W., Martinson, E., Omi, P.N., Shepperd, W., Zeller, K., 2003. Fire behavior, fuels treatments, and fire suppression on the Hayman Fire. U.S. Forest Service Gen. Tech. Rpt. RMRS-GTR-114, Rocky Mountain Research Station, Missoula, Montana, USA.

Fule, P.Z., Swetnam, T.W., Brown, P.M., Falk, D.A., Peterson, D.L., Allen, C.D., Aplet, G.H., Battaglia, M.A., Binkley, D., Farris, C., Keane, R.E., Margolis, E.Q., Grissino-Mayer, H., Miller, C., Hull Sieg, C., Skinner, C., Stephens, S.L., Taylor, A., 2014. Unsupported inferences of high severity fire in historical western United States dry forests: Response to Williams and Baker. Glob. Ecol. Biogeogr. 23, 825–830.

Fulé, P.Z., Korb, J.E., Wu, R., 2009. Changes in forest structure of a mixed conifer forest, southwestern Colorado, USA. For. Ecol. Manage. 258, 1200–1210.

Gibbons, P., van Bommel, L., Gill, A.M., Cary, G.J., Driscoll, D.A., Bradstock, R.A., Knight, E., Moritz, M.A., Stephens, S.L., Lindenmayer, D.B., 2012. Land management practices associated with house loss in wildfires. PLoS One 7, e29212.

Hagmann, R.K., Franklin, J.F., Johnson, K.N., 2013. Historical structure and composition of ponderosa pine and mixed-conifer forests in south-central Oregon. For. Ecol. Manage. 304, 492–504.

Haney, A., Apfelbaum, S., Burris, J.M., 2008. Thirty years of post-fire succession in a southern boreal forest bird community. Amer. Midland Nat. 159, 421–433.

Hanson, C.T., North, M.P., 2008. Postfire woodpecker foraging in salvage-logged and unlogged forests of the Sierra Nevada. Condor 110, 777–782.

Hanson, C.T., 2013. Pacific fisher habitat use of a heterogeneous post-fire and unburned landscape in the southern Sierra Nevada, California, USA. Open Forest Sci. J. 6, 24–30.

Hanson, C.T., 2014. Conservation concerns for Sierra Nevada birds associated with high-severity fire. Western Birds 45, 204–212.

Hanson, C.T., Odion, D.C., DellaSala, D.A., Baker, W.L., 2009. Overestimation of fire risk in the Northern Spotted Owl Recovery Plan. Conserv. Biol. 23, 1314–1319.

Hanson, C.T., Odion, D.C., 2014. Is fire severity increasing in the Sierra Nevada mountains, California, USA? Int. J. Wildland Fire 23, 1–8.

Hanson, C.T., Odion, D.C., 2015a. Historical forest conditions within the range of the Pacific Fisher and Spotted Owl in the central and southern Sierra Nevada, California, USA. Natural Areas Journal.

Hanson, C.T., Odion, D.C., 2015b. Sierra Nevada fire severity conclusions are robust to further analysis: a reply to Safford et al. Int. J. Wildland Fire 24, 294–295.

Hessburg, P.F., Salter, R.B., James, K.M., 2007. Re-examining fire severity relations in pre-management era mixed conifer forests: inferences from landscape patterns of forest structure. Landsc. Ecol. 22, 5–24.

Heyerdahl, E.K., Brubaker, L.B., Agee, J.K., 2001. Spatial controls of historical fire regimes: a multiscale example from the Interior West, USA. Ecology 82, 660–678.

Heyerdahl, E.K., Lertzman, K., Wong, C.M., 2012. Mixed-severity fire regimes in dry forests of southern interior British Columbia, Canada. Can. J. Forest Res. 42, 88–98.

Huckaby, L.S., Kaufmann, M.R., Stoker, J.M., Fornwalt, P.J., 2001. Landscape Patterns of Montane Forest Age Structure Relative to Fire History at Cheesman Lake in the Colorado Front Range. U.S.D.A. Forest Service Proceedings RMRS-P-22, Rocky Mountain Research Station, Missoula, Montana, USA.

Hutto, R.L., 1995. Composition of bird communities following stand-replacement fires in Northern Rocky Mountain (U.S.A.) conifer forests. Conserv. Biol. 9, 1041–1058.

Hutto, R.L., 2006. Toward meaningful snag-management guidelines for postfire salvage logging in North American conifer forests. Conserv. Biol. 20, 984–993.

Hutto, R.L., 2008. The ecological importance of severe wildfires: some like it hot. Ecol. Appl. 18, 1827–1834.

Iniguez, J.M., Swetnam, T.W., Baisan, C.H., 2009. Spatially and temporally variable fire regime on Rincon Mountain, Arizona, USA. Fire Ecol. 5, 3–21.

Jenkins, S.E., Hull Sieg, C., Anderson, D.E., Kaufman, D.S., Pearthree, P.A., 2011. Late Holocene geomorphic record of fire in ponderosa pine and mixed-conifer forests, Kendrick Mountain, northern Arizona, USA. Int. J. Wildland Fire 20, 125–141.

Johnson, E.A., Gutsell, S.L., 1994. Fire frequency models, methods and interpretations. Adv. Ecol. Res. 25, 239–283.

Kasischke, E.S., Turetsky, M.R., 2006. Recent changes in the fire regimes across the North American boreal region—Spatial and temporal patterns of burning across Canada and Alaska. Geophys. Res. Lett. 33, L09703.

Kaufmann, M.R., Regan, C.M., Brown, P.M., 2000. Heterogeneity in ponderosa pine/Douglas-fir forests: age and size structure in unlogged and logged landscapes of central Colorado. Can. J. Forest Res. 30, 698–711.

Keeley, J.E., Zedler, P.H., 2009. Large, high-intensity fire events in southern California shrublands: debunking the fine-grain age patch model. Ecol. Appl. 19, 69–94.

Kinney, A., 1900. Forest and water. Post Publishing Company, Los Angeles, California, USA.

Kitzberger, T., Araoz, E., Gowda, J., Mermoz, M., Morales, J., 2012. Decreases in fire spread probability with forest age promotes alternative community states, reduced resilience to climate variability and large fire regime shifts. Ecosystems 15, 97–112. http://dx.doi.org/10.1007/s10021-011-9494-y.

Klenner, W., Walton, R., Arsenault, A., Kremsater, L., 2008. Dry forests in the Southern Interior of British Columbia: Historical disturbances and implications for restoration and management. For. Ecol. Manage. 256, 1711–1722.

Kotliar, N.B., Hejl, S.J., Hutto, R.L., Saab, V.A., Melcher, C.P., McFadzen, M.E., 2002. Effects of fire and post-fire salvage logging on avian communities in conifer-dominated forests of the western United States. Stud. Avian Biol. 25, 49–64.

Lindenmayer, D.B., Foster, D.R., Franklin, J.F., Hunter, M.L., Noss, R.F., Schmiegelow, F.A., Perry, D., 2004. Salvage harvesting policies after natural disturbance. Science 303, 1303.

Malison, R.L., Baxter, C.V., 2010. The fire pulse: wildfire stimulates flux of aquatic prey to terrestrial habitats driving increases in riparian consumers. Can. J. Fish. Aquat. Sci. 67, 570–579.

Marcoux, H., Gergel, S.E., Daniels, L.D., 2013. Mixed-severity fire regimes: How well are they represented by existing fire-regime classification systems? Can. J. Forest Res. 43, 658–668.

Marlon, J.R., Bartlein, P.J., Gavin, D.G., Long, C.J., Anderson, R.S., Briles, C.E., Brown, K.J., Colombaroli, D., Hallett, D.J., Power, M.J., Scharf, E.A., Walsh, M.K., 2012. Long-term perspective on wildfires in the western USA. Proc. Natl. Acad. Sci. 109, E535–E543.

Mast, J.N., Veblen, T.T., Hodgson, M.E., 1997. Tree invasion within a pine/grassland ecotone: An approach with historic aerial photography and GIS modeling. For. Ecol. Manage. 93, 187–194.

Mast, J.N., Veblen, T.T., Linhart, Y.B., 1998. Disturbance and climatic influences on age structure of ponderosa pine at the pine/grassland ecotone, Colorado Front Range. J. Biogeogr. 25, 743–755.

Miller, J.D., Skinner, C.N., Safford, H.D., Knapp, E.E., Ramirez, C.M., 2012. Trends and causes of severity, size, and number of fires in northwestern California, USA. Ecol. Appl. 22, 184–203.

Minnich, R.A., Barbour, M.G., Burk, J.H., Sosa-Ramirez, J., 2000. Californian mixed-conifer forests under unmanaged fire regimes in the Sierra San Pedro Martir, Baja California, Mexico. J. Biogeogr. 27, 105–129.

Moritz, M.A., Batllori, E., Bradstock, R.A., Gill, A.M., Handmer, J., Hessburg, P.F., Leonard, J., McCaffrey, S., Odion, D.C., Schoennagel, T., Syphard, A.D., 2014. Learning to coexist with wildfire. Nature 515, 58–66.

Odion, D.C., Frost, E.J., Strittholt, J.R., Jiang, H., DellaSala, D.A., Moritz, M.A., 2004. Patterns of fire severity and forest conditions in the Klamath Mountains, northwestern California. Conserv. Biol. 18, 927–936.

Odion, D.C., Hanson, C.T., 2006. Fire severity in conifer forests of the Sierra Nevada, California. Ecosystems 9, 1177–1189.

Odion, D.C., Hanson, C.T., 2008. Fire severity in the Sierra Nevada revisited: conclusions robust to further analysis. Ecosystems 11, 12–15.

Odion, D.C., Moritz, M.A., DellaSala, D.A., 2010. Alternative community states maintained by fire in the Klamath Mountains, USA. J. Ecol. 98, 96–105.

Odion, D.C., Hanson, C.T., 2013. Projecting impacts of fire management on a biodiversity indicator in the Sierra Nevada and Cascades, USA: the Black-backed Woodpecker. Open For. Sci. J. 6, 14–23.

Odion, D.C., Hanson, C.T., Arsenault, A., Baker, W.L., DellaSala, D.A., Hutto, R.L., Klenner, W., Moritz, M.A., Sherriff, R.L., Veblen, T.T., Williams, M.A., 2014. Examining historical and current mixed-severity fire regimes in ponderosa pine and mixed-conifer forests of western North America. PLoS One 9, e87852.

Paritsis, J., Veblen, T.T., Holz, A., 2014. Positive fire feedbacks contribute to shifts from *Nothofagus pumilio* forests to fire-prone shrublands in Patagonia. J. Veget. Sci. 26, 89–101. http://dx.doi.org/10.1111/jvs.12225, Online.

Perry, D.A., Hessburg, P.F., Skinner, C.N., Spies, T.A., Stephens, S.L., Taylor, A.H., Franklin, J.F., McComb, B., Riegel, G., 2011. The ecology of mixed severity fire regimes in Washington, Oregon, and northern California. For. Ecol. Manage. 262, 703–717.

Perry, G.L.W., Wilmshurst, J.M., McGlone, M.S., McWethy, D.B., Whitlock, C., 2012. Explaining fire-driven landscape transformation during the Initial Burning Period of New Zealand's prehistory. Glob. Chang. Biol. 18, 1609–1621.

Pierce, J.L., Meyer, G.A., Jull, A.J.T., 2004. Fire-induced erosion and millennial-scale climate change in northern ponderosa pine forests. Nature 432, 87–90.

Platt, R.V., Schoennagel, T., 2009. An object-oriented approach to assessing changes in tree cover in the Colorado Front Range 1938–1999. For. Ecol. Manage. 258, 1342–1349.

Pyne, S.J., 1982. Fire in America: A Cultural History of Wildland and Rural Fire (Cycle of Fire). University of Washington Press, Seattle.

Raphael, M.G., Morrison, M.L., Yoder-Williams, M.P., 1987. Breeding bird populations during twenty-five years of postfire succession in the Sierra Nevada. The Condor 89, 614–626.

Romme, W.H., Despain, D.G., 1989. Historical perspective on the Yellowstone Fires of 1988. BioScience 39, 695–699.

Roos, C.I., Swetnam, T.W., 2012. A 1416-year reconstruction of annual, multidecadal, and centennial variability in area burned for ponderosa pine forests of the southern Colorado Plateau region, Southwest USA. The Holocene 22, 281–290.

Rota, C.T., 2013. Not all forests are disturbed equally: population dynamics and resource selection of Black-backed Woodpeckers in the Black Hills, South Dakota. Ph.D. Dissertation, University of Missouri-Columbia, MO.

Schieck, J., Song, S.J., 2006. Changes in bird communities throughout succession following fire and harvest in boreal forests of western North America: literature review and meta-analyses. Can. J. Forest Res. 36, 1299–1318.

Schoennagel, T., Sherriff, R.L., Veblen, T.T., 2011. Fire history and tree recruitment in the Colorado Front Range upper montane zone: implications for forest restoration. Ecol. Appl. 21, 2210–2222.

Sestrich, C.M., McMahon, T.E., Young, M.K., 2011. Influence of fire on native and nonnative salmonid populations and habitat in a western Montana basin. Trans. Am. Fish. Soc. 140, 136–146.

Shatford, J.P.A., Hibbs, D.E., Puettmann, K.J., 2007. Conifer regeneration after forest fire in the Klamath-Siskiyous: how much, how soon? J. Forestry 105, 139–146.

Sherriff, R.L., Veblen, T.T., 2006. Ecological effects of changes in fire regimes in *Pinus ponderosa* ecosystems in the Colorado Front Range. J. Veget. Sci. 17, 705–718.

Sherriff, R.L., Veblen, T.T., 2007. A spatially explicit reconstruction of historical fire occurrence in the Ponderosa pine zone of the Colorado Front Range. Ecosystems 9, 1342–1347.

Sherriff, R.L., Platt, R.V., Veblen, T.T., Schoennagel, T.L., Gartner, M.H., 2014. Historical, observed, and modeled wildfire severity in montane forests of the Colorado Front Range. PLoS One 9, e106971.

Shinneman, D.J., Baker, W.L., 1997. Nonequilibrium dynamics between catastrophic disturbances and old-growth forests in ponderosa pine landscapes of the Black Hills. Conserv. Biol. 11, 1276–1288.

Siegel, R.B., Tingley, M.W., Wilkerson, R.L., Bond, M.L., Howell, C.A., 2013. Assessing home range size and habitat needs of Black-backed Woodpeckers in California: Report for the 2011 and 2012 field seasons. Institute for Bird Populations.

Steel, Z.L., Safford, H.D., Viers, J.H., 2015. The fire frequency-severity relationship and the legacy of fire suppression in California's forests. Ecosphere 6, 8.

Stephens, S.L., Agee, J.K., Fule, P.Z., North, M.P., Romme, W.H., Swetnam, T.W., Turner, M.G., 2013. Managing forests and fire in changing climates. Science 342, 41–42.

Stephens, S.L., Ruth, L.W., 2005. Federal forest fire policy in the United States. Ecol. Appl. 15, 532–542.

Swanson, M.E., Franklin, J.F., Beschta, R.L., Crisafulli, C.M., DellaSala, D.A., Hutto, R.L., Lindenmayer, D.B., Swanson, F.J., 2011. The forgotten stage of forest succession: early-successional ecosystems on forest sites. Front. Ecol. Environ. 9, 117–125.

Taylor, A.H., Skinner, C.N., 1998. Fire history and landscape dynamics in a late-successional reserve, Klamath Mountains, California, USA. For. Ecol. Manage. 111, 285–301.

Thompson, I., Mackey, B., McNulty, S., Mosseler, A., 2009. Forest Resilience, Biodiversity, and Climate Change. A synthesis of the biodiversity/resilience/stability relationship in forest ecosystems. Secretariat of the Convention on Biological Diversity, Montreal. Technical Series no. 43, 67 pages.

van Wagtendonk, J.W., van Wagtendonk, K.A., Thode, A.E., 2012. Factors associated with the severity of intersecting fires in Yosemite National Park, California, USA. Fire Ecol. 8, 11–32.

Veblen, T.T., Kitzberger, T., Donnegan, J., 2000. Climatic and human influences on fire regimes in ponderosa pine forests in the Colorado Front Range. Ecol. Appl. 10, 1178–1195.

Veblen, T.T., Lorenz, D.C., 1986. Anthropogenic disturbance and recovery patterns in montane forests, Colorado Front Range. Phys. Geogr. 7, 1–24.

Veblen, T.T., Lorenz, D.C., 1991. The Colorado Front Range: A Century of Ecological Change. University of Utah Press, Salt Lake City, 186 pp.

Veblen, T.T., 2003. Historic range of variability of mountain forest ecosystems: concepts and applications. The Forestry Chronicle 79, 223–226.

Weaver, H.A., 1943. Fire as an ecological and silvicultural factor in the ponderosa pine region of the Pacific slope. J. Forestry 41, 7–15.

Westerling, A.L., Hidalgo, H.G., Cayan, D.R., Swetnam, T.W., 2006. Warming and earlier spring increases western US forest wildfire activity. Science 313, 940–943.

Whitlock, C., Marlon, J., Briles, C., Brunelle, A., Long, C., Bartlein, P., 2008. Long-term relations among fire, fuel, and climate in the north-western US based on lake-sediment studies. Int. J. Wildland Fire 17, 72–83.

Williams, J., 2012. Exploring the onset of high-impact mega-fires through a forest land management prism. For. Ecol. Manage. 294, 4–10.

Williams, M.A., Baker, W.L., 2010. Bias and error in using survey records for ponderosa pine landscape restoration. J. Biogeogr. 37, 707–721.

Williams, M.A., Baker, W.L., 2011. Testing the accuracy of new methods for reconstructing historical structure of forest landscapes using GLO survey data. Ecol. Monogr. 81, 63–88.

Williams, M.A., Baker, W.L., 2012a. Spatially extensive reconstructions show variable-severity fire and heterogeneous structure in historical western United States dry forests. Glob. Ecol. Biogeogr. 21, 1042–1052.

Williams, M.A., Baker, W.L., 2012b. Comparison of the higher-severity fire regime in historical (A. D. 1800s) and modern (A.D. 1984-2009) montane forests across 624,156 ha of the Colorado Front Range. Ecosystems 15, 832–847.

Williams, M.A., Baker, W.L., 2013. Variability of historical forest structure and fire across ponderosa pine landscapes of the Coconino Plateau and south rim of Grand Canyon National Park, Arizona, USA. Landsc. Ecol. 28, 297–310.

Williams, M.A., Baker, W.L., 2014. High-severity fire corroborated in historical dry forests of the western United States: Response to Fulé et al. Glob. Ecol. Biogeogr. 23, 831–835.

Wright, C.S., Agee, J.K., 2004. Fire and vegetation history in the eastern Cascade mountains, Washington. Ecol. Appl. 14, 443–459.

Chapter 2

Ecological and Biodiversity Benefits of Megafires

Dominick A. DellaSala[1] and Chad T. Hanson[2]
[1]*Geos Institute, Ashland, OR, USA,* [2]*John Muir Project of Earth Island Institute, Berkeley, CA, USA*

2.1 JUST WHAT ARE MEGAFIRES?

Under extreme weather conditions (e.g., droughts with high temperatures and wind), any wildfire has the potential to cause dramatic changes in plant and wildlife communities across large landscapes, providing renewed opportunities for colonization by species that are adapted to the newly created postfire environments. Despite fire-suppression efforts, megafires have proven impossible to control and result in substantial landscape-level changes with or without human intervention because they most often occur under extreme weather events. Negative perceptions about megafires and high costs of property damage have resulted in controversy over fire management among the public, land managers, government officials, and even some ecologists and conservation groups that typically view these fires as "catastrophes."

Interestingly, megafires are self-organizing natural forces that possess the sheer capacity to drive themselves into critical disturbance dynamics independent of initial conditions (Clar et al., 1996). For instance, very large fires are affected by—and, in turn, generate—their own weather patterns with fire plumes ascending to over 9000 m (Figure 2.1). The plumes then create downward pressure gradients with high winds that direct fire spread, fire line propagation, and fire intensity over very large areas.

Megafires qualitatively differ from smaller fires in that, in small fires, overall severity tends to be low and plant communities seldom change; in large fires, however, they may experience myriad and varying natural succession trajectories and heterogeneity at multiple spatial scales over large areas because the severity associated with larger fires tends to be quite variable (Abrams et al., 1985; Kotliar and Wiens, 1990; Wu and Loucks, 1995; Odion et al., 2004; Kasischke and Turetsky, 2006; Odion et al., 2010). Most megafires are ecologically beneficial because they result in high levels of beta (changes in species composition across fire severities) and alpha (within a particular fire severity patch) diversity (Whittaker, 1960),

FIGURE 2.1 Smoke plume from a large nineteenth century fire in the San Gabriel Mountains of southern California (Kinney, 1900).

including a prolific pulse of critically important forest legacies (e.g., snags, downed logs, montane chaparral patches) over large landscapes.

In the western United States, megafires represent up to one-third of area affected by fires that occurred from 1984 to 2010 (Stavros et al., 2014). In other regions they are less common, but when they do occur they contribute to disproportionately high burn acres (especially severe fire patches) compared with those of more frequent and less severe fires. Just how large a fire has to be to qualify as a megafire has been mainly related to socioeconomic criteria (e.g., costs of fire suppression, property damage, loss of life), although there have been attempts to describe these fires using statistical attributes related to fire size and frequency (Lin and Rinaldi, 2009). The popular press and decision makers have used the term "megafire" most often in a negative sense, particularly as economic costs have soared into tens to hundreds of millions of dollars per fire (see Chapter 12). The term also increasingly has been used in science policy articles (Attiwill and Binkley, 2013; Williams, 2013), most often in a command-and-control sense aimed at preventing fires from reaching megafire status.

With the recognition that a certain amount of subjectivity is inherent in describing such a broad concept, we recommend that the determination of megafires be based on the spatiotemporal characteristics of large fires using statistically derived criteria from multidecadal fire records at the regional scale. For instance, one criterion for megafire determination might be when an individual fire is >2 standard deviations above the average size of fires in a given region during a specified period (e.g., 25 years).

Statistical definitions can be cross-checked for biological relevance by comparing fire characteristics to the regenerative properties of fire-dependent (pyrogenic) species (as in Turner and Dale, 1998). Megafire determinations also need to be based on biologically meaningful time lines, given the slow (relative to human time lines) postfire change of abiotic factors (e.g., soils, microclimate) and the resilience of surviving or regenerating seed sources as part of the typical postfire successional stages (see Swanson et al., 2011; DellaSala et al., 2014). Moreover, megafire determinations should take into account the importance of large fires in closing the gap on regional fire deficits (Odion et al., 2014) and maintaining large patches of complex early seral forest—at more significant spatial and temporal scales—to provide habitat for rare and declining fire-dependent species (DellaSala et al., 2014; Hanson, 2014).

In this chapter we generally refer to a threshold value for megafires as landscape-scale fires of ~50,000 ha (500 km^2) (see, e.g., Keeley and Zedler, 2009) because this is most likely to be a biologically meaningful scale. Megafires may, however, occasionally occur over much larger areas, such as the 2002 Biscuit fire (~220,000 ha) in the Oregon portion of the Klamath-Siskiyou ecoregion (see Chapter 11); the 2002 Rodeo-Chedeski fire in Arizona (186,866 ha); the 2011 Wallow fire (217,741 ha) in the southwestern United States (Williams and Baker, 2012a); and the 2013 Rim fire (104,176 ha) in the western Sierra Nevada (USDA, 2014). Other regional megafires (three separate events over a 7-year period) have been as large as three million hectares, as occurred in southeastern Australia (Attiwill and Adams, 2013).

Large fires have been reoccurring in the western United States with regularity, but there has been no recent uptick in fire area, although in some years there have been multiple events (Figure 2.2). The largest ones (>50,000 ha), of interest herein, occur in forests/woodlands, shrubland, and grassland (herb) vegetation types (Figure 2.3).

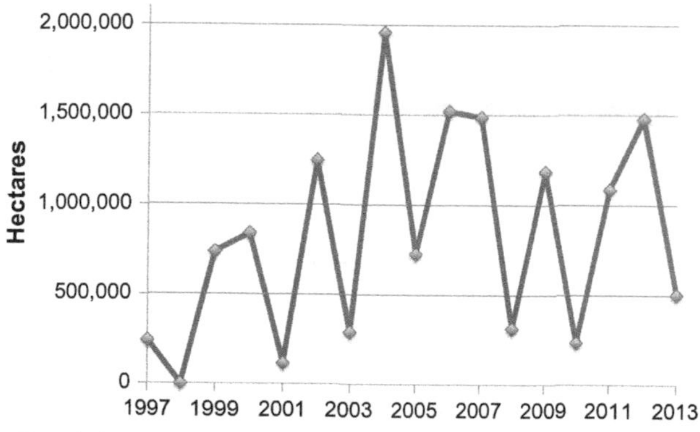

FIGURE 2.2 Cumulative area of large (>40,000 ha) fires (as defined by the National Interagency Fire Center) in the United States from 1997 to 2013. (http://www.nifc.gov/fireInfo/fireInfo_stats_lgFires.html; accessed January 17, 2015). Area burned over time is non-significant (Mann-Kendall trend test z=1.03, p=0.30)

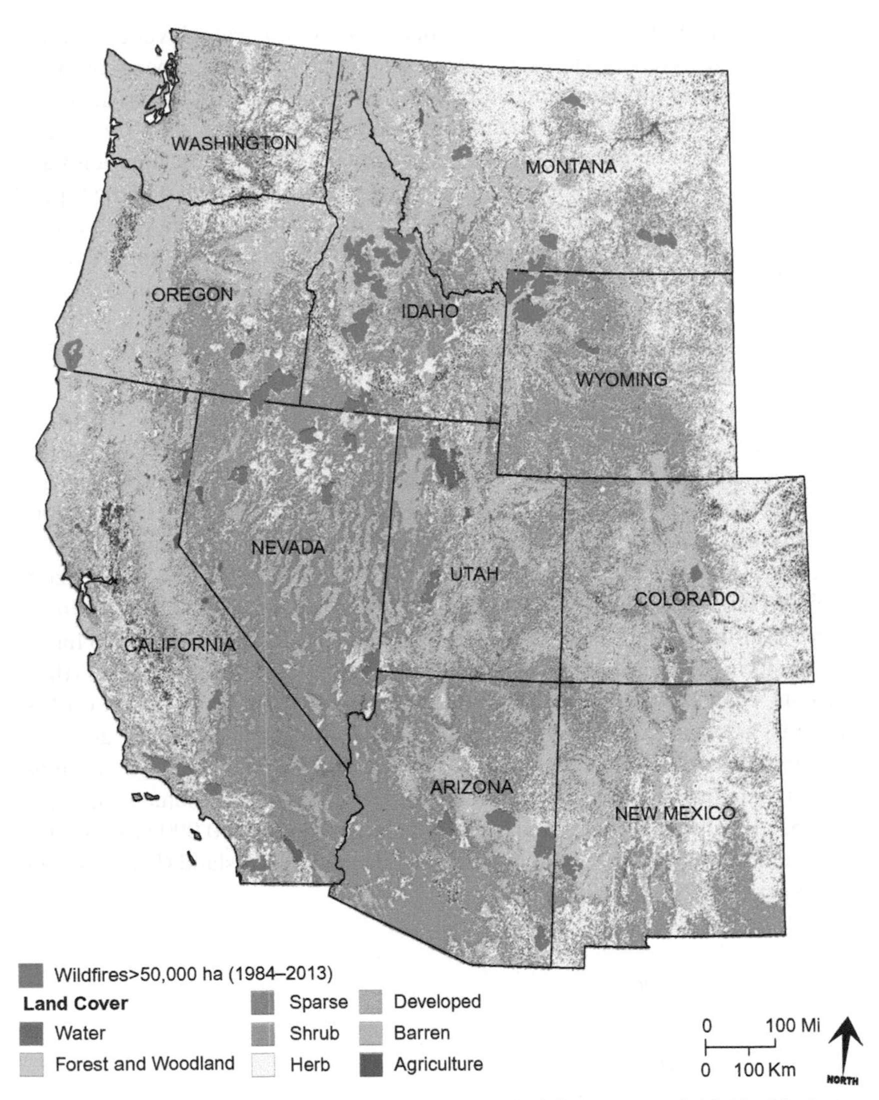

FIGURE 2.3 Megafires (>50,000 ha) in the western United States as recorded by the Monitoring Trends in Burn Severity data set of the US Forest Service from 1984 to 2013 (http://www.mtbs.gov/nationalregional/burnedarea.html; accessed January 15, 2015). Fire perimeters are displayed on top of existing vegetation lifeform category resampled to 90 m using the LANDFIRE program (http://www.landfire.gov/lf_mosaics.php; accessed January 14, 2015). Figure does not include Rim fire in the central Sierra Nevada. Figure prepared by J. Leonard, Geos Institute.

2.2 MEGAFIRES AS GLOBAL CHANGE AGENTS

In this chapter we describe some of the more general properties of megafires and why we believe they play an important role in maintaining biodiversity and ecosystem processes. We start with a global perspective and then zoom in on the western United States, where the term megafire ostensibly has its origin (Pyne, 2007). We also discuss whether megafires are currently increasing or have the capacity to do so under conditions of global climate change and implications to remedy the ongoing fire deficit in many regions of western North America and the world (Niklasson and Granström, 2004; Odion et al., 2014).

Naturally occurring megafires (or very large fires, as described by others) have been reported in many regions of the world. Notable examples include Portugal (Tedim et al., 2013); northern China and southeastern Siberia (Cahoon et al., 1994); southern France (Ganteaume and Jappiot, 2013); Greece (San-Miguel-Ayanz et al., 2013); the western United States (many examples [Stavros et al., 2014], most notably the greater Yellowstone ecosystem [Turner et al., 1998]); boreal forests of Canada, Alaska, and Russia (Kasischke and Bruhwiler, 2002); portions of Australia (Lindenmayer et al., 2010); and sub-Saharan Africa (Bird and Cali, 1998). Other human-caused megafires, such as those that have occurred in the Brazilian Amazon, are not considered ecologically driven events and therefore are not included here.

Based on global patterns of fire behavior, the occurrence of megafires is governed primarily by top-down drivers such as increases in the sea surface temperature (Skinner et al., 2006), the Pacific Decadal Oscillation (when positive; Morgan et al., 2008), El Niño-Southern Oscillation (Skinner et al., 2006), and midtropospheric surface-blocking events (high-pressure systems) during summer months (Johnson and Wowchuk, 1993). Local factors (terrain, vegetation) may contribute to the extent of a fire (Thompson and Spies, 2009; Ganteaume and Jappiot, 2013), but the occurrence and severity of a megafire are largely weather-related, top-down phenomena.

2.3 MEGAFIRES, LARGE SEVERE FIRE PATCHES, AND COMPLEX EARLY SERAL FORESTS

The relationship of megafires to native biodiversity warrants special attention by ecologists and land managers. Several studies have addressed this issue in various ways and for very large, intense fires covering ~20,000-50,000 ha. For example, after the fires of 1988, 2000, and 2003, burned areas in the Intermountain West were transformed into unique early-successional bird communities after the disturbance (Hutto 1995, Hutto, 2008, Smucker et al. 1995). In southwest Oregon's 2002 Biscuit fire, large high-severity fire patches and high-severity reburned areas supported high levels of native plant and bird species, with a richness and abundance equal to or greater than levels found in unburned mature forest (Donato et al., 2009; Fontaine et al., 2009). In the southeastern Sierra Nevada mountains of California, greater richness of bird species was reported in the ~61,000 ha McNally fire of 2002 than in adjacent, unburned mature/old conifer forest (Siegel and Wilkerson, 2005), similar to bird richness levels in the ~18,000 ha Donner fire of 1960 in the northern Sierra Nevada (Raphael et al., 1987). Likewise, in the northern Sierra Nevada, the highest total bird abundance was found in unlogged high-severity fire areas within the

~23,000 ha Storrie fire of 2000 (Burnett et al., 2010). A more recent investigation of moderate-/high-severity fire areas in the 104,178 ha Rim fire of 2013 in the central Sierra Nevada found high avian abundance and diversity just 1 year after the fire, including numerous species that were absent or nearly absent in adjacent unburned forest, concluding that these higher-severity areas in the Rim fire created "a rich habitat for early-successional birds that will sustain these rarer species on the Sierra landscape for years to come" (Fogg et al., 2014).

Megafires also influence habitat for rare and management-sensitive species such as the California spotted owl (*Strix occidentalis occidentalis*). This forest species mainly selected unlogged high-severity areas in the McNally fire area for foraging, more than lower fire severity areas and more than unburned mature/old forest (Bond et al., 2009). After the 90,265 ha Horseshoe 2 fire of 2011 in ponderosa pine (*Pinus ponderosa*) and mixed-conifer forests of Arizona, Mexican spotted owl (*Strix occidentalis lucida*) populations increased, and spotted owl reproduction increased as well, particularly in territories with higher levels of high-severity fire (Moors, 2012, 2013), likely due to an enhanced small mammal prey base (Ganey et al., 2014). Similarly, at one year post-fire, but before post-fire logging, Lee and Bond (2015) found 92% occupancy of historical California spotted owl territories in the Rim fire, which is higher than average annual occupancy in unburned mature/old forest, and pair occupancy was not reduced in territories with mostly high-severity fire effects. Some rare, sensitive

BOX 2.1 Pulse Versus Chronic Disturbance

Any large natural disturbance (e.g., landscape-level insect outbreaks, hurricanes, volcanic eruptions, big floods, coastal storms) triggers a pulse of biological activity in its aftermath (Turner and Dale, 1998). In the case of fires, a pulse disturbance occurs when a large fire of high intensity kills the majority of vegetation, leaving charred, fire-killed trees and other vegetation that may persist for decades to centuries as key structural elements in the new forest. Many terrestrial and aquatic species (see Chapters 3-6) depend on these postfire structural elements (Donato et al., 2012). By contrast, a chronic disturbance is a reoccurring one that can also repeatedly affect large landscapes. Most anthropogenic disturbances are chronic, accumulating in space and over time, and often exceed the capacity of disturbance-adapted species to regenerate. Anthropogenic chronic disturbances take place over short time lines that are outside historical bounds in terms of size and intensity, and they remove or inhibit, rather than create, biological legacies. As such, they may "flip" ecosystem dynamics to altered states that carry irreparable consequences to biodiversity (Paine et al., 1998; Lindenmayer et al., 2011). Examples include logging, the application of herbicides, and planting of nursery-stock conifers that typically follow large fires; reseeding of postfire areas with nonnative plants for erosion abatement, where nonnative plants then outcompete the native plant establishment (Beyers, 2004); roads and chronic sediment input into streams that impact aquatic species (Trombulak and Frissell, 2000; Colombaroli and Gavin, 2010); and cattle grazing following reseeding in postfire areas, which compacts fragile soils and favors exotic over native species (Beschta et al., 2013; also see Chapter 11).

FIGURE 2.4 An example of complex early seral forest created by a large, high-severity fire patch in the Sierra Nevada. Note the preponderance of biological legacies (snags, shrubs).

bat species also are most strongly associated with high-severity fire areas (Buchalski et al., 2013), and Pacific fishers (*Pekania pennanti*) (McNally fire) use unlogged moderate-/high-severity fire areas within dense, mature forest at levels comparable to their use of adjacent dense, old, unburned forest (Hanson, 2013). This is especially true for females (Hanson, 2015).

Megafires also trigger a "pulse" disturbance (Box 2.1) that results in the replenishment of complex early seral forests that support abundant wildlife communities. Complex early seral habitat is the recently established vegetation stage in which intense fires kill most of the overstory trees (Swanson et al., 2011; DellaSala et al., 2014). This stage of forest succession produces biologically rich patches populated by large numbers of large snags, fallen logs, montane chaparral (patches of native flowering and berry-producing shrubs), and natural conifer regeneration of variable density that link successional stages across a temporal gradient (Swanson et al., 2011). Notably, the legacy pulse provided by large fires "lifeboats" essential structural elements from the unburned forests to the recent fire-created one, which then continues over decades to centuries as a forest matures. In this fashion the precocity of complex early seral vegetation is evident in structural elements originating long before and immediately after the disturbance and persisting over time (Donato et al., 2012; see Figure 2.4). Dead trees, for instance, especially large ones, play a key role in complex early seral communities, and when they are removed via logging (which is most often the case), most fire-dependent bird species decline, and many disappear altogether (Hutto, 2008; Hanson, 2014; also see Chapter 11).

2.4 HISTORICAL EVIDENCE OF MEGAFIRES

In fire-adapted ecosystems around the world, there is historical evidence that megafires have occurred for millennia. For instance, in Victoria, Australia,

~2 million ha burned in the "Black Friday Bushfires" in 1939, much of which occurred in montane-*Eucalyptus* forests (Attiwill and Adams, 2013). In 1825 in the forests of New Brunswick, Canada, the Miramichi fire spanned ~1.2 million hectares (Wein and Moore, 1977), the Saquenay fire in Quebec grew to ~390,000 ha in 1870 (NYT, 1870), and the "Great Fire of 1910" burned over 1.2 million hectares in northeastern Washington, northern Idaho, and western Montana.[1] Remarkably, the Great Fire burned most of the fire area in just 2 days, killing 87 people, mostly firefighters (Pyne, 2008; Egan, 2010). Large fires were also quite common in the Pacific Northwest during warm phases of the Holocene and pre-Holocene (up to 20,000 years ago; Whitlock, 1992).

In the western United States, the historical significance of large fires has been recorded by the National Interagency Fire Center (Appendix **2.1**). Examination of historical records indicates effects of fire over large areas on property and loss of life, thereby understandably contributing to command-and-control responses. Determining whether today's megafires are uncharacteristic requires a comprehensive understanding of a region's biological and cultural history of fire over long time lines. Thus, the historical context of large, high-severity fire patches is especially pertinent to the question: Are megafires categorically a modern anomaly or are such fires within the historical envelope, before modern fire suppression, land management practices, and anthropogenic climate change?

As a regional example we provide some relevant historical accounts and fire history reconstruction studies to address the occurrence of very large, high-severity fire patches in ponderosa pine and mixed-conifer forests of the western United States before the effects of fire suppression and logging. Readers should note that we are now focusing on large patches of high-severity fire within the megafire "matrix" that is a mosaic of varying patch sizes and severities. As such, we grouped historical megafires geographically to show broad-level influences from past fires and for later discussion regarding what the future may hold for megafires.

Rocky Mountain Region

Perhaps the most well-known example of a historical megafire is the "Big Burn" of 1910 in the northern Rocky Mountains, a high-severity fire of ~1.2 million hectares that occurred under extreme fire-weather conditions in mostly remote, unlogged forest. The preburn forests included mostly lodgepole pine (*Pinus contorta*) and ponderosa pine/Douglas-fir (*Pseudotsuga menziesii*) (Leiberg, 1900a). To reiterate, the 1910 fire burned most of its landscape in just 2 days. But large, intense fires were not unusual for the region's pre-fire suppression era. For instance, high-severity fire patches hundreds to thousands of hectares in size were mapped by early US Geological Survey researchers in the

---

1 For historical photos go to: https://www.google.com/search?q=photos+of+Big+Blow+out+1910+fires&espv=2&biw=1280&bih=629&tbm=isch&tbo=u&source=univ&sa=X&ei=JmNQVOPzGKaf8QG544GgCg&ved=0CDIQ7Ak; accessed October 28, 2014.

late 1800s and early 1900s in ponderosa pine forest in similarly remote, unmanaged areas (Leiberg, 1900a).

In the central Rockies, within the ponderosa pine forests of Colorado's Front Range, numerous large, higher-severity fire patches (mixed-severity and higher-severity combined) were documented in the mid- to late 1800s (Williams and Baker, 2012b). Many of the higher-severity patches within the large fire perimeter were 1000 to 3000 ha; the maximum higher-severity patch size was 8331 ha. The mean and maximum historical high severity patch sizes were even larger than current patch sizes in this same area (Williams and Baker, 2012b). A recent analysis indicates that only a relatively minor proportion (16%) of the montane conifer forests of the Colorado Front Range have experienced an increase in fire severity such that they may exceed historical norms—and such areas are generally at the lowest elevations (Sherriff et al., 2014).

Eastern Cascades and Southern Cascades

High-severity fire patches of 1000-5000 and 5000-10,000 ha within very large fires in the mixed-conifer forests of the Eastern Cascades of Washington State have been described using historical accounts (Perry et al., 2011) and reconstructions based on field surveys from the mid- to late 1800s (Baker, 2012). Similarly, pioneering historical ecologists (Leiberg, 1903, pp. 273-275, plate XL) mapped and reported a single high-severity fire patch (~14,000 ha) in predominantly ponderosa pine forest near Mount Pitt, south of Crater Lake, in unlogged forests of the eastern Oregon Cascades. In the mixed-conifer and fir (*Abies* spp.) forests of the Southern Cascades of California, "widespread and high-severity fires" burned across vast areas, indicated by fire scars at numerous locations separated by >30 km and occurring in the same years (1829, 1864, 1889) during dry conditions (Bekker and Taylor, 2010).

Oregon Coast Range and Klamath Region

In the western Oregon Coast Range, the 1849 Yaquina and 1853 Nestucca fires each spanned ~200,000 ha (Gannett, 1902; Morris, 1934; see also Figure 2.5 below). In the Klamath region of southern Oregon, Leiberg (1900b) documented a high-severity fire area of ~24,000 ha (likely resulting from more than one fire) before the arrival of settlers and occurring in roughly equal areas of lodgepole and ponderosa pine.

Sierra Nevada

Leiberg (1902, plate VII) mapped numerous large, high-severity fire patches (defined as 75-100% mortality of timber volume) occurring in the 1800s, including many patches 1000-5000 ha and numerous 5000-10,000 ha patches, within unlogged mixed-conifer forests (Hanson, 2007). Leiberg noted that "a large proportion" of these patches occurred in the early part of the 1800s, before the arrival of settlers. High-severity fire patches were mapped only if

FIGURE 2.5 The Nestucca (a) and Yaquina (b) fires in the mid-1800s (Gannett, 1902).

Leiberg and his team were able to locate evidence of the previous stands (e.g., the remains of fire-killed overstory trees and large downed logs). He also estimated the age of the naturally regenerating stands following high-severity fire (Leiberg, 1902). Because Leiberg mapped high-severity fire patches that occurred over the course of the nineteenth century, some of these large patches could have resulted from fires occurring in different years or decades, but in most cases it would be exceedingly improbable that large, high-severity fire patches happened to occur immediately adjacent to previous large patches. Baker (2014), using General Land Office survey data from the mid- to late 1800s, found high-severity fire patches up to ~8000 and ~9000 ha in mixed-conifer and ponderosa pine forests of western Sierra Nevada.

Southwestern United States and Pacific Southwest

Fire history reconstructions have revealed contiguous areas with only high-severity fire effects over 500-1000 ha, and one >10,000 ha, within ponderosa pine forests in the Black Mesa and southeastern Mogollon Plateau areas of eastern-central and northern Arizona, respectively (Williams and Baker, 2012a). This was based on spatially extensive US General Land Office field plots from the mid- to late 1800s (Williams and Baker, 2012a) that were submitted to rigorous accuracy testing and have been found to be robust upon further examination and analysis (Williams and Baker, 2012a; Williams and Baker, 2014). Lang and Stewart (1910), in ponderosa pine forests of northern Arizona, observed "[v]ast denuded areas, charred stubs and fallen trunks and the general prevalence of blackened poles seem to indicate [fire] frequency and severity...."

In the mountains of southern California, in what is now the northern portion of the Cleveland National Forest, the Santiago Canyon fire occurred in 1889, covering 125,000-200,000 ha of chaparral and dry conifer forest; the size varies among accounts (Keeley and Zedler, 2009).

Black Hills

In ponderosa pine-dominated forests of the Black Hills of South Dakota, a single high-severity fire patch of ~19,000 ha was documented (Dodge, 1876). Just over two decades later, US Geological Survey researchers found "a young pine forest springing up" in this area (Graves, 1899, p. 146) and photographed and described additional large, high-severity fire areas.

2.5 MEGAFIRES AND LANDSCAPE HETEROGENEITY

There is much hyperbole, misunderstanding, and fear conveyed in the media's reports of megafires; these are based on unverified anecdotal representations from land managers, firefighters, timber industry officials, local community leaders, and politicians about overwhelmingly high-severity fire effects on forests or high-severity fire patches tens of thousands of hectares in size with no surviving trees and little or no potential for natural conifer regeneration, as in the case of the Rim fire of 2013 in the forests of the western Sierra Nevada of California (Cone, 2013; Jarvis, 2014). Once the smoke clears, however, the evidence generally indicates predominantly low- to moderate-severity fire effects over large areas, substantial heterogeneity within large, high-severity fire patches, and forests naturally regenerating in ways that promote a diversity of successional stages and plant communities (Table 2.1).

Moreover, based on recent observations of high-severity fire patches within megafires, substantial intrapatch heterogeneity occurs at multiple scales. Where high-severity fire effects occur, large overstory trees often survive in varying densities, and the overall basal area mortality can be 100% in some portions of patches but often can be ~60-80% as well (Table 2.2).

TABLE 2.1 High-Severity Fire Proportions in Recent Megafires in the Western United States

Fire/location	Size (hectares)	Year	High severity area (%)	Source
Rim (California)	104,178	2013	20% (52% low, 28% moderate)	www.mtbs.gov
Biscuit (Oregon)	~200,000	2002	29% (41% low, 30% moderate)	www.mtbs.gov
Rodeo-Chedeski (Arizona)	186,866	2002	37%	Thompson and Spies (2009) Williams and Baker (2012a)
Hayman (Colorado)	53,212	2002	22%	Williams and Baker (2012a)
Wallow (Arizona)	217,741	2011	16%	Williams and Baker (2012a)

TABLE 2.2 Basal Area Mortality for Relative Delta Normalized Burn Ratio (RdNBR) Higher-Severity Fire Thresholds of 574 and 800 in Klamath and Sierra Nevada plots

		Basal Area Mortality					
		Small Trees Included		*Trees ≥30 cm DBH*		*Trees ≥50 cm DBH*	
Region	**RdNBR**	*Mean % (SD)*	n	*Mean % (SD)*	n	*Mean % (SD)*	n
Klamath	574±100	60.9 (35.6)	18	51.8 (39.0)	16	48.1 (40.1)	16
	800±100	75.8 (24.8)	18	67.9 (33.4)	16	58.9 (37.1)	13
Sierra Nevada	574±50	60.9 (35.1)	67	51.0 (39.5)	58	41.1 (44.4)	43
	800±50	83.4 (27.2)	69	76.0 (35.5)	65	60.2 (46.2)	41

n represents the number of field validation plots.
Values are derived from US Forest Service field validation plot data and fire severity values from satellite imagery. Reproduced from Hanson et al. (2010).

In addition, patches of low- to moderate-severity fire, ranging from 0.1 ha to dozens of hectares, occur throughout large, high-severity fire patches, such as in the largest (>4000 ha) high-severity fire patches within mixed-conifer forests during the McNally fire (Buchalski et al., 2013; Hanson, 2013: Figure 1) and the

Biscuit fire (Halofsky et al., 2011), as well as the ~104,000 ha Rim fire that occurred on the western slopes of the central Sierra Nevada mountains in 2013.

Photos taken in September 2014 (Figure 2.6a and c) show "flushing," the production of new, green needles from surviving terminal buds 1 year after a fire where there was 100% initial mortality of foliage (Hanson and North, 2009) of ponderosa pines hundreds of meters into the interior of one of the largest high-severity fire patches (~1000 ha) of the Rim fire. Surveys by one of us (C.T.H.) indicate, on average, >20 surviving trees per hectare in this large, high-severity fire patch, with live trees variably distributed in clumps, generally across 0.1 to 10 ha. We also observed natural postfire conifer regeneration 1 year after the fire within the same large, high-severity fire patch in the Rim fire (Figure 2.6b and d). Our postfire surveys indicate, on average, ~250 naturally regenerating conifers per hectare >200 m into the interior of this patch, with ponderosa and sugar pine (*Pinus lambertiana*) dominating interior regeneration (Figure 2.7).

Recent research found that natural postfire conifer regeneration in large, high-severity fire patches is both vigorous and heterogeneous. Crotteau et al. (2013) found 715 naturally regenerated conifer seedlings per hectare in large, high-severity fire patches after the 2000 Storrie fire in the northern Sierra Nevada, and, in the same fire area, Cocking et al. (2014) found such high-severity fire patches to play a key role in the regeneration and maintenance of California black oak (*Quercus kelloggii*), comparable to the findings of Haire and McGarigal (2008) with regard to aspen (*Populus* sp.) and oak (*Quercus* spp.) in large, high-severity fire patches in forests of the southwestern United States. In mixed evergreen forests of southwestern Oregon and northwestern California, Shatford et al. (2007) found several hundred conifer seedlings per hectare, even when plots were ≥300 m into high-severity fire patches, and even where native shrub cover was very high to complete (i.e., the conifers grew up through the shrub cover). Additional conifer regeneration occurred in successive postfire years. Donato et al. (2009) made similar findings in large, high-severity fire patches of the Biscuit fire in the same region. One recent study by the US Forest Service found relatively little natural conifer regeneration in high-severity fire patches in the northern Sierra Nevada and advocated for postfire logging and plantation establishment (Collins and Roller, 2013); however, a visit to this area by one of us (C.T.H.) found that the study sites had generally been clearcut before the fires (and thus there was little or no conifer seed source even before the fires occurred) or were nearly pure black oak stands before the fires.

While natural postfire conifer regeneration can be relatively lower with increasing distance into large, high-severity fire patches (Haire and McGarigal, 2010), the irregular nature of high-severity fire patch boundaries leads to a surprisingly small proportion of the area of large, high-severity fire patches within megafires that is more than a few hundred meters from the nearest live-tree edge (see Figure 5 in Halofsky et al., 2011). In light of this, and

FIGURE 2.6 High-severity patches within the Rim Fire (2013) in the Sierra region, showing complex early seral forests and conifer establishment a year later. (Photos by Doug Bevington.)

Continued

FIGURE 2.6—Cont'd

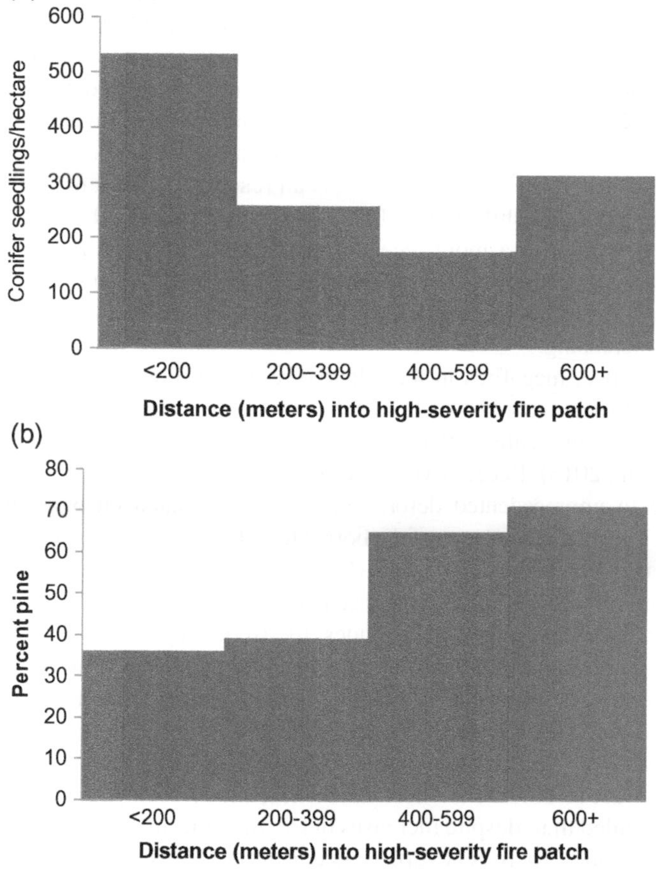

FIGURE 2.7 (a) Natural postfire conifer regeneration within a large, high-severity fire patch 1 year after the Rim Fire of 2013, in terms of seedling density and (b) the percentage of regeneration comprised of pine species (ponderosa and sugar pine) with increasing distance into the high-severity fire patch.

given the infrequency of megafires on the landscape scale, the spatially scarce interior areas of large, high-severity fire patches in megafires can play a vitally important role in maintaining complex early seral forest, dominated by native shrubs, for a relatively longer period of time before such habitat is replaced by conifer stands through natural succession. This can help to maintain populations of at-risk bird species associated with montane chaparral—many of which are declining as a result of fire suppression, postfire logging, and subsequent shrub removal and the establishment of artificial conifer plantations (Hanson, 2014).

2.6 ARE MEGAFIRES INCREASING?

Because megafires have received considerable attention, in part because of exurban sprawl into "firesheds" (areas where homes and structures abut fire-prone forests; Chapter 13), the public is naturally concerned about whether they are increasing in size or frequency. But determining whether megafires are increasing depends on the historical baseline chosen for comparisons with current conditions. The historical baseline is a primary factor in determinations of what may be considered characteristic, or "natural," versus uncharacteristic, or anthropogenic, with respect to any fire. In Chapter 9 we illustrate how more recent historical versus longer time lines can result in significant bias about fire increases, leading to shifting baseline perspectives (i.e., the baseline for comparisons is shifted to a more recent time line that may not reflect historical or evolutionary conditions under which fire-adapted communities evolved). The selection of the baseline, therefore, affects determinations about whether fires are increasing.

For instance, megafires in the Algarve region of Portugal, where there has been no historical record of such fires, have recently (within decades) been considered the "new reality" that is overwhelming local firefighting resources (Tedim et al., 2013). Recent (within decades) megafires in Amazonia have been attributed to unprecedented deforestation and associated changes in regional climates. The total annual area of the boreal forests of Canada and Alaska affected by fire has doubled since the 1960s (Kasischke and Turetsky, 2006), and some have correlated this increase with recent warming trends (Gillett et al., 2004). Moreover, over the past several decades, the frequency of fire years with $>1\%$ of the region burned in certain boreal ecozones increased from an average of 5 times per decade during the 1960s/1970s to an average of 13 times per decade during the 1980s/1990s. But are these time lines sufficient to declare today's megafires as truly "uncharacteristic?"

In contrast to studies of short time lines, those covering longer intervals have concluded that, despite increases in fire in some areas of Canada's forests in recent decades, there is nevertheless currently far less fire than historically, such that there is now only about one-fourth as much annual fire as there was circa 1850 (Bergeron et al., 2004). Similarly, a meta-analysis of landscape-scale fires in forested regions of the western United States indicated much more variability in fire severity and extent in the historical record than previously reported. This was based on multiple lines of evidence, including early surveys (1880s General Land Office surveys), stand reconstructions (based on plot-level data), and charcoal evidence of large fires from thousands of years ago (Odion et al., 2014). To the surprise of many, based on historical comparisons, these researchers documented a current high-severity fire deficit rather than a surplus, as often assumed.

Using General Land Office records and other evidence, high-severity fire deficits have been documented for the Eastern Cascades, Northern and Central Rocky Mountains, Klamath, Sierra Nevada, and the southwestern United States (Odion and Hanson, 2013; Baker, 2014; Hanson and Odion, 2014; Odion et al., 2014). A stand-age analysis indicated that there is currently one-half to one-fourth as much high-severity fire, depending on the region, as there was before the early twentieth century in mixed-conifer and ponderosa pine forests of the western United States (Odion et al., 2014). While there are some equivocal indications of increases in fire severity (e.g., increases in the area of high-severity fire, but not the proportion of high-severity fire effects) in some regions, such as portions of the southwestern United States and the southern Rockies (Dillon et al., 2011), most studies have found no increase in fire severity in most forested regions of the western United States (Hanson et al., 2009; Collins et al., 2009; Dillon et al., 2011; Miller et al., 2012; Hanson and Odion, 2014). One recent study reported an increase in fire severity in the Sierra Nevada (Miller and Safford, 2012), whereas a larger analysis found no such trend (Hanson and Odion, 2014; Hanson and Odion, 2015). Thus, whether the gaps between current and historical fire occurrence are closing in these regions is unclear, suggesting that actions to allow or facilitate more managed wildland fire (fires caused by lightning that are allowed to burn for ecological benefit) should be encouraged where ongoing fire deficits exist and fires can burn under safe conditions.

In Chapter 9 we discuss what the future of megafires might be in a changing climate that, based on regional, scaled-down climate change projections, may begin to close fire-area deficits over the coming decades with climate change. In most cases, however, postfire landscapes are not allowed to go through successional stages that result in high levels of biodiversity because they are so often logged after fires, followed by removal of native shrubs and the establishment of artificial conifer plantations (Chapter 11), putting numerous bird species at risk (Hanson, 2014). So, even though climate change may increase megafires, the ensuing postfire landscapes most often are degraded by postfire management, rather than enhanced by the fire. In fact, in general, the larger the megafire, the more severe the resource extraction feedback imposed, resulting in extensive and intensive landscape-scale degradation that is at least on par with the degradation of green forests (see Chapter 11).

2.7 LANGUAGE MATTERS

Because of the cultural fear of and misunderstanding about forest fires, as well as the climate of political and economic opportunism facilitated by postfire logging policies, having an objective dialogue about forest fires in general, and megafires in particular, is most difficult. The highly charged language so often used to describe fires, especially large, intense ones, exacerbates conflict. For a more ecologically literate public dialog to occur, and for forest conservation to be most effectively informed by current science, we submit that the vernacular of fire must become less charged and more objective. Currently, major newspapers and television stations, policymakers, and land managers commonly describe a particular forest fire as having "destroyed," "damaged," "devastated," "nuked," "razed," "moonscaped," or "consumed" the area within the fire's perimeter, or they describe a certain area of forest as having been "lost" to fire, or "deforested" by fire. As an alternative, we suggest that, at a minimum, neutral language, such as "affected," "spanned," or "covered" (e.g., "The fire affected over 50,000 ha" or "The fire perimeter now spans approximately 50,000 ha"), or language that reflects the now well-documented ecological benefits of large mixed-severity fires, such as "restored," be used. Fire "risk," "hazard," or "threat" would become fire "chance" or "probability," "hazardous fuels" would become "post-fire habitat," and "stand-destroying" or "stand-replacing" fire might instead become "stand-initiating" or "stand-renewing" fire. In this vein, especially in regions with ongoing fire deficits relative to the historical spatiotemporal extent of fire, a larger-than-typical fire year would no longer be a "bad" fire year or "the worst" fire year but, rather, would be an "above-average" fire year. Fundamentally unscientific terms such as "catastrophic wildfire" would no longer be used.

Until we change the way we think about, and describe, fire in our forests, efforts to provide greater protections for postfire habitat and the many rare and imperiled wildlife species associated with such habitat will continue to be at a disadvantage, as will efforts to encourage greater use of managed wildland fire in more remote forested areas. We believe that such a shift in our vernacular is more than warranted given the current state of scientific knowledge about the relative scarcity of postfire habitat and its great ecological importance to native biodiversity.

2.8 CONCLUSIONS

Postfire landscapes resulting from megafires that produce large, high-severity fire patches have become the forgotten seral stage (Swanson et al., 2011), devalued by most land managers, government officials, and even some scientists and conservation groups. We encourage conservationists and ecologists to refrain from calling the ensuing postfire landscape a "catastrophe," particularly given the fire deficit in some places and the ecological rarity of unlogged postfire landscapes. Megafires have been a top-down driver of ecological and evolutionary change for millennia and will continue to be a dominant natural force, despite command-and-control actions, because local weather effects mostly govern these fires. Whether they are increasing requires reconstruction of appropriate historical baselines, and more comprehensive analysis of the periodicity, scale, and severity of large fires over long time lines and spatial gradients, relative to those previously used to describe these fires. Megafires may become more of the new "norm" in some areas as a result of anthropogenic climate change that has the potential to drive more of these fires over shorter time lines, though the role of multidecadal climate cycles, such as ocean wind fluctuations, must always be taken into account (LAT, 2014).

We recommend that ecologists and land managers set up permanent plots in megafire areas to better understand long-term effects on fire-dependent communities in terms of their resilience to fire, particularly reburns. Land managers wishing to determine whether these fires are characteristic should use more comprehensive historical accounts to avoid shifting baseline perspectives that result in management actions that exacerbate widespread declines in species that are dependent on the pulse of complex early seral vegetation (Hanson, 2014). Megafires may be the only large pulse of biological legacies that a particular area receives for many decades, or even centuries, and the structural elements and landscape heterogeneity produced by these events is not re-created by management. Megafire pulses create levels of native biodiversity and wildlife abundance comparable to, or greater than, those found in unburned old forest.

By comparison, postfire management often is associated with chronic disturbances that operate outside the bounds of historical and evolutionary processes in terms of patch sizes, disturbance periodicity, and intensity (see Chapter 11). Such events are not replacements for landscapes generated after fire because they typically remove most biological legacies created by large fires. Thus, from the standpoint of pyrophilous communities, large postfire landscapes of high ecological integrity (unlogged) are newly recovered habitat areas rather than habitats "recovering" from an undisturbed state. Managers, scientists, and conservation groups wishing to maintain fire-dependent biodiversity should plan for these areas in reserve design and treat them on par with the more celebrated old-growth forests, given the comparability of biodiversity and rarity (DellaSala et al., 2014).

Managers also wishing to maintain biodiversity over large landscapes, such as in national parks, lands with wilderness character, and intact areas, should allow the occurrence of megafires to operate as a top-down ecosystem process via appropriate wildfire responses (e.g., let them burn under safe conditions, as in the case of US national parks and increasingly in some US national forests). Regardless of their immediate scenic change (green to charred), within 1 to 3 years high-severity fire patches in conifer forests become rich, colorful landscapes with an abundance of flowers, shrubs, snags, downed logs, natural conifer regeneration, and the sights and sounds of abundant and diverse wildlife associated with this habitat. Moreover, from the standpoint of fire-dependent communities, beauty is in the eye of the beholder. Postfire landscapes shaped by megafires are dynamic places that are constantly changing as a result of natural successional processes; they are not biological wastelands, as often claimed or assumed.

Government officials and local communities living in firesheds need to prepare for megafires because these events will always be with us and may increase in frequency and extent in some regions as long as we keep pumping greenhouse gases into the atmosphere and deforesting large regions that otherwise sequester and store carbon for long periods (see Chapter 10). Effective land-use planning (e.g., zoning) is needed to restrict exurban sprawl into firesheds and to limit increasing damage to human structures. That is, with more people and structures occupying firesheds, the prospects for even greater socioeconomic losses from megafires will escalate, triggering increased attempts at ecosystem-degrading command-and-control actions, unless proactive steps are taken to reduce fire risk in the home-ignition zone (see Chapter 13). Attempting to squelch small fires before they become megafires may succeed in nondrought years; however, this only perpetuates command-and-control ecosystem degradation given the loss of biological pulses. Once megafires do occur—generally under extreme weather conditions—they are self-reinforcing events that are extinguished when the weather changes (summer to fall, monsoonal summer rains) regardless of what we do to them. A fundamental shift in thinking is needed to change public attitudes toward megafires in recognition of the substantial ecological benefits to forest ecosystems and biodiversity, particularly where ongoing fire deficits are occurring and megafires are burning safely in the backcountry.

APPENDIX 2.1 FIRES OF HISTORICAL SIGNIFICANCE FROM RECORDS COMPILED BY THE NATIONAL INTERAGENCY FIRE CENTER (HTTP://WWW.NIFC.GOV/FIREINFO/FIREINFO_STATS_HISTSIGFIRES.HTML)

Date	Name	Location	Hectares	Significance
October 1825	Miramichi and Maine fires	New Brunswick and Maine	1.2 million	160 Lives lost
1845	Great Fire	Oregon	600,000	Large area burned
1849	Yaquina	Oregon	180,000	Large area burned

Continued

Date	Name	Location	Hectares	Significance
1868	Coos	Oregon	120,000	Large area burned
October 1871	Peshtigo	Wisconsin and Michigan	1,512,000	1500 Lives lost in Wisconsin
	Great Chicago	Illinois	Undetermined	250 Lives lost
				17,400 Structures destroyed
September 1881	Lower Michigan	Michigan	1,000,000	169 Lives lost
				3000 Structures destroyed
September 1894	Hinckley	Minnesota	64,000	418 Lives lost
	Wisconsin	Wisconsin	Several million	Undetermined; some lives lost
February 1898	Series of South Carolina fires	South Carolina	1.2 million	Unconfirmed report of 14 lives lost and numerous structures and sawmills destroyed
September 1902	Yacoult	Washington and Oregon	≥400,000	38 Lives lost
April 1903	Adirondack	New York	254,800	Large area burned
August 1910	Great Idaho	Idaho and Montana	1.2 million	85 Lives lost
October 1918	Cloquet-Moose Lake	Minnesota	480,000	450 Lives lost
				38 Communities destroyed
September 1923	Giant Berkley	California	Undetermined	624 Structures destroyed and 50 city blocks leveled

Continued

Date	Name	Location	Hectares	Significance
August 1933	Tillamook	Oregon	124,400	1 Life lost
				Same area burned again in 1939
October 1933	Griffith Park	California	Undetermined	29 Lives lost and 150 people injured
August 1937	Blackwater	Wyoming	Undetermined	15 Lives lost and 38 people injured
July 1939	Northern Nevada	Nevada	Undetermined	5 Lives lost
				First recorded firefighting fatality in a sage brush fuel type
October 1943	Hauser Creek	California	4000	11 US Marines killed and 72 injuries
				Fire was started by gunnery practice
October 1947	Maine	Maine	82,271	16 Lives lost
1949	Mann Gulch	Montana	1736	13 Smokejumpers killed
July 1953	Rattlesnake	California	Undetermined	15 Lives lost
1956	Inaja	California	17,200	11 Lives lost
November 1966	Loop	California	Undetermined	13 El Cariso Hotshots lost their lives
1967	Sundance	Idaho	22,400	Burned 20,000 ha in just 9 hrs
September 1970	Laguna	California	70,170	382 Structures destroyed
July 1972	Moccasin Mesa	New Mexico	1072	Fire suppression activities destroyed many

Continued

Date	Name	Location	Hectares	Significance
				archeological sites, which resulted in a national policy to include cultural resource oversight in wildland fires on federal lands
July 1976	Battlement Creek	Colorado	Undetermined	5 Lives lost
July 1977	Sycamore	California	322	234 Structures destroyed
November 1980	Panorama	California	9440	325 Structures destroyed
1985	Butte	Idaho	Undetermined	72 Firefighters deployed fire shelters for 1 to 2 hrs
1987	Siege of '87	California	256,000	Valuable timber lost in the Klamath and Stanislaus National Forests
1988	Yellowstone	Montana and Idaho	634,000	Large area burned
September 1988	Canyon Creek	Montana	100,000	Large area burned
June 1990	Painted Cave	California	1960	641 Structures destroyed
	Dude Fire	Arizona	9667	6 Lives lost
				63 Homes destroyed
October 1991	Oakland Hills	California	600	25 Lives lost and 2900 structures destroyed
August 1992	Foothills Fire	Idaho	102,800	1 Life lost

Continued

Date	Name	Location	Hectares	Significance
1993	Laguna Hills	California	6800	366 Structures destroyed in 6 hrs
July 1994	South Canyon Fire	Colorado	742	14 Lives lost
	Idaho City Complex	Idaho	61,600	1 Life lost
August 1995	Sunrise	Long Island	2000	Realization that the East can have fires similar to the West
August 1996	Cox Wells	Idaho	87,600	Largest fire of the year
June 1996	Millers Reach	Alaska	14,934	344 Structures destroyed
July 1997	Inowak	Alaska	244,000	Threatened 3 villages
1998	Volusia Complex	Florida	44,452	Thousands of people evacuated from several counties
1998	Flagler/St. John	Florida	37,862	Forced the evacuation of thousands of residents
August 1999	Dunn Glen Complex	Nevada	115,288	Largest fire of the year
August-November 1999	Big Bar Complex	California	56,379	Series of fires caused several evacuations during a 3.5-month period
September-November 1999	Kirk Complex	California	34,680	Hundreds of people were evacuated by this complex of fires that burned for almost 3 months

Continued

Date	Name	Location	Hectares	Significance
May 2000	Cerro Grande	New Mexico	19,060	Originally a prescribed fire, 235 structures destroyed and the Los Alamos National Laboratory damaged
July 2001	Thirtymile	Washington	3720	14 Fire shelters were deployed
				4 Lives lost
June 2002	Hayman	Colorado	54,400	600 Structures destroyed
	Rodeo-Chediski	Arizona	184,800	426 Structures destroyed
July 2003	Cramer	Idaho	5538	2 Lives lost
October 2003	Cedar	California	110,000	2400 Structures destroyed
				15 Lives lost
2004	Taylor Complex	Alaska	522,237	Alaska fires during 2004 burned over 2.54 million hectares
June 2005	Cave Creek Complex	Arizona	99,324	11 Structures destroyed
				Largest fire in the Sonoran Desert ever recorded
March 2006	East Amarillo Complex	Texas	362,898	80 Structures destroyed
				12 Lives lost
				Largest fire during the 2006 fire season
April 2007	Big Turnaround Complex	Georgia	155,207	Largest fire for the US Fish & Wildlife Service outside of Alaska

Continued

Date	Name	Location	Hectares	Significance
July 2007	Murphy Complex	Idaho	260,806	One of the largest fires in Idaho
2010	Long Butte	Idaho	120,000	
	Jefferson	Idaho	43,600	
2010	Four Mile Canyon	Colorado	2500	A wind-driven fire northwest of Boulder, Colorado, burned more than 170 structures and one fire engine
	Bastrop County Complex	Texas		1400 Residences burned in 3 days and two civilians were killed
June 2010	Schultz	Arizona	6000	Threatened hundreds of homes; a 12-year-old girl was tragically killed by flash floods that came out of the area burned by this fire
Jun 2011	Las Conchas	New Mexico	62,400	Threatened the Los Alamos National Laboratory
	Wallow	Arizona and New Mexico	215,200	Largest single fire ever recorded in the lower 48 states
August 2011	Pagami Creek	Minnesota	37,600	A significant 4-day wind event caused 32,800 ha to burn in late August and early September

Continued

Date	Name	Location	Hectares	Significance
May 2012	Whitewater-Baldy	New Mexico	119,138	Largest fire in New Mexico
June 2012	Waldo Canyon	Colorado	7579	346 Homes burned
	White Draw	South Dakota	3600	C-130 Airtanker crash kills four crewmembers
	Long Draw	Oregon	223,051	One of the largest fires in Oregon
June 2013	Yarnell Hill	Arizona	3360	19 Lives lost
August 2013	Rim	California	102,926	Largest fire in Sierra Nevada since formal records began in the 1930s

REFERENCES

Abrams, M.D., Sprugel, D.G., Dickman, D.I., 1985. Multiple successional pathways on recently disturbed jack pine sites in Michigan. For. Ecol. Manag. 10, 31–48.

Attiwill, P.M., Adams, M.A., 2013. Megafires, inquiries and politics in the eucalypt forests of Victoria, south-eastern Australia. For. Ecol. Manag. 294, 45–53.

Attiwill, P.M., Binkley, D., 2013. Exploring the megafire reality: a "Forest Ecology and Management" conference. For. Ecol. Manag. 294, 1–3.

Baker, W.L., 2012. Implications of spatially extensive historical data from surveys for restoring dry forests of Oregon's eastern Cascades. Ecosphere 3, Article 23.

Baker, W.L., 2014. Historical forest structure and fire in Sierran mixed-conifer forests reconstructed from General Land Office survey data. Ecosphere 5, Article 79.

Bekker, M.F., Taylor, A.H., 2010. Fire disturbance, forest structure, and stand dynamics in montane forest of the southern Cascades, Thousand Lakes Wilderness, California, USA. Ecoscience 17, 59–72.

Bergeron, Y., Gauthier, S., Flannigan, M., Kafka, V., 2004. Fire regimes at the transition between mixedwood and coniferous boreal forest in northwestern Quebec. Ecology 85, 1916–1932.

Beschta, R.L., Donahue, D.L., DellaSala, D.A., Rhodes, J.J., Karr, J.R., O'Brien, M.H., Fleischner, T.L., Williams, C.D., 2013. Adapting to climate change on western public lands: addressing the ecological effects of domestic, wild, and feral ungulates. Environ. Manag. 51, 474–491.

Beyers, J.L., 2004. Postfire seeding for erosion control: effectiveness and impacts on native plant communities. Conserv. Biol. 18, 947–956.

Bird, M.I., Cali, J.A., 1998. A million-year record of fire in sub-Saharan Africa. Nature 394, 767–769.

Bond, M.L., Lee, D.E., Siegel, R.B., Ward Jr., J.P., 2009. Habitat use and selection by California Spotted Owls in a postfire landscape. J. Wildl. Manag. 73, 1116–1124.

Buchalski, M.R., Fontaine, J.B., Heady III, P.A., Hayes, J.P., Frick, W.F., 2013. Bat response to differing fire severity in mixed-conifer forest, California, USA. PLoS One 8, e57884.

Burnett, R.D., Taillie, P., Seavy, N., 2010. Plumas Lassen Study 2009 Annual Report. U.S. Forest Service, Pacific Southwest Region, Vallejo, CA.

Cahoon, D.R., Stocks, B.J., Levine, J.S., Cofer, W.R., Pierson, J.M., 1994. Satellite analysis of the severe 1987 forest fires in northern China and southeastern Siberia. J. Geophys. Res. 99, 18627–18638.

Clar, S., Drossel, B., Schwabl, F., 1996. Forest fires and other examples of self-organized criticality. J. Phys. Condens. Matter 8, 6803.

Cocking, M.I., Varner, J.M., Knapp, E.E., 2014. Long-term effects of fire severity on oak-conifer dynamics in the southern Cascades. Ecol. Appl. 24, 94–107.

Collins, B.M., Roller, G.B., 2013. Early forest dynamics in stand-replacing fire patches in the northern Sierra Nevada, California, USA. Landsc. Ecol. 28, 1801–1813.

Collins, B.M., Miller, J.D., Thode, A.E., Kelly, M., van Wagtendonk, J.W., Stephens, S.L., 2009. Interactions among wildland fires in a long-established Sierra Nevada natural fire area. Ecosystems 12, 114–128.

Colombaroli, D., Gavin, D.G., 2010. Highly episodic fire and erosion regime over the past 2000 y in the Siskiyou Mountains, Oregon. Proc. Natl. Acad. Sci. U. S. A. 107, 18909–18914.

Cone, T., 2013. Nearly 40 Percent of Rim Fire Land a Moonscape. Associated Press news story, Fresno, California, USA, September 19, 2013.

Crotteau, J.S., Varner III, J.M., Ritchie, M.W., 2013. Post-fire regeneration across a fire severity gradient in the southern Cascades. For. Ecol. Manag. 287, 103–112.

DellaSala, D.A., Bond, M.L., Hanson, C.T., Hutto, R.L., Odion, D.C., 2014. Complex early seral forests of the Sierra Nevada: what are they and how can they be managed for ecological integrity? Natural Areas J. 34, 310–324.

Dillon, G.K., Holden, Z.A., Morgan, P., Crimmins, M.A., Heyerdahl, E.K., Luce, C.H., 2011. Both topography and climate affected forest and woodland burn severity in two regions of the western US, 1984 to 2006. Ecosphere. 2, Article 130.

Dodge, I.R., 1876. The Black Hills: A Minute Description of the Routes, Scenery, Soil, Climate, Timber, Gold, Geology, Zoology, Etc. James Miller Publisher, New York.

Donato, D.C., Fontaine, J.B., Robinson, W.D., Kauffman, J.B., Law, B.E., 2009. Vegetation response to a short interval between high-severity wildfires in a mixed-evergreen forest. J. Ecol. 97, 142–154.

Donato, D.C., Campbell, J.L., Franklin, J.F., 2012. Multiple successional pathways and precocity in forest development: can some forests be born complex? J. Veg. Sci. 23, 576–584.

Egan, T., 2010. The Big Burn: Teddy Roosevelt and the Fire that Saved America. Houghton Mifflin Harcourt, Boston.

Fogg, A., Burnett, R.D., Steel, Z.L., 2014. Short Term Changes in Avian Community Composition within the Sierra Nevada's Massive Rim Fire. Point Blue Conservation Science, Petaluma, California, USA.

Fontaine, J.B., Donato, D.C., Robinson, W.D., Law, B.E., Kauffman, J.B., 2009. Bird communities following high-severity fire: response to single and repeat fires in a mixed evergreen forest, Oregon, USA. For. Ecol. Manag. 257, 1496–1504.

Ganey, J.L., Kyle, S.C., Rawlinson, T.A., Apprill, D.L., Ward Jr., J.P., 2014. Relative abundance of small mammals in nest core areas and burned wintering areas of Mexican spotted owls in the Sacramento Mountains, New Mexico. The Wilson Journal of Ornithology 126, 47–52.

Gannett, H., 1902. The Forests of Oregon. U.S. Geological Survey, Government Printing Office, Washington, D.C..

Ganteaume, A., Jappiot, M., 2013. What causes large fires in southern France? For. Ecol. Manag. 294, 76–85.

Gillett, N.P., Weaver, A.J., Zwiers, F.W., Flannigan, M.D., 2004. Detecting the effect of climate change on Canadian forest fires. Geophys. Res. Lett. 31, L18211.

Graves, H.S., 1899. The Black Hills forest reserve. In: The Nineteenth Annual Report of the Survey, 1897-1898. Part V. Forest Reserves. U.S. Geological Survey, Washington, D.C, pp. 67–164.

Haire, S.L., McGarigal, K., 2008. Inhabitants of landscape scars: succession of woody plants after large, severe forest fires in Arizona and New Mexico. Southwest. Nat. 53, 146–161.

Haire, S.L., McGarigal, K., 2010. Effects of landscape patterns of fire severity on regenerating ponderosa pine forests (*Pinus ponderosa*) in New Mexico and Arizona, USA. Landsc. Ecol. 25, 1055–1069.

Halofsky, J.E., Donato, D.C., Hibbs, D.E., Campbell, J.L., Donaghy Cannon, M., Fontaine, J.B., Thompson, J.R., Anthony, R.G., Bormann, B.T., Kayes, L.J., Law, B.E., Peterson, D.L., Spies, T.A., 2011. Mixed-severity fire regimes: lessons and hypotheses from the Klamath-Siskiyou ecoregion. Ecosphere. 2, Article 40.

Hanson, C.T., 2007. Post-fire management of snag forest habitat in the Sierra Nevada. Ph.D. Dissertation, University of California at Davis, Davis, California, USA.

Hanson, C.T., 2013. Habitat use of Pacific fishers in a heterogeneous post-fire and unburned forest landscape on the Kern Plateau, Sierra Nevada, California. Open For. Sci. J. 6, 24–30.

Hanson, C.T., 2014. Conservation concerns for Sierra Nevada birds associated with high-severity fire. West. Birds 45, 204–212.

Hanson, C.T., 2015. Use of higher-severity fire areas by female Pacific fishers on the Kern Plateau, Sierra Nevada, California, USA. The Wildlife Society Bulletin (in press).

Hanson, C.T., North, M.P., 2009. Post-fire survival and flushing in three Sierra Nevada conifers with high initial crown scorch. Int. J. Wildland Fire 18, 857–864.

Hanson, C.T., Odion, D.C., 2014. Is fire severity increasing in the Sierra Nevada mountains, California, USA? Int. J. Wildland Fire 23, 1–8.

Hanson, C.T., Odion, D.C., 2015. Sierra Nevada fire severity conclusions are robust to further analysis: a reply to Safford et al. Intl J Wildland Fire 24, 294–295.

Hanson, C.T., Odion, D.C., DellaSala, D.A., Baker, W.L., 2009. Overestimation of fire risk in the Northern Spotted Owl Recovery Plan. Conserv. Biol. 23, 1314–1319.

Hanson, C.T., Odion, D.C., DellaSala, D.A., Baker, W.L., 2010. More-comprehensive recovery actions for Northern Spotted Owls in dry forests: reply to Spies et al. Conserv. Biol. 24, 334–337.

Hutto, R.L., 1995. Composition of bird communities following stand-replacement fires in Northern Rocky Mountain (U.S.A.) conifer forests. Cons. Bio. 9, 1041–1058.

Hutto, R.L., 2008. The ecological importance of severe wildfires: some like it hot. Ecol. Appl. 18, 1827–1834.

Jarvis, B., 2014. Scorched Earth: Extreme Wildfires may Mean Forests without Trees. Sierra Magazine, San Francisco, California, USA, January/February 2014.

Johnson, E.A., Wowchuk, D.R., 1993. Wildfires in the southern Canadian Rocky Mountains and their relationship to mid-trophospheric anomalies. Can. J. For. Res. 23, 1213–1222.

Kasischke, E.S., Bruhwiler, L.P., 2002. Emissions of carbon dioxide, carbon monoxide, and methane from boreal forest fires in 1998. J. Geophys. Res. 107, FFR2-1–FFR2-14.

Kasischke, E.S., Turetsky, M.R., 2006. Recent changes in the fire regimes across the North American boreal region—spatial and temporal patterns of burning across Canada and Alaska. Geophys. Res. Lett. 33, L09703.

Keeley, J.E., Zedler, P.H., 2009. Large, high-intensity fire events in southern California shrublands: debunking the fine-grain age patch model. Ecol. Appl. 19, 69–94.

Kinney, A., 1900. Forest and water. Post Publishing Company, Los Angeles, CA, USA.

Kotliar, N.B., Wiens, J.A., 1990. Multiple scales of patchiness and patch structure: a hierarchical framework for the study of heterogeneity. Oikos 59, 253–260.

Lang, D.M., Stewart, S.S., 1910. Reconnaissance of the Kaibab National Forest. Unpublished report. Northern Arizona University, Flagstaff.

LAT, 2014. West Coast Warming Is Linked to Natural Changes. Tony Barboza, Los Angeles Times, Los Angeles, California, USA, September 23, 2014.

Lee, D.E., Bond, M.L., 2015. Occupancy of California spotted owl sites following a large fire in the Sierra Nevada, California. The Condor 117, 228–236.

Leiberg, J.B., 1900a. Bitterroot forest reserve. In: USDI Geological Survey, Twentieth Annual Report to the Secretary of the Interior, 1898-99, Part V. Forest Reserves. US Government Printing Office, Washington, D.C., pp. 317–410.

Leiberg, J.B., 1900b. Cascade range forest reserve, Oregon, from township 28 south to township 37 south, inclusive; together with the Ashland Forest Reserve and adjacent forest regions from township 28 south to township 41 south, inclusive, and from range 2 west to range 14 east, Willamette Meridian, inclusive. In: U.S. Geological Survey Annual Report 21 (Part V), pp. 209–498.

Leiberg, J.B., 1902. Forest conditions in the northern Sierra Nevada, California. USDI Geological Survey, Professional Paper No. 8. U.S. Government Printing Office, Washington, D.C.

Leiberg, J.B., 1903. Southern part of Cascade Range Forest Reserve. In: Langille, H.D., Plummer, F.G., Dodwell, A., Rixon, T.F., Leiberg, J.B., Gannett, H. (Eds.), Forest Conditions in the Cascade Range Forest Reserve, Oregon. Professional Paper No. 9, Series H, Forestry, 6. Department of the Interior, US Geological Survey, Government Printing Office, Washington, D.C., pp. 229–289.

Lin, J., Rinaldi, S., 2009. A derivation of the statistical characteristics of forest fires. Ecol. Model. 220, 898–903.

Lindenmayer, D., Blair, D., McBurney, L., Banks, S., 2010. Forest Phoenix: How a Great Forest Recovers After Wildfire. CSIRO Publishing, Victoria, Australia.

Lindenmayer, D.B., Hobbs, R.J., Likens, G.E., Krebs, C.J., Banks, S.C., 2011. Newly discovered landscape traps produce regimes shifts in wet forests. Proc. Natl. Acad. Sci. U. S. A. 108, 15887–15891.

Miller, J.D., Safford, H., 2012. Trends in wildfire severity: 1984 to 2010 in the Sierra Nevada, Modoc Plateau, and southern Cascades, California, USA. Fire Ecol. 8, 41–57.

Miller, J.D., Skinner, C.N., Safford, H.D., Knapp, E.E., Ramirez, C.M., 2012. Trends and causes of severity, size, and number of fires in northwestern California, USA. Ecol. Appl. 22, 184–203.

Moors, A., 2012. Occupancy and reproductive success of Mexican spotted owls in the Chiricahua Mountains 2012. Report to the Coronado National Forest. Supervisor's office, Moors Wildlife Management Services, Globe, Arizona.

Moors, A., 2013. Occupancy and reproductive success of Mexican spotted owls in the Chiricahua Mountains 2013. Report to the Coronado National Forest. Supervisor's office, Moors Wildlife Management Services, Globe, Arizona.

Morgan, P., Heyerdahl, E.K., Gibson, C.E., 2008. Multi-season climate synchronized forest fires throughout the 20th-century, northern Rockies, USA. Ecology 89, 717–728.

Morris, W.G., 1934. Forest fires in western Oregon and western Washington. Or. Hist. Q. 35, 313–339.

Niklasson, M., Granström, A., 2004. Short facts on Swedish fires with emphasis on fire history. In: Sigurgeirsson, A., Jõgiste, K. (Eds.), Natural Disturbances Dynamics as Components of

Ecosystem Management Planning: Abstracts and Short Papers from the Workshop of the SNS Network.pp. 22–27, Geysir, Iceland, October 11-15, 2003.

NYT, 1870. The Great Saguenay Fire. New York Times, New York, USA, July 18, 1870.

Odion, D.C., Hanson, C.T., 2013. Projecting impacts of fire management on a biodiversity indicator in the Sierra Nevada and Cascades, USA: the Black-backed Woodpecker. Open For. Sci. J. 6, 14–23.

Odion, D.C., Frost, E.J., J.R., Strittholt, Jiang, H., DellaSala, D.A., Moritz, M.A., 2004. Patterns of fire severity and forest conditions in the western Klamath Mountains, California. Conserv. Biol. 18, 927–936.

Odion, D.C., Moritz, M.A., DellaSala, D.A., 2010. Alternative community states maintained by fire in the Klamath Mountains, USA. J. Ecol. 98, 96–105.

Odion, D.C., Hanson, C.T., Arsenault, A., Baker, W.L., DellaSala, D.A., Hutto, R.L., Klenner, W., Moritz, M.A., Sherriff, R.L., Veblen, T.T., Williams, M.A., 2014. Examining historical and current mixed-severity fire regimes in ponderosa pine and mixed-conifer forests of western North America. PLoS One 9, e87852.

Paine, R.T., Tegner, M.J., Johnson, E.A., 1998. Compounded perturbations yield ecological surprises. Ecosystems 1, 535–545.

Perry, D.A., Hessburg, P.F., Skinner, C.N., Spies, T.A., Stephens, S.L., Taylor, A.H., Franklin, J.F., McComb, B., Riegel, G., 2011. The ecology of mixed severity fire regimes in Washington, Oregon, and northern California. For. Ecol. Manag. 262, 703–717.

Pyne, S.J., 2007. Megaburning: the meaning of megafires and the means of management. In: Proceedings of the 4th International Wildland Fire Conference, Seville, Spain, May 13-17.

Pyne, S.J., 2008. Year of the Fires: The Story of the Great Fires of 1910. Mountain Press Publishing Company, Missoula, Montana. ISBN 978 0 87842 544 0, pp. 155–157, 175–176.

Raphael, M.G., Morrison, M.L., Yoder-Williams, M.P., 1987. Breeding bird populations during twenty-five years of postfire succession in the Sierra Nevada. Condor 89, 614–626.

San-Miguel-Ayanz, J., Manuel Moreno, J., Camia, A., 2013. Analysis of large fires in European Mediterranean landscapes: lessons learned and perspectives. For. Ecol. Manag. 294, 11–22.

Shatford, J.P.A., Hibbs, D.E., Puettmann, K.J., 2007. Conifer regeneration after forest fire in the Klamath-Siskiyous: how much, how soon? J. For. 2007, 139–146.

Sherriff, R.L., Platt, R.V., Veblen, T.T., Schoennagel, T.L., Gartner, M.H., 2014. Historical, observed, and modeled wildfire severity in montane forests of the Colorado Front Range. PLoS One 9, e106971.

Siegel, R.B., Wilkerson, R.L., 2005. Short- and Long-Term Effects of Stand Replacing Fire on a Sierra Nevada Bird Community. The Institute for Bird Populations, Point Reyes Station, CA.

Skinner, W.R., Shabbar, A., Flannigan, M.D., Logan, K., 2006. Large forest fires in Canada and the relationship to global sea surface temperatures. J. Geophys. Res. Atmos. 111, 27.

Smucker, K.M., Hutto, R.L., Steele, B.M., 2005. Changes in bird abundance after wildfire: importance of fire severity and time since fire. Ecol. Appl. 15, 1535–1549.

Stavros, E.N., Abatzoglou, J., Larkin, N.K., McKenzie, D., Steel, E.A., 2014. Climate and very large wildland fires in the contiguous western USA. Int. J. Wildland Fire 23 (7), 899–914. http://dx.doi.org/10.1071/WF13169.

Swanson, M.E., Franklin, J.F., Beschta, R.L., Crisafulli, C.M., DellaSala, D.A., Hutto, R.L., Lindenmayer, D., Swanson, F.J., 2011. The forgotten stage of forest succession: early successional ecosystems on forest sites. Front. Ecol. Environ. 9, 117–125.

Tedim, F., Remelgado, R., Borges, C., Carvalho, S., Martins, J., 2013. Exploring the occurrence of megafires in Portugal. For. Ecol. Manag. 294, 86–96.

Thompson, J.R., Spies, T.A., 2009. Vegetation and weather explain variation in crown damage within a large mixed-severity wildfire. For. Ecol. Manag. 258, 1684–1694.

Trombulak, S.C., Frissell, C.A., 2000. Review of ecological effects of roads on terrestrial and aquatic communities. Conserv. Biol. 14, 18–30.

Turner, M.G., Dale, V.H., 1998. Comparing large, infrequent disturbances: what have we learned? Ecosystems 1, 493–496.

Turner, M.G., Baker, W.L., Peterson, C.J., Peet, R.K., 1998. Factors influencing succession: lessons from large, infrequent natural disturbances. Ecosystems 1, 511–523.

USDA, 2014. Rim Fire Recovery Project, Final Vegetation Report. U.S. Department of Agriculture, Forest Service, Stanislaus National Forest, Sonora, California, USA.

Wein, R.W., Moore, J.M., 1977. Fire history and rotations in the New Brunswick Acadian Forest. Can. J. For. Res. 7, 285–294.

Whitlock, C., 1992. Vegetational and climatic history of the Pacific Northwest during the Last 20,000 years: implications for understanding present-day biodiversity. Northwest Environ. J. 8, 5–28.

Whittaker, R.H., 1960. Vegetation of the Siskiyou Mountains, Oregon and California. Ecol. Monogr. 30, 279–338.

Williams, J., 2013. Exploring the onset of high-impact megafires through a forest land management prism. For. Ecol. Manag. 294, 4–10.

Williams, M.A., Baker, W.L., 2012a. Spatially extensive reconstructions show variable-severity fire and heterogeneous structure in historical western United States dry forests. Glob. Ecol. Biogeogr. 21, 1042–1052.

Williams, M.A., Baker, W.L., 2012b. Comparison of the higher-severity fire regime in historical (A.D. 1800s) and modern (A.D. 1984-2009) montane forests across 624,156 ha of the Colorado Front Range. Ecosystems 15, 832–847.

Williams, M.A., Baker, W.L., 2014. High-severity fire corroborated in historical dry forests of the western United States: Response to Fule´ et al. Glob. Ecol. Biogeogr. 23, 831–835.

Wu, J., Loucks, O.L., 1995. From balance of nature to hierarchical patch dynamics: a paradigm shift in ecology. Q. Rev. Biol. 70, 439–466.

Chapter 3

Using Bird Ecology to Learn About the Benefits of Severe Fire

Richard L. Hutto[1], Monica L. Bond[2] and Dominick A. DellaSala[3]
[1]*Division of Biological Sciences, University of Montana, Missoula, MT, USA,* [2]*Wild Nature Institute, Hanover, NH, USA,* [3]*Geos Institute, Ashland, OR, USA*

3.1 INTRODUCTION

In this chapter we do not provide an encyclopedic review of the more than 450 published papers that describe some kind of effect of fire on birds. In other words, we are not systematically proceeding through a litany of fire effects on birds of southeast pine forests, California chaparral, Australian eucalypt forests, South African fynbos, and so forth. Instead, we have chosen to highlight underappreciated principles or lessons that emerge from selected studies of birds in ecosystems born of, and maintained by, mixed- to high-severity fire. Those lessons show how important and misunderstood basic fire ecology is when it comes to managing fire-dependent forest lands and shrublands, and the lessons apply to all fire-dependent ecosystems that have historically experienced severe fire—fires that are severe enough to stimulate an ecological succession of plant communities (as described in Chapter 1). We also focus our attention primarily on conifer forest ecosystems of the western United States because they undergo an amazing transformation following severe fire and because studies of these systems clearly reveal how birds evolved with, and now require, severe fire. Insight that emerges from the study of bird populations is overlooked in management circles worldwide. This is unfortunate because the insight one can gain by studying the ecology of individual bird species argues strongly that severe fire needs to be maintained in the landscape if we hope to maintain the integrity of most fire-dependent ecological systems.

Most studies of fire effects on birds are disappointingly "empty" because they are merely lists of birds that benefit from or are hurt by fire; they are not placed in the broader context of what a self-sustaining fire-dependent system looks like. To understand whether a particular change in abundance is "good" or "bad" requires insight into what ought to be, which requires an understanding of the patterns that occur under conditions that are as natural as possible for any given vegetation system. That, in turn, requires replicated study of what we can expect to find after "natural" fire in any given system. Thus, a study of the effects of, say, prescribed understory fire on birds is meaningless without knowing what a "natural" fire in that system would ordinarily produce. Many studies might show that bird species A increases after a prescribed fire, but is that a good thing? If bird species B increases after postfire salvage logging, is that a good thing? If bird diversity is higher in one fire treatment versus another, is that a good thing? For studies of fire effects to be useful, we need to address questions that inform management by tapping into a solid understanding of what constitutes a "natural" response to fire, and that requires knowing something about the fire regime under which a given system evolved. Only through distribution patterns and adaptations of individual species (not through effects on bird guilds or on diversity and similar composite metrics) can we begin to understand which kind of fire regime necessarily gave rise to specific patterns of habitat use and to adaptations that have evolved over millennia. Birds are excellent messengers; they carry all the information we need to reconstruct the historical conditions under which they evolved. All we have to do is listen.

3.2 INSIGHTS FROM BIRD STUDIES

Lesson 1: The Effects of Fire Are Context Dependent; Species Respond Differently to Different Fire Severities and Other Postfire Vegetation Conditions

One extremely important lesson that has emerged from studies of the fire effects on birds is that a given effect depends entirely on the vegetation type, the kind of fire, and the time since the fire (Recher and Christensen, 1981; Woinarski and Recher, 1997). For years, individual bird species have been labeled as "positive responders" or "negative responders" or "mixed responders" when, in fact, any species can be all of the above. The actual response of a bird species (or of any species) to fire, then, is dependent on context. The earliest papers on fire effects rarely provided details about the nature of the fire being studied, so the first attempt to conduct a meta-analysis based on a compilation of published results of fire effects (Kotliar et al., 2002) necessarily generated a lot of "mixed" responses by birds because some papers said a species was positively affected and others said the same species was negatively affected by fire. The seeming disagreement among studies was, in most cases, a simple result of researchers looking at different postfire vegetation conditions and times since fire. It was not until Smucker et al. (2005) separated their data into categories of fire severity and time since the fire that responses began to look much more consistent among studies that share a particular vegetation type, fire type, and time since the fire. As soon as one accounts for these factors, it becomes clear that the responses of most bird species are quite consistent and that most bird species benefit from severe fire (as we will more fully discuss below).

Time Since Fire

Species that benefit from severe fire are not only those that flourish during the first year or two following the disturbance event. The same can be said for species that are restricted to years 2-4, years 5-10, or even years 50-100 following severe fire. In fact, *most* plant and animal species are present only during a limited time period following a disturbance. Therefore, *most* plant and animal species in disturbance-based systems depend on disturbance to periodically create the conditions they need. Many bird species that thrive after fire have been mislabeled as species hurt by fire because studies of bird response to fire typically involve only a brief period of time soon after the fire. For example, although Williamson's sapsucker (*Sphyrapicus thyroideus*) was labeled a "mixed responder" and brown creeper (*Certhia americana*) a "negative responder" in the meta-analysis by Kotliar et al. (2002), and the change in house wren (*Troglodytes aedon*) abundance was labeled "insignificant" in a recently published study by Seavy and Alexander (2014), each of these species typically reaches its peak abundance several years after a fire, as revealed in an 11-year postfire study conducted after the Black Mountain fire, which burned near Missoula, Montana, in 2003 (Figure 3.1). Thus, each species clearly benefits from severe fire when viewed in the proper (and perhaps very restricted) time frame after fire.

By extending the duration of a postfire study beyond the first few years after a fire, most bird species reveal a unimodal response to time since fire, and most benefit from fire; they reveal a greater probability of detection in the burned forest at some point during that postfire period than in the same forest before fire or in the surrounding unburned forest (Taylor and Barmore, 1980; Reilly, 1991a, 2000; Taylor et al., 1997; Hannon and Drapeau, 2005; Saab et al., 2007; Chalmandrier et al., 2013; Hutto, 2015). These results force one to appreciate that if for a period of time after a fire conditions remain better than they are in very old plant communities near the end of the late seral stage of succession, then disturbance is periodically necessary to create the conditions needed by that species. Thus a species being "hurt" in the short term by fire is not evidence that fire is somehow "bad" for that species and that it would have been better off without fire. In fact, once a system is beyond the ideal postdisturbance time period for a species, the only way to periodically "restore" conditions needed by that species is to disturb the system with another severe fire and then wait for the appropriate time period following disturbance again. The lesson is this: one cannot assess the effects of fire on any plant or animal species without examining whether the species is restricted to a period of time preceding the oldest possible vegetation condition.

A necessary consequence of different species occurring at different points in time following fire (in association with changes in vegetation type and structure) is that we must embrace natural severe disturbance processes because they create starting points for the development of the full range of vegetation-age

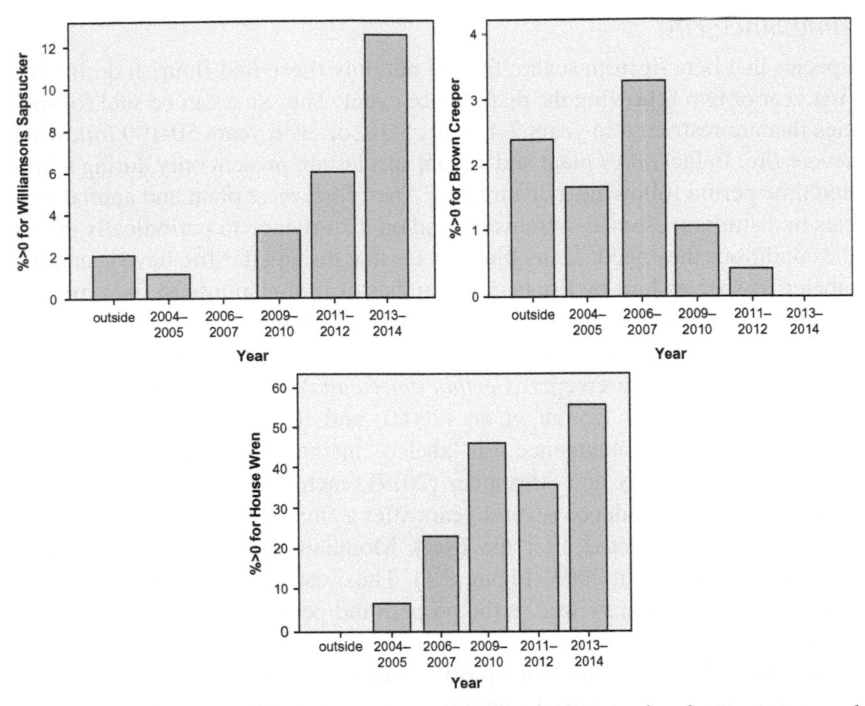

FIGURE 3.1 The probabilities of occurrence of Williamson's sapsucker, brown creeper, and house wren were significantly greater several years after the 2003 Black Mountain fire than they were either before the fire (as determined from survey data "outside" the burn perimeter in unburned, mixed-conifer forest of the same type) or during the first 2 years following the fire (R.L. Hutto, unpublished data; sample sizes exceed 150 point counts for each time period; $P < 0.05$, log linear analyses). Therefore, the benefit of severe fire for some species cannot be detected without restricting data collection to within a specific time period after the fire event.

categories, which, in turn, are needed for the maintenance of biological diversity (in particular beta diversity, the turnover in species number across gradients). Moreover, mixed-severity fires (which can result only from high-severity fire events) help provide a variety of kinds of starting points, which, in turn, also help maintain biological diversity (Smucker et al., 2005; Haney et al., 2008; Rush et al., 2012; Sitters et al., 2014; see also Chapters 4-6).

Old Growth

As already emphasized, most bird species clearly depend on severe fire to reset the clock, which stimulates development of the particular postdisturbance "age" to which they are best adapted. Still, many bird species are restricted in their habitat distribution to an end-of-the-line successional stage—they are dependent on old growth. There are also ecosystems (e.g., eucalyptus forests, chaparral) where severe fire is natural but where there are few, if any, early

FIGURE 3.2 Resprouting eucalyptus trees following a severe fire that burned through the area only months earlier. *(Photograph by Richard Hutto, taken in November 1999 near −34.284030°S, 150.725373°E in the tablelands above Wollongong, New South Wales, Australia.)*

fire-dependent bird species because many of the dominant plant species resprout, yielding a plant community structure and composition that "recovers" rapidly after fire (Figure 3.2). In these instances most bird species are associated with "mature" forms of those plant communities and would appear to do well if there were no fire at all (e.g., Taylor et al., 2012).

In all vegetation types that undergo plant succession following mixed- to high-severity fire, there will always be some bird species that depend on long-unburned vegetation. Therefore, discovering that those species are absent in the short term or "hurt" by fire is not unexpected, nor is it a necessarily a problem that needs to be addressed. The fact that fire temporarily removes large parts of a landscape from the pool of suitable conditions for those species is not a problem because the loss of suitable conditions is temporary, and there are usually nearby "refuges" of suitable conditions in places that have not burned for a long time (Bain et al., 2008; Leonard et al., 2014; Robinson et al., 2014; Winchell and Doherty, 2014). Natural systems exist as an ever-changing mosaic of different postfire ages—all vegetation ages are present at some point in space all the time. A significant problem emerges only when humans remove or degrade so much of the older vegetation through timber harvesting or land conversion that there is now a perceived risk of fire to those species that depend on older vegetation stands that are too few and far between. Understand clearly, however, that the absence of late-succession forest refuges is a problem that stems from excessive logging or development, not from the presence of fire per se.

Now that we are down to the last remaining old-growth forest remnants in California and Oregon, some believe that we should thin the forests around those remnants to protect them from fire. The effect of altering mature forest surrounding the last remaining old-growth remnants on the remnants themselves is, however, unknown. Moreover, as has been discussed in reference to eucalyptus forest systems, many old-growth forest patches are old precisely because they are situated in places that are relatively immune to severe fire (Bowman, 2000); the same is undoubtedly true of many old-growth mixed-conifer forest patches. Unburned forest patches surrounding unburned, old-growth forest patches also have been suggested to be important as dispersal corridors across which old-growth species may recolonize recently burned areas as succession proceeds toward later stages (Pyke et al., 1995; Robinson et al., 2014; Seidl et al., 2014). Therefore, proposals to thin the forest around remaining old-growth stands may be well intentioned but reflect a lack of appreciation for the resilience associated with plant communities born of, and maintained by, natural disturbance processes (a case in point is the spotted owl [*Strix occidentalis*]; see Box 3.1).

Postfire Vegetation Conditions

One must account not only for time since fire but also for fire severity and other forest conditions (e.g., vegetation composition and tree density) to adequately assess fire effects on animal species. Smucker et al. (2005) accounted for both time since fire and fire severity in an analysis of bird occurrence patterns following the Bitterroot fires of 2000 in Montana, and the results were profound.

BOX 3.1 Old-Growth Species and Severe Disturbance Events

There are a number of old-growth-dependent species in North American conifer forests, but severe fire may not pose anywhere near the threat to those species that one might suppose. Consider the spotted owl, one of the most iconic old-growth-dependent bird species in the Pacific Northwest, California, and Southwest (extending into northern Mexico). This federally listed threatened raptor typically nests, roosts, and forages in dense conifer and mixed-conifer-oak forests dominated by large (>50-cm diameter at breast height), older trees and peppered with big decadent snags and fallen logs. High levels of canopy cover (generally >60%) from overhead foliage is an important component of nesting and roosting stands; thus, spotted owls were long presumed to be seriously harmed where severe fire burned the forest canopy. Indeed, over the past several decades, most forest management efforts in the range of the spotted owl (a Forest Service management indicator species) has been driven by logging to prevent or reduce fire to "save" the owl, including the latest U.S. Fish & Wildlife Service recovery plans for the northern and Mexican spotted owls. Yet, the forests where the owl dwells have experienced mixed- and high-severity fire for millennia. So how do these birds actually respond when severe fire affects habitat within their home ranges?

Several studies have demonstrated that all three subspecies of spotted owl can survive and thrive (i.e., successfully reproduce) within territories that have experienced moderate- and high-severity fire (Bond et al., 2002; Jenness et al., 2004;

Continued

BOX 3.1 Old-Growth Species and Severe Disturbance Events—Cont'd

Roberts et al., 2011; Lee et al., 2012, 2013). Exceptionally high levels of severe fire in a nest stand can cause spotted owls to abandon that territory (Lee et al., 2013), but only a small fraction of sites ever exceed that threshold in any given fire. Moreover, a higher probability of abandonment after fire was documented only in a small geographical region where prefire forest patches were limited or isolated (Lee et al., 2013) and in areas that were logged after fire (Lee et al., 2012; Clark et al., 2013); reduced occupancy did not occur in unlogged areas where prefire forest cover was more abundant (Lee et al., 2012, 2013). For example, the year after the 2013 Rim Fire—one of the largest fires to occur in California within the past century—at least six pairs of California spotted owls (*S. occidentalis occidentalis*) were detected in sites where >70% of the "suitable habitat" around their nest stands burned at high severity. (At one occupied site severe fire burned 96% of the habitat!) Why do they stick around in burned territory? One study found California spotted owls selectively hunted (mostly for woodrats and gophers) in stands recently burned by severe fire when those burned forests were available to them and relatively near the nest or roost stand (Bond et al., 2009, 2013). Another study showed that during winter, Mexican spotted owls (*S. occidentalis lucida*) moved up to 14 km into burned forests where prey biomass was 2-6 times greater than in their breeding-season nesting areas (Ganey et al., 2014). Spotted owls are perch-and-pounce predators, so it is not surprising that they avoided foraging in areas that were logged after fire, as there were no longer any perch trees (Bond et al., 2009), nor is it surprising that postfire logging reduced site occupancy and survival rates (Clark et al., 2013; Lee et al., 2013). In these studies, spotted owls still preferred to nest and roost in green forests, underscoring the importance of unburned/low-severity refuges within the larger landscape mosaic of mixed-severity fire. Still, the point is that where severe fire is natural, even old-growth species can partake of its bounty. The spotted owl, too, is sending a message here: A natural fire regime provides a bedroom, nursery, and kitchen for even old-growth-dependent species, as long as the burned forest is left standing.

Despite this evidence, the U.S. Fish & Wildlife Service is now calling for aggressive, large-scale thinning in northern spotted owl habitat in dry forests as a means of reducing fire intensity (U.S. Fish and Wildlife Service, 2011). This "recovery" objective for the owl was developed over objections raised by scientists (Hanson et al., 2009, 2010) and professional societies such as The Wildlife Society and Society for Conservation Biology. Notably, Odion et al. (2014b) simulated changes in owl habitat over a four-decade period following fire and the kind of thinning proposed by federal land managers. The simulation study showed that thinning over large landscapes would remove 3.4-6.0 times more of their dense, late-successional habitat in the Klamath and dry Cascades, respectively, than forest fires would, even given a future increase in the amount of high-severity fire. Further, Baker (2015) documented that before extensive Euro-American settlement, mixed- and high-severity fires shaped dry forests in the Eastern Cascades of Oregon and provided important habitat for northern spotted owls there. These studies challenge the paradigm that severe fire is a serious threat to spotted owls, which evolved in landscapes shaped by such fire, and that extensive logging is needed to ameliorate this widely believed but overstated threat.

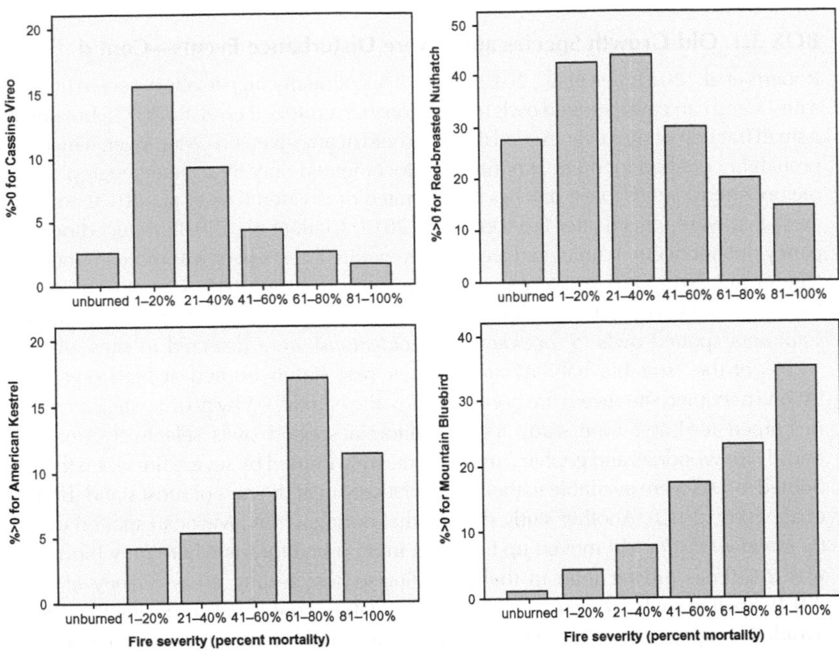

FIGURE 3.3 Example plots of the percentage occurrence of four mixed-conifer bird species in relation to fire severity in the first few years after fire. Data were drawn from 7043 survey points distributed across 110 different fires that burned since 1988 in western Montana. Sample sizes exceed 700 point counts per severity category. All patterns are significant ($P < 0.05$, log linear analyses). Note that each species is more abundant in burned than in unburned forest, and each is relatively abundant at a level of burn severity (percentage of tree mortality) that differs from that occupied by the other species. Scientific names for birds from top left clockwise to bottom right are: *Vireo cassinii, Sitta canadensis, Sialia currucoides, Falco sparverius.*

Once they accounted for fire severity alone, it became abundantly clear that many of the same bird species that had been labeled as "mixed responders" to fire by others (e.g., Kotliar et al., 2002) were not at all mixed in their response to fire. The importance of fire severity is strikingly apparent in even the simplest graphs of percentage occurrence across severity categories (Figure 3.3).

Lesson 2: Given the Appropriate Temporal and Vegetation Conditions, Most Bird Species Apparently Benefit from Severe Fire

After we combine information on the time since fire, fire severity, and perhaps one or two additional vegetation variables, most bird species apparently benefit from severe fire. For each species there is a particular combination of burned forest variables that creates ideal conditions for that species, as evidenced by an abundance that exceeds that in a long-unburned patch of the same vegetation type. Indeed, when Hutto and Patterson (2015) considered just two fire-context variables (time since fire and fire severity), they found 46 of 50 species to be more abundant in some combination of those two variables than in long-unburned stands (Figure 3.4). Thus, not only are most species relatively abundant in one burned forest condition or another, but the average point in space and time occupied by each species is also species specific (Figure 3.5).

As an introduction to some of the fascinating biology surrounding severely burned forests, consider the following bird species. The black-backed woodpecker (*Picoides arcticus*), American three-toed woodpecker (*Picoides dorsalis*), hairy woodpecker (*Picoides villosus*), northern flicker (*Colaptes auratus*), and Lewis's woodpecker (*Melanerpes lewis*) are all more abundant in severely burned than unburned mixed-conifer forest (see patterns of habitat occurrence for four of the five species in Figures 3.11 and 3.12) because of an abundance of food (beetle larvae and ants) and potential nest sites associated with standing

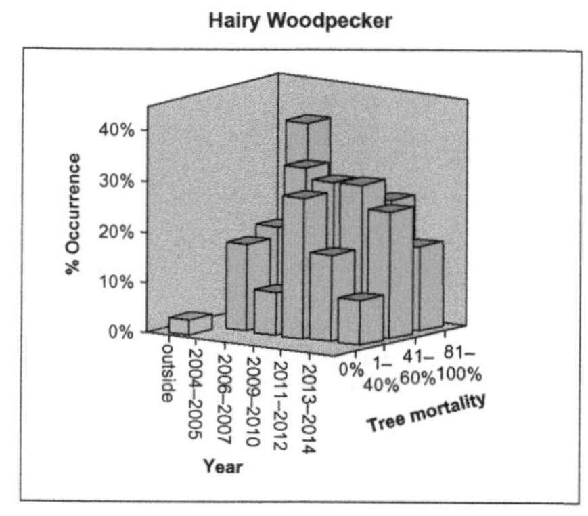

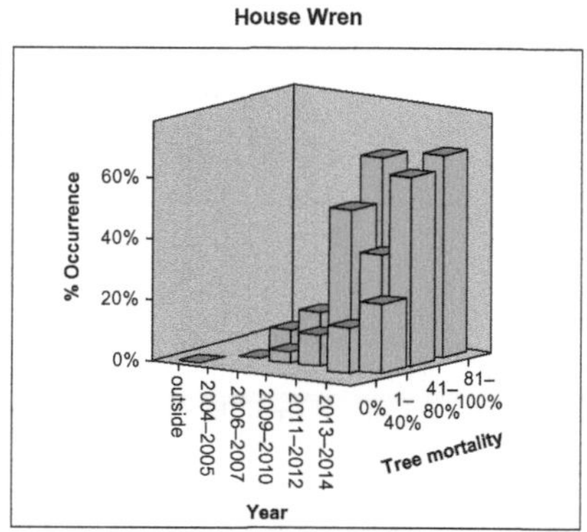

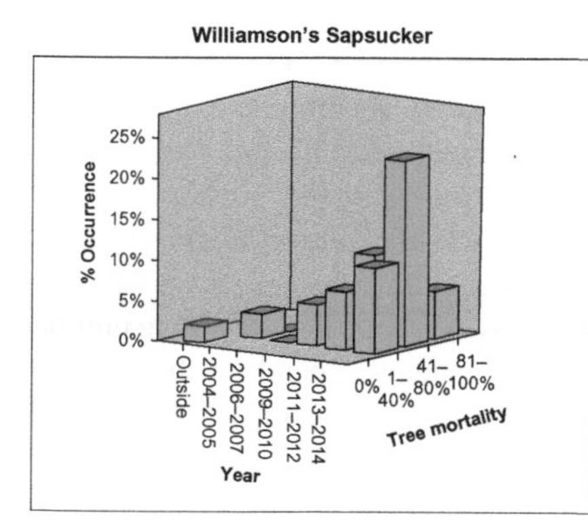

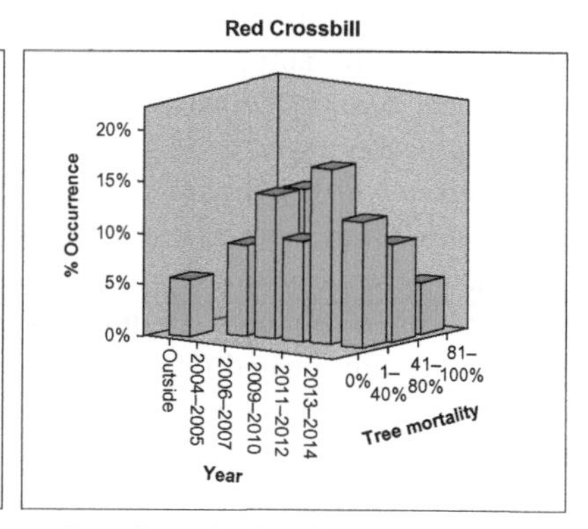

FIGURE 3.4 Example plots of percentage occurrence for various mixed-conifer bird species in relation to both time since fire and fire severity after the 2003 Black Mountain fire near Missoula, Montana (R.L. Hutto, unpublished; sample sizes exceed 35 point counts for each time-by-severity category; all patterns are significantly nonrandom as determined by log linear analyses [$P < 0.05$]). The examples were selected to illustrate that each species is more abundant in burned than in unburned forest (the occurrence rate in unburned forest shown in the first time period), and each is most abundant in a different combination of time since fire and burn severity (percentage of tree mortality).

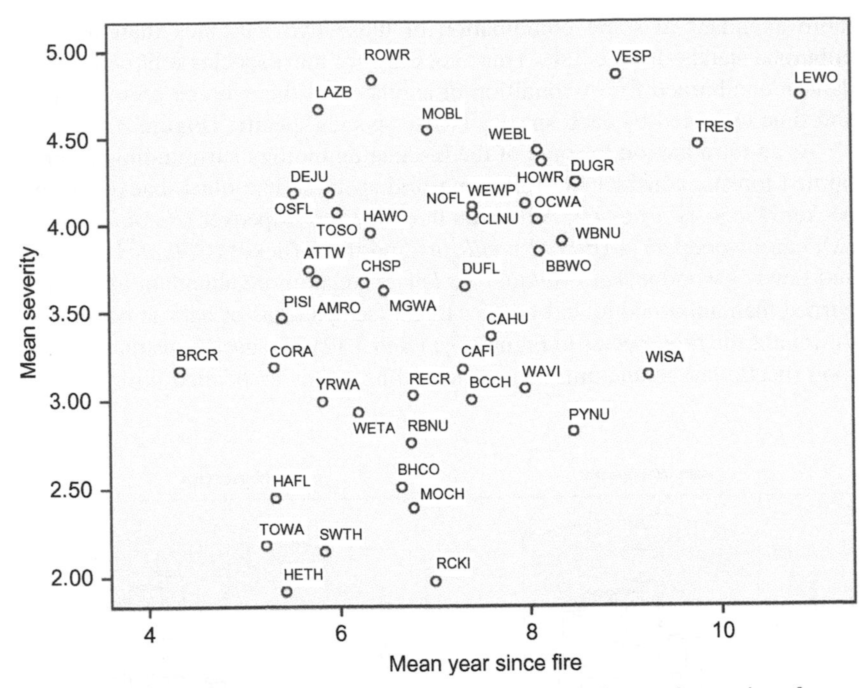

FIGURE 3.5 In combination, the mean time since fire and mean fire severity at points of occurrence for each of 46 (mnemonically coded) species differs from that of every other species. Mean values were calculated from the kind of data presented in Figure 3.4.

dead trees. The Williamson's sapsucker and olive-sided flycatcher (*Contopus cooperi*) find the abrupt edges between severely burned and unburned forest to be ideal nest locations (Figure 3.6). A host of secondary cavity-nesting and snag-nesting species (e.g., northern hawk owl [*Surnia ulula*], great gray owl [*Strix nebulosa*], mountain bluebird [*Sialia currucoides*], western bluebird [*Sialia mexicana*], house wren, and tree swallow [*Tachycineta bicolor*]) benefit from new forest openings, where they find a mature-forest legacy of already existing broken-top snags (Figure 3.7), where a disproportionately large number of nest sites are located (Hutto, 1995). These species depend on the kinds of snags that become common only after a forest reaches the mature- to old-growth stage and then burns in a severe fire. A variety of species (e.g., flammulated owl [*Psiloscops flammeolus*], mountain bluebird, Townsend's solitaire [*Myadestes townsendi*], and dark-eyed junco [*Junco hyemalis*]) make use of the cavities created by burned-out root wads or uprooted trees that happen to blow down in the first few years after severe fire (Figure 3.8). Many species (e.g., Clark's nutcracker [*Nucifraga columbiana*], Cassin's finch [*Haemorhous cassinii*], red crossbill (*Loxia curvirostra*], and pine siskin [*Spinus pinus*]) take advantage of seeds that are released or made available in cones that open after severe fire

FIGURE 3.6 Williamson's sapsucker (left) and olive-sided flycatcher (right) are known to nest disproportionately often near the abrupt edges between severely burned and unburned forest. *(Photographs by Richard Hutto (left) and Bruce Robertson (right))*.

FIGURE 3.7 Compared with burned trees with intact tops, broken-top snags that were already snags before the fire burned are used disproportionately more often as nest sites by cavity-nesting bird species. The black-backed woodpecker also roosts almost entirely in burned-out hollows, forked trunks, or other relatively unusual structures that create crevices in "deformed" snags that existed before the forest burned (Siegel et al., 2014). Pictured (left to right) are a young hairy woodpecker in its nest cavity, an American robin (*Turdus migratorius*) nest, and a northern flicker nest. The implications are profound—old-growth elements (snags) are really important to birds that depend on burned forest conditions, so burned, old-growth forests are as valuable to wildlife as unburned old-growth forests. *(Photographs by Richard Hutto.)*

FIGURE 3.8 The architecture of a burned forest becomes modified after trees begin to blow down in the first few years after a fire, and a number of bird species make use of the root wads as nest sites. A Townsend's solitaire nest is highlighted here. *(Photograph by Richard Hutto.)*

FIGURE 3.9 Few people seem to realize how important Clark's nutcrackers are as seed dispersers after severe fire in ponderosa pine forests. Pictured here are examples of a nutcracker extracting seeds from a ponderosa pine (*Pinus ponderosa*) cone that opened after fire (left) and a nutcracker with a throat pouch full of seeds in the scorched ground beneath a ponderosa pine canopy. *(Photographs by Richard Hutto.)*

(Figure 3.9). Still more bird species (e.g., calliope hummingbird [*Selasphorus calliope*], lazuli bunting [*Passerina amoena*], and MacGillivray's warbler [*Geothlypis tolmiei*]) use the shrub-dominated early seral stage for feeding and nesting and as display sites (Hutto, 2014).

Lesson 3: Not only Do Most Bird Species Benefit from Severe Fire, but Some also Appear to *Require* Severe Fire to Persist

The black-backed woodpecker has become an iconic indicator of severely burned forests because its distribution is nearly restricted to such conditions. Bent (1939) provided the first description of the unusual association between this woodpecker species and burned forests when he noted that Manly Hardy wrote to Major Bendire in 1895 about finding the woodpecker to be "... so abundant in fire-killed timber areas that I once shot the heads off six in a few minutes when short of material for a stew." This anecdote, reflecting the importance of severe fire, went largely unnoticed until the 1970s, when Dale Taylor undertook a study of birds in relation to time since fire in the Yellowstone and Grand Teton National Parks. His more systematic study uncovered the same remarkable pattern. Taylor was the first person to evaluate data drawn from a series of burned conifer forest stands of differing ages, and he found the appearance of the black-backed woodpecker to be restricted to the first few years after fire (Taylor and Barmore, 1980). A subsequent before-and-after fire study by Apfelbaum and Haney (1981) and studies of burned versus adjacent unburned forest by Niemi (1978), Pfister (1980), and Harris (1982) provided additional evidence that this bird species is strongly associated with burned forest conditions. Following the Rocky Mountain fires of 1988, Hutto (1995) conducted a more comprehensive study of the distribution of black-backed woodpeckers across a broad range of vegetation types. That study served to reinforce the notion that this species is an ideal indicator of severely burned mixed-conifer forest. More specifically, Hutto provided a meta-analysis of his own and already published bird survey data collected from burned forests and from more than a dozen unburned vegetation types; those data showed the black-backed woodpecker to be relatively restricted to burned forests. To address the potential problem of putting too much faith in distribution patterns derived from bird occurrence rates that were based on a variety of study durations and methods, Hutto subsequently coordinated the collection of standardized bird survey data from more than 18,000 points distributed across every major vegetation type in the U.S. Forest Service Northern Region. The results (Hutto, 2008) were strikingly similar to what earlier studies showed: one is hard pressed to find a black-backed woodpecker anywhere but in a recently burned forest (Figure 3.10).

Numerous studies (most published just in the past decade) provide additional detail that can help us better understand this remarkable association between the black-backed woodpecker and severely burned forests. Here we list some of the insights we have gained:

1. The magical appearance of woodpeckers within weeks of a fire (Blackford, 1955; Uxley, 2014) suggests that either smoke, or perhaps the fire or burned landscape itself, provides a stimulus for birds to colonize newly burned forests.

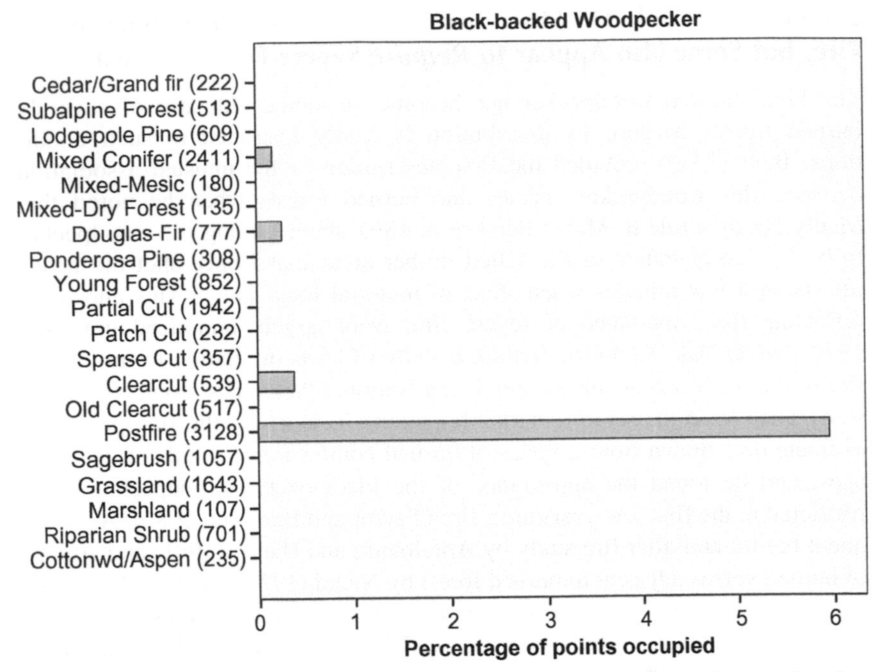

FIGURE 3.10 Histogram bars indicate the percentage of points (sample sizes in parentheses) at which the black-backed woodpecker was detected in each of 21 distinct vegetation types within northern Idaho and western Montana. The distribution is nonrandom ($X^2 = 559.43$; $df = 19$; $P < 0.0001$) and reveals that the black-backed woodpecker is highly specialized in its use of burned conifer forest. *(Data from Hutto (2008)).*

2. Breeding and nest densities increase more rapidly than expected on the basis of recruitment alone (Yunick, 1985; Youngman and Gayk, 2011), which suggests that the process of immigration after fire is significant.
3. Woodpecker diet, which is based mainly on wood-boring beetle larvae that feed almost exclusively on recently burned and killed trees (Murphy and Lehnhausen, 1998; Powell et al., 2002; Fayt et al., 2005), reflects the broad postfire change in animal community composition that accompanies severe fire.
4. The woodpecker's nonrandom use of forest patches containing dense, larger-diameter trees (Saab and Dudley, 1998; Saab et al., 2002, 2009; Nappi and Drapeau, 2011; Dudley et al., 2012; Seavy et al., 2012) that have burned at high rather than low severity (Schmiegelow et al., 2006; Koivula and Schmiegelow, 2007; Hanson and North, 2008; Hutto, 2008; Nappi and Drapeau, 2011; Youngman and Gayk, 2011; Siegel et al., 2013) is striking and consistent among studies.
5. The window of opportunity for occupancy by this species is not only soon after fire, but generally lasts only about a half-dozen years before the birds

(and the abundant native beetle populations) disappear (Taylor and Barmore, 1980; Apfelbaum and Haney, 1981; Murphy and Lehnhausen, 1998; Hoyt and Hannon, 2002; Saab et al., 2007; Nappi and Drapeau, 2009; Saracco et al., 2011).

6. The size of the home ranges of black-backed woodpeckers within burned forests are significantly smaller (indicating better quality habitat) than those outside burned forests (Rota et al., 2014b; Tingley et al., 2014). Even more telling is that nest success is significantly higher inside than outside burned forests (Nappi and Drapeau, 2009; Rota et al., 2014a).
7. Estimated population growth rates are insufficient to maintain a growing population outside burned forests (Rota et al., 2014a). Thus, although one could argue that low woodpecker densities in green-tree forests multiplied by a much larger unburned forest area might yield even more woodpeckers in green forests (Fogg et al., 2014), a sink area alone (no matter how large) can never yield a viable population of woodpeckers (Odion and Hanson, 2013).
8. The importance of severely burned forests as foraging locations for wintering black-backed woodpeckers is virtually unknown; the only detailed work so far (Kreisel and Stein, 1999) revealed densities that were an order of magnitude greater in burned than in unburned forests.

The biology surrounding this single bird species clearly reflects not only the ecological importance but also the necessity of severely burned forests, but major environmental organizations have yet to focus conservation efforts on burned forests (Schmiegelow et al., 2006), and management guidelines developed by state agencies to designate important wildlife habitats (e.g., https://www.dfg.ca.gov/biogeodata/cwhr/) do not even have burned conifer forests on their radar.

The distributional stronghold of the black-backed woodpecker might be considered to lie within the boreal forests of Canada, which nobody doubts are among the most severe-fire-dependent ecosystems in the world, but the bird's distribution south into the California Sierras and Rocky Mountains of the Intermountain West confirms that severe fires in those areas have been historically important as well. A North American forest bird species that is more narrowly restricted to a single forest condition does not exist; the black-backed woodpecker is the definition of a specialist. Everything about this bird species, including its distribution, territory size, breeding success, and even coloration pattern (which matches blackened trees), all indicate that this species needs expansive patches of severely burned forest to persist (Figure 3.11).

We have taken the liberty to provide extensive detail on this particular species because its ecological story carries significant management implications. Because public land managers have a responsibility to manage for the maintenance of all vertebrate species, finding even a single species that depends on severe fire should be enough to raise their awareness that severely burned

FIGURE 3.11 Black-backed woodpecker—a species that is relatively restricted in its distribution to severely burned forests. *(Photograph by Richard Hutto.)*

mixed-conifer forests provide necessary habitat as well. Thus the black-backed woodpecker is an ideal focal species for bringing attention to the fact that burned forest conditions are important to maintain in the landscape (DellaSala et al., 2014). The evolutionary history that has led to a strong association between burned forests and the woodpecker also raises questions about whether (as many assume) severe fires in mixed-conifer forests are really beyond the historical natural range of variation, whether we need to be thinning forests outside the wildland-urban interface to reduce fire severity, whether we need to be suppressing fire outside the wildland-urban interface, and whether we should "salvage" log trees (including important legacy trees; see Chapter 11) after fire. Yes, the story surrounding this focal species is important.

Bird Species in Other Regions That Seem to Require Severe Fire

Do any other bird species seem not only to benefit from but also to require severe fire to persist? The presence of a species in a specific environment and its absence elsewhere would be a clear indication that it depends on that particular environment. For species that occur across a range of environmental conditions, the places where they are relatively abundant are also likely to represent places that are required for population persistence because they persist in source areas and they are generally less abundant in, and their abundance is more variable through time in, more marginal areas (Pulliam, 1988; Sergio and Newton, 2003). Although the same level of biological detail that has been amassed for the black-backed woodpecker has not been collected for most other fire-associated bird species, the habitat distribution patterns of numerous bird species reveal that they are nowhere more abundant than in recently burned forests. For example, Hutto (1995) listed 15 species that were more abundant in recently burned forests than in any of 14 other vegetation types. Graphs generated from surveys conducted across an even broader range of vegetation types show just how striking these habitat distribution patterns can be: numerous species are nowhere more abundant than they are in severely burned forests (Hutto and Young, 1999) (Figure 3.12).

Many mixed-conifer bird species (e.g., black-backed woodpecker, American three-toed woodpecker, hairy woodpecker, northern flicker, olive-sided flycatcher, western wood-pewee [*Contopus sordidulus*], dusky flycatcher [*Empidonax oberholseri*], mountain bluebird, Townsend's solitaire, house wren, tree swallow, lazuli bunting, Clark's nutcracker, red crossbill) fall consistently into a short-term "benefit" category, as revealed either by some measure of abundance or nest success in studies of burned versus unburned or before versus after fire (Bock and Lynch, 1970; Bock et al., 1978; Taylor and Barmore, 1980; Apfelbaum and Haney, 1981; Raphael et al., 1987; Hutto, 1995; Kotliar et al., 2002; Hannah and Hoyt, 2004; Smucker et al., 2005; Mendelsohn et al., 2008; Seavy and Alexander, 2014). Even severely burned patches within conifer forests that we have come to associate with low-severity fire can provide critically important habitat for species like the buff-breasted flycatcher [*Moucherolle beige*] (Kirkpatrick et al., 2006; Conway and Kirkpatrick, 2007; Hutto et al., 2008).

One of the most celebrated examples of a fire specialist involves the federally endangered Kirtland's warbler (*Setophaga kirtlandii*). It occurs almost exclusively in young (5- to 23-year-old) jack pine (*Pinus banksiana*) forest historically created by severe fire (Walkinshaw, 1983). In addition, pairing success is significantly higher in burned than in unburned forests (98% vs. 58% success; Probst and Hayes, 1987). The need for severe fire is obvious not only because, historically, it must have taken severe fires to stimulate forest succession but also because of how its critically endangered population increased dramatically after a fire accidentally escaped within its breeding range (James and McCulloch, 1995). Managers have had difficulty trying to recreate conditions that mimic natural postfire conditions through the use of logging techniques (Probst and Donnerwright, 2003; Spaulding and Rothstein, 2009), and efforts to use these artificial means to maintain warbler populations miss the point. Conservation efforts should be directed toward maintaining severely burned forests, not toward finding a way around the natural fire disturbance process.

In Australia, where few species are thought to be restricted to recently burned shrubland or forest conditions, early colonists are viewed as generalists, and management concerns are focused on postfire decreases in late-succession specialists (Serong and Lill, 2012). Nevertheless, recent data from Lindenmayer

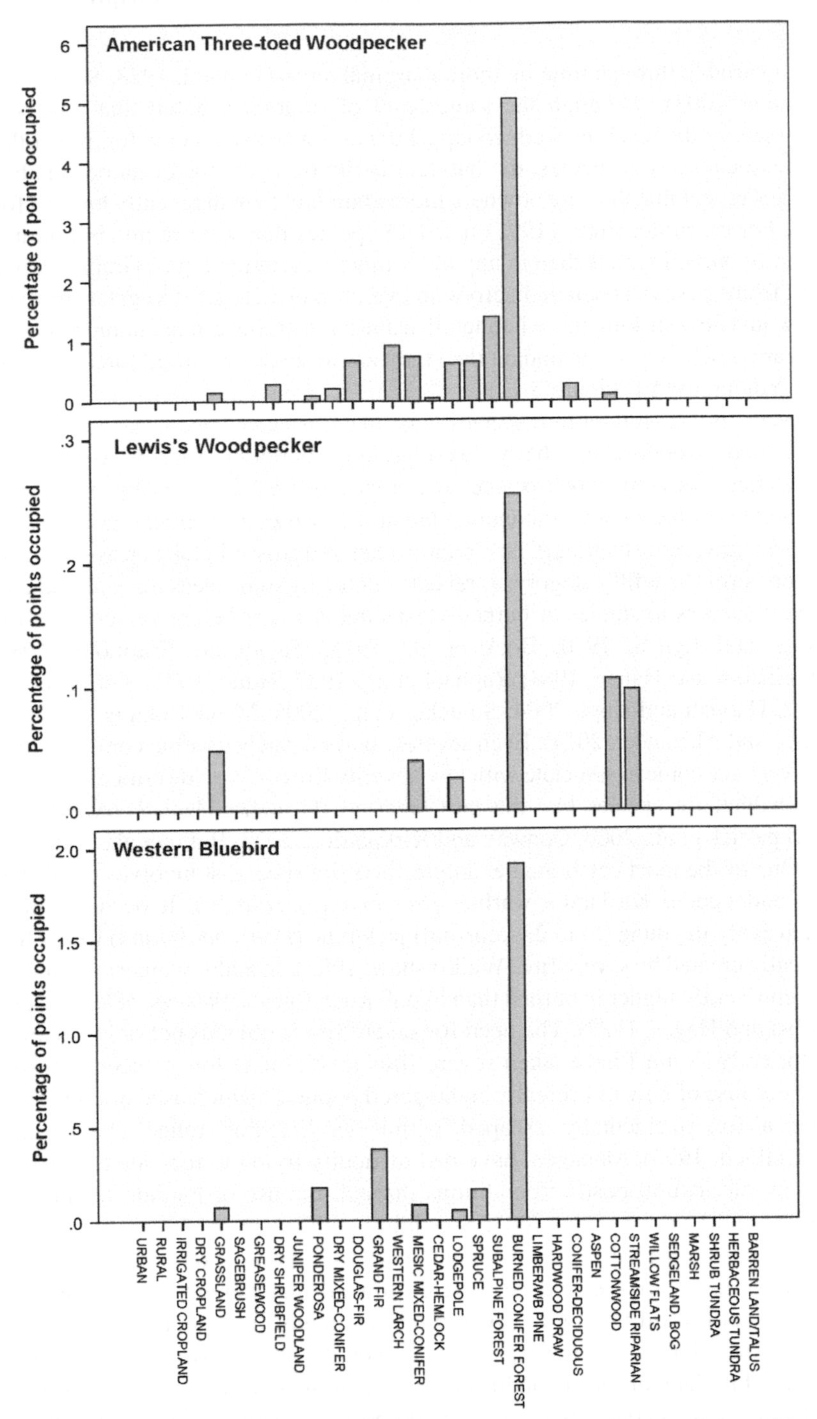

FIGURE 3.12 Several graphs depicting species that seem to be more abundant in burned forests than in any other vegetation type in the northern Rocky Mountains. Data were drawn from a subset of the Northern Region Landbird Monitoring Program database consisting of 20,000 survey points distributed across northern Idaho and western Montana.

et al. (2014) show that a number of bird species decline in abundance 1-2 years after moderate to severe fire but then return to levels comparable to, or *higher* than, those in unburned forests within 3 years following fire. Indeed, upon further inspection, we found that the superb fairywren (*Malurus cyaneus*), gray fantail (*Rhipidura albiscapa*), yellow-faced honeyeater (*Lichenostomus chrysops*), white-fronted honeyeater (*Purnella albifrons*), dusky robin (*Melanodryas vittata*), flame robin (*Petroica phoenicea*), willie wagtail (*Rhipidura leucophrys*), gray shrike-thrush (*Colluricincla harmonica*), varied sittella (*Daphoenositta chrysoptera*), apostlebird (*Struthidea cinerea*), white-browed scrubwren (*Sericornis frontalis*), brown thornbill (*Acanthiza pusilla*), spotted pardalote (*Pardalotus punctatus*), welcome swallow (*Hirundo neoxena*), dusky woodswallow (*Artamus cyanopterus*), black-faced woodswallow (*Artamus cinereus*), and silver-eye (*Zosterops lateralis*) each have been shown by one or more authors to be more abundant in severely burned than in long unburned, dry sclerophyll forests (Christensen and Kimber, 1975; McFarland, 1988; Reilly, 1991a,b, 2000; Turner, 1992; Taylor et al., 1997; Fisher, 2001; Leavesley et al., 2010; Recher and Davis, 2013; Lindenmayer et al., 2014). Thus many eucalyptus forest species also seem to require severe fire to create the early successional forest conditions within which they are most abundant, but most of those species are not restricted to conditions that occur during the first year or two after fire. In comparison with the dramatic change in bird species composition following severe fire in mixed-conifer forests, there is, in fact, a notable lack of turnover in bird species composition following severe fire in eucalyptus forests (compare before-and-after fire data from Australia and the western United States in Table 3.1). This difference in response to fire is presumably because eucalyptus trees resprout rapidly from epicormic shoots (Figure 3.2). Lindenmayer et al. (2014) also note that in montane ash forests, "... very rapid vegetation regeneration and canopy closure on severely burned sites ... may limit the influx of open-country birds and preclude the evolutionary development of early successional species" (p. 474). Nevertheless, the bird species listed above suggest that many may depend on slightly later stages of succession before the development of a fully mature forest and that a slightly different perspective might be needed to expose the ecological importance of severe fire to birds of Australian eucalypt forests.

Taken together, we hope we have provided enough ecological information derived from birds to solidify the notion that severe fire in most severe-fire-dependent shrublands and forests is both natural and necessary for maintenance of the ecological integrity of such systems.

Postfire Management Implications

Severe fire is natural and necessary in most—not relatively few—conifer forest types and in many other vegetation types worldwide as well (see Chapters 1 and 2). Current management practices designed to prevent fire,

TABLE 3.1 Probabilities of the occurrence of bird species in burned and unburned Australian eucalypt forests in the tablelands above Wollongong, New South Wales, and in burned and unburned mixed-conifer forests in western Montana (R.L. Hutto, unpublished data). Numbers of survey points are given in parentheses. Birds are ordered by the unburned-to-burned ratio of abundance, and species that are completely absent from or are significantly (Mann-Whitney *U* tests) less abundant in the opposite condition are highlighted in yellow. In both locations are bird species restricted to either early or later successional stages, but the amount of species turnover (degree of replacement of late with early succession specialists) is less pronounced after severe fire in Australia than after severe fire in the western United States

Australian eucalyptus forest			**Western North American mixed-conifer forest**		
Species	*unburned (n=39)*	*burned (n=35)*	*Species*	*unburned (n=1143)*	*burned (n=638)*
New Holland Honeyeater	0.161	0	Townsend's Warbler	0.4	0.03
Little Wattlebird	0.095	0	Solitary Vireo	0.238	0.021
Scarlet Robin	0.019	0	Golden-crowned Kinglet	0.235	0.021
Yellow-faced Honeyeater	0.040	0.231	Gray Jay	0.084	0.009
Painted Button-Quail	0	0.035	Pileated Woodpecker	0.052	0.006
Grey Shrike-thrush	0	0.058	Swainson's Thrush	0.43	0.062
Olive-backed Oriole	0	0.131	Varied Thrush	0.103	0.015
			White-breasted Nuthatch	0.017	0.003
			Black-capped Chickadee	0.053	0.012
			Red-breasted Nuthatch	0.591	0.145
			Ruby-crowned Kinglet	0.316	0.086

Hammond's Flycatcher	0.091	0.027
Hermit Thrush	0.048	0.015
Orange-crowned Warbler	0.098	0.036
Western Tanager	0.398	0.163
Mountain Chickadee	0.219	0.092
MacGillivray's Warbler	0.201	0.095
Yellow-rumped Warbler	0.521	0.249
Warbling Vireo	0.145	0.098
Clark's Nutcracker	0.022	0.047
Pine Siskin	0.111	0.257
Rufous Hummingbird	0.014	0.038
Northern Flicker	0.076	0.21
Calliope Hummingbird	0.01	0.03
Song Sparrow	0.004	0.015
Olive-sided flycatcher	0.025	0.107
Rufous-sided Towhee	0.01	0.044
Cassin's Finch	0.029	0.13
American Kestrel	0.003	0.015
Mourning Dove	0.004	0.021
Hairy Woodpecker	0.021	0.124
Three-toed Woodpecker	0.007	0.056
Northern Waterthrush	0.003	0.033

Continued

TABLE 3.1 Probabilities of the occurrence of bird species in burned and unburned Australian eucalypt forests in the tablelands above Wollongong, New South Wales, and in burned and unburned mixed-conifer forests in western Montana (R.L. Hutto, unpublished data). Numbers of survey points are given in parentheses. Birds are ordered by the unburned-to-burned ratio of abundance, and species that are completely absent from or are significantly (Mann-Whitney *U* tests) less abundant in the opposite condition are highlighted in yellow. In both locations are bird species restricted to either early or later successional stages, but the amount of species turnover (degree of replacement of late with early succession specialists) is less pronounced after severe fire in Australia than after severe fire in the western United States—Cont'd

Australian eucalyptus forest			Western North American mixed-conifer forest		
Species	*unburned (n=39)*	*burned (n=35)*	*Species*	*unburned (n=1143)*	*burned (n=638)*
			Green-tailed Towhee	0.001	0.012
			White-crowned Sparrow	0.002	0.027
			Lazuli Bunting	0.01	0.148
			House Wren	0.004	0.086
			Western Wood-pewee	0.003	0.104
			Mountain Bluebird	0.004	0.281
			American Robin	0.185	0.441
			Lincoln's Sparrow	0	0.015
			Tree Swallow	0	0.089
			Rock Wren	0	0.044
			Black-backed Woodpecker	0	0.05

suppress fire, mitigate fire severity, "restore" or "rehabilitate" burned forests after fire, and mimic the effects of severe fire are incompatible with the maintenance of ecosystem integrity (Chapter 13). Below we use results from bird research as evidence to support this statement, and we offer positive suggestions about what land managers could be doing differently.

Fire Prevention Should Be Focused on Human Population Centers

The dependence of so many bird (and many other plant and animal) species on conditions created by severe fire is clear. It necessarily follows that we cannot prevent fire and still retain anything close to a natural world. The obvious alternative is to focus prevention efforts toward population centers that are most at risk from severe fire so that fire can be left to periodically restore forest conditions elsewhere. Smokey Bear needs to refine his message so that it reflects a desire to save human lives and property, *not* a desire to save trees from fire in our wildlands (see Chapter 13).

Fire Suppression Should Be Focused on the Wildland-Urban Interface (or Fireshed)

Because many species depend on severe fire, it also necessarily follows that we should focus suppression efforts on areas immediately adjacent to human settlements (see Chapter 13). Wildland firefighters should serve primarily as support for firefighters who defend homes and human lives. Efforts to suppress fire beyond settled areas should be viewed as little more than efforts to save the forest from itself—forests need fire in the same way that they need sunlight and rain.

High-Severity Fires Beget Mixed-Severity Results

In contrast with high-severity fire, low-severity understory fires cannot create as broad a range of postfire conditions as severe fires can, nor can they stimulate the postfire process of ecological succession like a severe fire can. Therefore, managing for the maintenance of biodiversity requires more conscientious management for the maintenance of severe fires and the mixed-severity landscape effects that result from such fires (Nappi et al., 2010; Taylor et al., 2012).

Mitigate Fire Severity Through Thinning only Where such Fuel Reduction Is Appropriate

Because many species depend on severe fire, it necessarily follows that we should focus forest-thinning efforts in the wildland-urban interface and perhaps beyond that in what are basically artificial tree plantations that have resulted from past timber harvesting (see Odion et al., 2014a for review of this topic). The distributions of black-backed woodpeckers and many other fire-dependent plant and animal species make it abundantly clear that a reduction in fire severity is ecologically justified in only a very small proportion of vegetation types (Odion et al., 2014a; Sherriff et al., 2014). The presence of numerous fire-dependent species in most conifer forests throughout the American West (as illustrated by the abundance of bird research results considered in this chapter) is the strongest possible indication that the same forests have burned severely for millennia and are well within the historical range of natural variation.

The distribution of birds like the black-backed woodpecker and other fire-dependent plant and animal species, which blanket most of the forested land in the American West, are clearly at odds with claims (e.g., Haugo et al., 2015) that as much as 40% of public forested lands in parts of the United States are in need of restoration to prevent or mitigate the effects of severe fire. Lower-severity fires do not produce the mixed- and high-severity conditions needed by the most fire-dependent bird species, so efforts to mitigate fire severity in most places is incompatible with maintenance of the ecological integrity of most conifer forest systems (Odion et al., 2014a). So, what should we be doing differently? We could realize that modeled estimates indicating that our forests are in conditions that lie beyond the historical natural range of variation are just that—modeled estimates that rest strongly on many untested assumptions. We should always compare modeled results with insight gained by ecologists who can also draw strong inferences about historical conditions and, more specifically, about the kind of environments that necessarily led to adaptations of plants and animals—adaptations that reflect the distant past much more accurately than other methods commonly used to reconstruct natural fire regimes.

Postfire "Salvage" Logging in the Name of Restoration or Rehabilitation Is Always Inappropriate

Postfire "salvage" logging, seeding, planting, and shrub removal have overwhelmingly negative effects on natural systems (Lindenmayer et al., 2004; Lindenmayer and Noss, 2006; McIver and Starr, 2006; Swanson et al., 2011; DellaSala et al., 2014; Hanson, 2014), and birds have been instrumental in uncovering that fact. There is nothing as obvious to a birdwatcher as the negative effect of postfire salvage logging on the most fire-dependent birds (Uxley, 2014), and these anecdotal impressions are backed up by the strongest and most consistent scientific results ever published on any wildlife management issue (Hutto, 1995, 2006; Morissette et al., 2002; Nappi et al., 2004; Hutto and Gallo, 2006; Koivula and Schmiegelow, 2007; Hanson and North, 2008; Cahall and Hayes, 2009; Saab et al., 2009; Rost et al., 2013). One look at (Figure 3.13), or one walk through, a salvage-logged forest (also see Chapter 11) after knowing something about the biological wonder associated with a severely burned forest should be enough to convince any thinking person that there is no justification for this kind of land management activity.

It is bad enough that forests logged after fire are made unsuitable for black-backed woodpeckers and other early postfire specialists, but much worse is that postfire logging and shrub removal through mechanical or chemical means may also act as an "ecological trap" (Robertson and Hutto, 2006). This can occur

FIGURE 3.13 A vivid view of what can only be described as an ecological disaster following this postfire salvage logging operation, which took place after the 1988 Combination fire in Montana. *(Photograph by Richard Hutto.)*

when birds are attracted to burned areas that seem to be suitable and then those areas are suddenly transformed by logging or shrub removal into unsuitable habitat in an unnaturally rapid period of time. This is the most reasonable explanation for why black-backed woodpeckers are more abundant in dense, burned forests that are logged after fire than they are in burned forests that are logged before fire—birds are not attracted to the latter, where tree densities are too low and sizes are too small to provide suitable habitat, but they are attracted to the former before the trees are unexpectedly removed (Hutto, 2008). Similarly, the disproportionate use of recently logged, unburned, old-growth forests in Canada (Tremblay et al., 2009) suggests that black-backed woodpeckers sometimes make the best of a marginal situation, not that they "prefer" recently logged forests.

Although the ecological responses of birds to postfire salvage logging may differ among globally different ecosystems (Rost et al., 2012), there is absolutely no ecological justification for this kind of logging in the mixed-conifer forests of the western United States, nor is there an economic justification to salvage log after fire, because there are always better places to harvest timber without anywhere near the negative ecological consequences associated with postfire salvage logging. This is a matter of setting priorities for timber harvest, and burned forests should be at the bottom of the list. Burned forests not only provide unique ecological value, they also set the stage for the development of a variety of future forest conditions—conditions that are much more varied than those associated with development after artificial disturbance from logging. Forests have their own rules and timetables associated with the natural process of ecological succession, and we should embrace that variety and complexity. What could be done differently? Postfire rehabilitation should focus on roads, culverts, and other infrastructure issues, and nothing else. We need to recognize that new forest conditions get created after fire, and a disturbance-dependent forest does not need to be "fixed" after disturbance takes place.

We Can Do more Harm Than Good Trying to "Mimic" Nature

Prescribed burning, forest thinning, and the use of other forms of artificial disturbance in an effort to mimic nature are often poor substitutes for natural disturbance processes. Prescribed burning is usually done out of season, too frequently, and in a manner that is far too mild to have the necessary effects in most systems that evolved with fire (England, 1995; Tucker and Robinson, 2003; Penman and Towerton, 2008; Peters and Sala, 2008; Arkle and Pilliod, 2010; Rota et al., 2014a). Thinning forests in a manner thought to mimic disturbance effects is also likely to be problematic because natural disturbance (the process of fire itself) produces effects that cannot be emulated through artificial means (Schieck and Song, 2006; Reidy et al., 2014). Moreover, a thinned forest that subsequently burns in a natural fire event will not be suitable as postfire habitat for early postfire specialists because of the reduction in tree densities and sizes (Hutto, 2008). Finally, the use of forest thinning in the name of forest restoration is inappropriately applied to relatively mesic mixed-conifer forests that are unlikely to be in need of restoration, as indicated by a lack of posttreatment change in bird communities toward what one would expect if the forests were actually outside the historical range of natural variation (Hutto et al., 2014).

Except in the case of an endangered species, the worst management approach is one that focuses narrowly on creating *artificial* conditions needed by a single species. This is "single-species management," which is not the same thing as using a "management indicator approach." Management indicators are not meant to be tools that enable land managers to artificially modify land conditions to benefit a single species. Instead, a management indicator species should be used as an indication of a particular kind of "natural" condition that needs to be maintained on the landscape and as a check that the land condition is indeed acceptable to a species that requires such conditions. Even for an endangered species, we should always be thinking about maintaining the "natural" conditions that historically maintained its population. Thus although artificial tree plantations may provide conditions used by Kirtland's warbler (Spaulding and Rothstein, 2009), the bird historically nested beneath the canopy of young trees born of fire. Therefore we should create conditions safe enough to allow natural severe fire events to unfold throughout most of its historical range. As clearly stated in the Endangered Species Act (ESA, Section 2), "the purposes of this act are to provide a means whereby the *ecosystems upon which endangered species and threatened species depend* may be conserved . . ." (our italics). Conservation should be about the larger system (e.g., maintaining a fire disturbance-based jack pine forest system), not about finding a way to maintain a species through artificial means. Thus the black-backed woodpecker is an "indicator" or "focal species" that should be used to inform us about a critically important "natural" disturbance process and vegetation condition we need to maintain—severely burned forests and all the associated organisms that thrive within them.

What could we be doing differently? We need to trust that disturbance-dependent systems need severe disturbance (yes, that means a lot of tree death) to stimulate ecological succession in a manner that is indeed natural. We also need to appreciate that modeled *means and standard deviations* associated with measures of forest structure are not the same things as historical *ranges of variation* associated with the same measures. While some places have tree densities that exceed some estimated historical average value, it does not mean they fall outside the historical range of natural variation. Land managers need to relax in response to severe fire. As long as we can reduce the frequency of human-caused fires and remain safe during naturally ignited fire events, a management option that lets nature take its course will work just fine (Gill, 2001; Bradstock, 2008). In this context, noting that safety is best achieved through mechanical treatments in small areas immediately adjacent to structures (Cohen, 2000; Cohen and Stratton, 2008; Winter et al., 2009; Stockmann et al., 2010; Gibbons et al., 2012; Syphard et al., 2014), and not through mechanical treatments in more remote wildlands, is important. Given this fact, why treatments in relatively remote, publicly owned wildlands have become the tactic most commonly used to reduce wildfire risk is puzzling (Schoennagel et al., 2009).

Concluding Remarks

The most important ecological lessons we can take away from the bird research described in this chapter are that (1) many species have evolved to the point where they now require severe fire to create the conditions they need, and (2) even though some ecological systems may have departed significantly from what are believed to be historical conditions (e.g., tree plantations in the Pacific Northwest), birds are telling us (through their behavior and distribution patterns) that the vast majority of fire-dependent ecosystems are still well within the historical range of natural variation, are plenty "resilient," and are fully capable of proceeding quite naturally through the process of succession following a severe-fire event. Therefore, thinning forests in the name of restoration is largely unnecessary. If this were not true, the world would be full of places that experienced a severe fire disturbance and then underwent an unnatural transformation or "type conversion" following the disturbance event, never to return to what was there before disturbance. It is most telling that those kinds of places are rare indeed.

For those who would like to read, view, or hear more about the relationship between birds and severe fire, there are excellent children's books (e.g., Peluso, 2007; Collard, 2015); several informative videos, including a field trip that illustrated many of the patterns discussed here (listed in the Preface); and a

Fire Ecology Lab Facebook page (https://www.facebook.com/FireEcologyLab) devoted to building an appreciation for the role of severe fire in our forests.

REFERENCES

Apfelbaum, S., Haney, A., 1981. Bird populations before and after wildfire in a Great Lakes pine forest. Condor 83, 347–354.

Arkle, R.S., Pilliod, D.S., 2010. Prescribed fires as ecological surrogates for wildfires: a stream and riparian perspective. For. Ecol. Manag. 259, 893–903.

Bain, D.W., Baker, J.R., French, K.O., Whelan, R.J., 2008. Post-fire recovery of eastern bristlebirds (*Dasyornis brachypterus*) is context-dependent. Wildl. Res. 35, 44–49.

Baker, W.L., 2015. Historical Northern Spotted Owl habitat and old-growth dry forests maintained by mixed-severity wildfires. Landscape Ecol. 30, 665–666.

Bent, A.C., 1939. Life histories of North American woodpeckers. U.S. Natl. Mus. Bull. 174, 334.

Blackford, J.L., 1955. Woodpecker concentration in burned forest. Condor 57, 28–30.

Bock, C.E., Lynch, J.F., 1970. Breeding bird populations of burned and unburned conifer forest in the Sierra Nevada. Condor 72, 182–189.

Bock, C.E., Raphael, M., Bock, J.H., 1978. Changing avian community structure during early post-fire succession in the Sierra Nevada. Wilson Bull. 90, 119–123.

Bond, M.L., Gutierrez, R.J., Franklin, A.B., LaHaye, W.S., May, C.A., Seamans, M.E., 2002. Short-term effects of wildfires on spotted owl survival, site fidelity, mate fidelity, and reproductive success. Wilson Bull. 30, 1022–1028.

Bond, M.L., Lee, D.E., Siegel, R.B., Ward, J.P., 2009. Habitat use and selection by California Spotted Owls in a postfire landscape. J. Wildl. Manag. 73, 1116–1124.

Bond, M.L., Lee, D.E., Siegel, R.B., Tingley, M.W., 2013. Diet and home range size of California Spotted Owls in a burned forest. Western Birds 44, 114–126.

Bowman, D.M.J.S., 2000. Australian Rainforests: Islands of Green in a Land of Fire. Cambridge University Press, Cambridge, UK.

Bradstock, R.A., 2008. Effects of large fires on biodiversity in south-eastern Australia: disaster or template for diversity? Int. J. Wildland Fire 17, 809–822.

Cahall, R.E., Hayes, J.P., 2009. Influences of postfire salvage logging on forest birds in the Eastern Cascades, Oregon, USA. For. Ecol. Manag. 257, 1119–1128.

Chalmandrier, L., Midgley, G.F., Barnard, P., Sirami, C., 2013. Effects of time since fire on birds in a plant diversity hotspot. Acta Oecol. 49, 99–106.

Christensen, P., Kimber, P.C., 1975. Effects of prescribed burning on the flora and fauna of south-west Australian forests. Proc. Ecol. Soc. Aust. 9, 85–106.

Clark, D.A., Anthony, R.G., Andrews, L.S., 2013. Relationship between wildfire, salvage logging, and occupancy of nesting territories by northern spotted owls. J. Wildl. Manag. 77, 672–688.

Cohen, J.D., 2000. Preventing disaster: home ingnitability in the wildland-urban interface. J. For. 98, 15–21.

Cohen, J.D., Stratton, R.D., 2008. Home destruction examination: Grass Valley fire. USDA Forest Service R5-TP-026b. pp. 1–26.

Collard III, S.B., 2015. Fire Birds: Valuing Natural Wildfires and Burned Forests. Mountain Press for Bucking Horse Books, Missoula, MT.

Conway, C.J., Kirkpatrick, C., 2007. Effect of forest fire suppression on Buff-breasted Flycatchers. J. Wildl. Manag. 71, 445–457.

DellaSala, D.A., Bond, M.L., Hanson, C.T., Hutto, R.L., Odion, D.C., 2014. Complex early seral forests of the Sierra Nevada: what are they and how can they be managed for ecological integrity? Nat. Areas J. 34, 310–324.

Dudley, J.G., Saab, V.A., Hollenbeck, J.P., 2012. Foraging-habitat selection of Black-backed Woodpeckers in forest burns of southwestern Idaho. Condor 114, 348–357.

England, A.S., 1995. Avian community organization along a post-fire age gradient in California chaparral. Ph.D. Thesis, University of California-Davis.

Fayt, P., Machmer, M.M., Steeger, C., 2005. Regulation of spruce bark beetles by woodpeckers—a literature review. For. Ecol. Manag. 206, 1–14.

Fisher, A.M., 2001. Avifauna changes along a *Eucalyptus* regeneration gradient. EMU 101, 25–31.

Fogg, A.M., Roberts, L.J., Burnett, R.D., 2014. Occurrence patterns of Black-backed Woodpeckers in green forest of the Sierra Nevada Mountains, California, USA. Avian Conserv. Ecol. 9, 3.

Ganey, J.L., Kyle, S.C., Rawlinson, T.A., Apprill, D.L., Ward, J.P., 2014. Relative abundance of small mammals in nest core areas and burned wintering areas of Mexican Spotted Owls in the Sacramento Mountains. New Mexico. Wilson J. Ornith. 126, 47–52.

Gibbons, P., van Bommel, L., Gill, A.M., Cary, G.J., Driscoll, D.A., Bradstock, R.A., Knight, E., Moritz, M.A., Stephens, S.L., Lindenmayer, D.B., 2012. Land management practices associated with house loss in wildfires. PLoS One 7, e29212.

Gill, A.M., 2001. Economically destructive fires and biodiversity conservation: an Australian perspective. Conserv. Biol. 15, 1558–1560.

Haney, A., Apfelbaum, S., Burris, J.M., 2008. Thirty years of post-fire succession in a southern boreal forest bird community. Am. Midl. Nat. 159, 421–433.

Hannah, K.C., Hoyt, J.S., 2004. Northern Hawk Owls and recent burns: does burn age matter? Condor 106, 420–423.

Hannon, S.J., Drapeau, P., 2005. Bird responses to burning and logging in the boreal forest of Canada. Stud. Avian Biol. 30, 97–115.

Hanson, C.T., 2014. Conservation concerns for Sierra Nevada birds associated with high-severity fire. Western Birds 45, 204–212.

Hanson, C.T., North, M.P., 2008. Postfire woodpecker foraging in salvage-logged and unlogged forests of the Sierra Nevada. Condor 110, 777–782.

Hanson, C.T., Odion, D.C., DellaSala, D.A., Baker, W.L., 2009. Overestimation of fire risk in the Northern Spotted Owl Recovery Plan. Conserv. Biol. 23, 1314–1319.

Hanson, C.T., Odion, D.C., DellaSala, D.A., Baker, W.L., 2010. More-comprehensive recovery actions for Northern Spotted Owls in dry forests: reply to Spies et al. Conserv. Biol. 24, 334–337.

Harris, M.A., 1982. Habitat use among woodpeckers in forest burns. M.S. Thesis, University of Montana, Missoula, MT.

Haugo, R., Zanger, C., DeMeo, T., Ringo, C., Shlisky, A., Blankenship, K., Simpson, M., Mellen-McLean, K., Kertis, J., Stern, M., 2015. A new approach to evaluate forest structure restoration needs across Oregon and Washington, USA. For. Ecol. Manag. 335, 37–50.

Hoyt, J.S., Hannon, S.J., 2002. Habitat associations of Black-backed and Three-toed woodpeckers in the boreal forest of Alberta. Can. J. For. Res. 32, 1881–1888.

Hutto, R.L., 1995. Composition of bird communities following stand-replacement fires in northern Rocky Mountain (U.S.A.) conifer forests. Conserv. Biol. 9, 1041–1058.

Hutto, R.L., 2006. Toward meaningful snag-management guidelines for postfire salvage logging in North American conifer forests. Conserv. Biol. 20, 984–993.

Hutto, R.L., 2008. The ecological importance of severe wildfires: some like it hot. Ecol. Appl. 18, 1827–1834.

Hutto, R.L., 2014. Time budgets of male Calliope Hummingbirds on a dispersed lek. Wilson Journal of Ornithology 126, 121–128.

Hutto, R.L., Gallo, S.M., 2006. The effects of postfire salvage logging on cavity-nesting birds. Condor 108, 817–831.

Hutto, R.L., Young, J.S., 1999. Habitat relationships of landbirds in the Northern Region, USDA Forest Service. USDA Forest Service General Technical Report RMRS-GTR-32. pp. 1–72.

Hutto, R.L., Patterson, D.A., 2015. Hidden positive fire effects on birds exposed only after controlling for fire severity and time since fire. unpublished MS.

Hutto, R.L., Conway, C.J., Saab, V.A., Walters, J.R., 2008. What constitutes a natural fire regime? Insight from the ecology and distribution of coniferous forest birds in North America. Fire Ecol. 4, 115–132.

Hutto, R.L., Flesch, A.D., Fylling, M.A., 2014. A bird's-eye view of forest restoration: do changes reflect success? For. Ecol. Manag. 327, 1–9.

James, F.C., McCulloch, C.E., 1995. The strength of inferences about causes of trends in populations. In: Martin, T.E., Finch, D.M. (Eds.), Ecology and Management of Neotropical Migratory Birds: A Synthesis and Review of Critical Issues. Oxford University Press, New York, NY, pp. 40–51.

Jenness, J.S., Beier, P., Ganey, J.L., 2004. Associations between forest fire and Mexican Spotted Owls. For. Sci. 50, 765–772.

Kirkpatrick, C., Conway, C., Jones, P.B., 2006. Distribution and relative abundance of forest birds in relation to burn severity in southeastern Arizona. J. Wildl. Manag. 70, 1005–1012.

Koivula, M.J., Schmiegelow, F.K.A., 2007. Boreal woodpecker assemblages in recently burned forested landscapes in Alberta, Canada: effects of post-fire harvesting and burn severity. For. Ecol. Manag. 242, 606–618.

Kotliar, N.B., Hejl, S.J., Hutto, R.L., Saab, V.A., Melcher, C.P., McFadzen, M.E., 2002. Effects of fire and post-fire salvage logging on avian communities in conifer-dominated forests of the western United States. Stud. Avian Biol. 25, 49–64.

Kreisel, K.J., Stein, S.J., 1999. Bird use of burned and unburned coniferous forests during winter. Wilson Bull. 111, 243–250.

Leavesley, A.J., Cary, G.J., Edwards, G.P., Gill, A.M., 2010. The effect of fire on birds of mulga woodland in arid central Australia. Int. J. Wildland Fire 19, 949–960.

Lee, D.E., Bond, M.L., Siegel, R.B., 2012. Dynamics of breeding-season site occupancy of the California Spotted Owl in burned forests. Condor 114, 792–802.

Lee, D.E., Bond, M.L., Borchert, M.I., Tanner, R., 2013. Influence of fire and salvage logging on site occupancy of spotted owls in the San Bernardino and San Jacinto Mountains of Southern California. J. Wildl. Manag. 77, 1327–1341.

Leonard, S.W.J., Bennett, A.F., Clarke, M.F., 2014. Determinants of the occurrence of unburnt forest patches: potential biotic refuges within a large, intense wildfire in south-eastern Australia. For. Ecol. Manag. 314, 85–93.

Lindenmayer, D.B., Noss, R.F., 2006. Salvage logging, ecosystem processes, and biodiversity conservation. Conserv. Biol. 20, 949–958.

Lindenmayer, D.B., Foster, D.R., Franklin, J.F., Hunter, M.L., Noss, R.F., Schmiegelow, F.A., Perry, D., 2004. Salvage harvesting policies after natural disturbance. Science 303, 1303.

Lindenmayer, D.B., Blanchard, W., McBurney, L., Blair, D., Banks, S.C., Driscoll, D.A., Smith, A.L., Gill, A.M., 2014. Complex responses of birds to landscape-level fire extent, fire severity and environmental drivers. Divers. Distrib. 20, 467–477.

McFarland, D., 1988. The composition, microhabitat use and response to fire of the avifauna of subtropical heathlands in Cooloola National Park, Queensland. EMU 88, 249–257.

McIver, J.D., Starr, L., 2006. A literature review on the environmental effects of postfire logging. West. J. Appl. For. 16, 159–168.

Mendelsohn, M.B., Brehme, C.S., Rochester, C.J., Stokes, D.C., Hathaway, S.A., Fisher, R.N., 2008. Responses in bird communities to wildland fires in southern California. Fire Ecol. 4, 63–82.

Morissette, J.L., Cobb, T.P., Brigham, R.M., James, P.C., 2002. The response of boreal forest songbird communities to fire and post-fire harvesting. Can. J. For. Res. 32, 2169–2183.

Murphy, E.G., Lehnhausen, W.H., 1998. Density and foraging ecology of woodpeckers following a stand-replacement fire. J. Wildl. Manag. 62, 1359–1372.

Nappi, A., Drapeau, P., 2009. Reproductive success of the Black-backed Woodpecker (*Picoides arcticus*) in burned boreal forests: are burns source habitats? Biol. Conserv. 142, 1381–1391.

Nappi, A., Drapeau, P., 2011. Pre-fire forest conditions and fire severity as determinants of the quality of burned forests for deadwood-dependent species: the case of the black-backed woodpecker. Can. J. For. Res. 41, 994–1003.

Nappi, A., Drapeau, P., Savard, J.-P.L., 2004. Salvage logging after wildfire in the boreal forest: is it becoming a hot issue for wildlife? For. Chron. 80, 67–74.

Nappi, A., Drapeau, P., Saint-Germain, M., Angers, V.A., 2010. Effect of fire severity on long-term occupancy of burned boreal conifer forests saproxylic insects and wood-foraging birds. Int. J. Wildland Fire 19, 500–511.

Niemi, G.J., 1978. Breeding birds of burned and unburned areas in northern Minnesota. Loon 50, 73–84.

Odion, D.C., Hanson, C.T., 2013. Projecting impacts of fire management on a biodiversity indicator in the Sierra Nevada and Cascades, USA: the black-backed woodpecker. Open Forest Sci. J. 6, 14–23.

Odion, D.C., Hanson, C.T., Arsenault, A., Baker, W.L., DellaSala, D.A., Hutto, R.L., Klenner, W., Moritz, M.A., Sherriff, R.L., Veblen, T.T., Williams, M.A., 2014a. Examining historical and current mixed-severity fire regimes in ponderosa pine and mixed-conifer forests of western North America. PLoS One 9, e87852 (87851–87814).

Odion, D.C., Hanson, C.T., DellaSala, D.A., Baker, W.L., Bond, M.L., 2014b. Effects of fire and commercial thinning on future habitat of the Northern Spotted Owl. Open Ecol. J. 7, 37–51.

Peluso, B.A., 2007. The Charcoal Forest: How Fire Helps Animals and Plants. Mountain Press, Missoula, MT.

Penman, T.D., Towerton, A.L., 2008. Soil temperatures during autumn prescribed burning: implications for the germination of fire responsive species? Int. J. Wildland Fire 17, 572–578.

Peters, G., Sala, A., 2008. Reproductive output of ponderosa pine in response to thinning and prescribed burning in western Montana. Can. J. For. Res. 38, 844–850.

Pfister, A.R., 1980. Postfire avian ecology in Yellowstone National Park. M.S. Thesis, Washinton State University, Pullman, WA.

Powell, H.D.W., Hejl, S.J., Six, D.L., 2002. Measuring woodpecker food: a simple method for comparing wood-boring beetle abundance among fire-killed trees. J. Field Ornithol. 73, 130–140.

Probst, J.R., Donnerwright, D., 2003. Fire and shade effects on ground cover structure in Kirtland's Warbler habitat. Am. Midl. Nat. 149, 320–334.

Probst, J.R., Hayes, J.P., 1987. Pairing success of Kirtland's Warblers in marginal vs. suitable habitat. Auk 104, 234–241.

Pulliam, H.R., 1988. Sources, sinks, and population regulation. Am. Nat. 132, 652–661.

Pyke, G.H., Saillard, R., Smith, J., 1995. Abundance of Eastern Bristlebirds in relation to habitat and fire history. EMU 95, 106–110.

Raphael, M.G., Morrison, M.L., Yoder-Williams, M.P., 1987. Breeding bird populations during twenty-five years of post-fire succession in the Sierra Nevada. Condor 89, 614–626.

Recher, H.F., Christensen, P.E., 1981. Fire and the evolution of the Australian biota. In: Keast, A., Junk, D.W. (Eds.), Ecological Biogeography of Australia. D.W. Junk, The Hague, pp. 135–162.

Recher, H.F., Davis Jr., W.E., 2013. Response of birds to a wildfire in the Great Western Woodlands, Western Australia. Pac. Conserv. Biol. 19, 188–203.

Reidy, J.L., Thompson Iii, F.R., Kendrick, S.W., 2014. Breeding bird response to habitat and landscape factors across a gradient of savanna, woodland, and forest in the Missouri Ozarks. For. Ecol. Manag. 313, 34–46.

Reilly, P., 1991a. The effect of wildfire on bird populations in a Victorian coastal habitat. EMU 91, 100–106.

Reilly, P., 1991b. The effect of wildfire on bush bird populations in six Victorian coastal habitats. Corella 15, 134–142.

Reilly, P., 2000. Bird populations in a Victorian coastal habitat twelve years after a wildfire in 1983. EMU 100, 240–245.

Roberts, S.L., van Wangtendonk, J.W., Miles, A.K., Kelt, D.A., 2011. Effects of fire on spotted owl site occupancy in a late-successional forest. Biol. Conserv. 144, 610–619.

Robertson, B.A., Hutto, R.L., 2006. A framework for understanding ecological traps and an evaluation of existing ecological evidence. Ecology 87, 1075–1085.

Robinson, N.M., Leonard, S.W.J., Bennett, A.F., Clarke, M.F., 2014. Refuges for birds in fire-prone landscapes: the influence of fire severity and fire history on the distribution of forest birds. For. Ecol. Manag. 318, 110–121.

Rost, J., Clavero, M., Brotons, L., Pons, P., 2012. The effect of postfire salvage logging on bird communities in Mediterranean pine forests: the benefits for declining species. J. Appl. Ecol. 49, 644–651.

Rost, J., Hutto, R.L., Brotons, L., Pons, P., 2013. Comparing the effect of salvage logging on birds in the Mediterranean Basin and the Rocky Mountains: common patterns, different conservation implications. Biol. Conserv. 158, 7–13.

Rota, C.T., Millspaugh, J.J., Rumble, M.A., Lehman, C.P., Kesler, D.C., 2014a. The role of wildfire, prescribed fire, and mountain pine beetle infestations on the population dynamics of black-backed woodpeckers in the black hills, South Dakota. PLoS One 9, e94700.

Rota, C.T., Rumble, M.A., Millspaugh, J.J., Lehman, C.P., Kesler, D.C., 2014b. Space-use and habitat associations of Black-backed Woodpeckers (Picoides arcticus) occupying recently disturbed forests in the Black Hills, South Dakota. For. Ecol. Manag. 313, 161–168.

Rush, S., Klaus, N., Keyes, T., Petrick, J., Cooper, R., 2012. Fire severity has mixed benefits to breeding bird species in the southern Appalachians. For. Ecol. Manag. 263, 94–100.

Saab, V.A., Dudley, J.G., 1998. Responses of cavity-nesting birds to stand-replacement fire and salvage logging in ponderosa pine/Douglas-fir forests of southwestern Idaho. USDA Forest Service Research Paper RMRS-RP-11. pp. 1–17.

Saab, V., Brannon, R., Dudley, J., Donohoo, L., Vanderzanden, D., Johnson, V., Lachowski, H., 2002. Selection of fire-created snags at two spatial scales by cavity-nesting birds. USDA Forest Service General Technical Report PSW-GTR-181. pp. 835–848.

Saab, V.A., Russell, R.E., Dudley, J., 2007. Nest densities of cavity-nesting birds in relation to postfire salvage logging and time since wildfire. Condor 109, 97–108.

Saab, V.A., Russell, R.E., Dudley, J.G., 2009. Nest-site selection by cavity-nesting birds in relation to postfire salvage logging. For. Ecol. Manag. 257, 151–159.

Saracco, J.F., Siegel, R.B., Wilkerson, R.L., 2011. Occupancy modeling of Black-backed Woodpeckers on burned Sierra Nevada forests. Ecosphere 2, 1–17.

Schieck, J., Song, S.J., 2006. Changes in bird communities throughout succession following fire and harvest in boreal forests of western North America: literature review and meta-analyses. Can. J. For. Res. 36, 1299–1318.

Schmiegelow, F.K.A., Stepnisky, D.P., Stambaugh, C.A., Koivula, M., 2006. Reconciling salvage logging of boreal forests with a natural-disturbance management model. Conserv. Biol. 20, 971–983.

Schoennagel, T., Nelson, C.R., Theobald, D.M., Carnwath, G.C., Chapman, T.B., 2009. Implementation of National Fire Plan treatments near the wildland-urban interface in the western United States. Proc. Natl. Acad. Sci. 106, 10706–10711.

Seavy, N.E., Alexander, J.D., 2014. Songbird response to wildfire in mixed-conifer forest in southwestern Oregon. Int. J. Wildland Fire 23, 246–258.

Seavy, N.E., Burnett, R.D., Taille, P.J., 2012. Black-backed woodpecker nest-tree preference in burned forests of the Sierra Nevada, California. Wildl. Soc. Bull. 36, 722–728.

Seidl, R., Rammer, W., Spies, T.A., 2014. Disturbance legacies increase the resilience of forest ecosystem structure, composition, and functioning. Ecol. Appl. 24, 2063–2077.

Sergio, F., Newton, I., 2003. Occupancy as a measure of territory quality. J. Anim. Ecol. 72, 857–865.

Serong, M., Lill, A., 2012. Changes in bird assemblages during succession following disturbance in secondary wet forests in south-eastern Australia. EMU 112, 117–128.

Sherriff, R.L., Platt, R.V., Veblen, T.T., Schoennagel, T.L., Gartner, M.H., 2014. Historical, observed, and modeled wildfire severity in montane forests of the Colorado Front Range. PLoS One 9, e106971.

Siegel, R.B., Tingley, M.W., Wilkerson, R.L., Bond, M.L., Howell, C.A., 2013. Assessing Home Range Size and Habitat Needs of Black-Backed Woodpeckers in California. The Institute for Bird Populations, Point Reyes Station, California.

Siegel, R.B., Wilkerson, R.L., Tingley, M.W., Howell, C.A., 2014. Roost sites of the Black-backed Woodpecker in burned forest. Western Birds 45, 296–303.

Sitters, H., Christie, F.J., Di Stefano, J., Swan, M., Penman, T., Collins, P.C., York, A., 2014. Avian responses to the diversity and configuration of fire age classes and vegetation types across a rainfall gradient. For. Ecol. Manag. 318, 13–20.

Smucker, K.M., Hutto, R.L., Steele, B.M., 2005. Changes in bird abundance after wildfire: importance of fire severity and time since fire. Ecol. Appl. 15, 1535–1549.

Spaulding, S.E., Rothstein, D.E., 2009. How well does Kirtland's warbler management emulate the effects of natural disturbance on stand structure in Michigan jack pine forests? For. Ecol. Manag. 258, 2609–2618.

Stockmann, K., Burchfield, J., Calkin, D., Venn, T., 2010. Guiding preventative wildland fire mitigation policy and decisions with an economic modeling system. For. Policy Econ. 12, 147–154.

Swanson, M.E., Franklin, J.F., Beschta, R.L., Crisafulli, C.M., DellaSala, D.A., Hutto, R.L., Lindenmayer, D.B., Swanson, F.J., 2011. The forgotten stage of forest succession: early-successional ecosystems on forest sites. Front. Ecol. Environ. 9, 117–125.

Syphard, A.D., Brennan, T.J., Keeley, J.E., 2014. The role of defensible space for residential structure protection during wildfires. Int. J. Wildland Fire 23, 1165–1175.

Taylor, D.L., Barmore, W.J., 1980. Post-fire succession of avifauna in coniferous forests of Yellowstone and Grand Teton National Parks, Wyoming. In R.M. DeGraaf (Ed.). Workshop Proceedings: Management of Western Forests and Grasslands for Nongame Birds. USDA Forest Service General Technical Report INT-86, Ogden, UT. pp. 130–145.

Taylor, R., Duckworth, P., Johns, T., Warren, B., 1997. Succession in bird assemblages over a seven-year period in regrowth dry sclerophyll forest in south-east Tasmania. EMU 97, 220–230.

Taylor, R.S., Watson, S.J., Nimmo, D.G., Kelly, L.T., Bennett, A.F., Clarke, M.F., 2012. Landscape-scale effects of fire on bird assemblages: does pyrodiversity beget biodiversity? Divers. Distrib. 18, 519–529.

Tingley, M.W., Wilkerson, R.L., Bond, M.L., Howell, C.A., Siegel, R.B., 2014. Variation in home-range size of Black-backed Woodpeckers. Condor 116, 325–340.

Tremblay, J.A., Ibarzabal, J., Dussault, C., Savard, J.-P.L., 2009. Habitat requirements of breeding Black-backed Woodpeckers (*Picoides arcticus*) in managed, unburned boreal forest. Avian Conserv. Ecol. 4, 2.

Tucker Jr., J.W., Robinson, W.D., 2003. Influence of season and frequency of fire on Henslow's Sparrows (*Ammodramus henslowii*) wintering on gulf coast pitcher plant bogs. Auk 120, 96–106.

Turner, R.J., 1992. Effect of wildfire on birds at Weddin Mountain, New South Wales. Corella 16, 65–74.

U.S. Fish & Wildlife Service, 2011. Revised Recovery Plan for the Northern Spotted Owl (*Strix occidentalis caurina*). USFWS, Portland, Oregon.

Uxley, W., 2014. Firebird. Bird Watching J. October, 26–31.

Walkinshaw, L.H., 1983. Kirtland's Warbler, the Natural History of an Endangered Species. Cranbrook Institute of Science, Bloomfield Hills, MI.

Winchell, C.S., Doherty, P.F., 2014. Effects of habitat quality and wildfire on occupancy dynamics of Coastal California Gnatcatcher (*Polioptila californica californica*). Condor 116, 538–545.

Winter, G., McCaffrey, S., Vogt, C.A., 2009. The role of community policies in defensible space compliance. For. Policy Econ. 11, 570–578.

Woinarski, J.C.Z., Recher, H.F., 1997. Impact and response: a review of the effects of fire on the Australian avifauna. Pac. Conserv. Biol. 3, 183–205.

Youngman, J.A., Gayk, Z.G., 2011. High density nesting of Black-backed Woodpeckers (*Picoides arcticus*) in a post-fire Great Lakes jack pine forest. Wilson J. Ornithol. 123, 381–386.

Yunick, R.P., 1985. A review of recent irruptions of the Black-backed Woodpecker and Three- toed Woodpecker in eastern North America. J. Field Ornithol. 56, 138–152.

Chapter 4

Mammals and Mixed- and High-severity Fire

Monica L. Bond
Wild Nature Institute, Hanover, NH, USA

4.1 INTRODUCTION

Mammals are ecologically and economically important members of the landscapes in which they live. Large herbivores like deer (*Odocoileus* spp.) and elk (*Cervus elaphus*), and predators like bears (*Ursus* spp.) and wolves (*Canis lupus*), are highly conspicuous and well-known "flagship" mammal species, whereas rodents, bats, and mustelids are cryptic but no less important in their ecosystems. Many species have developed broad ecological tolerance from exposure to environmental variation and natural disturbances over long time periods (Lawler, 2003). However, widespread hunting and excessive habitat fragmentation of landscapes by modern-day humans are qualitatively and quantitatively different from the natural disturbances to which these mammals were exposed in the past (Spies and Turner, 1999), and they have resulted in contraction of historical ranges and population declines. In North America alone notable population declines include elk, grizzly bears (*Ursus arctos*), gray wolves, Canadian lynx (*Lynx canadensis*), bighorn sheep (*Ovis canadensis*), beaver (*Castor canadensis*), the larger species of forest mustelids, and several herteromyid rodents.

Mixed- and high-severity wildfire is a natural disturbance in many vegetation systems of North America, the Mediterranean, Australia, and Africa (see Chapters 1, 2, and 8). The effects of severe fire on organisms vary spatially and temporally, by habitat type, and by species, but how do these disturbances specifically impact mammals? As with any natural disturbance, some species are adversely affected ("fire-averse" species), others benefit ("fire-loving" or pyrophilous species), and still others have a neutral response to fires.

The dynamics of populations and communities of mammals after severe fire depend on factors such as the degree of ecological change, time since fire, size and spatial configuration of burned and unburned areas, extent of edge, isolation of habitat patches by urbanization and roads, and invasion of nonnative species (Smith, 2000; Shaffer and Laudenslayer, 2006; Arthur et al., 2012; Diffendorfer

The Ecological Importance of Mixed-Severity Fires: Nature's Phoenix

et al., 2012; Fontaine and Kennedy, 2012). In theory, mammalian populations should be stable and resilient across the landscape wherever prefire populations and critical habitats are not greatly reduced and/or fragmented by human activities, and where severe fires occur in a spatial and temporal pattern in which a species has evolved (Shaffer and Laudenslayer, 2006). The capability of fire-loving individuals to utilize severely burned areas or for fire-averse populations to recover after fire, however, can be compromised when prefire habitat fragmentation has resulted in small and/or isolated populations and where postfire management actions, such as logging of burned trees and use of herbicides and pesticides, adversely influence population dynamics and habitat use (see Chapter 11).

In this chapter I provide an overview of published studies about mammalian responses to mixed- and high-severity fires in forests, woodlands, shrublands, deserts, and grasslands around the world. I describe research on the effects of severe fire on four major taxonomic groups of mammals: bats, small mammals, carnivores, and ungulates. I emphasized peer-reviewed publications, particularly those with robust methodologies and analyses, because these are the accepted standard in science. I also used non-peer-reviewed data when necessary to supplement information from the peer-reviewed literature. I do not cite every published study but instead provide a balanced overview of severe-fire effects on these taxa. I encourage readers to investigate further the scientific literature on habitat use and population responses of mammals to severe fire because the state of the science is constantly evolving.

Few studies have documented direct effects of fire on wildlife (e.g., mortality from asphyxiation, heat stress, burning, or physiological stress; however, see Singer et al., 1989), but wildlife biologists generally agree that direct mortality from fire is typically very low and does not significantly influence populations (Smith, 2000). Thus, I focus here on the indirect responses to severe fire, such as postfire occupancy, abundance or density, survival, reproduction, and use of habitat (e.g., breeding, resting, foraging). I define "significant effects" according to the generally accepted scientific definition of statistical significance (i.e., at the 0.05 probability level). I exclude studies that simulated or modeled fires, choosing instead to focus on observations of real systems responding to severe wildfire.

Appendix 4.1 is a summary of published studies by mammalian taxa and directional response to severe wildfire (negative, neutral, positive) over three time periods after fire. I present results from studies comparing unburned habitats with high-severity burn from wildfire (rather than prescribed fire) and without the confounding effect of postfire logging. For small mammals, only species with enough detections to determine directional response were included in the appendix.

4.2 BATS

Bats perform unique and critical ecosystem services by consuming vast quantities of insects, thereby transferring nutrients, most notably nitrogen, from foraging to roosting areas via their feces (Gruver and Keinath, 2006). Bats are predators of adult mosquitoes and thus play an important role in controlling mosquito populations and reducing disease transmission (Reiskind and Wund, 2009). Further, nectar-feeding bats are primary pollinators of many plant species throughout the world (Molina-Freaner and Eguiarte, 2003).

The current literature on the effects of fire on bats strongly suggests that mixed- and high-severity fires are explicitly beneficial. In a study comparing the relative activity of six phonic groups of mostly rare and sensitive bat species across unburned and moderate- and high-severity burned mixed-conifer stands 1 year after fire in the southern Sierra Nevada, bat activity in burned areas was equivalent to or greater than activity in unburned areas for all groups based on echolocation frequencies (Buchalski et al., 2013). Indeed, two of the phonic groups showed a positive response to high-severity fire but a neutral response to moderate-severity fire, demonstrating the importance of severity-specific responses. The positive response to mixed- and high-severity fire by bats mirrors findings for a range of bird species (see Chapter 3) and provides evidence of a long evolutionary relationship between bats and severe fire.

Several studies have documented how roosting bats use basal hollows of large trees (Gellman and Zielinski, 1996; Zielinski and Gellman, 1999; Fellers and Pierson, 2002; Mazurek, 2004). (Figure 4.1) Basal hollows are cavities formed by repeated fire scarring and healing (Zielinski and Gellman, 1999). For bats that roost in basal hollows of large trees, high-severity fire may destroy or reduce the longevity of existing roost trees, but it also creates new roost trees. In addition, fire creates gaps in the canopy that increase the amount of solar radiation reaching the subcanopy where bats roost. These warmer temperatures may facilitate thermoregulation (Brigham et al., 1997; Boyles and Aubrey, 2006) and are particularly beneficial to reproductive females because increased temperatures are associated with greater fetal and neonate growth (Brigham et al., 1997; Johnson et al., 2009). Finally, high-severity fire creates a "pulse" of insect prey (e.g., aquatic insects (Malison and Baxter, 2010), moths, beetles, and flies (Schwab, 2006)), as well as new natural edge habitat that provides novel foraging opportunities (Fellers and Pierson, 2002).

Comparisons of food web components between unburned watersheds and areas of low- and high-severity fires 5 years after fire in Douglas-fir (*Pseudotsuga menziesii*) and ponderosa pine (*Pinus ponderosa*) forests in central Idaho showed high insect biomass in heavily burned areas and correspondingly high bat detection rates (Malison and Baxter, 2010). Notably, high-severity sites had almost five times more biomass of zoobenthic insects and more than three times the number of emerging adult aquatic insects than low-severity sites (and twice as many as unburned areas). The frequency of bat echolocation calls also was significantly greater at high-severity sites than at unburned sites, because aquatic insects emerging from streams into the terrestrial environment are an important food source for bats. In a review of the responses of stream benthic macroinvertebrates to fire, Minshall (2003) concluded that "results for macroinvertebrates generally support the belief that fire and similar natural disturbance events are not detrimental to the sustained

FIGURE 4.1 Basal hollows in large trees are created by periodic fire scarring and healing, creating important roost sites for bats. A Townsend's big-eared bat (*Corynorhinus townsendii*) roost tree in a coast redwood (*Sequoia sempervirens*) in Grizzly Creek State Park, northern California. *(Photo by M.J. Mazurek (2015).)*

maintenance of diverse and productive aquatic ecosystems (i.e., those found in undisturbed forests)" (p. 159). While individual taxa respond differently to the physical changes in stream structure and short-term and long-term postfire changes in vegetation, Minshall noted that streams are inherently unstable and dynamic environments in which disturbance, including high-severity fire, is a regular occurrence, and many species are opportunistic and can shift food resources in response to fire.

In mid-elevation forests burned at mixed and high severity in western Montana, Schwab (2006) characterized roost sites and sampled potential prey sources for two forest-dwelling, insectivorous bat species, the little brown bat (*Myotis lucifugus*) and the long-eared myotis (*Myotis evotis*). These species roosted in larger-diameter snags (standing dead trees) in high-density stands of fire-killed trees. Proximity to perennial streams also was important in roost site selection for these two species in burned forests. Wildland fire apparently created an abundance of roosting sites and insect prey for bats. Although the abundance of Lepidoptera (moths) and Trichoptera (caddis flies) was similar in burned and unburned forests, the abundance of Diptera (flies) and Coleoptera (beetles) was significantly higher in burned forests. Overall, the median capture rate of all insects in the burn was 1.78 times higher than the median capture rate in unburned forests– but there was considerable variability in the composition and abundance of particular species. Eight of the 11 orders of insects were more abundant in burned sites. In addition, beetles, flies, and caddis flies were significantly more abundant in burned than unburned sites in the first year after fire, although they decreased significantly the second year after fire. Thus, retention of burned trees the first year is important for insectivorous bats. In fact, removing burned trees decreased mammalian (and avian) predation on the abundance of insects that occurred 1 year after fire. Snags in unburned forests can be recruited from existing green trees, but in severely burned forests postfire logging eliminates both existing and future snags for nearly a century because few trees are available for snag recruitment until large-diameter trees have regrown (Schwab, 2006).

As with many bird species, mixed- and high-severity fire in forest ecosystems likely enhances foraging opportunities for bats (Buchalski et al., 2013). Many insect species inhabiting coniferous forests are highly evolved to exploit severely burned forests and are aptly termed "pyrophilous." Certain beetle species in particular are strongly attracted to highly burned forests. Saint-Germain et al. (2004) noted that, "some insect groups have adapted to recurrent forest fires by evolving sensory organs and life strategies that allow them to exploit these high quality habitats efficiently. Pyrophilous Buprestids of the genera *Oxypteris* and *Merimna* and the Cerambycid *Arhopalus tristis* (F.) have been shown to respond physiologically to smoke and/or heat generated by fire, and use them as signals leading toward the newly created habitat ... Several other Coleoptera species uncommon in mature forests congregate in exceptionally high densities in burned stands" (p. 583).

In a study of fire-loving beetle communities in a large fire that burned boreal black spruce (*Picea mariana*) forest in Quebec, Canada, more than half of the 86 taxa captured were restricted to burned stands (Saint-Germain et al., 2004). Moreover, total captures and species richness were higher in burned stands, especially the oldest severely burned forests. Captures were significantly lower the second year after the fire for all burned stands, indicating that the utility of burned forests for these beetles is greatest in the first year following fire.

Insects utilizing dead trees occur at much lower abundances in low-severity sites, which by definition have far fewer fire-killed trees than high-severity sites. Malison and Baxter (2010) stated that, "our results suggest that high severity fires do not play the same ecological role as low severity fires and allowing high severity fires to burn (rather than suppressing them) in certain forest types could be important in maintaining ecosystem function" (p. 577). Similarly, in his severely burned study site, Schwab (2006) noted, "26% of all [insect] families captured were restricted to sites within the burn suggesting

BOX 4.1

(1) Bats preferentially roost and forage in burned forests.
(2) High-severity fire creates a superabundance of native insect prey.
(3) Bats select denser stands of fire-killed trees for roosting in burned forests and forage significantly more in forests burned by high-severity fire than in unburned and low-severity fire-affected forests.
(4) Large burned trees for roosting have significant positive benefits for bats.
(5) Postfire logging removes roost trees, reduces the abundance of prey, and reduces habitat suitability for bats.

a unique environment created only after fire." Thus, ecological changes caused by mixed- and high-severity fires cannot be mimicked by low-severity prescribed burns (also see Chapter 13 for similar discussion) (Box 4.1).

4.3 SMALL MAMMALS

Small mammals are critically important to ecosystems because they can influence vegetation structure and composition by dispersing seeds and ectomycorrhizal fungi and by aerating soils (Maser et al., 1978). They also provide an essential prey base for carnivores, and the distribution of small mammals can affect the use of space and the habitat selection of their predators (Carey et al., 1992; Ward et al., 1998). Small mammals have comparatively small home ranges and therefore are quite sensitive to habitat change, making them good biological indicators (Haim and Izhaki, 1994). Small mammal assemblages include rodents and insectivores of the families Soricidae (shrews), Talpidae (moles), Aplodontidae (mountain beavers), Sciuridae (squirrels, chipmunks, and marmots), Geomyidae (gophers), Heteromyidae (pocket mice and kangaroo rats), superfamily Muroidea (voles, mice, and woodrats), and Dipodidae (jumping mice). Larger-bodied small mammals include rodents in the Castoridae (beaver) and Erethizontidae (porcupine) families, as well as lagomorphs (pika, hares, and rabbits), and Australian and American marsupials (Marsupialia).

The occupation of severely burned areas by small mammals is related to regrowth of the vegetation structure with which various species are associated (Torre and Díaz, 2004; Lee and Tietje, 2005; Vamstad and Rotenberry, 2010; Diffendorfer et al., 2012; Kelly et al., 2012; Borchert and Borchert, 2013), as well as with seed and insect production and availability (Coppeto et al., 2006), and cavities created by woodpeckers in snags (Tarbill, 2010). I discuss fire effects on small mammals according to habitat type but give special attention to the deer mouse (*Peromyscus maniculatus*)—an exceptionally "fire-loving" species—in its own section. (Figure 4.2)

FIGURE 4.2 Deer mice increase after severe fire in a variety of habitats. A deer mouse captured two years after forest dominated by Douglas-fir with some lodgepole pine (*Pinus contorta*), western larch (*Larix occidentalis*), and ponderosa pine burned severely in the 2005 Tarkio Fire, Montana. *(Photo by Rafal Zwolak (2005).)*

Chaparral and Coastal Sage Scrub

The chaparral and coastal sage scrub vegetation types in central and southern California support an exceptionally rich diversity of rodents that are well-adapted to a regime of periodic, very-high-intensity fire (see Chapter 7). Many studies have examined small-mammal communities after both prescribed and wildfire in these vegetative types. During intense fires, some individuals among small, less vagile animals may suffer mortality, but many others survive in rock crevices, riparian areas, large downed logs, and underground burrows where temperatures remain cool and the air clean (Chew et al., 1959; Quinn, 1979; Lawrence, 1966; Wirtz, 1995; Smith, 2000). Following fire, small-mammal communities change over time (Diffendorfer et al., 2012; Arthur et al., 2012; Borchert and Borchert, 2013) and space (Schwilk and Keeley, 1998), depending on the vegetation associations of the various species. Species preferring open habitat, including pocket mice (*Chaetodipus* spp.), California voles (*Microtus californicus*), harvest mice (*Reithrodontomys megalotis*), and, especially, kangaroo rats (*Dipodomys* spp.) and deer mice can increase quite dramatically and quickly after severe shrubland fire. Over a period of several years, as shrubs resprout and grow denser and as different food sources become available, small-mammal species preferring a shrubby overstory, including woodrats (*Neotoma* spp.), California mice (*Peromyscus californicus*), brush mice (*Peromyscus boylii*), and cactus mice (*Peromyscus eremicus*), increase in number (Cook, 1959; Wirtz, 1977; Price and Waser, 1984; Brehme et al., 2011; Borchert and Borchert, 2013). Compared with unburned chaparral and grassland, severely burned chaparral had the highest rodent diversity 4 years after a high-intensity wildfire near Mount Laguna in San Diego County (Lillywhite, 1977). Published data are not currently available for lagomorphs in chaparral wildfires, but prescribed burning of chamise (*Adenostoma fasciculatum*) chaparral in northern California increased black-tailed jackrabbit (*Lepus californicus*) densities by 500-1000% the year following fire (Howard, 1995).

Forests

Forests offer important habitats for small mammals, especially shrews, mice, tree voles, and squirrels. Mixed- and high-severity fire in forested habitats can have pronounced effects on small-mammal populations by creating or transforming habitat structures such as live and dead trees, shrubs, and coarse woody debris. While some studies have shown that severely burned conifer forests in North America support fewer individuals of some rodents and insectivores immediately after fire compared with adjacent unburned sites (e.g., pinyon mice [*Peromyscus truei*; Borchert et al., 2014] and masked shrews (*Sorex cinereus*) and southern red-backed voles [*Myodes gapperi*; Zwolak and Forsman, 2007]), numbers begin to rebound several years after fire, often by individuals surviving in unburned refuges within the larger burn perimeter. Northern red-backed voles (*Myodes rutilus*), considered old-growth specialists, began repopulating an intense burn in boreal Alaska from surrounding unburned forest and started reproducing 3 years thereafter (West, 1982).

Unburned refuges and vegetation changes over time also mediate postfire mammal population dynamics in other forests types, notably *Eucalyptus* forests in Australia. Numbers of bush rats (*Rattus fuscipes*) and agile antechinus (*Antechinus agilis*) were reduced compared with populations in adjacent unburned forests 6 months after severe fire in a mountain ash (*Eucalyptus regnans*) forest, but the population in the burned area was composed of residual animals that had survived the fire rather than animals recolonizing from adjacent forests (Banks et al., 2011). Long-term studies are especially useful because responses relative to time since fire can be quantified. One study examined marsupial population dynamics over a 28-year period following severe wildfire in a southeastern Australia *Eucalyptus* forest reserve (Arthur et al., 2012). Bandicoots (*Isoodon obesulus* and *Perameles nasuta*) increased immediately following the fire, peaked 15 years later, and then declined, associated with an increase and decline of shrub cover. The potoroo (*Potorous tridactylus*) population was similar before and immediately after the fire but began to increase a decade later as tree cover increased. Wombats (*Vombatus ursinus*) exhibited a stable population trend for the first decade after the fire, then slowly declined along with a decline in ground litter cover. Finally, larger macropods (eastern gray kangaroo [*Macropus giganteus*], red-necked wallaby [*Macropus rufogriseus*], and swamp wallaby [*Wallabia biocolor*]) remained at high densities after the fire then declined a decade later as vegetation cover increased.

Rabbits and hares are associated with shrubs and small conifers that provide cover (Ream, 1981; Howard, 1995). Severe fire temporarily eliminates this habitat structure, but it quickly returns as the vegetation regrows, stimulated by intense fire. Snowshoe hares (*Lepus americanus*) in a boreal forest in Alberta, Canada, moved out of intensely burned sites to surrounding habitat immediately after fire but returned the second summer after the fire when shrubs resprouted, and the postfire population trajectory increased above prefire numbers (Keith and Surrendi, 1971).

Tree squirrels, including Douglas squirrels (*Tamiasciurus douglasii*) and northern flying squirrels (*Glaucomys sabrinus*), typically are associated with late-successional coniferous forests in California and the Pacific Northwest in the United States (Carey, 2000); thus they may be adversely affected by intense fire (Zwolak and Forsman, 2007), but few data currently are available to refute or support this hypothesis. Chipmunks and ground squirrels can occupy forests after severe fire where shrubs provide cover and food (Borchert et al., 2014). Townsend's chipmunks (*Neotamias townsendii*) were abundant in early seral forests with dense shrub cover (Campbell and Donato, 2014). Gray-collared chipmunks (*Tamias cinereicollis*) and least chipmunks (*Tamias minimus*) showed no significant response to wildfire in ponderosa pine forests of the southwestern United States (Converse et al., 2006), and the proportion and composition of two chipmunk species, *Tamias amoenus* and *Tamias ruficaudus*, did not differ between severely burned and unburned conifer forest in Montana (Zwolak and Forsman, 2007).

The increase in the availability, amount, and quality of forage for herbivorous small mammals is an important determinant of the post-severe-fire community. In plots recently burned by large, intense wildfires in a Mediterranean pine-oak woodland in Spain, the abundance of small mammals—mostly mice and shrews—was higher than expected based on vegetation characteristics alone (Torre and Díaz, 2004). The authors attributed small-mammal increases to large quantities of seeds and seedlings in burned sites.

Deserts

The role of severe fire and its effects on small mammals in desert grasslands is somewhat controversial (Killgore et al., 2009; Vamstad and Rotenberry, 2010). Most desert systems are not adapted to frequent fire because many species of long-lived perennial desert plants have low recruitment rates and long life spans and lack the ability to resprout. Fire size and frequency in some areas has increased recently because of the invasion of exotic grasses from livestock grazing (Brooks, 2000) and other causes (Burbidge and McKenzie, 1989). In general, most research shows a lack of significant long-term effects of intense fire on the abundance of desert small mammals, although fire can alter community composition. Similar to shrub types in southern California, rodents in the family Heteromyidae increased following a large, intense wildfire in a perennial grassland in southeastern Arizona, whereas species in the family Cricetidae declined immediately after fire, began increasing 4 years after fire, and returned to prefire levels by the sixth year (Bock et al., 2011). Rodent abundance and species richness were no different between burned and unburned plots after wildfires in Joshua tree (*Yucca bevifolia*) woodlands of the Mojave Desert in the American Southwest (Vamstad and Rotenberry, 2010). Merriam's kangaroo rat (*Dipodomys merriami*) dominated the burned sites. As postfire vegetation changed from annuals to subshrubs and then to long-lived perennials, however, the composition of rodent species changed and the diversity of rodents increased over time.

Habitat type is important to fire effects in deserts. In Australia, wildfires in stony desert habitats with sparse grasses have less effect on habitat structure and small mammals than wildfires in sandy desert habitats with denser hummock grass spinifex (*Triodia* spp.) (Pastro et al., 2014). For example, an intense wildfire did not affect the total abundance and species richness of small mammals in the stony (gibber) desert in central Australia, although some species increased and others decreased immediately following fire (Letnic et al., 2013). By contrast, 9 months after intense wildfire in a spinifex grassland in the same region, small-mammal diversity declined compared with before the fire and with prescribed burned areas, although the abundance of animals captured was similar (Pastro et al., 2011). Data were unavailable from wildfires, but hare (*Lepus* spp.) abundance increased by 300% after prescribed burning in East African savanna grasslands (Ogen-Odoi and Dilworth, 1984).

Deer Mice

In North America, generalist deer mice are often the most abundant rodent after severe fire in a variety of vegetation types (Borchert et al., 2014). This species responds strongly and positively to high-intensity fire in both shrubland and conifer forests. Deer mice increased significantly over time in moderately and severely burned mixed-conifer forests in the San Bernardino Mountains of southern California over a 5-year period after fire (Borchert et al., 2014). During 2 years subsequent to intense fire, deer mice were invariably the most numerous species in burned study sites in a Douglas-fir-Western larch forest in Montana (Zwolak and Forsman, 2008). Converse et al. (2006) attributed increased abundance of deer mice after wildfire in southwestern United States ponderosa pine forests to greater seed production or detectability of seeds after fire.

Dramatic increases in deer mice in severely burned conifer forests were not simply a result of colonization of the burn by animals from surrounding unburned forests. When population densities were low, the vast majority of individually ear-tagged deer mice were found in forest areas after severe fire, and mice appeared regularly in unburned forests only when population densities were high (Zwolak and Forsman, 2008). This finding indicated that severely burned forest was *preferred* deer mouse habitat and that the postfire population increase was intrinsic to the burn; thus the burn itself was a source habitat.

BOX 4.2

(1) After intense wildfire, small-mammal communities are dynamic and associated with vegetation structure at different successional stages.
(2) Intense fire may increase the availability and abundance of seeds and seedlings for herbivorous small mammals.
(3) Unburned refuges and time since fire are important determinants of small-mammal communities following intense fire.
(4) The richness and abundance of small-mammal species is high following intense fire in chaparral and coastal sage scrub communities of southern California. Heteromyid rodents and deer mice often dominate severely burned shrublands, and heteromyids dominate postburn desert grasslands.
(5) Some small-mammal species decrease shortly after intense fire in North American conifer forests, but they can recover to prefire levels within 1 to several years after fire. Deer mice dramatically increase following intense fire.

Overall, these observations from small-mammal studies in mixed- and severely burned shrublands, forests, and grasslands underscore the important roles played by high-severity fire patches, unburned refuges within a fire area, and the time since fire in population dynamics after severe fire (Box 4.2).

4.4 CARNIVORES

Carnivores are critically important "top-down" regulators of ecosystem processes. Elimination of top carnivores unleashes a cascade of adverse effects, including relaxation of predation as a selective force on prey species, spread of disease, explosions of herbivore populations, and subsequent reproductive failure and local extinction of some plants, birds, herptiles, and rodents (Crooks and Soulé, 1999; Terborgh et al., 2001). Soulé Large carnivores include ursids (bears), canids (wolves), and larger felids (puma, lions, and jaguars). Medium-sized carnivores, or "mesocarnivores," include canids (coyotes and foxes), Procyonidae (ringtails and raccoons), mustelids (wolverine, marten, fisher, weasels, mink, and badgers), Mephitidae (skunks), and smaller felids (lynx and bobcats). Currently published research on carnivores in mixed and severe wildfires is limited primarily to forested habitats.

Mesocarnivores and Large Cats

Many mesocarnivores are associated with forested habitats. Some are habitat generalists, whereas others are forest specialists, riparian associates, or semi-aquatic (Buskirk and Zielinski, 2003). Martens (*Martes* spp.) occur in dense coniferous or deciduous forests across the northern hemisphere. They also regularly use severely burned habitats. Some evidence suggests martens use burns only when postfire trees are not logged. For instance, stone marten (*Martes foina*) were not detected in an intensely burned but extensively postfire-logged Aleppo pine (*Pinus halepensis*) forest in Greece the second and third years after wildfire and logging (Birtsas et al., 2012). These martens were found only in Turkish red pine (*Pinus brutia*) forests burned by wildfire 9 years earlier and not in nearby unburned forests (Soyumert et al., 2010). In coniferous forests of the Alaskan taiga, resident and transient American martens (*Martes americanus*) were captured in a 6-year-old unlogged burn more often than in an island of unburned mature forest surrounded by the burn (Paragi et al., 1996). The authors did not quantify burn severity in their study area but described fire-affected sites as having portions of "severe" burn, and most of the vegetation was in early to mid-seral stages, with dead, fire-scarred trees still standing, consistent with mixed- and high-severity fire. There was no age difference between martens trapped live in the mature forests and those trapped in the burn, and marten foraging intensity was greatest in the recently burned area (Paragi et al., 1996). Conversely, another study found martens avoided stands of boreal forests burned from 2 to 20 years prior (Gosse et al., 2005), but the study did not quantify or describe burn severity nor specify whether the burned forest was logged.

Larger cousins to the marten, fisher (*Martes pennanti* or *Pekania pennanti*) are rare mesocarnivores associated with dense, mature, boreal and mixed conifer-hardwood forests of North America (Powell and Zielinski, 1994). A recent study in the southern Sierra Nevada, however, used scat sampling to detect fisher habitat preferences and demonstrated that the species used denser, mature forests that had experienced moderate- and high-severity fire 10 and 12 years prior and that were not logged after fire (Hanson, 2013) (Figure 4.3). It is likely that both martens and fishers use severely burned forests for foraging rather than denning. These results provide intriguing evidence that even old-forest specialist species are adapted to and can exploit postfire conditions in regions where mixed- and high-severity fire is natural (see Chapter 3, Box 3.1: spotted owls).

Foxes apparently prefer severely burned forest areas over unburned areas, but they may be less tied to forest structure than martens and fishers and thus less sensitive to postfire logging. Red fox (*Vulpes vulpes*) in Turkish red pine forests were detected more often in the 9-year-old unlogged wildfire area (Soyumert et al., 2010); in postfire-logged Aleppo pine forests in Greece, red foxes were detected most often in severely burned areas, rather than moderately and unburned areas (Birtsas et al., 2012). In 3 of 4 years after intense wildfire in mixed-conifer forests of the San Bernardino Mountains in southern California, gray foxes (*Urocyon cinereoargenteus*) were detected more often in mixed-severity burned over unburned areas, and in two of the years no foxes at all were captured in the unburned area, but coyote (*Canis latrans*) were detected more often in unburned forests (Borchert, 2012). Both gray fox and coyote scats were more numerous in areas burned by intense wildfire than in unburned areas

FIGURE 4.3 Representative foraging location based upon global positioning system coordinates for a confirmed female Pacific fisher scat detection site several hundred meters into the interior of the largest high-severity fire patch (>5000 ha) in the McNally Fire of 2002, Sequoia National Forest, California. *(Photo by Chad Hanson (2014).)*

2 years after fire in interior chaparral, Madrean evergreen woodland, and ponderosa pine forest in Arizona (Cunningham et al., 2006).

Striped skunk (*Mephitis mephitis*), ringtail (*Bassariscus astutus*), and raccoon (*Procyon lotor*) were photocaptured only in mixed-conifer forests in southern California burned by high-intensity fire, but each were photographed only once (Borchert, 2012). Bobcat (*Lynx rufus*) were photocaptured in similar numbers in severely burned and unburned forest, but captures in the burned area decreased over time over the 4 years of the study. Finally, mountain lion (*Puma concolor*) were photocaptured more often in severely burned forest, but the overall sample was small (four lion in burned areas, one lion in unburned areas).

Bears

Although grizzly bears are flexible in the habitats they use, in British Columbia, Canada, radio-collared grizzly bears strongly selected open forest burned by wildfires 50-70 years earlier at high elevations because these sites supported prolific huckleberries (McLellan and Hovey, 2001) (see Box 4.3). Wildfire also promotes the regeneration of whitebark pine (*Pinus albicaulis*) seeds, another important food source for bears (Kunkel, 2003). Wildfire is not equivalent to logging, as regenerating timber harvests were rarely used by bears in any season (McLellan and Hovey, 2001).

One study compared the demographics and physiology of black bears (*Ursus americanus*) occupying burns of two ages, 13 and 35 years old, in spruce

(*Picea* spp.) and aspen (*Populus tremuloides*) forests of the Kenai Peninsula of Alaska (Schwartz and Franzmann, 1991). The authors did not specify burn intensity, but they noted that 5% of the older burn was logged after fire for "improvement" of moose (*Alces alces*) habitat, and they pointed out that the more recent fire burned at a greater intensity than the older fire. The density of bears and the percentage of cubs born were similar between the two sites, but all age groups of bears were significantly larger in the recent burn area. Bears in the older burn area consumed more cranberries (Box 4.3), whereas the number of moose calves consumed per bear was much larger in the recent burn area, likely explaining the larger size of the bears. Females in the recent burn area also produced litters at a younger age and had a shorter interval between weaning of yearlings than females in the older burn area. Moreover, cub survival was significantly higher in the recent burn area. The vigor of black bear populations was associated with moose abundance, which was significantly enhanced in the 13-year-old fire area.

Another study compared the demography of a population of black bears in interior chaparral, Madrean evergreen woodland, and ponderosa pine forest, burned by high-intensity wildfire for 3 years after fire using (1) the population in a nearby unburned site for 3 years and (2) results from earlier demographic research on the fire site from 20 years earlier, conducted over a 6-year period (Cunningham and Ballard, 2004). The sex ratio at the 3-year-old burned site was more skewed toward males than in either the unburned reference site or 20 years before the burn. The authors presumed that the fire had reduced the adult female population; however, it is also possible that the female population already had been reduced in the 20 years before the fire occurred, when the population was

BOX 4.3 Seed Dispersal by Carnivores

Fleshy fruits are an important component of the diet of many carnivores, especially during certain seasons when other resources are scarce. Indeed, the germination of many seeds is facilitated by passage through the carnivore gut because it removes the fruit pericarp and scarifies the seed coat (Herrera, 1989). Carnivores are important dispersers of seeds because they have relatively large home ranges and long gut retention times, thus spreading the seeds far from the parent plant. This may be an important mechanism whereby early seral habitats are seeded. For example, in experimental and field tests in severely burned Aleppo pine forest in Spain, Rost et al. (2012) demonstrated that carnivores, including red fox, stone marten, and European badger (*Meles meles*), were important dispersers of Mediterranean hackberry (*Celtis australis*) seeds into the burned areas. These carnivores traveled long distances into the fire area, dispersing seeds more than 1 km from the parent plant. Moreover, seeds collected from scat (i.e., that had passed through the gut) in the burned study area had a significantly greater germination rate than unscarified seeds, both in the greenhouse and in the field.

BOX 4.4

(1) Grizzly bears use areas burned by intense wildfire because of increases in berry production, although results from studies of the effects of intense fire on black bear demographics are equivocal.
(2) Martens and fisher are mesocarnivores that are dense, mature forest specialists for denning and resting but use severely burned forests that were not logged after fire, most likely for foraging.
(3) Foxes regularly use severely burned forests (regardless of postfire logging for one Mediterranean species), but results from research on coyotes are equivocal.
(4) Carnivores are important dispersers of seeds deep into severely burned forest areas.

not monitored. Indeed, an alternative scenario could be that the population of both adult females and males had been declining at Four Peaks before fire, and the fire actually attracted males to the site, who have larger home ranges, thus skewing the sex ratio.

The above study reported complete reproductive failure in the 3 years after fire at the burned site compared with 36% of cubs surviving to 1 year of age on the unburned control site (Cunningham and Ballard, 2004). More cubs had survived to year 1 at the burned site 20 years before the fire. During the 1970s, however, complete reproductive failure also occurred in the absence of fire during 3 of the 6 years of study. Thus years of complete reproductive failure in that study area were not unusual. Overall, reproductive success was lowest in the burned forest compared with the same site 20 years before fire and an unburned reference site, suggesting the possibility of negative short-term effects of high-intensity fire on black bear reproduction. The mortality of adult bears from hunting, however, was 2.5 times higher in the fire area than in the unburned area (Cunningham et al., 2001), which would be expected to influence cub survival, potentially confounding results. The overall density of black bears in the fire area was higher than prefire densities in the area (Cunningham et al., 2001) (Box 4.4).

4.5 UNGULATES

As major herbivorous components of ecosystems, ungulates can act as keystone species with profound effects on vegetation development and productivity in forests, woodlands, and grassland ecosystems throughout the world (Hobbs, 1996; Wisdom et al., 2006). Hobbs (1996) stated, "ungulates are not merely outputs of ecosystems, they may also serve as important regulators of ecosystem processes at several scales of time and space" (p. 695). Ungulates, Hobbs further noted, are "important agents of environmental change, acting to create spatial heterogeneity, accelerate successional processes, and control the switching of ecosystems between alternative states." Ungulates regulate nitrogen cycling and influence plant size and morphology (Singer et al., 2003). Because grazing and browsing by ungulates affects the biomass, structure, and type of vegetation available to burn, these animals can actually regulate the dynamics of fire (Hobbs, 1996; Wisdom et al., 2006).

Episodic disturbance agents such as fire strongly interact with ungulate herbivory over space and time. For example, removal of fine fuels by ungulate grazers may reduce the frequency of ground fires but can increase crown fires by enhancing the development of ladder trees, especially when combined with a relatively long absence of fire (Hobbs, 1996). Further, postfire plant regeneration provides forage species that are highly palatable to ungulates, which attracts ungulates to burned areas, where they influence vegetation regrowth after fire (Canon et al., 1987; Wan et al, 2014). Moose rapidly immigrated to burned areas after a large wildfire in mixed coniferous-deciduous forests of northern Minnesota (Peek, 1974). In fact, fire size can moderate the adverse effects of ungulate herbivory on vegetation recovery. Compared with small fires, large fires "swamp" the effects of ungulate herbivory, for example, by providing sufficient new grass production to offset browsing, and enabling woody species such as aspen (*Populus sp.*) to grow to tree height (Biggs et al., 2010). In intensively burned ponderosa pine, mixed-conifer, and spruce-fir forests of northern New Mexico, elk selectively foraged on grasses over shrubs (Biggs et al., 2010). In 25 wildfires throughout five national forests in Utah, larger areas of aspen forest that burned with greater severity had the highest growth potential for aspen regeneration, and these high burn-severity conditions stimulated defensive chemicals in plants that lowered the levels of damage done by ungulate browsing (Wan et al., 2014). Wan et al. noted that "this effect may be particularly strong if amplified over large post-fire landscapes by saturating the browse capacity of the ungulate community." (See Box 4.5).

Positive effects of high-severity fire on ungulates likely are most pronounced in vegetation types that are most adapted to high-intensity fires, such as aspen forests and shrublands. Mountain or bighorn sheep selected intensely burned shrublands up to 15 years after fire in Montana (DeCesare and Pletscher, 2006) and in southern California mountains (Bleich et al., 2008). Wildfire increased the carrying capacity of southern California desert bighorn sheep (*Ovis canadensis nelsoni*) in the San Gabriel Mountains, dramatically increasing the number of animals in this endangered population (Holl et al., 2004). A large natural fire on the eastern slopes of the Sierra Nevada mountains in California improved the winter range of Sierra Nevada bighorn sheep (*Ovis canadensis sierrae*) by increasing green forage availability, shifting diet composition to include more forbs, and possibly decreasing predation risk from mountain lions by increasing visibility (Greene et al., 2012). Overall, large, high-severity fire in bighorn sheep shrubland/forest habitats increases forage quality and availability as well as visual openness, which is critical because several populations are listed as endangered.

Studies investigating the impact of fire on mule deer (*Odocoileus hemionus*), a common herbivore in the western United States, indicate that populations tend to increase after severe fire, especially in chaparral communities. In a review of the literature on ungulate responses to fire, Smith (2000) reported mule deer density in intensely burned chaparral was more than twice as high as that in mature chaparral in California, and it increased 400% the first year after high-intensity fire in chamise chaparral. Density then decreased each year afterward until preburn levels were reached 5-12 years later. Chamise chaparral burned by a large wildfire in California had more deer use per square mile than unburned chamise chaparral (Bendell, 1974). In northern coastal California, mule deer densities in chaparral burned by high-intensity wildfire the year before were four times greater than in unburned chaparral (Taber and Dasmann, 1957). Because the fire described in this study was relatively small, deer may have moved from one area to another rather than actually increasing the population via higher birth rates. Similarly, mule deer in central coastal California strongly preferred burned habitat, with a 400% increase in the density of deer in prescribe-burned chaparral near oak woodlands, relative to preburn density, by the second growing season (Klinger et al., 1989). Here the increase in the use of burned chaparral was attributed to movements of deer from adjacent oak woodlands rather than an intrinsic increase in population size. Heavy use of prescribe-burned chamise chaparral by mule deer was reported in the San Jacinto Mountains of southern California (Roberts and Tiller, 1985).

Other studies documented postfire increases in the number of mule deer in conifer forests. Visual observations of 543 mule deer indicated a preference for burned over unburned Douglas-fir/ninebark (*Physocarpus sp.*) (*Physocarpus* spp.) and burned ponderosa pine/bluebunch wheatgrass (*Pseudoroegneria spicata*) habitat types during winter and spring in the Selway-Bitterroot Wilderness of Idaho, although the authors did not specifically define the burn severity of sites used by deer (Keay and Peek, 1980). Two other studies that documented increases in mule deer in burned forests hypothesized that postfire logging removes protective cover, a critical habitat element for mule deer. Significantly more deer droppings were located in pinyon-juniper woodlands of Arizona burned by high-intensity fire 13 years earlier than in adjacent unburned areas (McCulloch, 1969). The author surmised that the standing forest of dead trees and fallen trunks provided some cover for deer from predators. Both mule deer and elk used intensely burned lodgepole pine forests at two sites in Wyoming significantly more than paired clearcut sites of the same ages (9 and 5 years old), based on fecal pellet counts (Davis, 1977). Davis (1977, p. 787) stated: "Deer and elk use was greater in burned areas with standing dead timber than in clearcut areas without it. In the Sierra Madre study area, the burned and clearcut plots both had the same number of plant species present, and they both had standing dead timber. However, the burned plot with much more standing dead timber had more deer and elk use. Fire opened up the canopy allowing light to enter, stimulating growth of forage plants, while the dead trees left standing provided good protective cover" (see Figure 4.4).

FIGURE 4.4 Mule deer respond positively to high-severity fire in forests. In this photo, mule deer forage on fresh vegetation growing in the first post-fire year following the Rim fire of 2013 on the Stanislaus National Forest, central Sierra Nevada. *(Photo by Chad Hanson (2014).)*

Available studies generally report increases in the reproductive rates and body condition of female mule deer in burned habitats. The reproductive rate was 1.32 fawns per doe in the first year after wildfire in northern coastal California, compared with 0.77 fawns per doe in unburned chaparral (Taber and Dasmann, 1957). After 3 years, the reproductive rate of deer at the burned site declined to that of deer in the unburned site. Chamise chaparral burned by a large wildfire produced heavier deer, and does had a higher frequency of ovulation, gave birth to more fawns, and wintered in better condition than does in dense, unburned chamise (Bendell, 1974). Another study, however, documented no difference in fawn-to-doe ratios between burned and unburned chaparral interspersed with oak woodlands in central California (Klinger et al., 1989).

Foraging studies indicate that mule deer populations in chaparral habitats burned by high-intensity fire often increase as a result of the increased availability of browse. *Ceanothus*—a high-quality food for ungulates (Hobbs, 1996)—is abundant after fire because it reproduces from seed that is scarified by burning (Smith, 2000). Thimbleberry (*Rubus parviflorus*) also generally increases after fire (Smith, 2000). Moreover, fire can increase the palatability of foliage for deer as well as the crude protein content (Smith, 2000). The improved quantity and quality of browse may be related to the fire-caused increase in available nutrients in the soil. As such, deer populations often benefit from the increased food production and nutritional value of their food in recently burned areas. Length and surface enlargement factor of papillae (the surface area within the intestine for absorbing nutrients) of necropsied mule deer were greater in those from high-intensity burned than unburned ponderosa pine habitat in the southern Black Hills of South Dakota (Zimmerman et al., 2006). These physiological factors indicate higher forage quality, such as greater concentration of volatile fatty acids. The authors concluded that fire was beneficial at the mucosal level for mule deer: the increase in forage quality from burning caused a rapid change in papillary morphology, allowing the deer to take up more nutrients.

Lichens in boreal habitats are preferred winter forage for caribou (*Rangifer tarandus*), yet large wildfires that depleted lichens had no effect on home-range size, range fidelity, or the survival and fecundity of woodland caribou (*Rangifer tarandus caribou*) in Alberta, Canada (Dalerum et al., 2007). Caribou avoided foraging in burned compared with unburned areas (Dalerum et al., 2007; Joly et al., 2010), although burn severity was not quantified, and some of the fires occurred 50 years before study. Lichens are significantly reduced by wildfire and take decades to recover to prefire abundance (Joly et al., 2010) (Box 4.5).

BOX 4.5

(1) Ungulates interact strongly with episodic disturbances. Many are attracted to severely burned areas because of increased forage palatability and availability, where in turn they influence vegetation regrowth.
(2) Elk, bighorn sheep, and mule deer generally increase after intense fire in shrublands and forests.
(3) The larger the area of high-severity fire, the lower the adverse impact on regrowth of aspen forests from ungulate herbivory.
(4) Caribou may be adversely affected when intense fire reduces lichen used for winter forage.

4.6 MANAGEMENT AND CONSERVATION RELEVANCE

The abundance of certain mammal species after fire has direct benefits to land managers in the form of irreplaceable ecosystem and economic services. Bats are voracious predators of insects—many of them consume crop and forest pests—and as such are important regulators of insect populations, including disease-carrying mosquitoes (Reiskind and Wund, 2009). Bats are also critical pollinators of many plants (Molina-Freaner and Eguiarte, 2003). The loss of bats in North America could cost the economy $3.7 billion per year in agricultural losses alone (Boyles et al., 2011). Small mammals aerate the soil and, along with many carnivores, are important dispersers of seeds and fungi (Maser et al., 1978; Rost et al., 2012). Large carnivores are top-down regulators of smaller carnivores and ungulates, and are vital to the health and function of natural ecosystems. Ungulates help to cycle nitrogen and provide big-game hunting opportunities and food for humans. Indeed, in 2001 alone, hunting of ungulates and large carnivores in the United States contributed to approximately $25 billion in retail sales and $17 billion in salaries and wages and employed 575,000 people (IAFWA, 2002). These animals include mule deer, bighorn sheep, moose, elk, and bear, all of which use or thrive within heavily burned habitats.

As described here, a great many mammals benefit from mixed- and high-severity fire and play essential roles in postfire ecosystem dynamics. Land managers rarely weigh these benefits when evaluating the impacts of large fires of mixed- and high-severity, however, thus undervaluing their ecological and economic importance. The vital ecosystem services of mammals in postfire areas should be quantified and carefully considered when planning potentially harmful management activities such as postfire logging and common management activities following postfire logging, such as the application of herbicides and rodenticides.

4.7 CONCLUSIONS

The extraordinary abundance and diversity of mammals using (e.g., American marten, Pacific fisher, grizzly bear) and even thriving (e.g., deer mice, kangaroo rats, bats, mule deer, elk, bighorn sheep) in severely burned grassland, shrubland, and forested habitats is an important indicator of the high habitat suitability of these areas. Prescribed burning does not provide the expected gains in biological diversity for a range of mammal, reptile, bird, and plant taxa (Pastro et al., 2014). Only large, severe wildfires create significant ecological changes associated with increases in fire-loving species, and, as demonstrated herein, only larger fires can "swamp" the effects of ungulate herbivory on postfire vegetation. Mixed- and high-severity fires globally have unique ecological value that must be weighed against the dominant paradigm that such natural disturbance events are "catastrophic" (Zwolak and Foresman, 2008; also see Chapters 1, 2, and 13). Mammals and other wildlife using intensely burned forests provide myriad ecological services that benefit people and ecosystems alike.

APPENDIX 4.1 THE NUMBER OF STUDIES BY TAXA SHOWING DIRECTIONAL RESPONSE (NEGATIVE, NEUTRAL, OR POSITIVE) TO SEVERE WILDFIRE OVER THREE TIME PERIODS FOLLOWING FIRE. STUDIES CITED INCLUDE UNBURNED AREAS COMPARED TO SEVERELY BURNED AREAS WITH NO POST-FIRE LOGGING, AND EXCLUDED PRESCRIBED BURNS. FOR SMALL MAMMALS, ONLY SPECIES WITH ENOUGH DETECTIONS TO DETERMINE DIRECTIONAL RESPONSE WERE REPORTED.

	1-5 yr post-fire			6-10 yr post-fire			>10 yr post-fire		
	Negative	*Neutral*	*Positive*	*Negative*	*Neutral*	*Positive*	*Negative*	*Neutral*	*Positive*
Bats[1]		1	3						
Small Mammals[2]									
Masked shrew	1								
White-toothed shrew			1						1
Tamias spp.		4							
Pacific kangaroo rat		1	2			1			
Dulzura kangaroo rat			2						
Merriam's kangaroo rat			1			1			1
California pocket mouse		1	1						
San Diego pocket mouse	1	2							
Bush rat	1								
Long-haired rat	1								

Continued

	1-5 yr post-fire			6-10 yr post-fire			>10 yr post-fire		
	Negative	*Neutral*	*Positive*	*Negative*	*Neutral*	*Positive*	*Negative*	*Neutral*	*Positive*
Red-backed vole	2								
California vole		1	2						
Canyon mouse	1			1			1		
Brush mouse		1	1						
Deer mouse		2	5		1				
California mouse	3		1			1			
Cactus mouse	1	1	1						
Pinyon mouse	1		1	1					
Harvest mouse	2	1	1						
Desert woodrat	2								
Big-eared woodrat	1	1							
Snowshoe hare			1						
Antechinus	1								
Potoroo		1				1			1
Bandicoot			1			1	1		
Wombat		1			1		1		
Macrocarps (3 spp)		1		1			1		

Carnivores[3]						
American marten				1		
Stone marten				1		
Pacific fisher				1		
Gray fox		2				
Red fox				1		
Black bear	1					1
Grizzly bear						1
Ungulates[4]						
Caribou	2		2		2	
Moose		1				
Bighorn sheep		3		2		2
Elk		1				
Mule deer		4		1		1

[1]*Bat citations: Schwab 2006, Malison and Baxter 2010, Buchalski et al. 2013. Bats are categorized by phonic groups: of 6 phonic groups in 3 studies, 5 phonic groups showed positive response and 1 showed neutral response.*
[2]*Small mammal citations: Keith and Surrendi 1971, Cook 1959, Wirtz 1977, West 1982, Price and Waser 1984, Torre and Díaz 2004, Converse et al. 2006, Zwolak and Forsman 2007, Zwolak and Forsman 2008, Vamstad and Rotenberry 2010, Brehme et al. 2011, Banks et al. 2011, Arthur et al. 2012, Borchert and Borchert 2013, Letnic et al. 2013, Borchert et al. 2014*
[3]*Carnivore citations: Paragi et al. 1996, McLellan and Hovey 2001, Cunningham and Ballard 2004, Cunningham et al. 2006, Soyumert et al. 2010, Borchert 2012, Hanson 2013*
[4]*Ungulate citations: Taber and Dasmann 1957, McCulloch 1969, Bendell 1974, Peek 1974, Davis 1977, Keay and Peek 1980, Smith 2000, Holl et al. 2004, Dalerum et al. 2007, Bleich et al. 2008, Biggs et al. 2010, Joly et al. 2010, Greene et al. 2012*

Studies cited include unburned areas compared with severely burned areas with no postfire logging; they exclude prescribed burns. For small mammals, only species with enough detections to determine directional response are reported.

REFERENCES

Arthur, A.D., Catling, P.C., Reid, A., 2012. Relative influence of habitat structure, species interactions and rainfall on the post-fire population dynamics of ground-dwelling vertebrates. Austral Ecol. 37, 958–970.

Banks, S.C., Dujardin, M., McBurney, L., Blair, D., Barker, M., Lindenmayer, D.B., 2011. Starting points for small mammal population recovery after wildfire: recolonisation or residual populations? Oikos 120, 26–37.

Bendell, J.F., 1974. Effects of fire on birds and mammals. In: Kozlowski, T.T., Ahlgren, C.E. (Eds.), Fire and Ecosystems. Academic Press, New York, NY, USA, pp. 73–138.

Biggs, J.R., VanLeeuwen, D.M., Holechek, J.L., Valdez, R., 2010. Multi-scale analyses of habitat use by elk following wildfire. Northwest Sci. 84, 20–32.

Birtsas, P., Sokos, C., Exadacylos, S., 2012. Carnivores in burned and adjacent unburned areas in a Mediterranean ecosystem. Mammalia 76, 407–415.

Bleich, V.C., Johnson, H.E., Holl, S.A., Konde, L., Torres, S.G., Krausman, P.R., 2008. Fire history in a chaparral ecosystem: Implications for conservation of a native ungulate. Rangel. Ecol. Manage. 61, 571–579.

Bock, C.E., Jones, Z.F., Kennedy, L.J., Block, J.H., 2011. Response of rodents to wildfire and livestock grazing in an Arizona desert grassland. Am. Midl. Nat. 166, 126–138.

Borchert, M.I., 2012. Mammalian carnivore use of a high-severity burn in conifer forests in the San Bernardino Mountains of southern California, USA. Hystrix Ital. J. Mammal. 23, 50–56.

Borchert, M.I., Borchert, S., 2013. Small mammal use of the burn perimeter following a chaparral wildfire in southern California. Bull. South. Calif. Acad. Sci. 112, 63–73.

Borchert, M.I.D., Farr, P., Rimbenieks-Negrete, M.A., Pawlowski, M.N., 2014. Responses of small mammals to wildfire in a mixed conifer forest in the San Bernardino Mountains, California. Bull. South. Calif. Acad. Sci. 113, 81–95.

Boyles, J.G., Aubrey, D.P., 2006. Managing forests with prescribed fire: Implications for a cavity-dwelling bat species. For. Ecol. Manag. 222, 108–115.

Boyles, J.G., Cryan, P.M., McCracken, G.F., Kunz, T.H., 2011. Economic importance of bats in agriculture. Science 332, 41–42.

Brehme, C.S., Clark, D.R., Rochester, C.J., Fisher, R.N., 2011. Wildfires alter rodent community structure across four vegetation types in southern California, USA. Fire Ecol. 7, 81–98.

Brigham, R.M., Vonhof, M.J., Barclay, R.M.R., Gwilliam, J.C., 1997. Roosting behavior and roost-site preferences of forest-dwelling California bats (*Myotis californicus*). J. Mammal. 78, 1231–1239.

Brooks, M.L., 2000. Competition between alien annual grasses and native annual plants in the Mojave Desert. Am. Midl. Nat. 144, 92–108.

Buchalski, M.R., Fontaine, J.B., Heady III, P.A., Hayes, J.P., Frick, W.F., 2013. Bat response to differing fire severity in mixed-conifer forest California, USA. PLoS One 8 (3), e57884. http://dx.doi.org/10.1371/journal.pone.0057884.

Burbidge, A.A., McKenzie, N.L., 1989. Patterns in the modern decline of Western Australia's vertebrate fauna: causes and conservation implications. Biol. Conserv. 50, 143–198.

Buskirk, S.W., Zielinski, W.J., 2003. Small and mid-sized carnivores. In: Zabel, C.J., Anthony, R.G. (Eds.), Mammal Community Dynamics: Management and Conservation in the Coniferous Forests of Western North America. Cambridge University Press, New York, NY, USA, pp. 207–249.

Campbell, J.L., Donato, D.C., 2014. Trait-based approaches to linking vegetation and food webs in early-seral forests of the Pacific Northwest. For. Ecol. Manag. 324, 172–178.

Canon, S.K., Urness, P.J., DeByle, N.V., 1987. Habitat selection, foraging behavior, and dietary nutrition of elk in burned aspen forest. J. Range Manag. 40, 433–438.

Carey, A.B., 2000. Effects of new forest management strategies on squirrel populations. Ecol. Apps. 10, 248–257.

Carey, A.B., Horton, S.P., Biswell, B.L., 1992. Northern spotted owls: Influence of prey base and landscape character. Ecol. Monogr. 62, 223–250.

Chew, R.M., Butterworth, B.B., Grechman, R., 1959. The effects of fire on the small mammal populations of the chaparral. J. Mammal. 40, 253.

Converse, S.J., White, G.C., Block, W.M., 2006. Small mammal responses to thinning and wildfire in ponderosa pine-dominated forests of the southwestern United States. J. Wildl. Manag. 70, 1711–1722.

Cook, S.F., 1959. The effects of fire on a population of small rodents. Ecology 40, 102–108.

Coppeto, S.A., Kelt, D.A., Van Vuren, D.H., Wilson, J.A., Bigelow, S., 2006. Habitat associations of small mammals at two spatial scales in the northern Sierra Nevada. J. Mammal. 87, 402–413.

Crooks, K.R., Soulé, M.E., 1999. Mesopredator release and avifaunal extinctions in a fragmented system. Nature 400, 563–566.

Cunningham, S.C., Ballard, W.B., 2004. Effects of wildfire on black bear demographics in central Arizona. Wildl. Soc. Bull. 32, 928–937.

Cunningham, S.C., Monroe, L.M., Kirkendall, L., Ticer, C.L., 2001. Effects of the catastrophic Lone fire on low, medium, and high mobility wildlife species. Technical Guidance Bulletin No. 5, Arizona Game and Fish Department, Phoenix, Arizona, USA.

Cunningham, S.C., Kirkendall, L., Ballard, W., 2006. Gray fox and coyote abundance and diet responses after a wildfire in central Arizona. Western North American Nat. 66, 169–180.

Dalerum, F., Boutin, S., Dunford, J.S., 2007. Wildfire effects on home range size and fidelity of boreal caribou in Alberta, Canada. Can. J. Zool. 85, 26–32.

Davis, P.R., 1977. Cervid response to forest fire and clearcutting in southeastern Wyoming. J. Wildl. Manag. 41, 785–788.

DeCesare, N.J., Pletscher, D.H., 2006. Movements, connectivity, and resource selection of Rocky Mountain bighorn sheep. J. Mammal. 87, 531–538.

Diffendorfer, J.G.M., Fleming, S., Tremor, W. Spencer, Beyers, J.L., 2012. The role of fire severity, distance from fire perimeter and vegetation on post-fire recovery of small-mammal communities in chaparral. Int. J. Wildlife Fire 21, 436–448. http://dx.doi.org/10.1071/WF10060.

Fellers, G.M., Pierson, E.D., 2002. Habitat use and foraging behavior of Townsend's big-eared bat (*Corynorhinus townsendii*) in coastal California. J. Mammal. 83, 167–177.

Fontaine, J.B., Kennedy, P.L., 2012. Meta-analysis of avian and small-mammal response to fire severity and fire surrogate treatments in U.S. fire-prone forests. Ecol. Appl. 22, 1547–1561.

Gellman, S.T., Zielinski, W.J., 1996. Use of bats of old-growth redwood hollows on the north coast of California. J. Mammal. 77, 255–265.

Gosse, J.W., Cox, R., Avery, S.W., 2005. Home-range characteristics and habitat use by American martens in eastern Newfoundland. J. Mammal. 86, 1156–1163.

Greene, L., Hebblewhite, M., Stephenson, T.R., 2012. Short-term vegetation response to wildfire in the eastern Sierra Nevada: implications for recovering an endangered ungulate. J. Arid Environ. 87, 118–128.

Gruver, J. C., Keinath, D.A., 2006. Townsend's Big-eared Bat (*Corynorhinus townsendii*): a technical conservation assessment. USDA Forest Service, Rocky Mountain Region. http://www.fs.fed.us/r2/projects/scp/assessments/townsendsbigearedbat.pdf.

Haim, A., Izhaki, I., 1994. Changes in rodent community during recovery from fire—relevance to conservation. Biodivers. Conserv. 3, 573–585.

Hanson, C.T., 2013. Habitat use of Pacific fishers in a heterogeneous post-fire and unburned forest landscape on the Kern Plateau, Sierra Nevada, California. Open For. Sci. J. 6, 24–30.

Herrera, C.M., 1989. Frugivory and seed dispersal by carnivorous mammals, and associated fruit characteristics, in undisturbed Mediterranean habitats. Oikos 55, 250–262.

Hobbs, N.T., 1996. Modification of ecosystems by ungulates. J. Wildl. Manag. 60, 695–713.

Holl, S.A., Bleich, V.C., Torres, S.T., 2004. Population dynamics of bighorn sheep in the San Gabriel Mountains, California, 1967-2002. Wildl. Soc. Bull. 32, 412–426.

Howard, J. 1995. *Lepus californicus* (black-tailed jackrabbit). In Fire Effects Information System, [Online]. U.S. Department of Agriculture, Forest Service, Rocky Mountain Research Station, Fire Sciences Laboratory (Producer). Available: http://www.fs.fed.us/database/feis/animals/mammal/leca/all.html.

International Association of Fish and Wildlife Agencies (IAFWA), 2002. Economic Importance of Hunting in America. IAFWA, Washington, DC.

Johnson, J.B., Edwards, J.W., Ford, W.M., Gates, J.E., 2009. Roost tree selection by northern myotis (*Myotis septentrionalis*) maternity colonies following prescribed fire in a Central Appalachian Mountains hardwood forest. For. Ecol. Manag. 258, 233–242.

Joly, K., Chapin III, F.S., Klein, D.R., 2010. Winter habitat selection by caribou in relation to lichen abundance, wildfires, grazing, and landscape characteristics in northwest Alaska. Ecoscience 17, 321–333.

Keay, J.A., Peek, J.M., 1980. Relationships between fires and winter habitat of deer in Idaho. J. Wildl. Manag. 44, 372–380.

Keith, L.B., Surrendi, D.C., 1971. Effects of fire on a snowshoe hare population. J. Wildl. Manag. 35, 16–26.

Kelly, L.T., Nimmo, D.G., Spence-Bailey, L.M., Taylor, R.S., Watson, S.J., Clarke, M.F., Bennett, A.F., 2012. Managing fire mosaics for small mammal conservation: a landscape perspective. J. Appl. Ecol. 49, 412–421.

Killgore, A., Jackson, E., Whitford, W.G., 2009. Fire in Chihuahuan Desert grassland: Short-term effects on vegetation, small mammal populations, and faunal pedoturbation. J. Arid Environ. 73, 1029–1034.

Klinger, R.C.M., Kutilek, J., Shellhammer, H.S., 1989. Population responses of black-tailed deer to prescribed burning. J. Wildl. Manag. 53, 863–871.

Kunkel, K.E., 2003. Ecology, conservation, and restoration f large carnivores in western North America. In: Zabel, C.J., Anthony, R.G. (Eds.), Mammal Community Dynamics: Management and Conservation in the Coniferous Forests of Western North America. Cambridge University Press, New York, NY, USA, pp. 250–295.

Lawler, T.E., 2003. Faunal composition and distribution of mammals in western coniferous forests. In: Zabel, C.J., Anthony, R.G. (Eds.), Mammal Community Dynamics: Management and Conservation in the Coniferous Forests of Western North America. Cambridge University Press, New York, NY, USA, pp. 41–80.

Lawrence, G.E., 1966. Ecology of vertebrate animals in relation to chaparral fire in the Sierra Nevada foothills. Ecology 47, 278–291.

Lee, D.E., Tietje, W.D., 2005. Dusky-footed woodrat demography and prescribed fire in a California oak woodland. J. Wildl. Manag. 69, 760–769.

Letnic, M., Tischler, M., Gordon, C., 2013. Desert small mammal responses to wildfire and predation in the aftermath of a La Niña driven resource pulse. Austral Ecol. 38, 841–849.

Lillywhite, H.B., 1977. Effects of chaparral conversion on small vertebrates in southern California. Biol. Conserv. 11, 171–184.

Malison, R.L., Baxter, C.V., 2010. The fire pulse: wildfire stimulates flux of aquatic prey to terrestrial habitats driving increases in riparian consumers. Can. J. Fish. Aquat. Sci. 67, 570–579.

Maser, C., Trappe, J.M., Nussbaum, R.A., 1978. Fungal-small mammal interrelationships with emphasis on Oregon coniferous forests. Ecology 59, 799–809.

Mazurek, M.J., 2004. A maternity roost of Townsend's big-eared bats (*Corynorhinus townsendii*) in coast redwood basal hollows in northwestern California. Northwest. Nat. 85, 60–62.

McCulloch, C.Y., 1969. Some effects of wildfire on deer habitat in pinyon-juniper woodland. J. Wildl. Manag. 33, 778–784.

McLellan, B.N., Hovey, F.W., 2001. Habitats selected by grizzly bears in a multiple use landscape. J. Wildl. Manag. 65, 92–99.

Minshall, G.W., 2003. Responses of stream benthic macroinvertebrates to fire. For. Ecol. Manag. 178, 155–161.

Molina-Freaner, F., Eguiarte, L.E., 2003. The pollination biology of two paniculate agaves (Agavaceae) from northwestern Mexico: contrasting roles of bats as pollinators. Am. J. Bot. 90, 1016–1024.

Ogen-Odoi, A.A., Dilworth, T.G., 1984. Effects of grassland burning on the savanna hare-predator relationship in Uganda. Afr. J. Ecol. 22, 101–106.

Paragi, T.F., Johnson, W.N., Katnik, D.D., Magoun, A.J., 1996. Marten selection of postfire seres in the Alaskan taiga. Can. J. Zool. 74, 2226–2237.

Pastro, L.A., Dickman, C.R., Letnic, M., 2011. Burning for biodiversity or burning biodiversity? Prescribed burn vs. wildfire impacts on plants, lizards, and mammals. Ecol. Appl. 21, 3238–3253.

Pastro, L.A., Dickman, C.R., Letnic, M., 2014. Fire type and hemisphere determine the effects of fire on the alpha and beta diversity of vertebrates: a global meta-analysis. Glob. Ecol. Biogeogr. 23, 1146–1156.

Peek, J.M., 1974. Initial response of moose to a forest fire in northeastern Minnesota. Am. Midl. Nat. 91, 435–438.

Powell, R.A., Zielinski, W.J., 1994. Fisher. In: Ruggiero, L.F., Aubry, K.B., Buskirk, S.W., Zielinski, W.J. (Eds.), The Scientific Basis for Conserving Forest Carnivores: American Marten, Fisher, Lynx, and Wolverine. USDA Forest Service General Technical Report RM-254. Rocky Mountain Forest and Range Experimental Station, Fort Collins, CO.

Price, M.V., Waser, N.M., 1984. On the relative abundance of species: postfire changes in a coastal sage scrub rodent community. Ecology 65, 1161–1169.

Quinn, R.D., 1979. Effects of fire on small mammals in the chaparral. In: Koch, D.L. (Ed.), Cal-Neva Wildlife Transactions. Western Section of the Wildlife Society, Smartsville, California.

Ream, C.H., 1981. The effects of fire and other disturbances on small mammals and their predators: an annotated bibliography. USDA Forest Service, Intermountain Forest and Range Experiment Station General Technical Report INT-106.

Reiskind, M.H., Wund, M.A., 2009. Experimental assessment of the impacts of northern long-eared bats on ovipositing *Culex* (Diptera: Culcidae) mosquitoes. J. Med. Entomol. 46, 1037–1044.

Roberts, T.A., Tiller, R.L., 1985. Mule deer and cattle responses to a prescribed burn. Wildl. Soc. Bull. 13, 248–252.

Rost, J., Pons, P., Bas, J.M., 2012. Seed dispersal by carnivorous mammals into burnt forests: an opportunity for non-indigenous and cultivated plant species. Basic Appl. Ecol. 13, 623–630.

Saint-Germain, M., Drapeau, P., Hébert, C., 2004. Comparison of Coleoptera assemblages from a recently burned and unburned black spruce forests of northeastern North America. Biol. Conserv. 118, 583–592.

Schwab, N.A., 2006. Roost-site selection and potential prey sources after wildland fire for two insectivorous bat species (*Myotis evotis* and *Myotis lucifugus*) in mid-elevation forests of western Montana. M.S. thesis, University of Montana.

Schwartz, C.C., Franzmann, A.W., 1991. Interrelationship of black bears to moose and forest succession in the northern coniferous forest. Wildl. Monogr. 113, 1–58.

Schwilk, D.W., Keeley, J.E., 1998. Rodent populations after a large wildfire in California chaparral and coastal sage scrub. Southwest. Nat. 43, 480–483.

Shaffer, K.E., Laudenslayer Jr., W.F., 2006. Fire and animal interactions. In: Sugihara, N.G., van Wagtendonk, J., Shaffer, K.E., Fites-Kaufman, J., Thode, A.E. (Eds.), Fire in California's Ecosystems. University of California Press, Berkeley and Los Angeles, CA, pp. 118–143.

Singer, F.J., Schreier, W., Oppenheim, J., Garton, E.O., 1989. Fire impact on Yellowstone. Bioscience 39, 716–722.

Singer, F.J., Wang, G., Hobbs, N.T., 2003. The role of ungulates and large predators on plant communities and ecosystem processes in western national parks. In: Zabel, C.J., Anthony, R.G. (Eds.), Mammal Community Dynamics: Management and Conservation in the Coniferous Forests of Western North America. Cambridge University Press, New York, NY, USA, pp. 444–486.

Smith, J.K., 2000. Wildland fire in ecosystems: effects of fire on fauna. USDA Forest Service, Rocky Mountain Research Station General Technical Report RMRS-GTR-42-vol 1, Ogden, UT.

Soyumert, A., Tavşanoğlu, C., Macar, O., Kaynaş, B.Y., Gürkan, B., 2010. Presence of large and -medium-sized mammals in a burned pine forest in southwestern Turkey. Hystrix Ital. J. Mammal. 21, 97–102.

Spies, T.A., Turner, M.G., 1999. Dynamic forest mosaics. In: Hunter Jr., M.L. (Ed.), Maintaining Biodiversity in Forest Ecosystems. Cambridge University Press, Cambridge, UK, pp. 95–160.

Taber, R.D., Dasmann, R.F., 1957. The dynamics of three natural populations of the deer *Odocoileus hemionus columbianus*. Ecology 38, 233–246.

Tarbill, G.L., 2010. Nest site selection and influence of woodpeckers on recovery in a burned forest of the Sierra Nevada. Master's Thesis, California State University, Sacramento.

Terborgh, J., Lopez, L., Nunez, P., Rao, M., Shahabuddin, G., Orihuela, G., Riveros, M., Ascanio, R., Adler, G.H., Lambert, T.D., Balbas, L., 2001. Ecological meltdown in predator-free forest fragments. Science 294, 1923–1925.

Torre, I., Díaz, M., 2004. Small mammal abundance in Mediterranean post-fire habitats: A role for predators? Acta Oecol. 25, 137–143.

Vamstad, M.S., Rotenberry, J.T., 2010. Effects of fire on vegetation and small mammal communities in a Mojave Desert Joshua tree woodland. J. Arid Environ. 74, 1309–1318.

Wan, H.Y., Olson, A.C., Muncey, K.D., St. Clair, S.B., 2014. Legacy effects of fire size and severity on forest regeneration, recruitment, and wildlife activity in aspen forests. For. Ecol. Manag. 329, 59–68.

Ward Jr., J.P., Gutiérrez, R.J., Noon, B.R., 1998. Habitat selection by northern spotted owls: the consequences of prey selection and distribution. Condor 100, 79–92.

West, S.D., 1982. Dynamics of colonization and abundance in Central Alaskan populations of the northern red-backed vole, *Clethrionomys rutilus*. J. Mammal. 63, 128–143.

Wirtz, W.O., 1977. Vertebrate post-fire succession. In: Mooney, H.A., Conrad, C.E., technical coordinators (Eds.), Proceedings of the Symposium on Environmental Consequences of Fire and

Fuel Management in Mediterranean Ecosystems. USDA Forest Service General Technical Report WO-3.

Wirtz, W.O., 1995. Responses of rodent populations to wildfire and prescribed fire in southern California chaparral. In: Keeley, J.E., Scott, T. (Eds.), Brushfires in California: Ecology and Resource Management. International Association of Wildlife Fire, Fairfield, Washington, USA, pp. 63–67.

Wisdom, M.J., Vavra, M., Boyd, J.M., Hemstrom, M.A., Ager, A.A., Johnson, B.K., 2006. Understanding ungulate herbivory-episodic disturbance effects on vegetation dynamics: knowledge gaps and management needs. Wildl. Soc. Bull. 34, 283–292.

Zielinski, W.J., Gellman, S.T., 1999. Bat use of remnant old-growth redwood stands. Conserv. Biol. 13, 160–167.

Zimmerman, T.J., Jenks, J.A., Leslie Jr., D.M., 2006. Gastrointestinal morphology of female white-tailed and mule deer: effects of fire, reproduction, and feeding type. J. Mammal. 87, 598–605.

Zwolak, R., Forsman, K.R., 2007. Effects of a stand-replacing fire on small-mammal communities in montane forest. Can. J. Zool. 85, 815–822.

Zwolak, R., Forsman, K.R., 2008. Deer mouse demography in burned and unburned forest: no evidence for source-sink dynamics. Can. J. Zool. 86, 83–91.

Chapter 5

Stream-Riparian Ecosystems and Mixed- and High-Severity Fire

Breeanne K. Jackson[1], S. Mažeika P. Sullivan[1], Colden V. Baxter[2] and Rachel L. Malison[3]

[1]*School of Environment and Natural Resources, The Ohio State University, Columbus, OH, USA,* [2]*Stream Ecology Center, Department of Biological Sciences, Idaho State University, Pocatello, ID, USA,* [3]*Norwegian Institute for Nature Research, Trondheim, Norway*

5.1 DEFINING WILDFIRE SEVERITY AND STREAM-RIPARIAN BIOTIC RESPONSES

Wildfire is an important natural disturbance that has consequences for both structural and functional characteristics of riparian and stream ecosystems (Resh et al., 1988; Gresswell, 1999; Verkaik et al., 2013a). More than 20 years of studies now point to a diverse array of responses by stream-riparian organisms and ecosystems to wildfire. Ecological responses vary along gradients of fire characteristics, including severity, extent, frequency, time since disturbance, and hydrological context (Agee, 1993; Arkle et al., 2010; Romme et al., 2011), among others. Although high-severity fire can result in major changes to stream and riparian areas, including erosion and sedimentation, opening of the riparian canopy, inputs of large wood to the stream channel, and changes in water temperature and chemistry, low-severity fire may have little to no effect (Jackson and Sullivan, 2009; Arkle and Pilliod, 2010; Malison and Baxter, 2010a; Jackson et al., 2012) (Figure 5.1). Stream-riparian biota respond both directly to wildfire as well as indirectly via wildfire-induced changes in physical habitat (Arkle et al., 2010). Land managers often work to keep high-severity fire out of riparian zones using a suite of techniques, including fuel reduction (removal of trees and understory vegetation through mechanical thinning and/or prescribed fire) and suppression (Stone et al., 2010). However, stream and riparian organisms often are highly adapted to disturbances, including floods, drought, and wildfire (Dwire and Kauffman, 2003; Naiman et al., 2005), and dynamic fire regimes that operate over time and space may be important in maintaining the integrity and biodiversity of linked stream-riparian ecosystems (Bisson et al., 2003).

The Ecological Importance of Mixed-Severity Fires: Nature's Phoenix

FIGURE 5.1 Wildfire can be conceptualized along a gradient of fire severity, from unburned to high-severity burned. In riparian zones, wildfire severity is typically determined by assessing changes in both the tree canopy as well as understory vegetation. Low-severity wildfires are commonly characterized by intact riparian canopy and patchy and incomplete burning of understory vegetation, whereas high-severity wildfires typically burn the canopy and remove most if not all understory vegetation. The photos presented here represent unburned to high-severity burned riparian zones along low-order streams in Yosemite National Park (top) and the River of No Return Wilderness in central Idaho (bottom) between 3 and 11 years after fire. They illustrate some of the common responses to wildfire, including erosion and inputs of large wood (Grouse Creek, low-severity burned 3 years earlier; Buena Vista Creek, high-severity burned patch (foreground) 11 years earlier with moderate-severity burned area in the background; Tamarack Creek, high-severity burned 3 years earlier). Spatial patterns of fire severity can be highly heterogeneous in riparian zones; therefore differences between low-, moderate-, and high-severity burned areas are often difficult to distinguish, especially as time since fire increases. *(Photos by Breeanne K. Jackson (top row) and Rachel L. Malison (bottom row)).*

This chapter focuses on the effects of wildfire, across a gradient of severity, on organisms and processes in linked stream-riparian ecosystems. To address the range of wildfire effects, we concentrate on probable influences of wildfire on both abiotic and biotic characteristics across multiple levels of ecological organization (from individuals and populations to communities and ecosystems). Rather than presenting a complete review of the literature, we describe in relative depth examples of responses associated with each level of ecological organization. We also focus our discussion on the influences of wildfire severity and how these may vary over time, drawing principally on empirical evidence from the North American West, where much science, as well as resource management uncertainty and public dialogue, has been centered on the costs and benefits of wildfire (Pyne, 1997, 2004; Hutto, 2008). Moreover, we afford particular attention to fire-food web dynamics because food webs are a valuable window into the structure, function, and productivity of linked stream-riparian ecosystems (Wallace et al., 1997; Power and Dietrich, 2002; Baxter et al., 2005) and can provide spatially and temporally integrated perspectives on the effects of wildfire (e.g., Mihuc and Minshall, 2005). We conclude with a broad discussion of the potential importance of high-severity wildfire for biodiversity, conservation, and management of stream-riparian ecosystems.

Importance of Stream-Riparian Ecosystems

Even though aquatic ecosystems make up only about 2% of terrestrial landscapes, they are disproportionately relied on by humans for numerous natural resources (Postel and Carpenter, 1997). Streams and riparian areas act as conduits, reservoirs, and purification systems for fresh water (Sweeney et al., 2004). Riparian zones sustain unique communities of organisms, contributing >50%, on average, to regional species richness values (Sabo et al., 2005), and a disproportionate number of threatened and endangered species rely on aquatic and riparian habitats (Carrier and Czech, 1996), as do many organisms that provide food, medicine, and fiber to humans. In addition, these areas are valued as scenic and used for recreation.

The influence of wildfire as an agent of natural selection has resulted in a suite of organisms that exhibit apparent adaptations that make them resistant or resilient to wildfire, and riparian and aquatic organisms are no exception. Because riparian zones are transitional areas (or ecotones) between aquatic and terrestrial habitats, a diverse array of animals are associated with riparian corridors, ranging from aquatic (fish, benthic invertebrates) to amphibious (frogs, salamanders) to terrestrial (riparian birds, mammals, and reptiles), each exhibiting responses to wildfire that vary across gradients of fire severity (Box 5.1, Figure 5.2).

Despite their importance, riparian areas have been degraded worldwide, and in some regions the majority of riparian zones have been lost altogether. For

BOX 5.1

Examples of stream-riparian animals that may benefit from high-severity wildfire

(1) Immediate impacts may be negative, but stream invertebrate abundance and biomass frequently increase in the short to midterm following fire (Minshall, 2003; Verkaik et al., 2013a), and the production of emerging adult insects (i.e., aquatic insects that emerge from the water as winged adults) can increase as well (Mellon et al., 2008; Malison and Baxter, 2010a). Such increases may be accompanied by reductions in species diversity and dominance by insects that are habitat and trophic generalists, are drift-dispersers, and have multivoltine (having multiple generations per year) life cycles (e.g., Chironomidae, Baetidae) (Mihuc and Minshall, 1995; Minshall et al., 2001b). Climate and hydrologic context following wildfire may mediate mid- to longer-term impacts, however; for instance, Rugenski and Minshall (2014) reported increases in both invertebrate biomass and diversity in wilderness streams of Idaho more than 5 years following severe wildfire during a period of time characterized by reduced peaks in spring floods.

(2) Despite a long-standing assumption that high-severity wildfire has negative effects on stream fishes, in many cases immediate effects on fishes seem slight or recovery of populations occurs rapidly (Rieman et al., 1997; Sestrich et al., 2011), and there is mounting evidence of numerous indirect, positive effects on fish populations that may follow severe wildfire. For instance, the pulse in invertebrate production that can follow severe wildfire (Malison and Baxter, 2010a; also see Chapter 6) may provide increased food resources to fish. Even when wildfire is followed by scouring debris flows that may, at least temporarily, extirpate fish from a local stream reach (Howell, 2006), the combination of increased downstream transport of sediment and large wood that creates and maintains essential habitat (Bigelow et al., 2007), and increased export of drifting invertebrate prey from such tributaries (Harris et al. In press), may lead to net positive effects on fishes in recipient habitats. The pulse of natural erosion/sedimentation that can occur soon after high-severity fire can be associated with increases in native fish populations by ≥ 3 years after fire (Sestrich et al., 2011), possibly partly a result of enhanced spawning grounds.

(3) Streams and their adjacent riparian zones provide important foraging habitat for insectivorous bats (Seidman and Zabel, 2001; Russo and Jones, 2003; Fukui et al., 2006), where aquatic insects that emerge from streams as adults can comprise the majority of bat diets (Belwood and Fenton, 1976; Swift et al., 1985). The combination of increased emergence of stream insects and removal of the riparian canopy following high-severity fire may provide bats with better foraging conditions (Malison and Baxter, 2010b; Buchalski et al., 2013) (see Box 5.2 for additional details and Chapter 4 for a similar discussion of bat use of burned areas).

(4) Many birds that principally occupy riparian areas also rely on trees burned by fire (i.e., snags) for nesting cavities. For example, in the western United States, Lewis's woodpeckers (*Melanerpes lewis*), a cavity-nester and an aerial insectivore common in riparian zones, have been called "burn specialists" because they tend to be abundant in both recent (2-4 years after fire) and older (10-25 years after fire) high-severity burns (Linder and Anderson, 1998; Vierling and Saab, 2004). Lewis's woodpeckers and other aerial insectivorous birds can also benefit from increases in emergent insects and other aerial insect prey (e.g., Bagne and Purcell, 2011) following high-severity fires (see Chapter 3).

FIGURE 5.2 Common stream and riparian organisms that may interact with high-severity wildfire (clockwise from top left): Baetidae mayfly, larval form; Baetidae mayfly, adult form; Tetragnathidae spider; cutthroat trout (*Oncorhynchus clarkii*); western red bat (*Lasiurus blossevillii*); Sierra garter snake (*Thamnophis couchii*); and, at center, Lewis's woodpecker. *(Illustrations by Madeleine Ledford.)*

example, in California's Central Valley, approximately 99% of historic riparian zones have vanished as a result of land-use changes (Khorram and Katibah, 1984). These impairments are largely caused by a legacy of ecosystem degradation, fragmentation, and loss, as well as the expansion of nonnative species. In addition, although wildfire may be less frequent in riparian versus upland areas, fire disturbance may be more severe in riparian areas given the greater accumulation of fuel that may occur between wildfire events (Everett et al., 2003). Within this context, wildfire is generally viewed with a mix of concern and optimism. On the one hand, there are concerns about the implications of higher water temperatures and increased erosion and sedimentation for conservation of sensitive species and protection of ecosystem services. On the other hand, wildfire can be important in both maintaining biodiversity and ecosystem function (e.g., Arkle and Pilliod, 2010) and has been investigated as a potential restoration technique (e.g., Blank et al., 2003).

5.2 STREAM-RIPARIAN AREAS AND WILDFIRE SEVERITY

Although wildfire occurs across landscape types (e.g., forests, grasslands, deserts), understanding its role in shaping stream-riparian ecosystems is particularly critical given the important ecosystem services they provide. Notably, stream-riparian ecosystems differ from upland environments in moisture regime, topography, microclimate, vegetation, soils, and productivity (reviewed by Pettit and Naiman, 2007), and these differences can influence characteristics of wildfire. Fire severity in riparian zones is influenced by a number of factors including aspect, valley entrenchment, structure and composition of riparian vegetation, and stream size (Van de Water and North, 2011). The latter is of particular importance because wide riparian zones, characterized by a cooler, wetter microclimate, can act as a buffer against wildfire and therefore as a refuge for fire-sensitive species (Pettit and Naiman, 2007). Conversely, steep and highly entrenched streams with narrow riparian zones often are characterized by more dense fuels than their adjacent upland forests (Van de Water and North, 2011). In these cases, stream drainages can act as conduits for fire. For instance, there is evidence from montane ecosystems that riparian zones burn with equal or even greater frequency than upland forests (Van de Water and North, 2010, 2011) and that the extent of fire in riparian zones is highly correlated with the extent of fire in the uplands (Arkle and Pilliod, 2010).

5.3 TIME SINCE FIRE MATTERS

In addition to varying with wildfire severity, responses of both organisms and ecosystem processes may differ in the short term (days to months) versus longer term (years to decades) following fire. Immediate and short-term (days to 1 year following wildfire) changes in riparian systems caused directly by wildfire may be short-lived, but their effects can persist over longer time periods. Direct effects of fire on soils and vegetation, for example, can influence the quantity and quality of water in these systems long after the fire (Shakesby and Doerr, 2006). Interactions between wildfire and flooding generally result in patchy and temporally variable responses that can persist for months to decades (Pettit and Naiman, 2007). For example, the first rain event following wildfire can be of particular importance in determining to what extent erosion and sedimentation occur. In the midterm (often described as 2-10 years after fire), there may be an increase in primary productivity both within the riparian zone (plants) and in the stream (benthic algae and macrophytes) as a result of increased light penetration. Conversely, stream reaches burned by low-severity fire may not differ from unburned streams in these respects within the short to midterm (Jackson and Sullivan, 2009; Malison and Baxter, 2010a). Over the long term (>10 years), stream-riparian responses to high-severity wildfire are generally irregular and do not necessarily follow a direct succession. In addition, wildfire regimes are largely driven by climatic factors that vary greatly from year to year, resulting in stochastic fire-return intervals (Agee, 1993). Thus the long-term consequences of wildfire for stream-riparian organisms and ecosystem processes can be highly idiosyncratic and difficult to predict (see details in Section 5.5).

5.4 SPATIAL SCALE MATTERS

Another important dimension determining stream-riparian ecosystem responses to wildfire severity is spatial scale. At the local scale (i.e., 10^1-10^2 m), fire can effect processes such as removal of canopy (Jackson and Sullivan, 2009), mobilization of nutrients, erosion and sedimentation (Wondzell and King, 2003), and channel stability (Benda et al., 2003). At the catchment scale, fire can influence the timing and magnitude of runoff (Meyer and Pierce, 2003); the composition and structure of upland and riparian vegetation species (Dwire and Kauffman, 2003; Jackson and Sullivan, 2009); the local climate (Rambo and North, 2008); and the selection of habitat by organisms such as birds (Saab, 1999), bats (Malison and Baxter, 2010b; Buchalski et al., 2013), and fishes (Rieman and Clayton, 1997; Dunham et al., 2003). Patches of open canopy, large-wood accumulation, sedimentation, and bank erosion that shift over time create habitat mosaics that can result in nonlinear responses by aquatic and riparian organisms (Arkle et al., 2010). This highlights an important question of scale when it comes to assessing the effects of wildfire of varying severities on stream-riparian areas. Nearly all studies attempting to assess the effects of wildfire on stream-riparian ecosystems have been focused at relatively small spatial scales and over relatively short time periods; understanding the cumulative effects of wildfire will require investigations of patterns that propagate through stream networks over longer periods of time (Benda et al., 2004; Burton, 2005). The importance of riparian areas as conduits for organisms and refuges for biodiversity (Sabo et al., 2005), combined with the upstream-to-downstream connectivity quintessential to stream ecosystems (Hynes, 1975; Freeman et al., 2007), suggest that riparian responses to wildfire have implications that extend from riverscapes to landscapes. Regardless, the lack of investigations across spatial scales points to an important uncertainty regarding our attempts at synthesis presented below. Studies are needed to address this gap in understanding the effects of wildfire severity.

5.5 RESPONSES TO A GRADIENT OF WILDFIRE SEVERITY: EVIDENCE FROM THE NORTH AMERICAN WEST

Responses to wildfire severity can be grouped into abiotic (physical and chemical) and biotic (individual organisms, populations, communities, and ecosystems).

Physical Responses

In areas of low water volume, stream temperature can increase by several degrees during and immediately following (days to weeks after) high-severity wildfire (Hitt, 2003). Over longer time periods (months to years), the loss of riparian vegetation and reorganization of the streambed resulting from postfire shifts in channel geomorphology following severe wildfire can result in alterations to the heat budget of streams. Loss of shade and increased solar radiation result in higher stream temperatures (Dwire and Kauffman, 2003; Pettit and Naiman, 2007). The magnitude of temperature change is influenced by the severity of the fire, the total length of stream exposed, changes in riparian vegetation, and the degree of channel reorganization; some streams show little response and others warm considerably (Royer and Minshall, 1997; Dunham et al., 2007). Isaak et al. (2010) compiled a temperature database for a 2500 km river network in central Idaho to evaluate the effects of climate change and wildfire on stream temperatures. They found that within wildfire perimeters, stream temperature increases were 2-3 times greater than basin averages, with radiation accounting for 50% of the warming.

Physical responses of streams and riparian zones, such as alterations in hydrology and channel morphology, tend to be persistent effects of wildfire, with immediate responses that can last for decades after the fire event. For example, significant erosion and deposition of fine sediments in stream channels frequently follow high-severity fire (Wondzell and King, 2003). High-severity fire often consumes a significant portion of aboveground vegetation in the riparian zone and adjacent side slopes (Dwire and Kauffman, 2003). In addition, consumption of the litter layer and obstructions to overland water runoff, such as downed logs, conversion of organic material to small-particle ash, and the development of hydrophobic soils (DeBano, 2000; Doerr et al., 2003), can collectively contribute to reduced infiltration capacity of soils and the potential for increased overland flow, surface erosion, scouring of stream channels, and deposition of fine sediments (Wondzell and King, 2003; Shakesby and Doerr, 2006; Vila-Escale et al., 2007). Under some circumstances, wildfire may be followed by debris flows—liquefied landslides that reorganize channels, export large wood, and can scour streambeds to bedrock (Miller et al., 2003; Wondzell and King, 2003; May, 2007). These and other physical disturbances that can accompany high-severity wildfire may extend and change the trajectory of the postfire recovery of stream ecosystems. Whereas the local effects of the wildfire-debris flow combination may lead to simplification of in-stream structure and morphology that may exert negative effects on some stream organisms, this process also delivers sediment, wood, organic matter, and nutrients important to the complexity and character of downstream habitats (e.g., Benda et al., 2003; Harris et al. In press).

5.6 CHEMICAL RESPONSES

High-severity wildfires that consume the forest floor can considerably alter the magnitude and timing of overland flows, erosion, and solute delivery to streams (Williams and Melack, 1997; Seibert et al., 2010). Nutrients, contaminants, and organic compounds become concentrated after fire and can bind to fine sediments, thus increasing their transport into streams and elevating exposure to fishes and aquatic invertebrates (Malmon et al., 2007). Partial combustion of riparian vegetation into ash that increases soil ammonium concentrations and results in increased stream nitrogen concentrations (Wan et al., 2001) is also a common in-stream response to wildfire (Minshall et al., 1997; Williams and Melack, 1997; Bladon et al., 2008). Patterns of stream phosphorus concentrations following wildfire are less consistent, with evidence largely pointing to a brief (often returning to prefire conditions within a few weeks to a few months) but marked increase (e.g., Spencer and Hauer, 1991; Hauer and Spencer, 1998; Earl and Blinn, 2003) or to no change (Minshall et al., 1997; Stephens et al., 2004). Overall, increases in nutrient delivery from the upland, combined with greater light penetration and higher temperature, may prompt elevated in-stream primary productivity, with consequences for communities and food webs (Betts and Jones, 2009: see "Food-Web Dynamics," below).

In contrast to physical responses, chemical responses to wildfire generally have shorter-lived consequences (Minshall et al., 2003), largely because annual runoff often increases in the first couple years following fire (Moody and Martin, 2001). For example, Hall and Lombardozzi (2008) found that the Hayman Fire, one of the largest wildfires in Colorado history (>50-70% of the burn area was classified as moderate- to high-severity fire), altered water temperature and dissolved oxygen concentrations, as well as concentrations of nitrate, phosphate, and mineral salts, in stream water over the 2-year postburn period. Because of the variability in climate, local topography, and burn characteristics, among other factors, chemical responses to moderate- and high-severity fires can be highly variable; some streams return to baseline conditions within weeks following fire (Earl and Blinn, 2003), whereas other streams (or chemical constituents) show changes for multiple years (Hauer and Spencer, 1998; Mast and Clow, 2008). Effects of low-severity fires on stream chemistry seem to be slight and typically do not persist beyond the first year (Stephens et al., 2004; Bêche et al., 2005). Though results of most studies suggest fire-driven shifts in chemistry are relatively ephemeral, such work has focused on the expected, pulsed delivery of materials from the land that follows fire. By contrast, and unlike research in the forested uplands (e.g., Smithwick et al., 2005; Koyama et al., 2010), there has been virtually no investigation of the mid- to long-term changes in biogeochemical processes that may accompany the more persistent changes in stream conditions or the biota that occupy riparian soils and streambed sediments.

Immediate Effects on Individuals

Responses to wildfire at the level of individual organisms are largely behavioral and physiological, and they occur in immediate to short time periods following fire. Highly mobile animals, such as birds and mammals, can move to unaffected areas away from high temperatures and smoke. Some terrestrial animals, including large ungulates like elk (*Cervaus elaphus*) and bison (*Bison bison*), have been observed taking refuge in streams while wildfires are actively burning the upland (Allred et al., 2013). Though amphibians with in-stream life cycle stages can lack the mobility to survive or move long distances in response to physical or chemical changes that might accompany wildfire (Gresswell, 1999; Pilliod et al., 2003), some species can move to wet areas and/or burrow to avoid high temperatures during wildfire. Although specific examples from the North American West are sparse, in Australia, the anuran *Hyperolius nitidulus* can detect the sound of wildfire and seek refuge in wet areas (Grafe et al., 2002); American toads (*Bufo americanus*) were found partially buried in mud following a prescribed fire in Iowa (Pilliod et al., 2003); and Vogl (1973) discovered partially burned leopard frogs (*Rana sphenocephala*) and bullfrogs (*R. catesbiana*) in a wetland following a fire in Florida. There is some evidence that lethal temperatures and/or changes in stream water chemistry during or shortly after wildfire can lead to mortality of fishes and benthic invertebrates (Bozek and Young, 1994; Rinne, 1996; Rieman and Clayton, 1997; Howell, 2006). Even the direct, immediate effects of wildfire on benthic invertebrates are often negligible (Minshall, 2003), however, and these effects on fishes can also be quite variable. In some cases, fishes can be temporarily extirpated by the direct effects of high-severity fire (e.g., increased temperature, dissolved gases), especially in smaller streams (Dunham et al., 2003), but they often recover within weeks to months (Sestrich et al., 2011).

In-Stream Biotic Response: Populations and Communities

The impact of wildfire on benthic invertebrate communities varies with fire severity and over time (Minshall, 2003). Following the first large postfire runoff, invertebrate richness may decline. Communities may recover to prefire conditions 1-2 years following wildfire, but a common pattern observed in many settings is that community composition shifts toward an increase in the relative abundance of disturbance-adapted taxa (Mihuc and Minshall, 1995; Minshall, 2003; Verkaik et al., 2013a). For example, in the short term following the Mortar Creek Fire in central Idaho, disturbance-adapted taxa were more dominant, but total taxa richness converged with that of reference streams toward the end of a 10-year study (Minshall et al., 2001a). In streams in the same region, Malison and Baxter (2010a) found that benthic insect assemblage composition continued to vary with fire severity 5 years following wildfire, and stream reaches that experienced high-severity fire had the greatest biomass of insects like midges (Chironomidae) and Baetidae mayflies. Invertebrate communities in streams may also shift following fire in terms of dominant feeding traits. Many disturbance-adapted taxa that flourish after fire are also feeding generalists (Mihuc and Minshall, 1995), but, again, changes may be influenced by the state of riparian vegetation. For instance, following the Jesusita Fire in southern California, shredders (which frequently rely on shredding leaves that fall into streams) were more abundant in streams draining unburned basins than those that burned but retained a riparian canopy, and they were completely absent from basins where the riparian canopy was removed by fire (collector/filterer insects dominated in these streams) (Cooper et al., 2014). Therefore, although species composition and feeding guild representation within benthic invertebrate communities vary along gradients of fire severity and extent (Arkle et al., 2010), in the years to decades following fire the number of taxa may remain fairly consistent (Verkaik et al., 2013a).

Negative effects on fishes may last for months to years following high-severity fire if elevated stream temperatures create stressful conditions, as can be the case for salmonids near the southern margin of their range (e.g., Beakes et al., 2014), or if stream reaches also are influenced by debris flows that may occur after wildfire (Dunham et al., 2007). Fishes adapted to cold-water habitats may be particularly sensitive to elevated temperatures that can occur during and in the few years following high-severity wildfires (Sestrich et al., 2011) and can last 10 years or more (Gresswell, 1999; Isaak et al., 2010; Sestrich et al., 2011). High water temperatures after fire have been linked to reduced density of salmonids in the American Southwest (Rinne, 1996), Rocky Mountains (Isaak et al., 2010), and California (Beakes et al., 2014) because these fishes are especially dependent on cold water for spawning and juvenile rearing. For example, Sestrich et al. (2011) found that pools in reaches of a western Montana stream burned by high-severity fire were 2-6 °C warmer in summer months in the year following fire and that the density of native westslope cutthroat trout (*Oncorhynchus clarki lewisii*) was negatively correlated with the percentage of the catchment burned with moderate- or high-severity fire at 1 year after fire. This may be a result of increased bioenergetic demand, as demonstrated in the study by Beakes et al. (2014), who found that steelhead trout (*O. mykiss*) biomass was reduced in pools under canopy gaps 1 year after a fire in southern California. On the other hand, from 1 to 3 years after fire, Sestrich et al. (2011) found higher proportions of moderate- and high-severity fire were associated with increases in populations of native fishes.

Shifts in temperature associated with climate change may exacerbate short-term spikes in water temperature caused by fire, which may have additional consequences for fish populations. Stream temperature increases as a result of climate change resulted in the reduction of spawning and rearing habitat for bull trout (*Salvelinus confluentus*) by 8% to 16% each decade in one study area in central Idaho; fire, at least temporarily, contributed to reductions in spawning/rearing habitat, though population levels were not recorded, so the extent to which or whether fire actually reduced bull trout in burned areas is unknown (Isaak et al., 2010). Because recovery of fish populations following fire may be influenced by recolonization of burned stream reaches from nearby unburned or low-severity burned reaches, drainage connectivity, ecosystem size, and timing of life-history events such as spawning may interact with fire severity to influence population recovery (reviewed by Dunham et al., 2003). For example, endangered gila trout (*O. gilae*) in New Mexico are especially vulnerable because they live in small, isolated streams currently experiencing a frequent high-intensity fire regime (Propst et al., 1992).

Evidence regarding potential effects on the composition of fish assemblages is relatively sparse. Fishes like endangered salmonids in streams of the North American West have relatively specific habitat needs, and if they are occurring in systems close to the edge of their range or those that are already degraded and fragmented, then they may be the most vulnerable to fire or associated disturbance and warming temperatures (Dunham et al., 2003; Isaak et al., 2010; Beakes et al., 2014). On the other hand, direct investigations of fish community composition responses have been few and the results rather equivocal. For instance, studies in Idaho and Montana found little evidence of persistent, negative effects of even severe wildfire on salmonid fish assemblages (e.g., Neville et al., 2009; Sestrich et al., 2011). That effects of severe wildfire might facilitate invasions of nonnative fishes has also been posited (Dunham et al., 2003). In streams of western Montana, Sestrich et al. (2011) found no evidence of increases in abundance of or invasion by eastern brook trout (*Salvilinus fontinalis*) after wildfire, but this hypothesis must be more widely tested.

Scouring flows that can result from high-severity wildfire can temporarily extirpate invertebrates, fishes, and amphibians (Verkaik et al., 2013a). For example, Vieira et al. (2004) found that the first 100-year flood event following the 1996 Dome Fire in New Mexico reduced benthic invertebrate density to near zero. Within a year, however, benthic invertebrate density recovered to prefire levels, largely because of recolonization by those that disperse as larvae. Many fishes require relatively stable bed conditions with specific sediment class sizes for spawning; therefore, depending on the timing of fires, floods, and spawning, fish may be more or less affected by erosion and sedimentation. Amphibian populations like the California newt (*Taricha torosa*) are similarly affected if preferred oviposition sites are filled with sediment (Gamradt and Kats, 1997).

Over longer time scales (multiple years to decades) following wildfire, responses by stream organisms can be quite divergent, and this variation may be associated with the severity of wildfire and the trajectory of both stream and riparian habitat recovery. As described above, recovery within months to a few years is commonly observed among both invertebrate and fish populations, and debris flows in the early postfire time period may later be associated with increased native fish populations (Sestrich et al., 2011). Over longer time periods (multiple years), however, both groups may exhibit more interannual variability in stream reaches that have experienced high-severity wildfire than in those that burned with low-severity wildfire (Arkle and Pilliod, 2010). Annual variability of populations is largely linked to floods and droughts (Robinson et al., 2000; Arkle et al., 2010; Verkaik et al., 2013b). For example, in the Big Creek watershed of central Idaho, benthic invertebrate populations fluctuated annually; shifts were correlated with a combination of sediment, large wood, riparian cover, and benthic organic matter along a gradient of fire severity (Arkle et al., 2010), suggesting an interaction between fire severity and flooding in driving benthic invertebrate variability. On the other hand, in settings where drought is prevalent and accompanies wildfire, the effects of stream drying on aquatic organisms may outweigh most variation associated with fire severity, as has been observed in Mediterranean and Australian stream ecosystems (Verkaik et al., 2013b; Verkaik et al. In press). In any case, understanding the net consequences for stream organisms will likely require investigations that encompass not only different time scales but also responses at the scale of entire stream networks—studies that, thus far, are lacking.

Riparian Community and Ecosystem Responses

Because of the pervasive influence of riparian plant composition and structure on a host of ecosystem responses, the influence of wildfire on riparian plant communities has received broad attention and highlights the importance of disturbance for driving the composition and structure of stream-riparian communities. Riparian plants are highly adapted to disturbance (Naiman et al., 2005). In most cases this disturbance is flooding, and in certain biomes it is drought and fire (Pettit and Naiman, 2007). Therefore, riparian plant species often possess distinct life-history traits such as stump sprouting, seed banks, and clonal regeneration that allow them to withstand fire or recover quickly following even severe wildfire. For example, plants in riparian forests often exhibit higher foliar moisture content than upland plants, even within the same species (Agee et al., 2002), which can result in patches of lower fire severity and lower plant mortality (Kauffman and Martin, 1989, 1990). Tree species common to riparian areas in mountainous areas of the North American West, such as ponderosa pine (*Pinus ponderosa*), western larch (*Larix occidentalis*), and coast redwood (*Sequoia sempervirens*), have thick bark that protects them from mortality during low-intensity ground fires (Miller, 2000). Low- and moderate-severity fire can stimulate clonal regeneration of quaking aspen (*Populous tremuloides*) (Jones and DeByle, 1985; Romme et al., 1997; Bartos and Campbell, 1998), and aspen trees that are top-killed by high-severity fire are stimulated to produce numerous root suckers (Schier, 1973; Keyser et al., 2005). Many riparian shrubs, including alder (*Alnus* spp.), birch (*Betula* spp.), currant (*Ribes* spp.), rose (*Rosa* spp.), and snowberry (*Symphoricarpos* spp.), sprout from stumps, root crowns, and belowground stems following fire (Adams et al., 1982; Stickney, 1986; Miller, 2000; Kobziar and McBride, 2006) (Figure 5.3).

FIGURE 5.3 Riparian vegetation at Goat Creek, a tributary of Big Creek in central Idaho, 10 days after a moderate-severity fire (August 2006). In the foreground (and throughout the background), water birch (*Betula occidentalis*) sprouting among stumps can be seen.

Species composition and taxonomic richness of riparian vegetation varies along a gradient of low- to high-severity wildfire, and the trajectory of community response tends to not follow a predictable succession. One year following a prescribed fire in the Sierra Nevada, Bêche et al. (2005) observed a reduction in species richness of riparian vegetation. Five years following the Diamond Peak wildfire in central Idaho, however, Jackson and Sullivan (2009) found species richness of riparian vegetation did not vary across stream reaches characterized as unburned, low-severity burned, and high-severity burned. In addition, that study found riparian vegetation community composition did not differ between unburned and low-severity burned reaches, whereas high-severity burned reaches exhibited greater relative density of sun-loving species like blue elderberry (*Sambucus cerulean*) and red raspberry (*Rubus ideaus*). Moreover, herbaceous cover within high-severity burned reaches was dominated by invasive cheat grass (*Bromus tectorum*), which has been associated with increased fire frequency and rate of spread (Mack and D'Antonio, 1998).

The structure and composition of riparian vegetation is thought to be closely linked to recolonization dynamics of riparian invertebrates following high-severity wildfire, both in terms of species composition and when they recolonize. Bess et al. (2002) found that the total number of riparian arthropod species was similar before ($n=80$) and 9 months after ($n=79$) a high-severity fire in New Mexico. Of the original 80 species, 30 had not recovered, but 29 species that had not been recorded in the 3 years before the fire had appeared. Similarly, Jackson et al. (2012) found that 5 years after wildfire the taxonomic composition of terrestrial invertebrates falling into streams differed between those that flowed through reaches burned by high-severity wildfire versus those that had experienced low-severity wildfire. However, these investigators observed the total number of taxa was consistent across burn types. Therefore, although species turnover seems to be common following high-severity wildfire, richness may remain similar.

Given that vegetation and habitat structure are critical factors that drive habitat selection in birds, wildfire in riparian zones can have substantial influences on bird communities as well. Some guilds, such as aerial insectivores, have been found to generally favor burned areas (Kotliar et al., 2002; Russell et al., 2009), potentially as a response to improved foraging conditions following reduced canopy cover and increases in flying arthropods and emergent aquatic insects. Cavity nesters (i.e., those that nest in sheltered chambers versus open-cup nests) also are thought to respond positively to wildfire, in part because the dense stands of snags (dead trees) created by wildfire provide important nesting sites (Saab and Powell, 2005; Saab et al., 2009), although time since fire and fire severity influence these patterns (Chapter 4). Some bird species specialize in habitats burned by high-severity fires (Hutto, 2008; refer to Box 5.1). In managed forests of the Sierra Nevada of California, riparian-associated birds increased in abundance <6 years following low-severity prescribed fire (Bagne and Purcell, 2011). On the other hand, ground-nesting red-faced warblers (*Cardellina rubrifrons*) and yellow-eyed juncos (*Junco phaeonotus*) avoided nesting in riparian areas burned by low-severity fire 1-2 years following fire in southern Arizona (Kirkpatrick and Conway, 2010). The timing of fire may also be a particularly important factor for birds; spring burns, for example, can interfere with breeding activities (Kruse and Piehl, 1986). Thus, responses of bird communities to wildfire seem highly variable, benefitting some groups more than others.

Primary and Secondary Production

To understand the effects of wildfire on ecosystem processes, considering its effects on rates of both primary and secondary production is critical. At present, inferences regarding these effects must largely be drawn from studies that have measured indices of productivity, principally snapshots of the biomass of producers like streambed algae or consumers like invertebrates. Algal biomass frequently increases over the short to midterm after severe wildfire, likely because of increased light penetration into streams through canopy gaps combined with altered nutrient inputs (Robinson et al., 1994; Minshall et al., 1997; Spencer et al., 2003). One year following the Jesusita Fire in southern California, pools and riffles where the riparian canopy had been removed by fire exhibited 85%

more cover by filamentous microalgae than unburned or reaches with intact canopy (Cooper et al., 2014). In the 2 years following the Diamond Peak Fire in central Idaho, chlorophyll-*a* values and periphyton ash-free dry mass was significantly higher in burned streams (Rugenski and Minshall, 2014). Similarly, in the only direct measure of primary productivity of which we are aware, during the year following a fire in Alaska, Betts and Jones (2009) observed rates of gross primary productivity of aquatic periphyton were double those in unburned sites. If elevated temperatures or light inputs persist (as may occur if fires are severe and riparian canopies remain open), this pulse of aquatic productivity may endure as well, but this remains to be evaluated.

If wildfire leads to increased in-stream primary productivity, this may in turn contribute to higher rates of secondary production by benthic invertebrates, and such responses may be mediated by fire severity. For example, Malison and Baxter (2010b) found that benthic invertebrate biomass was fivefold greater in stream reaches that had been burned by high-severity wildfire 5 years earlier compared with low-severity burned sites. Based on the same study they reported that rates of emerging adult insects produced in reaches burned with high-severity fire were three times higher than those in unburned reaches or those burned with low-severity fire (Malison and Baxter, 2010a). Although a similar pattern of elevated emergence was observed following fire in Washington (Mellon et al., 2008), the generality of these observations has not been evaluated, and, remarkably, no study to date has measured annual rates of invertebrate production in response to wildfire.

Food-Web Dynamics

Food webs in streams and riparian zones are linked to one another via the bidirectional fluxes of materials and organisms. If increases in primary and secondary productivity do follow severe wildfire this may have far-reaching consequences for organisms at higher trophic levels in stream-riparian food webs (what Malison and Baxter, 2010b refer to as a "fire pulse"). For example, fishes have been shown to selectively forage at the confluence of mainstem rivers and smaller tributaries that have been burned by high-severity wildfire in the past 5-10 years. Presumably this is a result of greater export of benthic invertebrate prey originating from those tributaries (Koetsier et al., 2007); a recent study showed that, indeed, tributaries disturbed by fire and associated debris flows export more invertebrate prey than those that were unburned, and fish exhibit a preference for confluences within these disturbed streams (Harris, 2013). In addition, as emergent adults, stream insects are heavily relied on as food resources for riparian consumers like birds, bats, and spiders (reviewed by Baxter et al., 2005). For example, in central Idaho, Malison and Baxter (2010a, 2010b) not only observed amplified emergence from sites that burned with high severity, they also found that the abundance of riparian web-building spiders from the family Tetragnathidae was two times higher in reaches burned by high-severity fire. Conversely, Jackson and Sullivan (In press) found that Tetragnathidae density was not significantly different in stream reaches of Yosemite National Park (Sierra Nevada, California) affected by low-severity fire compared to those affected by high-severity fire within the past 3-15 years. This result was largely linked to climate; sites that experienced more annual precipitation supported a greater density of riparian spiders.

From the riparian zone to the stream, high-severity wildfire can alter the magnitude, composition, and timing of inputs of leaf litter and terrestrial invertebrates. Jackson et al. (2012) found that leaf litter inputs (dry weight) to streams 5 years after the Diamond Peak wildfire in central Idaho were 2-6 times greater in unburned reaches and 1.5-2 times greater in low-severity burned reaches compared with high-severity burned reaches where the riparian canopy was removed. In addition, inputs of terrestrial invertebrates were as much as four times greater to unburned reaches and two times greater to low-severity burned reaches compared with high-severity burned reaches. The importance of terrestrial invertebrates as prey items for fish has been demonstrated in detail (Allan, 1981; Wipfli, 1997; Piccolo and Wipfli, 2002; Allan et al., 2003; Carpenter et al., 2005), and high-severity wildfire may alter these subsidies. It seems, however, that synchronized stimulation of in-stream primary and secondary productivity by high-severity wildfire combined with changes to habitat structure can result in a net neutral, or even beneficial, effect on in-stream and riparian consumers (Box 5.2, Figure 5.4).

BOX 5.2

The effects of wildfire on riparian habitat are expected to influence insectivorous bat distributions, foraging, and population dynamics. In forested systems, rivers create spatial gaps in dense forest vegetation, allowing echolocating bats to forage effectively directly over river channels, with comparatively low activity within or beneath the forest canopy (Power et al., 2004; Ober and Hayes, 2008). Riparian trees and snags also provide important roosting habitats for multiple riparian bat species (Brack, 1983; Fleming et al., 2013). Fire-induced changes in riparian and bottomland vegetation structure therefore could have significant effects on both bat habitat and energetics. For example, Buchalski et al. (2013) found that, in mixed-conifer forests of California, some bat species preferentially select moderate- and high-severity burned areas for foraging, likely facilitated by reduced vegetation density and increased availability of prey and roosts after fire.

Rivers and their adjacent riparian zones also provide important foraging habitat for insectivorous bats (Seidman and Zabel, 2001; Russo and Jones, 2003; Fukui et al., 2006), where aquatic insects that emerge from the stream as adults can comprise the majority of bat diets (Belwood and Fenton, 1976; Swift et al., 1985). Higher aquatic insect availability is often implicated as the mechanism driving observations of higher rates of foraging activity within riverine landscapes compared with upland habitats (Swift et al., 1985; Brigham et al., 1992). For example, Fukui et al. (2006) showed that bat activity along a Japanese stream significantly decreased after aquatic insect emergence was experimentally reduced. Hagen and Sabo (2012) found that seasonal river drying in extremely arid climates resulted in the disappearance of both aquatic insects and bats. Thus, because wildfire has the potential to profoundly alter aquatic insect emergence, terrestrial consumers such as riparian insectivorous bats (Sabo and Power, 2002; Paetzold et al., 2005) may also be affected. For instance, Malison and Baxter (2010b) observed the greatest number of bat echolocation calls at stream sites influenced 5 years earlier by high-severity wildfire, suggesting that fires of different severity may have different effects on stream-riparian food webs via fire-induced changes in stream secondary productivity and subsequent aquatic insect emergence. Food availability also has been shown to be related to individual health: It can mediate stress levels in bats with seasonally fluctuating resources (e.g., aerial insects) (Lewanzik et al., 2012). Thus, although the exact nature of the responses may be species-specific, high-severity wildfire is expected to have strong effects on riparian bats through both direct and indirect mechanisms.

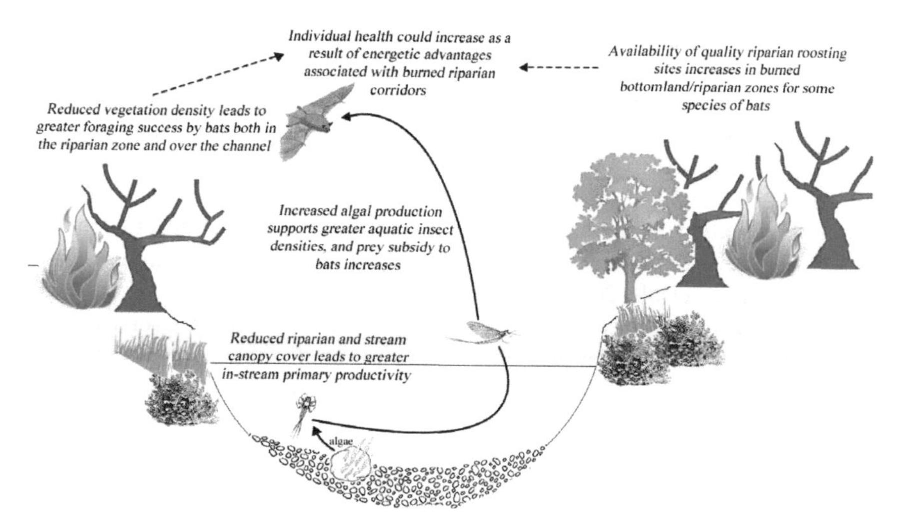

FIGURE 5.4 Potential effects of high-severity wildfire on aerial insectivorous bats in riparian corridors in the short to midterm (1-10 years following fire, although in some cases longer) under open canopy conditions. Solid arrows represent food web pathways; dashed lines represent indirect effects of wildfire via changes in habitat.

The dynamics of linked stream-riparian food webs integrate wildfire effects over both short and long time scales and across communities and ecosystems. The importance of riparian leaf litter, woody debris, and other plant material entering the stream from the riparian zone has a long history of study and appreciation (Vannote et al., 1980; Gregory et al., 1991). Aquatic sources of energy moving into riparian zones (e.g., via adult aquatic insect emergence, through flood pulses, and through movements of other organisms) and terrestrial-aquatic feedback loops also has received increasing attention (reviewed by Baxter et al., 2005). However, the role of wildfire with respect to these linkages is just starting to be described (Spencer et al., 2003; Malison and Baxter, 2010b; Jackson et al., 2012).

5.7 BIODIVERSITY, CONSERVATION, AND MANAGEMENT

Wildfire creates heterogeneous habitat conditions in space and time that may be important to the maintenance of native biodiversity and to the function of stream-riparian ecosystems. As we have summarized, whereas some organisms and processes may be negatively influenced by severe wildfire (at least on short time scales), many seem to be resilient over longer time periods. Indeed, stream-riparian ecosystems often are characterized as "dynamic mosaic[s] of spatial elements and ecological processes" (Ward et al., 2002). The creation and transformation of habitat patches and the facilitation of ecological functions in streams and rivers is largely driven by disturbance: foremost flooding, but also drought, ecosystem engineers (e.g., beaver (*Castor canadensis*)), and severe wildfire, as well as a host of human disturbances (e.g., dams, land use, and climate change). The importance of flooding for the maintenance of ecological function and biodiversity of stream-riparian ecosystems has been demonstrated in detail and can result in high habitat and species turnover compared with those in other ecosystems (Sullivan and Watzin, 2009; Tockner et al., 2010). For example, flooding creates and maintains a spatial mosaic of habitats that in turn foster diverse webs of interacting species (Junk et al., 1989; Stanford et al., 2005; Bellmore et al., 2014), and homogenization of flow regimes by dams and other means has been shown to greatly reduce global biodiversity (Poff et al., 2007). Similarly, because fire regimes are predictable over evolutionary timescales, it seems likely that alterations in the magnitude, frequency, timing, and extent of historic fire regimes will have consequences for stream-riparian biodiversity. Indeed, as has been the case for terrestrial ecosystems, a scientific consensus regarding the importance and, in many instances, benefits of wildfire to stream-riparian organisms and ecosystems seems to be emerging (Chapters 3 and 4), but at present there are at least two challenges to reaching such generalities that deserve consideration.

First, there remains a prevailing assumption that, although wildfire may have some ecological benefits, management of low-severity wildfire (or prescribed fire in its stead) should be preferred because it could represent a "goldilocks" condition, wherein such benefits might accrue while avoiding the perceived risks of high-severity fire. As we have summarized, the ecological effects (including various potential benefits) of high-severity wildfire in stream-riparian ecosystems do not seem to be mimicked by either low-severity wildfire or prescribed burning (e.g., Jackson and Sullivan, 2009; Arkle and Pilliod, 2010; Malison and Baxter, 2010b), which calls into question this assumption. On the other hand, interactions between severe wildfire and other sources of natural (flooding, drought, natural impoundments like debris jams) and anthropogenic disturbance (invasive species, postfire logging, channel alteration and impoundments, introduction of nutrients and contaminants) might have cumulative or even exponential effects on stream-riparian ecosystems, as have been described for other combinations of multiple stressors (Ormerod et al., 2010). For instance, connectivity in riverscapes (Fausch et al., 2002) is likely important in mediating the local effects of severe wildfire on communities of native organisms; if waterways are disconnected by large patches of unsuitable habitat, organisms may not be able to redistribute following wildfire (Gresswell, 1999; Dunham et al., 2003). Studies that explicitly evaluate how the sign and magnitude of responses to high-severity wildfire may differ with scale and in the context of other environmental stressors are needed.

The second, and perhaps even more difficult, challenge is that, just as science is beginning to provide some understanding of the ways in which wildfire of varying severity may affect ecological function, the entire context for such relationships is being altered by a changing climate. In western North America, climate change has been linked to recent increases in wildfire frequency and extent (Whitlock et al., 2003; Whitlock, 2004; Westerling et al., 2006, 2011a, 2011b) (but see Chapter 9). This has been accompanied by changes in the trajectory of regional vegetation states (Allen and Breshears, 1998; van Mantgem et al., 2009), and some have hypothesized that many of the assumptions regarding the resilience upon which existing fire ecology paradigms rest may now be poorly founded (see Davis et al., 2013 for a review). This highlights the need to understand how wildfire characteristics and recovery patterns of terrestrial vegetation over time mediate responses of stream-riparian ecosystems. Yet, proportionately few studies explicitly evaluate how these might be influenced by its severity, and most investigations have focused on short-term responses, with far fewer studies of mid- to longer-term dynamics (i.e., >2-3 years after fire; Romme et al., 2011; Rugenski and Minshall, 2014). Such investigations are needed to inform the adaptive management of landscapes and riverscapes under a changing climate.

A variety of management actions has been designed and used to mitigate the effects of severe wildfire on stream-riparian ecosystems, but the impact of these mitigation efforts is not always positive. For example, the use of prescribed fire as a tool to manage riparian ecosystem condition is increasing

(Stone et al., 2010), but because prescribed fires typically differ from wildfires in severity, timing, frequency, and extent (McIver et al., 2013), their influence on riparian and aquatic systems remains an open question (Boerner et al., 2008; Arkle and Pilliod, 2010). In addition, methods used during fire suppression efforts can have negative effects on stream-riparian ecosystems. For example, the use of fire retardants around aquatic systems has led to the mortality of aquatic organisms (Gaikowski et al., 1996; Buhl and Hamilton, 2000; Gimenez et al., 2004) and is therefore banned by firefighting agencies, but construction of fire lines within drainages continues. In some cases fire lines can facilitate the introduction of invasive species and be a significant source of chronic sediment delivery to streams following wildfires (reviewed by Beschta et al., 2004 and Karr et al., 2004).

Postfire management has the potential to be more disruptive to stream-riparian ecosystems and have longer-lasting consequences than high-severity wildfire itself (Beschta et al., 2004; Karr et al., 2004); therefore, any postfire management that does not mitigate the effects of suppression activities should be avoided, including planting with nonnative seeds, construction of debris dams, and postfire logging. Debris dams often are insufficient at ameliorating soil erosion and end up in stream channels following storms, where they impede the movement of organisms and disrupt flow. Mechanical disruption of soils, which often occurs as a result of postfire logging, increases chronic erosion and the deposition of fine sediments (McIver and Starr, 2001; McIver and McNeil, 2006), and soil compaction in forests can persist for 50-80 years (Quigley and Arbelbide, 1997), which may exceed the duration of effects from high-severity wildfire. Even dead vegetation provides soil stability; snags are important habitat for riparian organisms, and large wood is a significant and ecologically important structural element of stream-riparian ecosystems (Gregory et al., 2003). Thus postfire logging may reduce the quality of stream-riparian habitat in multiple ways. Whereas postfire management should be used with caution, prefire restoration of stream-riparian ecosystems might reduce potential negative effects of severe wildfire (Beschta et al., 2004); such efforts might include surfacing, stabilizing, and removing legacy roads; discouraging grazing in riparian zones; and restoring fluvial connectivity.

Finally, as we have described in this chapter, the effects of wildfire on stream-riparian ecosystems operate over gradients of severity, space, and time and across levels of ecological organization. For example, although there are likely to be winners and losers at the individual and population levels in the short term and over relatively small spatial scales, community- and ecosystem-level responses seem to be more neutral or positive, are longer lived, and tend to operate at relatively larger spatial scales. Therefore, management of stream-riparian ecosystems in landscapes that experience high-severity fire will benefit from a holistic perspective that takes into account heterogeneous responses over space and time.

5.8 CONCLUSIONS

Wildfire plays an important role in shaping the structural and functional characteristics of stream- riparian ecosystems. Though these ecosystems represent a small portion of the total landscape, they are a disproportionately vital source of natural resources, and understanding how wildfire may influence them and the ecosystem services they provide is critical. Riparian forests differ from upland forests in moisture regimes, microclimate, soils, topography, and vegetation and can act as conduits of wildfire, burning with equal or even greater frequency than upland forests. Many of the diverse organisms associated with riparian corridors, ranging from fishes and salamanders to birds and reptiles, are generally adapted to disturbance. Given that wildfire in particular has been a historic source of disturbance in many of these ecosystems, it is likely to be a key driver of biodiversity. Wildfire can impact streams and riparian zones at the individual, population, community, and ecosystem levels. Its effects vary with fire severity, time since fire, and the spatial scale of the fire. High-severity wildfire can have very different effects than low-severity wildfire, suggesting that a mosaic of fires of different severity may be necessary to maintain ecosystem function. Immediate and short-lived impacts such as increased stream temperatures can have negative effects on individual organisms and populations, although evidence generally suggests that their recovery is rapid, and there may be countervailing positive effects such as increased food availability for fishes, bats, and birds in aquatic-riparian environments. Long-term effects of wildfire are mediated by climate and can be irregular because of variation in site-specific physical characteristics. In addition, the impact of wildfire may depend not only on fire severity but also on the spatial scale (both total extent and patchiness) and timing of the fire, as well as the degree of hydrologic connectivity. Streams and riparian zones are highly connected pathways in landscapes; disruption of this continuity or impairment or loss of other natural features (e.g., modified riparian zones) as a result of human activities could lead to more detrimental effects of fire in these contexts. Because of the linked nature of stream-riparian ecosystems, and the disturbance-adapted organisms and food webs that characterize them, the role of wildfire in these ecosystems is likely essential to managing biodiversity and conservation across the landscape. Further research is needed in the following areas to better understand and predict the effects of fire severity in stream-riparian ecosystems and inform management: (1) investigations of fire-severity effects over larger spatial scales and longer time periods that integrate fire extent, patchiness, and continuity; (2) descriptions of interaction effects between fire and other sources of both natural and anthropogenic disturbance; (3) analyses of the ability of prescribed fires to emulate wildfires and provide ecosystem benefits; and (4) longer-term studies that integrate changes in fire regimes, vegetation, precipitation, and temperature resulting from climate change.

REFERENCES

Adams, D.E., Anderson, R.C., Collins, S.L., 1982. Differential response of woody and herbaceous species to summer and winter burning in an Oklahoma grassland. Southwest Nat. 27, 55–61.

Agee, J.K., 1993. Fire Ecology of the Pacific Northwest Forests. Island Press, Washington, D.C.

Agee, J.K., Wright, C.S., Williamson, N., Huff, M.H., 2002. Foliar moisture content of Pacific Northwest vegetation and its relation to wildland fire behavior. For. Ecol. Manag. 167, 57–66.

Allan, J.D., 1981. Determinants of diet of brook trout (*Salvelinus fontinalis*) in a mountain stream. Can. J. Fish. Aquat. Sci. 38, 184–192.

Allan, J.D., Wipfli, M.S., Caouette, J.P., Prussian, A., Rogers, J., 2003. Influence of streamside vegetation on inputs of terrestrial invertebrates to salmonid food webs. Can. J. Fish. Aquat. Sci. 60, 309–320.

Allen, C.D., Breshears, D.D., 1998. Drought-induced shift of a forest-woodland ecotone: rapid landscape response to climate variation. Proc. Natl. Acad. Sci. U. S. A. 95, 14839–14842.

Allred, B.W., Fuhlendorf, S.D., Hovick, T.J., Elmore, R.D., Engle, D.M., Joern, A., 2013. Conservation implications of native and introduced ungulates in a changing climate. Glob. Chang. Biol. 19, 1875–1883.

Arkle, R.S., Pilliod, D.S., 2010. Prescribed fires as ecological surrogates for wildfires: a stream and riparian perspective. For. Ecol. Manag. 259, 893–903.

Arkle, R.S., Pilliod, D.S., Strickler, K., 2010. Fire, flow and dynamic equilibrium in stream macroinvertebrate communities. Freshw. Biol. 55, 299–314.

Bagne, K.E., Purcell, K.L., 2011. Short-term responses of birds to prescribed fire in fire-suppressed forests of California. J. Wildl. Manag. 75, 1051–1060.

Bartos, D.L., Campbell Jr., R.B., 1998. Decline of quaking aspen in the interior west, examples from Utah. Rangelands 20, 17–24.

Baxter, C.V., Fausch, K.D., Saunders, W.C., 2005. Tangled webs: reciprocal flows of invertebrate prey link stream and riparian zones. Freshw. Biol. 50, 201–220.

Beakes, M.P., Moore, J.W., Hayes, S.A., Sogard, S.M., 2014. Wildfire and the effects of shifting stream temperature on salmonids. Ecosphere. 5, Article 63.

Bêche, L.A., Stephens, S.L., Resh, V.H., 2005. Effects of prescribed fire on a Sierra Nevada (California, USA) stream and its riparian zone. For. Ecol. Manag. 218, 37–59.

Bellmore, J.R., Baxter, C.V., Connolly, P.J., 2014. Spatial complexity reduces interaction strength in the meta-food web of a river floodplain mosaic. Ecology 96, 274–283.

Belwood, J.J., Fenton, M.B., 1976. Variation in diet of *Myotis lucifugus* (Chiroptera-vespertilionidae). Can. J. Zool. 54, 1674–1678.

Benda, L., Miller, D., Bigelow, P., Andras, K., 2003. Effects of post-wildfire erosion on channel environments, Boise River, Idaho. For. Ecol. Manag. 178, 105–119.

Benda, L., Poff, N.L., Miller, D., Dunne, T., Reeves, G., Pess, G., Pollock, M., 2004. The network dynamic hypothesis: how channel networks structure riverine habitats. Bioscience 54, 413–427.

Beschta, R.L., Rhodes, J.J., Kauffman, J.B., Gresswell, R.E., Minshall, G.W., Karr, J.R., Perry, D.A., Hauer, F.R., Frissell, C.A., 2004. Postfire management on forested public lands of the Western United States. Conserv. Biol. 18, 957–969.

Bess, E.C., Parmenter, R.R., McCoy, S., Molles, M.C., 2002. Responses of a riparian forest-floor arthropod community to wildfire in the middle Rio Grande Valley, New Mexico. Environ. Entomol. 31, 774–784.

Betts, E.F., Jones Jr., J.B., 2009. Impact of wildfire on stream nutrient chemistry and ecosystem metabolism in boreal forest catchments of interior Alaska. Arct. Antarct. Alp. Res. 41, 407–417.

Bigelow, P.E., Benda, L.E., Miller, D.J., Burnett, K.M., 2007. On debris flows, river networks, and the spatial structure of channel morphology. For. Sci. 53, 220–238.

Bisson, P.A., Rieman, B.E., Luce, C., Hessburg, P.F., Lee, D.C., Kershner, J.L., Reeves, G.H., Gresswell, R.E., 2003. Fire and aquatic ecosystems of the western USA: current knowledge and key questions. For. Ecol. Manag. 178, 213–229.

Bladon, K.D., Silins, U., Wagner, M.J., Stone, M., Emelko, M.B., Mendoza, C.A., Devito, K.J., Boon, S., 2008. Wildfire impacts on nitrogen concentration and production from headwater streams in southern Alberta's Rocky Mountains. Can. J. For. Res. 38, 2359–2371.

Blank, R.R., Chambers, J.C., Zamudio, D., 2003. Restoring riparian corridors with fire: effects on soil and vegetation. J. Range Manag. 56, 388–396.

Boerner, R.E.J., Huang, J., Hart, S.C., 2008. Impacts of fire and fire surrogate treatments on ecosystem nitrogen storage patterns: similarities and differences between forests of eastern and western North America. Can. J. For. Res. 38, 3056–3070.

Bozek, M.A., Young, M.K., 1994. Fish mortality resulting from delayed-effects of fire in the greater Yellowstone ecosystem. Great Basin Nat. 54, 91–95.

Brack Jr., V.W., 1983. The nonhibernating ecology of bats in Indiana with emphasis on the endangered Indiana bat, *Myotis Sodalis*. Ph.D. dissertation. Purdue University, West Lafayette, Indiana, USA.

Brigham, R.M., Aldridge, H., Mackey, R.L., 1992. Variation in habitat use and prey selection by Yuma bats, *Myotis yumanensis*. J. Mammal. 73, 640–645.

Buchalski, M.R., Fontaine, J.B., Heady, P.A., Hayes, J.P., Frick, W.F., 2013. Bat response to differing fire severity in mixed-conifer forest California, USA. PLoS One 8, e57884.

Buhl, K.J., Hamilton, S.J., 2000. Acute toxicity of fire-control chemicals, nitrogenous chemicals, and surfactants to rainbow trout. Trans. Am. Fish. Soc. 129, 408–418.

Burton, T.A., 2005. Fish and stream habitat risks from uncharacteristic wildfire: observations from 17 years of fire-related disturbances on the Boise National Forest, Idaho. For. Ecol. Manag. 211, 140–149.

Carpenter, S.R., Cole, J.J., Pace, M.L., Van de Bogert, M., Bade, D.L., Bastviken, D., Gille, C.M., Hodgson, J.R., Kitchell, J.F., Kritzberg, E.S., 2005. Ecosystem subsidies: terrestrial support of aqautic food webs from ^{13}C addition to contrasting lakes. Ecology 86 (10), 2737–2750.

Carrier, W.D., Czech, B., 1996. Threatened and endangered wildlife and livestock interactions. In: Krausman, P.R. (Ed.), Rangeland Wildlife. Society for Rangeland Management, Denver, CO, pp. 39–50.

Cooper, S.D., Page, H.M., Wiseman, S.W., Klose, K., Bennet, D., Even, T., Sadro, S., Nelson, C.E., Dudley, T.L., 2014. Physicochemical and biological responses of streams to wildfire severity in riparian zones. Freshw. Biol. e12523.

Davis, J.M., Baxter, C.V., Rosi-Marshall, E.J., Pierce, J.L., Crosby, B.T., 2013. Anticipating stream ecosystem responses to climate change: toward predictions that incorporate effects via land-water linkages. Ecosystems 16, 909–922.

DeBano, L.F., 2000. The role of fire and soil heating on water repellency in wildland environments: a review. J. Hydrol. 231, 195–206.

Doerr, S.H., Ferreira, A.J.D., Walsh, R.P.D., Shakesby, R.A., Leighton-Boyce, G., Coelho, C.O.A., 2003. Soil water repellency as a potential parameter in rainfall-runoff modelling: experimental evidence at point to catchment scales from Portugal. Hydrol. Process. 17, 363–377.

Dunham, J.B., Young, M.K., Gresswell, R.E., Rieman, B.E., 2003. Effects of fire on fish populations: landscape perspectives on peristance of native fishes and nonnative fish invasion. For. Ecol. Manag. 178, 183–196.

Dunham, J.B., Rosenberger, A.E., Luce, C.H., Rieman, B.E., 2007. Influences of wildfire and channel reorganization on spatial and temporal variation in stream temperature and distribution of fish and amphibians. Ecosystems 10, 335–346.

Dwire, K.A., Kauffman, J.B., 2003. Fire and riparian ecosystems in landscapes of the western USA. For. Ecol. Manag. 178, 61–73.

Earl, S.R., Blinn, D.W., 2003. Effects of wildfire ash on water chemistry and biota in South-Western USA streams. Freshw. Biol. 48, 1015–1030.

Everett, R., Schellhaas, R., Ohlson, P., Spurbeck, D., Keenum, D., 2003. Continuity in fire disturbance between riparian and adjacent sideslope Douglas-fir forest. For. Ecol. Manag. 175, 31–48.

Fausch, K.D., Torgersen, C.E., Baxter, C.V., Li, H.W., 2002. Landscapes to riverscapes: bridging the gap between research and conservation of stream fishes. Bioscience 52, 483–498.

Fleming, H.L., Jones, J.C., Belant, J.L., Richardson, D.M., 2013. Multi-scale roost site selection by Rafinesque's Big-eared Bat (*Corynorhinus rafinesquii*) and Southeastern Myotis (*Myotis austroriparius*) in Mississippi. Am. Midl. Nat. 169, 43–55.

Freeman, M.C., Pringle, C.M., Jackson, C.R., 2007. Hydrologic connectivity and the contribution of stream headwaters to ecological integrity at regional scales. J. Am. Water Resour. Assoc. 43, 5–14.

Fukui, D., Murakami, M., Nakano, S., Aoi, T., 2006. Effect of emergent aquatic insects on bat foraging in a riparian forest. J. Anim. Ecol. 75, 1252–1258.

Gaikowski, M.P., Hamilton, S.J., Buhl, K.J., McDonald, S.F., Summers, C.H., 1996. Acute toxicity of three fire-retardant and two fire-suppressant foam formulations to the early life stages of rainbow trout (*Oncorhynchus mykiss*). Environ. Toxicol. Chem. 15, 1365–1374.

Gamradt, S.C., Kats, L.B., 1997. Impact of chaparral wildfire-induced sedimentation on oviposition of stream-breeding California newts (*Taricha torosa*). Oecologia 110, 546–549.

Gimenez, A., Pastor, E., Zarate, L., Planas, E., Arnaldos, J., 2004. Long-term forest fire retardants: a review of quality, effectiveness, application and environmental considerations. Int. J. Wildland Fire 13, 1–15.

Grafe, T.U., Dobler, S., Linsenmair, K.E., 2002. Frogs flee from the sound of fire. Proc. R. Soc. B 269, 999–1003.

Gregory, S.V., Swanson, F.J., McKee, W.A., Cummins, K.W., 1991. An ecosystem perspective of riparian zones; focus on links between land and water. Bioscience 41, 540–552.

Gregory, S.V., Gurnell, A.M., Boyer, K.L., 2003. The ecology and management of wood in world rivers. American Fisheries Society, Bethesda, Maryland.

Gresswell, R.E., 1999. Fire and aquatic ecosystems in forested biomes of North America. Trans. Am. Fish. Soc. 128, 193–221.

Hagen, E.M., Sabo, J.L., 2012. Influence of river drying and insect availability on bat activity along the San Pedro River, Arizona (USA). J. Arid Environ. 84, 1–8.

Hall, S.J., Lombardozzi, D., 2008. Short-term effects of wildfire on montane stream ecosystems in the Southern Rocky Mountains: one and two years post-burn. Western North American Nat. 68, 453–462.

Harris, H.E., Baxter, C.V., Davis, J.M., In press. Debris flows mediate effects of wildfire on magnitude and composition of tributary subsidies to main-stem habitats. Freshwater Science.

Harris, H.E., 2013. Disturbance Cascade: How fire and debris flows affect headwater linkages to downstream and riparian ecosystems. M.S. thesis: Department of Biological Sciences, Idaho State University, Pocatello, ID, USA.

Hauer, F.R., Spencer, C.N., 1998. Phosphorus and nitrogen dynamics in streams associated with wildfire: a study of immediate and longterm effects. Int. J. Wildland Fire 8, 183–198.

Hitt, N.P., 2003. Immediate effects of wildfire on stream temperature. J. Freshw. Ecol. 18, 171–173.

Howell, P.J., 2006. Effects of wildfire and subsequent hydrologic events on fish distribution and abundance in tributaries of north fork John Day River. N. Am. J. Fish Manag. 26, 983–994.

Hutto, R.L., 2008. The ecological importance of severe wildfire: some like it hot. Ecol. Appl. 18, 1827–1834.

Hynes, H.B.N., 1975. The stream and its valley. Verh. Internat. Verein. Limnol. 19, 1–15.

Isaak, D.J., Luce, C.H., Rieman, B.E., Nagel, D.E., Peterson, E.E., Horan, D.L., Parkes, S., Chandler, G.L., 2010. Effects of climate change and wildfire on stream temperatures and salmonid thermal habitat in a mountain river network. Ecol. Appl. 20, 1350–1371.

Jackson, B.K., Sullivan, S.M.P., 2009. Influence of fire severity on riparian vegetation heterogeneity in an Idaho, U.S.A. wilderness. For. Ecol. Manag. 259, 24–32.

Jackson, B.K., Sullivan, S.M.P., In press. Responses of riparian tetragnathid spiders to wildfire in forested ecosystems of the California Mediterranean climate region, USA. Freshwater Science.

Jackson, B.K., Sullivan, S.M.P., Malison, R.L., 2012. Wildfire severity mediates fluxes of plant material and terrestrial invertebrates to mountain streams. For. Ecol. Manag. 278, 27–34.

Jones, J.R., DeByle, N.V., 1985. Fire. In: DeByle, N.V., Winokur, R.P. (Eds.), Aspen: ecology and management in the western United States. USDA Forest Service, Ogden, UT, pp. 77–81.

Junk, W.J., Bayley, P.B., Sparks, R.E., 1989. The flood pulse concept in river-floodplain systems. In: International Large River Symposium, pp. 110–127.

Karr, J.R., Rhodes, J.J., Minshall, G.W., Hauer, F.R., Beschta, R.L., Frissell, C.A., Perry, D.A., 2004. The effects of postfire salvage logging on aquatic ecosystems in the American West. Bioscience 54, 1029–1033.

Kauffman, J.B., Martin, R.E., 1989. Fire behavior, fuel consumption, and forest-floor changes following prescribed understory fires in Sierra-Nevada mixed conifer forests. Can. J. For. Res. 19, 455–462.

Kauffman, J.B., Martin, R.E., 1990. Sprouting shrub response to different seasons and fuel consumtion levels of prescribed fire in Sierra-Nevada mixed conifer ecosystems. For. Sci. 36, 748–764.

Keyser, T.L., Smith, F.W., Shepperd, W.D., 2005. Trembling aspen response to a mixed-severity wildfire in the Black Hills, South Dakota, USA. Can. J. For. Res. 35, 2679–2684.

Khorram, S., Katibah, E.F., 1984. Vegetation and land cover mapping of the middle fork of the Feather River watershed using landsat digital data. For. Sci. 30, 248–258.

Kirkpatrick, C., Conway, C.J., 2010. Importance of montane riparian forest and influence of wildfire on nest-site selection of ground-nesting birds. J. Wildl. Manag. 74, 729–738.

Kobziar, L.N., McBride, J.R., 2006. Wildfire burn patterns and riparian vegetation response along two Sierra Nevada streams. For. Ecol. Manag. 222, 254–265.

Koetsier, P., Tuckett, Q., White, J., 2007. Present effects of past wildfires on the diets of stream fish. Western North American Nat. 67, 429–438.

Kotliar, N.B., Heji, S.J., Hutto, R.L., Saab, V.A., Melcher, C.P., McFadzen, M.E., 2002. Effects of fire and post-fire salvage logging on avian communities in conifer-dominated forests of the western United states. Studies in Avian Biology. 25, 49–64.

Koyama, A., Kavanagh, K.L., Stephan, K., 2010. Wildfire effects on soil gross nitrogen transformation rates in coniferous forests of central Idaho, USA. Ecosystems 13, 1112–1126.

Kruse, A., Piehl, J., 1986. The impact of prescribed burning on groud-nesting birds. In: Clambey, G., Pemble, R. (Eds.), The prairie: past present and future. Proceedings of the 9th North American Prairie Conference, Moorhead, MN. Center for Environmental Studies, Tri-College University, Fargo, North Dakota, USA.

Lewanzik, D., Kelm, D.H., Greiner, S., Dehnhard, M., Voigt, C.C., 2012. Ecological correlates of cortisol levels in two bat species with contrasting feeding habits. Gen. Comp. Endocrinol. 177, 104–112.

Linder, K.A., Anderson, S.H., 1998. Nesting habitat of Lewis' Woodpeckers in southeastern Wyoming. J. Field Ornithol. 69, 109–116.

Mack, M.C., D'Antonio, C.M., 1998. Impacts of biological invasions on disturbance regimes. Trends Ecol. Evol. 13, 195–198.

Malison, R.L., Baxter, C.V., 2010a. Effects of wildfire of varying severity on benthic stream insect assemblages and emergence. J. N. Am. Benthol. Soc. 29, 1324–1338.

Malison, R.L., Baxter, C.V., 2010b. The fire pulse: wildfire stimulates flux of aquatic prey to terrestrial habitats driving increases in riparian consumers. Can. J. Fish. Aquat. Sci. 67, 570–579.

Malmon, D.V., Renean, S.L., Katzman, D., Lavine, A., Lyman, J., 2007. Suspended sediment transport in an ephemeral stream following wildfire. J. Geophys. Res. Earth Surf. 112, FO2006.

Mast, M.A., Clow, D.W., 2008. Effects of 2003 wildfires on stream chemistry in Glacier National Park, Montana. Hydrol. Process. 22, 5013–5023.

May, C., 2007. Sediment and wood routing in steep headwater streams: an overview of geomorphic processes and their topographic signatures. For. Sci. 53, 119–130.

McIver, J.D., McNeil, R., 2006. Soil disturbance and hill-slope sediment transport after logging of a severely burned site in northeastern Oregon. West. J. Appl. For. 21, 123–133.

McIver, J.D., Starr, L., 2001. A literature review on the environmental effects of postfire logging. West. J. Appl. For. 16, 159–168.

McIver, J.D., Stephens, S.L., Agee, J.K., Barbour, J., Boerner, R.E.J., Edminster, C.B., Erickson, K.L., Farris, K.L., Fettig, C.J., Fiedler, C.E., Haase, S., Hart, S.C., Keeley, J.E., Knapp, E.E., Lehmkuhl, J.F., Moghaddas, J.J., Otrosina, W., Outcalt, K.W., Schwilk, D.W., Skinner, C.N., Waldrop, T.A., Weatherspoon, C.P., Yaussy, D.A., Youngblood, A., Zack, S., 2013. Ecological effects of alternative fuel-reduction treatments: highlights of the National Fire and Fire Surrogate study (FFS). Int. J. Wildland Fire 22, 63–82.

Mellon, C.D., Wipfli, M.S., Li, J.L., 2008. Effects of forest fire on headwater stream macroinvertebrate communities in eastern Washington, USA. Freshw. Biol. 53, 2331–2343.

Meyer, G.A., Pierce, J.L., 2003. Climatic controls on fire-induced sediment pulses in Yellowstone National Park and central Idaho: a long-term perspective. For. Ecol. Manag. 178, 89–104.

Mihuc, T.B., Minshall, G.W., 1995. Trophic generalists vs trophic specialists—implications for food-web dynamics in post-fire streams. Ecology 76, 2361–2372.

Mihuc, T.B., Minshall, G.W., 2005. The trophic basis of reference and post-fire stream food webs 10 years after wildfire in Yellowstone National Park. Aquat. Sci. 67, 541–548.

Miller, M., 2000. Fire autecology. In: Brown, J.K., Smith, J.K. (Eds.), Wildland Fire and Ecosystems: Effects of Fire on Flora. USDA Forest Service, Ogden, UT, pp. 9–34, General Technical Report RMRS-GTR-412-vol2.

Miller, D., Luce, C., Benda, L., 2003. Time, space, and episodicity of physical disturbance in streams. For. Ecol. Manag. 178, 121–140.

Minshall, G.W., 2003. Responses of stream benthic macroinvertebrates to fire. For. Ecol. Manag. 178, 155–161.

Minshall, G.W., Robinson, C.T., Lawrence, D.E., 1997. Postfire responses of lotic ecosystems in Yellowstone National Park, USA. Can. J. Fish. Aquat. Sci. 54, 2509–2525.

Minshall, G.F., Brock, J.T., Andrews, D.A., Robinson, C.T., 2001a. Water quality, substratum and biotic responses of five central Idaho (USA) streams during the first year following the Mortar Creek fire. Int. J. Wildland Fire 10, 185–199.

Minshall, G.W., Robinson, C.T., Lawrence, D.E., Andrews, D.A., Brock, J.T., 2001b. Benthic macroinvertebrate assemblages in five central Idaho (USA) streams over a 10-year period following disturbance by wildfire. Int. J. Wildland Fire 10, 201–213.

Minshall, G.W., Bowman, K.E., Rugenski, B.A., Relyea, C., 2003. Monitoring of Streams in the Payette National Forest 1988-2003: Big Creek and South Fork Salmon Tributaries Pre- and Post-Fire. Payette National Forest, Idaho, USGS.

Moody, J.A., Martin, D.A., 2001. Post-fire, rainfall intensity-peak discharge relations for three mountainous watersheds in the western USA. Hydrol. Process. 15, 2981–2993.

Naiman, R.J., Decamps, H., McClain, M.E., 2005. Riparia: Ecology, Conservation, and Management of Streamside Communities. Elsevier Academic Press, Burlington, MA.

Neville, H., Dunham, J., Rosenberger, A., Umek, J., Nelson, B., 2009. Influences of wildfire, habitat size, and connectivity on trout in headwater streams revealed by patterns of genetic diversity. Trans. Am. Fish. Soc. 138, 1314–1327.

Ober, H.K., Hayes, J.P., 2008. Influence of vegetation on bat use of riparian areas at multiple spatial scales. J. Wildl. Manag. 72, 396–404.

Ormerod, S.J., Dobson, M., Hildrew, A.G., Townsend, C.R., 2010. Multiple stressors in freshwater ecosystems. Freshw. Biol. 55, 1–4.

Paetzold, A.C., Schubert, C.J., Tockner, K., 2005. Aquatic-terrestrial linkages along a braided river: riparian arthropods feeding on aquatic insects. Ecosystems 8, 748–758.

Pettit, N.E., Naiman, R.J., 2007. Fire in the riparian zone: characteristics and ecological consequences. Ecosystems 10, 673–688.

Piccolo, J.J., Wipfli, M.S., 2002. Does red alder (*Alnus rubra*) in upland riparian forests elevate macroinvertebrate and detritus export from headwater streams to downstream habitats in southeastern Alaska? Can. J. Fish. Aquat. Sci. 59, 503–513.

Pilliod, D.S., Bury, R.B., Hyde, E.J., Pearl, C.A., Corn, P.S., 2003. Fire and amphibians in North America. For. Ecol. Manag. 178, 163–181.

Poff, N.L., Olden, J.D., Merritt, D.M., Pepin, D.M., 2007. Homogenization of regional river dynamics by dams and global biodiversity implications. Proc. Natl. Acad. Sci. U. S. A. 104, 5732–5737.

Postel, S., Carpenter, S., 1997. Freshwater ecosystem services. In: Daily, G. (Ed.), Nature's Services: Societal Dependence on Natural Ecosystems. Island Press, Washington, D.C, pp. 195–214.

Power, M.E., Dietrich, W.E., 2002. Food webs in river networks. Ecol. Res. 17, 451–471.

Power, M.E., Rainey, W.E., Parker, M.S., Sabo, J.L., Smyth, A., Khandwala, S., Finlay, J.C., McNeely, F.C., Marsee, K., Anderson, C., 2004. River to watershed subsidies in an old-growth conifer forest. In: Polis, G.A., Power, M.E., Huxel, G.R. (Eds.), Food Webs at the Landscape Level. The University of Chicago Press, Chicago, IL, USA, pp. 217–240.

Propst, D.L., Stefferud, J.A., Turner, P.R., 1992. Conservation and status of Gila trout, *Oncorhynchus gilae*. Southwest. Nat. 37, 117–125.

Pyne, S.J., 1997. America's Fires: Management on Wildlands and Forests. Forest History Society, Durham, North Carolina.

Pyne, S.J., 2004. Tending Fire: Coping with America's Wildland Fires. Island Press, Washington.

Quigley, T.M., Arbelbide, S.J., 1997. An assessment of Ecosystem Components in the Interior Columbia Basin and Portions of the Klamath and Great Basins. USDA Forest Service, Portland, Oregon, USA, General Technical Report PNW-GTR-405.

Rambo, T.R., North, M.P., 2008. Spatial and temporal variability of canopy microclimate in a Sierra Nevada riparian forest. Northwest Sci. 82, 259–268.

Resh, V.H., Brown, A.V., Covich, A.P., Gurtz, M.E., Li, H.W., Minshall, G.W., Reice, S.R., Sheldon, A.L., Wallace, J.B., Wissmar, R.C., 1988. The role of disturbance in stream ecology. J. N. Am. Benthol. Soc. 7, 433–455.

Rieman, B.E., Clayton, J., 1997. Wildfire and native fish: issues of forest health and conservation of sensitive species. Fisheries 22, 6–15.

Rieman, B.E., Lee, D., Chandler, G., Meyer, D., 1997. Does wildfire threaten extinctions for salmonids? Responses of redband trout and bull trout following recent large fires on the Boise National Forest. In: Symposium on Fire Effects on Threatened and Endangered Species and Habitats. International Association of Wildland Fire, Fairfield, pp. 47–57.

Rinne, J.N., 1996. Short-term effects of wildfire on fishes and aquatic macroinvertebrates in the southwestern United States. N. Am. J. Fish Manag. 16, 653–658.

Robinson, C.T., Rushforth, S.R., Minshall, G.W., 1994. Diatom asssemblages of streams influenced by wildfire. J. Phycol. 30, 209–216.

Robinson, C.T., Minshall, G.W., Royer, T.V., 2000. Inter-annual patterns in macroinvertebrate communities of wilderness streams in Idaho, USA. Hydrobiologia 421, 187–198.

Romme, W.H., Turner, M.G., Gardner, R.H., Hargrove, W.W., Tuskan, G.A., Despain, D.G., Renkin, R.A., 1997. A rare episode of sexual reproduction in Aspen (*Populus tremuloides Michx*) following the 1988 Yellowstone fires. Natural Areas J. 17, 17–25.

Romme, W.H., Boyce, M.S., Gresswell, R., Merrill, E.H., Minshall, G.W., Whitlock, C., Turner, M.G., 2011. Twenty years after the 1988 Yellowstone fires: lessons about disturbance and ecosystems. Ecosystems 14, 1196–1215.

Royer, T.V., Minshall, G.W., 1997. Temperature patterns in small streams following wildfire. Arch. Hydrobiol. 140, 237–242.

Rugenski, A.T., Minshall, G.W., 2014. Climate-moderated responses to wildfire by macroinvertebrates and basal food resources in montane wilderness streams. Ecosphere. 5, Article 25.

Russell, R.E., Royle, J.A., Saab, V.A., Lehmkuhl, J.F., Block, W.M., Sauer, J.R., 2009. Modeling the effects of environmental disturbance on wildlife communities: avian responses to prescribed fire. Ecol. Appl. 19, 1253–1263.

Russo, D., Jones, G., 2003. Use of foraging habitats by bats in a Mediterranean area determined by acoustic surveys: conservation implications. Ecography 26, 197–209.

Saab, V., 1999. Importance of spatial scale to habitat use by breeding birds in riparian forests: a hierarchical analysis. Ecol. Appl. 9, 135–151.

Saab, V.A., Powell, H.D.W., 2005. Fire and avian ecology in North America: process influencing pattern. Stud. Avian Biol. 30, 1–13.

Saab, V.A., Russell, R.E., Dudley, J.G., 2009. Nest-site selection by cavity-nesting birds in relation to postfire salvage logging. For. Ecol. Manag. 257, 151–159.

Sabo, J.L., Power, M.E., 2002. Numerical response of lizards to aquatic insects and short-term consequences for terrestrial prey. Ecology 83, 3023–3036.

Sabo, J.L., Sponseller, R.A., Dixon, M., Gade, K., Harms, T., Heffernan, J., Jani, A., Katz, G., Soykan, C.U., Watts, J., Welter, J., 2005. Riparian zones increase regional species richness by harboring different, not more, species. Ecology 86, 56–62.

Schier, G.A., 1973. Origin and development of aspen root suckers. Can. J. For. Res. 3, 45–53.

Seibert, J., McDonnell, J.J., Woodsmith, R.D., 2010. Effects of wildfire on catchment runoff response: a modelling approach to detect changes in snow-dominated forested catchments. Hydrol. Res. 41, 378–390.

Seidman, V.M., Zabel, C.J., 2001. Bat activity along intermittent streams in Northwestern California. J. Mammal. 82, 738–747.

Sestrich, C.M., McMahon, T.E., Young, M.K., 2011. Influence of fire on native and nonnative salmonid populations and habitat in a Western Montana Basin. Trans. Am. Fish. Soc. 140, 136–146.

Shakesby, R.A., Doerr, S.H., 2006. Wildfire as a hydrological and geomorphological agent. Earth Sci. Rev. 74, 269–307.

Smithwick, E.A.H., Turner, M.G., Mack, M.C., Chapin, F.S., 2005. Postfire soil N cycling in northern conifer forests affected by severe, stand-replacing wildfires. Ecosystems 8, 163–181.

Spencer, C.N., Hauer, F.R., 1991. Phosphorus and nitrogen dynamics in streams during a wildfire. J. N. Am. Benthol. Soc. 10, 24–30.

Spencer, C.N., Gabel, K.O., Hauer, F.R., 2003. Wildfire effects on stream food webs and nutrient dynamics in Glacier National Park, USA. For. Ecol. Manag. 178, 141–153.

Stanford, J.A., Lorang, M.S., Hauer, F.R., 2005. The shifting habitat mosaic of river ecosystems. Verh. Internat. Verein. Limnol. 29, 123–136.

Stephens, S.L., Meixner, t., Poth, M., McGurk, B., Payne, D., 2004. Prescribed fire, soils, and stream water chemistry in a watershed in the Lake Tahoe Basin, California. Int. J. Wildland Fire 13, 27–35.

Stickney, P.F., 1986. First decade plant succession following the sundance forest fire. General Technical Report GTR-INT-197:26, USDA Forest Service, Northern Idaho.

Stone, K.R., Pilliod, D.S., Dwire, K.A., Rhoades, C.C., Wollrab, S.P., Young, M.K., 2010. Fuel reduction management practices in riparian areas of the western USA. Environ. Manag. 46, 91–100.

Sullivan, S.M.P., Watzin, M.C., 2009. Stream-floodplain connectivity and fish assemblage diversity in the Champlain Valley, Vermont, U.S.A. J. Fish Biol. 74, 1394–1418.

Sweeney, B.W., Bott, T.L., Jackson, J.K., Kaplan, L.A., Newbold, J.D., Standley, L.J., Hession, W.C., Horwitz, R.J., 2004. Riparian deforestation, stream narrowing, and loss of stream ecosystem services. Proc. Natl. Acad. Sci. U. S. A. 101, 14132–14137.

Swift, S.M., Racey, P.A., Avery, M.I., 1985. Feeding ecology of *Pipestrellus pipistrellus* (Chiroptera: Vespertilionidae) during pregnancy and lactation. II. Diet. J. Anim. Ecol. 54, 217–225.

Tockner, K., Lorang, M.S., Stanford, J.A., 2010. River floodplains are model ecosystems to test general hydrogeomorphic and ecological concepts. River Res. Appl. 26, 76–86.

Van de Water, K., North, M., 2010. Fire history of coniferous riparian forests in the Sierra Nevada. For. Ecol. Manag. 260, 384–395.

Van de Water, K., North, M., 2011. Stand structure, fuel loads, and fire behavior in riparian and upland forests, Sierra Nevada Mountains, USA; a comparison of current and reconstructed conditions. For. Ecol. Manag. 262, 215–228.

Van Mantgem, P.J., Stephenson, N.L., Byrne, J.C., Daniels, L.D., Franklin, J.F., Fule, P.Z., Harmon, M.E., Larson, A.J., Smith, J.M., Taylor, A.H., Veblen, T.T., 2009. Widespread increase of tree mortality rates in the western United States. Science 323, 521–524.

Vannote, R.L., Minshall, G.W., Cummins, K.W., Sedell, J.R., Cushing, C.E., 1980. The river continuum concept. Can. J. Fish. Aquat. Sci. 37, 130–137.

Verkaik, I., Rieradevall, M., Cooper, S.D., Melack, J.M., Dudley, T.L., Prat, N., 2013a. Fire as a disturbance in Mediterranean climate streams. Hydrobiologia 719, 353–382.

Verkaik, I., Vila-Escale, M., Rieradevall, M., Prat, N., 2013b. Seasonal drought plays a stronger role than wildfire in shaping macroinvertebrate communities of Mediterranean streams. Int. Rev. Hydrobiol. 98, 271–283.

Verkaik, I., Vila-Escale, M., Rieradevall, M., Baxter, C.V., Lake, P.S., Minshall, G.F., Reich, P., Prat, N., In press. Stream macroinvertebrate community responses to fire: are they the same in different fire-prone biogeographic regions? Freshwater Science.

Vieira, N.K., Clements, W.H., Guevara, L.S., Jacobs, B.F., 2004. Resistance and resilience of stream insect communities to repeated hydrologic disturbances after a wildfire. Freshw. Biol. 49, 1243–1259.

Vierling, K.T., Saab, V., 2004. Spatial scaling of woodpecker habitat selection in post-fire environments. Ecol. Soc. Am. Annu. Meet. Abstr. 89, 523.

Vila-Escale, M., Vegas-Vilarrubia, T., Prat, N., 2007. Release of polycyclic aromatic compounds into a Mediterranean creek (Catalonia, NE Spain) after a forest fire. Water Res. 41, 2171–2179.

Vogl, R.J., 1973. Effects of fire on plants and animals of a Florida wetland. Am. Midl. Nat. 89, 334–347.

Wallace, J.B., Eggert, S.L., Meyer, J.L., Webster, J.R., 1997. Multiple trophic levels of a forested stream linked to terrestrial litter inputs. Science 277, 102–104.

Wan, S., Hui, D., Luo, Y., 2001. Fire effects on nitrogen pools and dynamics in terrestrial ecosystems: a meta-analysis. Ecol. Appl. 11, 1349–1365.

Ward, J.V., Tockner, K., Arscott, D.B., Claret, C., 2002. Riverine landscape diversity. Freshw. Biol. 47, 517–539.

Westerling, A.L., Hidalgo, H.G., Cayan, D.R., Swetnam, T.W., 2006. Warming and earlier spring increase western US forest wildfire activity. Science 313, 940–943.

Westerling, A.L., Bryant, B.P., Preisler, H.K., Holmes, T.P., Hidalgo, H.G., Das, T., Shrestha, S.R., 2011a. Climate change and growth scenarios for California wildfire. Clim. Chang. 109, 445–463.

Westerling, A.L., Turner, M.G., Smithwick, E.A.H., Romme, W.H., Ryan, M.G., 2011b. Continued warming could transform Greater Yellowstone fire regimes by mid-21st century. Proc. Natl. Acad. Sci. U. S. A. 108, 13165–13170.

Whitlock, C., 2004. Forests, fires and climate. Nature 432, 28–29.

Whitlock, C., Shafer, S.L., Marlon, J., 2003. The role of climate and vegetation change in shaping past and future fire regimes in the northwestern US and the implications for ecosystem management. For. Ecol. Manag. 178, 5–21.

Williams, M.R., Melack, J.M., 1997. Effects of prescribed burning and drought on the solute chemistry of mixed-conifer forest streams of the Sierra Nevada, California. Biogeochemistry 39, 225–253.

Wipfli, M.S., 1997. Terrestrial invertebrates as salmonid prey and nitrogen sources in streams: contrasting old-growth and young-growth riparian forests in southeastern Alaska, USA. Can. J. Fish. Aquat. Sci. 54, 1259–1269.

Wondzell, S.M., King, J.G., 2003. Postfire erosional processes in the Pacific Northwest and Rocky Mountain regions. For. Ecol. Manag. 178, 75–87.

Chapter 6

Bark Beetles and High-Severity Fires in Rocky Mountain Subalpine Forests

Dominik Kulakowski[1] **and Thomas T. Veblen**[2]
[1]*Graduate School of Geography, Clark University, Worcester, MA, USA*
[2]*Department of Geography, University of Colorado-Boulder, Boulder, CO, USA*

6.1 FIRE, BEETLES, AND THEIR INTERACTIONS

For millennia the forests of the Rocky Mountains in the western United States have been shaped by wildfires (Romme and Despain, 1989, Sibold et al., 2006), outbreaks of insects (Eisenhart and Veblen, 2000, Jarvis and Kulakowski, 2015), and the potential interactions between these disturbances (Veblen et al., 1994) (Figure 6.1). The subalpine forests in this region are dominated by lodgepole pine (*Pinus contorta*) at lower elevations and Engelmann spruce (*Picea engelmannii*) and subalpine fir (*Abies lasiocarpa*) at higher elevations. These ecosystems are shaped by large, high-severity fires; outbreaks of mountain pine beetle (*Dendroctonus ponderosae*) in lodgepole pine forests and spruce beetle (*Dendroctonus rufipennis*) in spruce-fir forests; as well as the interactions between these disturbances. The potential role of modern fire suppression in altering stand conditions of these high-elevation forests is debatable, but the fact that these forests were historically shaped by fires that killed most canopy trees over extensive areas (i.e., patches >1000 ha) at long intervals (>100 years) is not disputed (Romme and Despain, 1989, Sibold et al., 2006). The focus of this chapter is subalpine forests characterized by lodgepole pine, mixed spruce, and subalpine fir, but we also touch on upper montane forests where Douglas-fir (*Pseudotsuga menziesii*) either form relatively pure stands or can be found mixed with the tree species typical of the subalpine zone. In these upper montane forests high-severity fires are characteristic, as are outbreaks of Douglas fir bark beetle (*Dendroctonus pseudotsugae*) (Harvey et al., 2013, Sherriff et al., 2014). This chapter deals specifically with interactions between fire and *Dendroctonus* bark beetles, but we also briefly mention other biotic and abiotic causes of massive tree mortality to place the main theme of the chapter into an appropriate context.

FIGURE 6.1 The forests of the Rocky Mountains in the western United States are shaped by wildfires (background), outbreaks of insects (foreground), and the potential interactions between these disturbances. *(Photo: D. Kulakowski)*

Disturbances such as wildfires and insect outbreaks interact with underlying environmental variability to determine the spatial and temporal heterogeneity of forest landscapes (White and Pickett, 1985) (Figure 6.2). Even large and severe disturbances do not homogenize the landscape but rather promote spatial heterogeneity as a result of spatial variability of disturbance severity and surviving residuals (Turner, 2010). The resulting patterns affect subsequent ecological processes including postdisturbance regeneration and susceptibility to subsequent disturbances (Turner, 2010), as well as biodiversity (Box 6.1). However, the importance of predisturbance conditions on susceptibility to disturbances

FIGURE 6.2 Wildfires and insect outbreaks interact with underlying environmental variability to determine the spatial and temporal heterogeneity of forest landscapes. *(Photo: D. Kulakowski)*

BOX 6.1 Bark Beetle Outbreaks and Biodiversity

Endemic and epidemic bark beetle outbreaks are important sources of structural heterogeneity and biodiversity in the conifer forests of western North America. Bark beetles are parts of many forest food webs and can be associated with a large number of organisms (Dahlsten, 1982). They can be hosts for parasites and food for a variety of animals, including spiders, birds, and other beetles (Koplin and Baldwin, 1970). The actual effect of any particular bark beetle outbreak on subsequent biodiversity depends on the initial forest conditions, the intensity of the outbreak, and the types of organisms considered.

Bark beetles can have far-reaching effects on ecological structures and biodiversity, which, when considered across scales from individual trees to entire landscapes, reveal their important roles as ecosystem engineers. At the scale of individual beetle galleries, they establish and maintain a microflora of fungi and bacteria that create a complex web of biosynthetic interactions affecting tree resistance and success of beetle attack. By reducing tree resistance, beetle attack creates opportunities for a wide diversity of saprogenic competitors (Raffa et al., 2008). Bark beetles themselves are an important food source for a diverse group of arthropods and vertebrates, including birds such as woodpeckers that are highly adapted to digging out larvae of woodboring insects. In general, a bark beetle outbreak initializes a release of resources that, in the short term, promotes the growth of populations of insectivorous birds (Saab et al., 2014). Overall, approximately twice as many bird species have increased, as opposed to decreased, in forests with bark beetle outbreaks (Saab et al., 2014). The longer- term impact of large outbreaks on avian diversity has not been widely studied but is likely to depend on the amount of tree mortality and the rate of recovery of unattacked host conifers as well as nonhost trees.

At the scale of forest stands, tree mortality caused by bark beetles increases structural heterogeneity through the creation of canopy gaps and enhanced growth of understory plants, which is likely to create a favorable habitat for many invertebrates and vertebrates. Outbreaks create snags that may be used by various birds and mammals, including woodpeckers, owls, hawks, wrens, warblers, bats, squirrels, American martens (*Martes americana*), and lynx (*Lynx canadensis*). Populations of cavity-nesting birds often increase following bark beetle outbreaks (Saab et al., 2014).

At a landscape scale, beetle outbreaks are likely to alter biodiversity through the creation of more diverse patch configurations and edge effects favoring some wildlife species. Wildlife associated with early seral habitats, such as deer and elk, are expected to be favorably influenced by an outbreak once there has been enough time for understory resources to respond to the creation of canopy openings (Saab et al., 2014). The consequences of a beetle outbreak for biodiversity at the scale of large stands and landscapes depend both on the intensity of the outbreak and on the preoutbreak forest landscape structure. For example, large areas of monotypic lodgepole pine stands that originated after late nineteenth- or early twentieth-century fires are typically low in structural diversity compared with landscapes with patches of beetle-killed trees. By contrast, a spruce beetle outbreak affecting old stands of Engelmann spruce and subalpine fir, which are already characterized by abundant standing dead and fallen large trees, would result in a less

Continued

BOX 6.1 Bark Beetle Outbreaks and Biodiversity—Cont'd

extreme alteration of stand conditions that may have a smaller impact on biodiversity.

Another particularly important effect of beetle outbreaks on ecosystem structure and biodiversity is evident in riparian habitats of mountain streams (Jackson and Wohl, 2015; see Chapter 5). Beetle-killed trees contribute to the recruitment of large, coarse, woody debris into riparian areas and stream systems, which exerts important beneficial influences on the storage of sediment and organic matter and on river and floodplain habitat for numerous animal species, including trout. Compared with timber harvesting, which can remove all riparian wood and severely deplete subsequent in-stream wood recruitment, beetle outbreaks provide a source of in-stream wood loads for decades following a beetle outbreak.

decreases as the intensity of disturbance increases. Therefore, how disturbances and their interactions determine landscape heterogeneity is likely to change as climatically driven disturbances become more intense under climate change.

Two-way interactions between fires and bark beetle outbreaks occur, whereby outbreaks can affect subsequent fires and fires can affect subsequent outbreaks. Fire regimes are fundamentally a function of fuels, weather, and topography. Because outbreaks can result in visible and abundant dead trees, expecting this change in fuels to affect fire regimes is intuitive. Conversely, because the time since the last severe fire strongly influences stand structure and tree size, which affect susceptibility to bark beetles, expecting fire regimes to affect susceptibility to outbreaks is also intuitive. These apparently simple relationships are made complex and nuanced by various sources of spatial heterogeneity that affect probabilities of fire and outbreaks, as well as rates of forest development following both of these disturbances. The topic of disturbance interactions includes questions of (1) linked disturbances (how the occurrence or severity of one disturbance affects the occurrence or severity of a subsequent disturbance) and (2) compounded disturbances (how two or more disturbances that occur in relatively short succession affect the overall disturbance intensity and postdisturbance development). In this chapter we focus primarily on linked beetle-fire disturbances but briefly address compounded effects because they are important in how disturbance interactions create spatial heterogeneity. This chapter deals mainly with the ecological interactions between wildfire and bark beetle outbreaks, but it is worth noting that these interactions are also central to questions of forest management in the Rocky Mountains and elsewhere. Management related to forest disturbances often aims to promote either resistance (i.e., reduced probability or magnitude of disturbance) or resilience (i.e., capacity of a forest to develop after a disturbance without tipping into an alternate state) (DeRose and Long, 2014). Therefore, understanding the ecological relationships between outbreaks and fires is fundamental to developing management strategies for increased resistance and resilience.

The potential influences of bark beetle outbreaks on subsequent fire hazard have been the subject of a lively public, political, and scientific debate; an emerging body of scientific research; and several substantive literature reviews (Parker et al. 2006; Romme et al., 1986, Jenkins et al., 2008, 2012, 2014; Kaufmann et al. 2008; Simard et al., 2008, Hicke et al., 2012a, Black et al., 2013). A lack of consensus on potential relationships between bark beetle outbreaks and actual or potential wildfire activity in the published literature has led to confusion among scientists, resource managers, and the general public. Some of this confusion and debate may be attributable to the specific research questions posed, the parameters selected to measure potential effects on fire regimes, and the type and initial conditions of ecosystems considered. In this chapter we examine the current state of knowledge about interactions between bark beetle outbreaks and fires in lodgepole pine and spruce-fir forests of the Rocky Mountains, with particular focus on high-severity fires and the role of disturbance interactions in creating landscape heterogeneity.

6.2 HOW DO OUTBREAKS AFFECT SUBSEQUENT HIGH-SEVERITY FIRES?

Methodological Considerations

Studies that address the question of how bark beetle outbreaks potentially affect subsequent wildfires typically focus on fire hazard (the fuel complex, including the type, volume, and arrangement of fuels that determine the ease of ignition and resistance to control regardless of the fuel type's weather-influenced moisture content); fire risk (the chance that a fire might start based on all causative agents, including fuel hazard, ignition source, and weather); or fire behavior (including flame length, rate of spread, or other measures of the fire) (Hardy 2005). Most studies have used research designs that can be broadly classified as (1) field experiments; (2) fire behavior modeling; and (3) retrospective case studies of actual fire events in beetle-affected forests. To date, field experiments in which fire behavior is intensely monitored in stands with actual or simulated bark beetle outbreaks have been rare and limited to a few locations in western Canada (Schroeder and Mooney, 2012). While these studies have yielded useful comparative observations of fire behavior in stands with differing fuel properties, the broader implications of such experimental results are severely limited by the weather conditions under which the burns are implemented and the narrow range of initial conditions (e.g., severity of beetle kill, time since beetle kill, and stand conditions before beetle kill) included in experimental burns. Although experimental burns comparing simulated beetle kill with control stands provide insights for improving fire behavior models, they have not contributed significantly to the more general question of whether outbreaks significantly affect fire regime parameters at broad spatial scales and over a range of time periods.

Given the paucity of experimental studies, this chapter necessarily focuses on studies using the other two methodological approaches: wildfire behavior modeling and retrospective case studies. Wildfire behavior modeling is applied across a range of spatial and temporal scales primarily for planning the management of a wildfire incident. In predicting wildland fire behavior, managers are particularly in need of a basis for predicting ignitions, rate of fire spread, energy released and associated flame-front dimensions, perimeter and area of fire growth, and closely related phenomena such as torching, crowning, and spotting (Alexander and Cruz, 2013). Wildland fire behavior models are commonly classified as empirical versus physics-based models. The former are applied in an operational decision-making context, whereas the latter are developed primarily for enhancing theoretical understanding of fire propagation.

Several empirical modeling systems have coupled the surface fire spread model by Rothermel (1972) with criteria and models of crown fire initiation and rate of crown fire propagation. Operational fire behavior models have been widely used to assess the effectiveness of fuel treatments. As pointed out by Cruz and Alexander (2010), however, simulations of the onset of crowning and rate of spread of active crown fire in conifer forests in the US West using these modeling systems exhibit significant underprediction bias. For example, standard fire behavior models such as NEXUS (Scott and Reinhardt, 2001) underpredict crown fire behavior because unrealistically high wind speeds are required for the onset and spread of crown fires in comparison with observed fire behavior (also see Chapter 13). Despite recognition of the shortcomings of operational fire models, they continue to be applied to assess the effectiveness of thinning treatments and to gauge the effects of forest insect infestations on crown fire potential. When applying operational fire behavior models to the latter objective, a near void of empirical evidence relating fuel characteristics and foliage flammability representative of different stages of bark beetle attack conditions has been widely recognized (Jolly et al., 2012). Thus, nearly all studies have been forced to use unverified models of foliage (live and dead) to address the effects of bark beetle outbreaks on potential fire behavior. Because stand-scale models assign a single average set of conditions to the entire stand, they cannot fully consider the significant amount of fine-scale heterogeneity in fuel moisture resulting from neighboring living and beetle-killed trees (Jolly et al., 2012). Although physics-based fire models can accommodate fine-scale variability in tree mortality, they are computationally demanding and have been applied to the assessment of the effects of bark beetles on potential fire behavior in only a few instances (Hoffman et al., 2012).

A series of studies using operational fire behavior models has led to a generalized expectation of the nature of changes in stand flammability following a bark beetle outbreak (Figure 6.3a). Operational fire behavior models have been used in the major research focus on how outbreaks may affect subsequent high-severity fires by altering fuel quantity and quality. Empirical research at scales from individual needles to stands has consistently shown major changes in the quality and arrangement of fuels following outbreaks. Each of these changes,

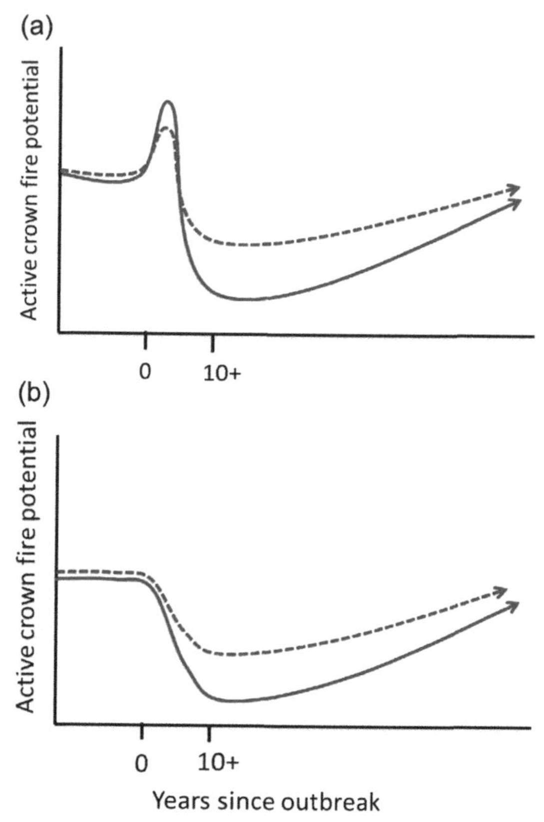

FIGURE 6.3 Active crown fire potential in lodgepole pine (solid lines) and spruce-fir stands (dashed lines) expected from modeling (a) and empirical (b) studies. Because of the structural differences between typical lodgepole pine and spruce-fir stands in the Rocky Mountains, the magnitude of structural changes caused by spruce beetle outbreaks in spruce-fir forests is likely to be less than that associated with mountain pine beetle outbreaks in lodgepole pine forests.

their magnitude, and their timing can potentially affect various aspects of fire regimes—sometimes in a contradictory manner. As a result, the effects of outbreaks on subsequent fires are contingent on a number of factors, including the intensity of the outbreak and the time since the outbreak.

Although changes in fuels following outbreaks are feasible to quantify, correctly understanding and accurately modeling their implications for actual fire regimes has proved to be much more elusive, in part because of important limitations in available fire models that compromise their effectiveness in characterizing fire behavior in beetle-affected forests (Jenkins et al., 2012, Page et al., 2014a,b). Consequently, much of what is known about the cumulative effects of outbreaks on fires is based on retrospective studies that have examined the occurrence of actual fires as they are related to preceding outbreaks. One major advantage of such studies is that they are free of the conceptual limitations of existing fire models and instead depict the actual relationships between outbreaks and fires. These studies also come with important limitations, however, including a limited number and variety of initial forest conditions, variable and sometimes unknown fire weather, and covariation among key variables—all of which can make disentangling complex ecological relationships and identifying underlying mechanisms difficult. As a result, how outbreaks affect fire behavior (e.g., fire line intensity, rate of spread) is poorly understood, whereas the effect of outbreaks on some fire regime parameters such as fire extent, frequency, probability of occurrence, and severity at landscape or broader spatial scales can be addressed more feasibly by retrospective studies.

Retrospective studies have generally relied on either field or remote sensing methods to examine how outbreaks affect fire attributes (Table 6.1). While field methods yield much more accurate data on disturbance severity, the required labor intensity results in the actual sampling of only relatively small areas, even though sample points can be distributed over much larger landscapes. By contrast, remote sensing provides contiguous data on disturbance extent and severity, but it is less accurate because of the relatively low spatial resolution of data and difficulties in remotely detecting signals of disturbances. Both methodological approaches are also vulnerable to unknown preoutbreak variability of stand structure, differences in burning conditions resulting from changing local fire weather during the event, as well as other variability that is not incorporated into a research design but that can affect fire behavior.

We stress that fire behavior modeling and retrospective approaches used to gauge effects of bark beetle outbreaks on subsequent wildfire activity differ fundamentally in the types of questions they are appropriate for addressing. Clearly, understanding fire behavior is important because of its ecological effects and its relevance for resistance of fire to control (Page et al., 2013b). That notwithstanding, in the context of understanding high-severity fire regimes, the central question is whether, and how, outbreaks may fundamentally or cumulatively alter those fire regimes.

Lodgepole Pine Forests

Outbreaks of mountain pine beetle (MPB) clearly alter the fuel structures of lodgepole pine forests (Page and Jenkins, 2007; Klutsch et al. 2009; Simard et al., 2011). Immediately following an MPB attack, important changes in foliar moisture, starch, sugar levels, fiber, and crude fat affect the flammability of lodgepole pine needles and can shorten ignition times for red needles as opposed to live green needles on unattacked trees (Jolly et al., 2012, Page et al., 2012). But how these changes scale up to affect high-severity fires at stand and landscape scales is complex and continues to be an active topic of research. Modeling studies have suggested that the potential of active crown fire may increase immediately after an outbreak, depending on the fire intensity generated by surface fuels (Hoffman et al., 2013). However, important differences in how foliar

TABLE 6.1 Characteristics of Retrospective Studies that Have Examined How Outbreaks Affect Various Fire Parameters

Authors	Location	Forest	Scale[a] (km^2)	Main Methods	Fire Parameters
Turner et al. (1999)	Yellowstone	Lodgepole	1 s	Field	Extent and severity
Kulakowski and Jarvis (2011)	Colorado	Lodgepole	1 s	Field	Occurrence
Harvey et al. (2013)	Yellowstone	Douglas fir	1 s	Field	Severity
Harvey et al. (2014a)	Northern Rockies	Lodgepole	1 s	Field	Severity
Harvey et al. (2014b)	Yellowstone	Lodgepole	1 s	Field	Severity
Kulakowski et al. (2003)	Colorado	Spruce	10 s	Field	Extent
Bigler et al. (2005)	Colorado	Spruce	10 s	Remote sensing	Extent and severity
Lynch et al. (2006)	Yellowstone	Lodgepole	100 s	Remote sensing	Occurrence
Kulakowski and Veblen (2007)	Colorado	Spruce and lodgepole	100 s	Remote sensing	Extent and severity
Renkin and Despain (1992)	Yellowstone	Lodgepole	1000s	Documentary	Occurrence and severity
Bebi et al. (2003)	Colorado	Spruce	1000s	Remote sensing	Occurrence

[a] *For field-based studies the scale indicates the total area sampled, even if samples were distributed across a much larger area.*

moisture content (FMC) on recently killed trees versus other fine fuels respond to changing environmental conditions can hamper accurate modeling of fire behavior (Page et al., 2013a,b, 2014a,b).

Limitations of fire behavior models mean that corresponding results are tentative and suggestive, particularly given the paucity of field experimentation to validate fire behavior models (Cruz and Alexander, 2010). Nevertheless, fire modeling studies of beetle-affected stands are useful for gaining insight into and developing hypotheses about possible consequences of outbreaks on fires. Schoennagel et al. (2012) predicted some effects of fuel moisture on fire behavior under certain weather scenarios. For example, active crown fire was modeled to be more probable at lower wind speeds and less extremely dry conditions assuming lower canopy fuel moisture in red and grey stages of MPB compared with the green stage (Schoennagel et al., 2012). Later in the outbreak more open canopies and high loads of large surface fuels resulting from tree fall were suggested to increase surface fireline intensities, possibly facilitating active crown fire at lower wind speeds (Schoennagel et al., 2012). However, Schoennagel et al. (2012) also suggested that if transition to crown fire occurs (outside the stand or within the stand via ladder fuels or wind gusts), active crown fire can be sustained at similar wind speeds, implying observed fire behavior may not be qualitatively different among MPB stages under extreme burning conditions. In sum, the probability of crown fire is likely to be similar across MPB stages and is characteristic of lodgepole pine forests where extremely dry, gusty weather conditions are key factors in determining fire behavior.

While much has been learned about how outbreaks affect fuels, how quickly canopy bulk density (CBD) decreases following tree mortality and how this varies with biophysical setting is perhaps the most important issue that remains poorly understood. CBD is important to fire regimes and fire behavior because it is an indicator of the amount and continuity of canopy fuels that are available to burn and carry a fire. Some conceptual frameworks suggest that CBD remains unchanged initially following an outbreak (Jenkins et al., 2012, Hicke et al., 2012a), but the limited empirical data that directly address this issue indicate that CBD decreases shortly after tree mortality (Simard et al., 2011). The issue of how quickly CBD decreases following outbreaks necessarily hinges on the synchrony of beetle attack and tree death within a stand, as well as site conditions. In other words, the question can be expressed as whether, and to what degree, the so-called red phase of outbreaks (characterized by beetle-killed trees with red needles) is homogeneous in timing. After a substantial reduction of stand-scale FMC, stand-level CBD would be most likely to remain unchanged following outbreaks if 100% of trees in a stand were killed during the initial year of the outbreak and if site moisture, temperature, and wind conditions promoted retention of dead needles. Retention of needles would be expected to be especially short at relatively dry, warm, and windy sites. If an outbreak lasts several years within a given stand (as it normally does; Schmid and Amman 1992), then it becomes increasingly likely that some trees would lose their needles before other trees are killed, effectively reducing CBD before major reductions in stand-scale FMC. Similarly, lower-severity outbreaks would affect fuels less than higher-severity outbreaks.

Empirical field observations support the notion that FMC and CBD decrease approximately simultaneously. After an MPB outbreak in Yellowstone

National Park, reduced canopy moisture content was coupled with reduced CBD (Simard et al., 2011). Based on these data, simulation models of fire behavior predicted that under intermediate wind conditions (40 to 60 km/h), the probability of active crown fire in stands recently affected by beetles would be lower than in stands not affected by beetles (Simard et al., 2011). In addition, if winds were below 40 km/h or above 60 km/h, stand structure would have little effect on fire behavior. Although the canopy is drier immediately after an outbreak, this does not translate to an increase in fire risk, likely because of the overriding effect of reduced CBD. Decreased CBD is predicted to continue being important in latter stages of outbreaks (5-60 years after) and lead to a consequential lower risk of active crown fire (Jenkins et al., 2008).

Given important uncertainties with modeling predictions of fire behavior after outbreaks, studies that have empirically examined actual high-severity fires have generally found that wildfire activity does not substantially increase, even immediately following MPB outbreaks (Figure 6.3b). For example, ongoing outbreaks of MPB (and spruce beetle) had no detectable effect on the extent or the severity of fire in a large complex of two major fires that burned over 10,000 ha in 2002 in northwestern Colorado, possibly because changes in fuels resulting from the outbreaks may have been overridden by climatic conditions (Kulakowski and Veblen, 2007). Similarly, an experimental burn in jack pine (*Pinus banksiana*) forests in Alberta indicated little difference in rates of canopy fire spread and weather threshold for crown involvement between stands with simulated effects of MPB versus control stands (Schroeder and Mooney, 2012). The effects of MPB on fire severity in the Northern Rockies have been shown to be contingent on burning conditions (Harvey et al., 2014a). Outbreaks in the red stage did not affect any measure of fire severity, except that under extreme burning conditions MPB outbreaks were associated with increased charring on trees that were presumably dead before the fire (Harvey et al., 2014a).

Studies of interactions among disturbances can be hampered by colinearity among stand attributes and a lack of constancy in the many variables expected to affect fire behavior in beetle-affected versus beetle-unaffected stands. For example, compared with stands with lower MPB-caused mortality, more lodgepole stands in which >50% of susceptible trees were killed in the preceding 5 to 15 years burned at high severity in 1988 in Yellowstone (Turner et al., 1999). Relatively old stands, however, which because of differences in fuel structures were inherently more likely to burn at a high severity than younger stands, were also more affected by MPB (Renkin and Despain, 1992). Therefore, whether the differences in the spatial patterns of the severe fires in Yellowstone were primarily the consequences of the outbreak or of prefire stand structure that was unrelated to the outbreak is not clear (Simard et al., 2012).

Other studies have reported that MPB outbreaks have negligible or mixed results on subsequent fire regimes. In Yellowstone National Park, stands affected by MPB outbreaks burned at a lower severity in the Robinson Fire in 1994 compared with adjacent stands that were not affected by MPBs before that fire (Omi, 1997). But stands affected by outbreak 13 to 16 years before the 1988 Yellowstone fires were slightly (about 11%) more likely to burn compared with stands unaffected by beetles (Lynch et al., 2006). By contrast, stands that were affected by outbreak 5 to 8 years before the 1988 fires were no more likely to burn compared with unaffected stands (Lynch et al., 2006). Over longer time periods in Colorado, the occurrence of fires in lodgepole pine forests has been shown to be unrelated to preceding MPB outbreaks but strongly associated with drought (Kulakowski and Jarvis, 2011).

Recent research in the Rockies in the northern United States has used field sampling of prefire stand conditions attributable to beetle kill and numerous field-based measures of severity in recent fires in lodgepole pine and mixed conifer forests (Harvey et al., 2013, 2014a,b). These studies have shown that field-based measurements of fire severity (i.e., fire-killed basal area, number of trees killed by fire, char height, and percentage of bole scorched) were primarily driven by burning conditions and topographic position. In lodgepole pine stands in the gray stage of MPB outbreak, fire severity was not related to outbreak severity under moderate burning conditions, except increased charring on trees that were presumably dead before the fire (Harvey et al., 2014a). But under extreme burning conditions, several measures of fire severity, including fire-caused mortality, increased with outbreak severity, possibly because of increased fireline intensity (Harvey et al., 2014a). In this study, however, outbreak severity in stands in the gray stage was relatively low (0-56% beetle-killed basal area). In fact, it was lower than the severity in stands in the red stage of outbreak in this study and also lower than other studies that found fire severity to be unrelated to outbreak in the gray stage (Simard et al., 2011, Schoennagel et al., 2012, Harvey et al., 2014b)). Further research is needed in stands in the gray stage following low-to moderate-severity outbreaks. Given the characteristic high-severity fires in lodgepole pine, however, fire behavior may not be qualitatively different among stages of MPB outbreak under extreme burning conditions (Schoennagel et al., 2012).

In addition to changing fuels, outbreaks also have been suggested to affect fire regimes by increasing lightning strikes because the number of standing dead trees increases in the years to decades following outbreaks. Although some evidence 1 and 6 years following a MPB outbreak in British Columbia has been presented to support this hypothesis, the influence of MPBs on ignitions has been reported to be less important than that of temperature and precipitation (Bourbonnais et al., 2014). That outbreaks could affect fire ignitions by indirectly affecting climate is also theoretically possible. Following an MPB outbreak in British Columbia, outgoing sensible and radiative heat fluxes increased enough to potentially modify local atmospheric processes, cloud cover, and precipitation (Maness et al. 2012). To date, however, no published study has found fire regimes actually to be influenced by these mechanisms. Although important questions remain, available studies indicate that outbreaks of MPB in lodgepole pine forests have not resulted in observable increases in fire risk, extent, or severity.

Spruce-Fir Forests

The spruce-fir forests of the Rocky Mountains are similar to those dominated by lodgepole pine in that they dominate the subalpine landscape, are relatively dense, and are characterized by infrequent high-severity fires and outbreaks of bark beetles. There are also a number of important differences between these two forest types that influence interactions between outbreaks and wildfires. Key differences are that spruce-fir forests are more mesic and include a large component of tree species not susceptible to the most important bark beetle. In addition, spruce-fir forests are represented by a much higher proportion of older (e.g., >200 years old) stands relative to lodgepole pine forests that are represented overwhelmingly by younger (e.g., 100-200 years old) stands in the Rocky Mountain region (Veblen, 1986, Veblen and Donnegan 2006). This difference in stand age distributions reflects the longer mean fire return intervals typical of the higher-elevation spruce-fir forests (Sibold et al., 2006). The result is that, compared with younger postfire stands of lodgepole pine, old stands of spruce-fir have a high component of standing dead large trees as a result of cumulative mortality from many causes over long time periods, even in the absence of a spruce beetle outbreak (Veblen, 1986, Veblen et al., 1991). Old spruce-fir stands also typically have much more biomass in the understory compared with younger lodgepole pine stands. The net effect of these structural differences is that the magnitude of structural changes caused by spruce beetle outbreak in old spruce-fir forests is likely to be less than that associated with MPB in younger postfire lodgepole pine forests (Figure 6.3a and b).

Nevertheless, similar to MPB outbreaks in lodgepole pine forests, important changes in foliar moisture and chemistry occur in Engelmann spruce following attack by spruce beetles. Immediately after a spruce beetle attack, FMC can be lower, proportions of lignin and cellulose can be higher, and proportions of carbohydrate-based compounds can be lower compared with that in green needles on unattacked trees, each of which can result in increased flammability of the foliage (Page et al., 2014a,b), which may lead to expectations of increased risk of active crown fire immediately following outbreaks (Figure 6.3a). However, any increase in crown flammability is short-lived because foliage on killed trees drops soon after a mass attack. Furthermore, as with MPB outbreaks, any increase in fire hazard would be contingent on the intensity and within-stand synchrony of an outbreak. For decades after high-severity spruce beetle outbreaks, models predict reduced probability of active crown fire because of persistent decreased CBD (Jenkins et al., 2008, DeRose and Long, 2009).

How changes in foliar moisture and chemistry relate to the actual risk of active crown fire and fire behavior has not yet been definitively established, in part because of the limitations of existing fire models and uncertainty about retention of dead needles on killed trees. Given these limitations and knowledge gaps, the results of retrospective studies of actual fires following outbreaks are important to understanding the consequences of beetle activity for fire risk and severity. After a 1940s spruce beetle outbreak that killed most canopy spruce over thousands of hectares in western Colorado, there was no increase in the numbers of fires over the period from 1950 to 1990 compared with unaffected subalpine forests (Bebi et al., 2003). Likewise, stands affected by beetles were unaffected by a low-severity fire that spread through adjacent forests several years after the outbreak subsided (Kulakowski et al., 2003), possibly because of increased moisture on the forest floor following the outbreak, which may have contributed to a proliferation of mesic understory plants (Reid, 1989). In fact, these beetle-affected stands did not burn until the extreme drought of 2002, during which large severe fires affected extensive areas of Colorado, including some spruce-fir stands that had been affected by the 1940s outbreak. The high-severity fires during that extreme drought were substantially affected by neither the 1940s outbreak (Bigler et al., 2005) nor by ongoing spruce beetle outbreaks (Kulakowski and Veblen, 2007). In sum, outbreaks of spruce beetle seem to have little or no effect on the occurrence or severity of fires in spruce-fir forests, primarily because high-severity fires in these forests depend on infrequent, severe droughts (e.g., Schoennagel et al., 2007). Under such extreme weather conditions, changes in fuels resulting from bark beetle outbreaks have only a minor, if any, effect on fire risk.

Why the Apparent Conflict Between Modeling and Observational Results?

As noted previously, fire behavior modeling and retrospective approaches are appropriate for addressing different sets of questions about how bark beetle outbreaks may affect subsequent wildfire activity. Fire behavior models are particularly useful in revealing insights about how beetle-killed fuels might result in uncharacteristic fire behavior at a stand scale that is, in turn, of fundamental importance to firefighter safety. Taken as a whole and extrapolated to larger landscapes, however, modeling-based studies lead to the expectation that, in general, beetle-killed forests should exhibit altered fire behavior (Figure 6.3a). Nevertheless, the pattern emerging from an admittedly small number of observational (retrospective) studies of wildfire activity in beetle-affected forests does not support that expectation (Figure 6.3b). The apparent discrepancy between expectations derived from fire behavior models and observational studies may have numerous, non-mutually exclusive explanations. Most important, these explanations include fine-scale heterogeneity in infestation severity and synchrony or overriding effects of topography and fire weather on patterns of burning. For example, fire weather, especially in high-elevation subalpine forests, strongly influences fire severity (Harvey et al., 2014a,b), fire occurrence (Turner et al., 1999), fire intensity (Bessie and Johnson, 1995), fire spread (Coen, 2005), fire ignition (Bourbonnais et al., 2014), and crown fire behavior (Simard et al., 2011, Schoennagel et al., 2012). During extreme fire weather that promotes high fire activity in subalpine forests, fuels are likely dry enough to promote extensive burning regardless of alterations to fuels as a result of bark beetle infestation. While fire behavior models are essential in conceptualizing potentially important driving factors of wildfire activity, particularly at a stand scale, their lack of validation by field experiments (Cruz and Alexander, 2010) limits their suitability for addressing questions of how bark beetle outbreaks may affect fire occurrence, extent, and severity across larger landscapes.

6.3 HOW DO HIGH-SEVERITY FIRES AFFECT SUBSEQUENT OUTBREAKS?

Fires and other disturbances can affect susceptibility to subsequent outbreaks as well as other disturbances by creating long-lasting legacies of forest structure. The effect of severe stand-replacing fires on subsequent outbreaks of bark beetles has been particularly clear as beetles preferentially attack larger trees and stands in advanced stages of development (Schmid and Frye 1977).

Lodgepole Pine Forests

Tree size is a major determinant of susceptibility to MPBs. Small-diameter trees provide less phloem to support beetle populations, and greater subcortical cooling of small trees contributes to higher mortality of beetles during winter (Safranyik, 2004). Therefore, small-diameter trees have historically been less susceptible to MPB attack, even during outbreaks that may kill most nearby canopy-size trees.

Because the high-severity fires characteristic of lodgepole pine forests in the US Rocky Mountains (Sibold et al., 2006; Veblen and Donnegan, 2006) result in postfire cohorts of small, young trees with thin bark, they can reduce stand susceptibility to MPBs (Kulakowski et al., 2012). However, reduced susceptibility of younger postfire stands was most pronounced for a 1940s/1950s outbreak, less so for a 1980s outbreak, and did not hold true for a 2000s/2010s outbreak. There are alternate but not mutually exclusive explanations for the varying relationship between severe fires and susceptibility to MPBs over the past century.

One possible explanation is that stand age no longer affects susceptibility to outbreak after stands reach a threshold age of >100-150 years (Taylor and Carroll 2004). Another possible explanation is related to the theoretical expectation that tree and stand attributes before a disturbance become less important in determining susceptibility to disturbance as the intensity of that disturbance increases. The warm and dry climate of the 2000s contributed to a high-intensity outbreak and likely stressed host trees and thereby reduced tree resistance to beetle attack (Safranyik 2004). Thus the susceptibility of younger stands during the 2000s/2010s outbreak may not have been reduced by preceding fires because of the very high intensity of that outbreak (Chapman et al., 2012) compared with the 1980s outbreak, which was much less intense (Smith et al., 2012).

Spruce-Fir Forests

Similar to dynamics in lodgepole pine forests, tree size is important in determining susceptibility of spruce to spruce beetle outbreaks because beetles preferentially attack larger trees and stands with structures associated with the latter stages of development (Schmid and Frye 1977). Consequently, young (<80 years) postfire stands of Engelmann spruce in Colorado have been less susceptible to attack by spruce beetle in the nineteenth (Kulakowski and Veblen, 2006) and twentieth centuries (Veblen et al., 1994, Bebi et al., 2003, Kulakowski et al., 2003), even when landscape-level outbreaks killed most large spruce in surrounding stands.

In contrast to the definitive influence of preceding high-severity fires on spruce beetle outbreaks over the past centuries, high-severity fires that occurred >100 years ago did not strongly influence stand structural traits linked to the 2002-2012 bark beetle outbreak (Hart et al., 2014b). This apparent decoupling may be associated with a threshold in stand age, beyond which stand structure is no longer critical in determining susceptibility to outbreak. Alternatively, climate-driven increases in outbreak intensity may be diminishing the importance of tree and stand attributes for outbreaks (Hart et al., 2014b) because warm and dry conditions promote larger beetle populations (Bentz et al. 2010) and decrease tree resistance to beetle attack (Mattson and Haack 1987, Hart et al., 2014a). This hypothesis is supported by numerous small-diameter and suppressed trees that were affected by spruce beetle in the twenty-first century before any eventual host saturation, suggesting that tree-level constraints have been relaxed in comparison with previous outbreaks (Hart et al., 2014b).

Nonbeetle Causes of Mortality

While outbreaks *of Dendroctonus* bark beetles in the Rocky Mountains have received most of the attention in research and management over the past 20 years, recognizing that there is widespread evidence of increasing tree mortality that may be attributed to a variety of biotic and abiotic factors is important. Late twentieth- and early twenty-first-century increases in tree mortality, manifested either as gradual increases in background tree mortality rates or pulses of forest die-off, have been widely documented across the western United States and associated with rising regional temperatures (van Mantgem et al., 2009, Williams et al., 2013). Some, but not all, of these increases in tree mortality are associated with lethal insects. Since the mid-1990s the forests of Colorado have experienced profound pulses of tree mortality coincident with warmer temperatures and episodes of reduced precipitation that have affected all the common tree species of the subalpine forests (Bigler et al., 2007, Worrall et al., 2010, Colorado State Forest Service, 2012, Hanna and Kulakowski 2012; Smith et al., 2015). Sudden and massive mortality of conifers since the mid-1990s in Colorado is well documented in relation to outbreaks of bark beetles: primarily MPB affecting lodgepole pine, limber pine (*Pinus flexilis*), and ponderosa pine (*P. ponderosa*) and spruce bark beetle affecting Engelmann spruce (Chapman et al., 2012, Colorado State Forest Service, 2012). Less well documented is the extensive mortality of subalpine fir attributed to western balsam bark beetle (*Dryocoetes confusus*) and fungal pathogens (Negron and Popp, 2009; Colorado State Forest Service, 2012). However, increases in background tree mortality in the absence of evidence of bark beetle infestation are also evident in subalpine forests in Colorado (Bigler et al., 2007, van Mantgem et al., 2009, Anderegg et al., 2014, Smith et al., 2015). Thus, increased tree mortality related to broad-scale warming but directly mediated by factors other than *Dendroctonus* beetles seems to be altering forest conditions in general across the Rocky Mountain region and focuses our attention on the need to better understand the complex factors that will affect forest resilience under continued climate warming. The potential influence of this mortality on fire regimes is informed by research on how outbreaks affect fire regimes, but it remains a priority area of research in and of itself.

6.4 HOW ARE INTERACTING FIRES AND BARK BEETLES AFFECTING FOREST RESILIENCE IN THE CONTEXT OF CLIMATE CHANGE?

Although thoroughly considering the nuanced links between outbreaks and wildfires is important, recognizing that the increase in both outbreaks (Raffa et al., 2008, Hart et al., 2014a) and wildfires (Dennison et al., 2014) is primarily driven by climatic warming may be as or perhaps even more important. Thus, although outbreaks are not increasing fire frequency or severity, the fact that both disturbances are increasing simultaneously as a result of a common driver leads to an increasing probability that forests will be affected by both of these disturbances in short succession. The occurrence of multiple disturbances in relatively short succession combined with a warming climate potentially may overcome the resilience of particular forest ecosystems and usher in transitions to alternative stable states (Buma and Wessman, 2011). The implications of such compounded disturbances on future forest conditions and resilience hinge on important differences in initial forest conditions, spatial heterogeneity in a biophysical setting, magnitude of the disturbance, and details of climate change (see Box 6.2).

The few studies that have examined the effects of compounded disturbances by outbreaks and fires so far have not reported major differences in regeneration

BOX 6.2 Bark Beetle Outbreaks and the Carbon Cycle

By killing many trees over large areas, bark beetle outbreaks can affect carbon uptake, sequestration, and release. Initially, killed trees stop taking up carbon as photosynthesis ceases; over the following decades, as trees decompose, carbon is released into the atmosphere and the soil (Hicke et al., 2012b). The rate of carbon uptake during postoutbreak stand regeneration depends on the stand conditions at the time of the outbreak, the postdisturbance environment, and the effect of disturbance interactions.

Kurz et al. (2008) modeled extensive and severe bark beetle outbreaks in Canada and hypothesized the effects of outbreaks on the carbon balance to be large and approximately 75% of the average carbon emitted as a result of fire in the largest outbreak year, but they did not measure carbon uptake from stand regeneration. Stinson et al. (2011) found that these forests were a carbon sink (net positive carbon sequestration) in 1990-2008, despite widespread beetle outbreaks and fires. Studies that have focused on postdisturbance regeneration over smaller areas have found that beetle-disturbed stands are carbon sources in the early years (Brown et al., 2010) but often recover to predisturbance levels over 10 to 25 years (Romme et al., 1986, Pfeifer et al., 2011) and in as little as 3-4 years as a result of increased productivity and growth of surviving trees and understory vegetation (Brown et al., 2010, Brown et al., 2012). Ghimire et al. (2015) recently combined postdisturbance forest regrowth trajectories derived from forest inventory data, a process-based carbon cycle model tracking decomposition, and aerial detection survey data to quantify the impact of outbreaks across the western United States. This study reported modeling results that predicted the amount of net carbon release to be large, but somewhat lower than that reported in Canada, likely because of differences in the assumptions of underlying models as well as in outbreaks dynamics between the two regions.

Bark beetle outbreaks clearly have, and will continue to have, important effects on the carbon cycle. However, important challenges remain in understanding the magnitude and duration of these effects, including how tree physiology controls ecosystem carbon fluxes (Frank et al., 2014). Specifically, accurately depicting carbon dynamics following current and future outbreaks depends on an improved understanding of forest regeneration following compounded disturbances and under potentially unfavorable climatic conditions.

compared with regeneration following only fire. For example, regeneration after severe fires in lodgepole pine in the northern Rockies was not affected by prefire MPB outbreaks (Harvey et al., 2014a). Likewise, regeneration after severe fire in 2002 in spruce-fir forests in western Colorado was not affected by spruce beetle outbreak in the 1940s (Kulakowski et al., 2013). Regeneration in most spruce-fir forests was unusually low, however, possibly because of generally unfavorable climatic conditions in the twenty-first century. This general scarcity of regeneration made investigating the consequences of compounded disturbances difficult, and it may be another indicator of overarching climatic influences on disturbance interactions. In another example, regeneration following high-severity fires in Douglas fir forests was generally low in contrast to that in stands burned by low-severity fires, except where those low-severity fires were preceded by Douglas-fir beetles (Harvey et al., 2013).

Tree species vary in how they are affected by compounded disturbances (Buma and Wessman, 2012, Kulakowski et al., 2013). Therefore, compounded disturbances have the potential to fundamentally change forest composition and future forest trajectories. Compounded disturbances can alter post-disturbance regeneration by either reducing seed source or by increasing the intensity of the secondary disturbance (Kulakowski and Veblen, 2007), which in turn may negatively influence soil and other microenvironmental conditions (Fonturbel et al. 2011). These two influences may be of minimal consequence for species that reproduce vegetatively, such as quaking aspen (*Populus tremuloides*), which has been shown to dominate initial regeneration following compounded disturbances, sometimes in stands dominated by conifers before the initial disturbance (Kulakowski et al., 2013). If compounded disturbances become more common under future climate scenarios, quaking aspen and other species that reproduce vegetatively may be favored over those that reproduce exclusively from seed. However, the susceptibility of aspen to climate-induced die-off (Anderegg et al., 2012; Hanna and Kulakowski 2012), as well as predicted reductions in its area of suitable habitat under scenarios of climate warming (Rehfeldt et al., 2009), lead to considerable uncertainty about how compounded disturbances under a varying climate will affect future forest trajectories.

6.5 CONCLUSIONS

In lodgepole pine and spruce-fir forests of the Rocky Mountains, effects of bark beetle outbreaks on fuels are complex. The magnitude and heterogeneity of changes in fuels following outbreaks vary with stand structure before the outbreak, the severity of the outbreak, and the rate of the outbreak (Jenkins et al., 2014). Important questions about how biophysical setting may affect the timing and duration of post-outbreak stages and about the importance of within-stand outbreak synchrony on fuel configurations remain. Still, much has been learned about how outbreaks affect fuels. Importantly, bark beetle outbreaks not only reduce FMC, modify foliar chemistry, and increase the volume of dead wood, which can promote several aspects of wildfires; but outbreaks also reduce canopy density, which lowers the amount of available fuel and can thereby decrease the probability of active crown fires and the likelihood of large, severe fires (Simard et al., 2011, Schoennagel et al., 2012). The relative importance of these contradictory effects during different phases of outbreaks continues to be an active area of research. Empirical studies of fuels immediately following outbreaks (Simard et al., 2011) and retrospective studies of fires in forests recently affected by outbreaks (Kulakowski and Veblen, 2007, Harvey et al., 2014a,b) suggest that outbreaks do not substantially increase—and may actually decrease—the risk of high-severity fires, even during and immediately following outbreaks. Likewise, there is general scientific agreement that the risk of active, high-severity crown fires decreases in the years to decades following outbreaks because of reduced CBD (Romme et al. 2006; Jenkins et al., 2008, 2012, 2014; Kaufmann et al. 2008; Simard et al., 2008, Hicke et al., 2012a, Black et al., 2013) (Figure 6.3b). Although this chapter has mainly focused on lodgepole pine and spruce forests, we note that similar conclusions are emerging for other forest types. For example, in upper montane forests dominated by Douglas-fir, prefire outbreak severity did not increase any measure of fire severity (Harvey et al., 2013).

Given the primacy of the importance of weather and climate in overall fire risk across broad regions and forest types (Dennison et al., 2014), any effects of bark beetle outbreaks on fire regimes should be considered in the context of climatic variability and the relative importance of climate versus changes in fuels associated with outbreaks. Research conducted thus far has consistently indicated that weather and climate are more important than the effects of outbreaks in determining fire risk and behavior in Rocky Mountain subalpine forests (Kulakowski and Jarvis, 2011, Schoennagel et al., 2012, Harvey et al., 2013, 2014a,b).

Severe wildfires, such as those that have been burning across western North America over recent decades, have been shown to decrease subsequent susceptibility to bark beetle outbreaks in lodgepole pine (Kulakowski et al., 2012) and spruce-fir forests (Veblen et al., 1994, Kulakowski et al., 2003, Bebi et al., 2003). This implies that climate-driven increases in wildfires have the potential to buffer ecosystems against the risk or severity of future bark beetle outbreaks. However, recent research also suggests that the modulating effect of fires on susceptibility to outbreaks may be contingent on current and future climate influences on beetle populations and tree resistances (Kulakowski et al., 2012, Hart et al., 2014b). Furthermore, prior occurrence of a large severe bark beetle outbreak may sufficiently deplete host populations to reduce the probability of the occurrence of a subsequent outbreak for periods of 70 years or more (Hart et al., 2015). The consequences of these interactions under future climate scenarios can be examined through simulation modeling. For example, under a weather scenario of sufficiently dry climate in the future, reduction of host tree populations may lower the probability of bark beetle outbreaks in some habitats (Temperli et al., in press).

As various forest disturbances become more frequent and extensive, understanding how multiple disturbances interact is of increasing importance. While additional questions remain, the best available science indicates that outbreaks of bark beetles do not increase the risk of high-severity fires in lodgepole pine and spruce-fir forests of the Rocky Mountains. Furthermore, the effects of outbreaks are much less important to fire risk than are weather and climate. By contrast, severe wildfires have been shown to reduce subsequent susceptibility to outbreaks in both forest types. The current state of knowledge does not support the common assumption that increases in bark beetle activity have resulted in increased wildfire activity. Therefore policy discussions should focus on societal adaptation to the effects of the underlying driving factor of increased tree mortality from insects and from burning: climate warming.

ACKNOWLEDGMENTS

Some of the research included in this review was supported by National Science Foundation Awards (no. 1262687 and 1262691). The authors thank R. Andrus, N. Gill, and N. Mietkiewicz for helpful comments on this work.

REFERENCES

Alexander, M.E., Cruz, M.G., 2013. Are the applications of wildland fire behaviour models getting ahead of their evaluation again? Environ. Model. Softw. 41, 65–71.

Anderegg, W.R.L., Berry, J.A., Smith, D.D., Sperry, J.S., Anderegg, L.D.L., Field, C.B., 2012. The roles of hydraulic and carbon stress in a widespread climate-induced forest die-off. Proc. Natl. Acad. Sci. U. S. A. 109, 233–237.

Anderegg, W.R.L., Anderegg, L.D.L., Berry, J.A., Field, C.B., 2014. Loss of whole-tree hydraulic conductance during severe drought and multi-year forest die-off. Oecologia 175, 11–23.

Bebi, P., Kulakowski, D., Veblen, T.T., 2003. Interactions between fire and spruce beetles in a subalpine Rocky Mountain forest landscape. Ecology 84, 362–371.

Bentz, B.J., Régnière, J., Fettig, C.J., Hansen, E.M., Hayes, J.L., Hicke, J.A., Kelsey, R.G., Negrón, J.F., Seybold, S.J., 2010. Climate change and bark beetles of the Western United States and Canada: direct and indirect effects. BioScience 60, 602–613.

Bessie, W.C., Johnson, E.A., 1995. The relative importance of fuels and weather on fire behavior in subalpine forests. Ecology 76, 747.

Bigler, C., Kulakowski, D., Veblen, T.T., 2005. Multiple disturbance interactions and drought influence fire severity in Rocky Mountain subalpine forests. Ecology 86, 3018–3029.

Bigler, C., Gavin, D., Gunning, C., Veblen, T.T., 2007. Drought induces lagged tree mortality in a subalpine forest in the Rocky Mountains. Oikos 116, 1983–1994.

Black, S.H., Kulakowski, D., Noon, B.R., DellaSala, D.A., 2013. Do bark beetle outbreaks increase wildfire risks in the central U.S. Rocky Mountains? Implications from recent research. Nat. Area J. 33, 59–65.

Bourbonnais, M.L., Nelson, T.A., Wulder, M.A., 2014. Geographic analysis of the impacts of mountain pine beetle infestation on forest fire ignition. Can. Geogr. 58, 188–202.

Brown, M., Black, T.A., Nesic, Z., Foord, V.N., Spittlehouse, D.L., Fredeen, A.L., Grant, N.J., Burton, P.J., Trofymow, J.A., 2010. Impact of mountain pine beetle on the net ecosystem production of lodgepole pine stands in British Columbia. Agric. For. Meteorol. 150, 254–264.

Brown, M.G., Black, T.A., Nesic, Z., Fredeen, A.L., Foord, V.N., Spittlehouse, D.L., Bowler, R., Burton, P.J., Trofymow, J.A., Grant, N.J., Lessard, D., 2012. The carbon balance of two lodgepole pine stands recovering from mountain pine beetle attack in British Columbia. Agric. For. Meteorol. 153, 82–93.

Buma, B., Wessman, C.A., 2011. Disturbance interactions can impact resilience mechanisms of forests. Ecosphere. 2, Article 64.

Buma, B., Wessman, C.A., 2012. Differential species responses to compounded perturbations and implications for landscape heterogeneity and resilience. For. Ecol. Manag. 266, 25–33.

Chapman, T., Veblen, T.T., Schoennagel, T., 2012. Spatio-temporal patterns of mountain pine beetle activity in the southern Rocky Mountains. Ecology 93, 2175–2185.

Coen, J.L., 2005. Simulation of the Big Elk Fire using coupled atmosphere-fire modeling. Int. J. Wildland Fire 14, 49–59.

Colorado State Forest Service, 2012. Report on the health of Colorado's forests: forest stewardship through active management (http://csfs.colostate.edu/pdfs/137233-forestreport-12-www.pdf).

Cruz, M.G., Alexander, M.E., 2010. Assessing crown fire potential in coniferous forests of western North America: a critique of current approaches and recent simulation studies. Int. J. Wildland Fire 19, 377–398.

Dahlsten, D.L., 1982. Relationships between bark beetles and their natural enemies. In: Mitton, J.B., Sturgeon, K.B. (Eds.), Bark Beetles in North American Conifers. University of Texas Press, Austin, pp. 140–182.

Dennison, P.E., Brewer, S.C., Arnold, J.D., Moritz, M.A., 2014. Large wildfire trends in the western United States, 1984–2011. Geophys. Res. Lett. 41, 2928–2933.

DeRose, R.J., Long, J.N., 2009. Wildfire and spruce beetle outbreak: simulation of interacting disturbances in the central Rocky Mountains. Ecoscience 16, 28–38.

DeRose, R.J., Long, J.N., 2014. Resistance and resilience: a conceptual framework for silviculture. Forest Sci. 60 (6), 1205–1212.

Eisenhart, K., Veblen, T.T., 2000. Dendrochronological detection of spruce bark beetle outbreaks in northwestern Colorado. Can. J. For. Res. 30, 1788–1798.

Fonturbel, M.T., Vega, J.A., Perez-Gorostiaga, P., Fernandez, C., Alonso, M., Cuinas, P., Jimenez, E., 2011. Effects of soil burn severity on germination and initial establishment of mari-time pine seedlings, under greenhouse conditions, in two contrasting experimentally burned soils. Int. J. Wildland Fire 20, 209–222.

Frank, J.M., Massman, W.L., Ewers, B.E., Huckaby, L.S., Negrón, J.F., 2014. Ecosystem CO2/H2O fluxes are explained by hydraulically limited gas exchange during tree mortality from spruce bark beetles. J. Geophys. Res. Biogeosci. 119, 1195–1215.

Ghimire, B., Williams, C.A., Collatz, G.J., Vanderhoof, M., Rogan, J., Kulakowski, D., Masek, J.G., 2015. Large carbon release from bark beetle outbreaks across Western United States imposes climate feedback. Global Change Biol. http://dx.doi.org/10.1111/gcb.12933.

Hanna, P., Kulakowski, D., 2012. The influences of climate on aspen dieback. For. Ecol. Manag. 274, 91–98.

Hardy, C.C., 2005. Wildland fire hazard and risk: problems, definitions, and context. For. Ecol. Manag. 211, 73–83.

Hart, S.J., Veblen, T.T., Eisenhart, K.S., Jarvis, D., Kulakowski, D., 2014a. Drought induces spruce beetle (*Dendroctonus rufipennis*) outbreaks across northwestern Colorado. Ecology 95, 930–939.

Hart, S.J., Veblen, T.T., Kulakowski, D., 2014b. Do tree and stand-level attributes determine susceptibility of spruce-fir forests to spruce beetle outbreaks? For. Ecol. Manag. 318, 44–53.

Hart, S.J., Veblen, T.T., Mietkiewicz, N., Kulakowski, D., 2015. Negative feedbacks on bark beetle outbreaks: Widespread and severe spruce beetle infestation restricts subsequent infestation. PLoS One, in press.

Harvey, B.J., Donato, D.C., Romme, W.H., Turner, M.G., 2013. Influence of recent bark beetle outbreak on fire severity and postfire tree regeneration in montane Douglas-fir forests. Ecology 94, 2475–2486.

Harvey, B.J., Donato, D.C., Turner, M.G., 2014a. Recent mountain pine beetle outbreaks, wildfire severity, and postfire tree regeneration in the US Northern Rockies. In: Proceedings of the National Academy of Sciences of the United States of America www.pnas.org/cgi/doi/10.1073/pnas.1411346111.

Harvey, B.J., Donato, D.C., Romme, W.H., Turner, M.G., 2014b. Fire severity and tree regeneration following bark beetle outbreaks: the role of outbreak stage and burning conditions. Ecol. Appl. 24, 1608–1625.

Hicke, J.A., Johnson, M.C., Hayes, J.L., Preisler, H.K., 2012a. Effects of bark beetle-caused tree mortality on wildfire. For. Ecol. Manag. 271, 81–90.

Hicke, J.A., Allen, C.D., Desai, A.R., Dietze, M.C., Hall, R.J., Hogg, E.H., Kashian, D.M., Moore, D., Raffa, K.F., Sturrock, R.N., Vogelmann, J., 2012b. Effects of biotic disturbances on forest carbon cycling in the United States and Canada. Glob. Chang. Biol. 18, 7–34.

Hoffman, C.M., Morgan, P., Mell, W., Parsons, R., Strand, E., Cook, S., 2012. Numerical simulation of crown fire hazard following bark beetle-caused mortality in lodgepole pine forests. For. Sci. 58, 178–188.

Hoffman, C.M., Morgan, P., Mell, W., Parsons, R., Strand, E., Cook, S., 2013. Surface fire intensity influences simulated crown fire behavior in lodgepole pine forests with recent mountain pine beetle-caused tree mortality. For. Sci. 59, 390–399.

Jackson, K.J., Wohl, E., 2015. Instream wood loads in montane forest streams of the Colorado Front Range, USA. Geomorphology 234, 161–170.

Jarvis, D., Kulakowski, D., 2015. Long-term history and synchrony of mountain pine beetle outbreaks. J. Biogeogr. http://dx.doi.org/10.1111/jbi.12489.

Jenkins, M.J., Hebertson, E., Page, W., Jorgensen, C.A., 2008. Bark beetles, fuels, fires and implications for forest management in the Intermountain West. For. Ecol. Manag. 254, 16–34.

Jenkins, M.J., Page, W.G., Hebertson, E.G., Alexander, M.E., 2012. Fuels and fire behavior dynamics in bark beetle-attacked forests in Western North America and implications for fire management. For. Ecol. Manag. 275, 23–34.

Jenkins, M.J., Runyon, J.B., Fettig, C.J., Page, W.G., Bentz, B.J., 2014. Interactions among the mountain pine beetle, fires, and fuels. For. Sci. 60, 489–501.

Jolly, W.M., Parsons, R.A., Hadlow, A.M., Cohn, G.M., McAllister, S.S., Popp, J.B., Hubbard, R.M., Negron, J.F., 2012. Relationships between moisture, chemistry, and ignition of Pinus contorta needles during the early stages of mountain pine beetle attack. For. Ecol. Manag. 269, 52–59.

Kaufmann, M.R., Aplet, G.H., Babler, M., Baker, W.L., Bentz, B., Harrington, M., Hawkes, B.C., Huckaby, L.S., Jenkins, M.J., Kashian, D.M., Keane, R.E., Kulakowski, D., McHugh, C., Negron, J., Popp, J., Romme, W.H., Schoennagel, T., Shepperd, W., Smith, F.W., Sutherland, E.K., Tinker, D., Veblen, T.T., 2008. The Status of Our Scientific Understanding of Lodgepole Pine and Mountain Pine Beetles - A Focus on Forest Ecology and Fire Behavior. The Nature Conservancy, Arlington, VA. GFI technical report 2008–2.

Klutsch, J.G., Negron, J.F., Costello, S.L., Rhoades, C.C., West, D.R., Popp, J., Caissie, R., 2009. Stand characteristics and downed woody debris accumulations associated with a mountain pine beetle (Dendroctonus ponderosae Hopkins) outbreak in Colorado. For. Ecol. Manag. 258, 641–649.

Koplin, J.R., Baldwin, P.H., 1970. Woodpecker predation on an endemic population of Engelmann spruce beetles. Am. Midl. Nat. 83, 510–515.

Kulakowski, D., Jarvis, D., 2011. The influence of mountain pine beetle outbreaks on severe wildfires in northwestern Colorado and southern Wyoming: a look at the past century. For. Ecol. Manag. 261, 1686–1696.

Kulakowski, D., Veblen, T.T., 2006. The effect of fires on susceptibility of subalpine forests to a 19th century spruce beetle outbreak in western Colorado. Can. J. For. Res. 36, 2974–2982.

Kulakowski, D., Veblen, T.T., 2007. Effect of prior disturbances on the extent and severity of wildfire in Colorado subalpine forests. Ecology 88, 759–769.

Kulakowski, D., Veblen, T.T., Bebi, P., 2003. Effects of fire and spruce beetle outbreak legacies on the disturbance regime of a subalpine forest in Colorado. J. Biogeogr. 30, 1445–1456.

Kulakowski, D., Jarvis, D., Veblen, T.T., Smith, J., 2012. Stand-replacing fires reduce susceptibility to mountain pine beetle outbreaks in Colorado. J. Biogeogr. 39, 2052–2060.

Kulakowski, D., Matthews, C., Jarvis, D., Veblen, T.T., 2013. Compounded disturbances in subalpine forests in western Colorado favor future dominance by quaking aspen (Populus tremuloides). J. Veg. Sci. 24, 168–176.

Kurz, W.A., Dymond, C.C., Stinson, G., Rampley, G.J., Neilson, E.T., Carroll, A.L., Ebata, T., Safranyik, L., 2008. Mountain pine beetle and forest carbon feedback to climate change. Nature 452, 987–990.

Lynch, H.J., Renkin, R.A., Crabtree, R.L., Moorcroft, P.R., 2006. The influence of previous mountain pine beetle (Dendroctonus ponderosae) activity on the 1988 Yellowstone fires. Ecosystems 9, 1318–1327.

Maness, H., Kushner, P.J., Fung, I., 2012. Summertime climate response to mountain pine beetle disturbance in British Columbia. Nat. Geosci. 6, 65–70.

Mattson, W.J., Haack, R.A., 1987. The role of drought in outbreaks of plant-eating insects. BioScience 27 (2), 110–119.

Negron, J.F., Popp, J.B., 2009. The flight periodicity, attack patterns, and life history of Dryocoetes confusus Swaine (Coleoptera: Curculionida: Scolytinae), the western basalm bark beetle, in north central Colorado. West. N. Am. Nat. 69 (4), 447–458.

Omi, P.N., 1997. Final Report: fuels modification to reduce large fire probability. Submitted to US Department of Interior, Fire Research Committee, Colorado State University, Fort Collins.

Page, W.G., Jenkins, M.J., 2007. Mountain pine beetle-induced changes to selected lodgepole pine fuel complexes within the intermountain region. For. Sci. 53, 507–518.

Page, W.G., Jenkins, M.L., Runyon, J.B., 2012. Mountain pine beetle attack alters the chemistry and flammability of lodgepole pine foliage. Can. J. For. Res. 42, 1631–1647.

Page, W.G., Jenkins, M.J., Alexander, M.E., 2013a. Foliar moisture content variations in lodgepole pine over the diurnal cycle during the red stage of mountain pine beetle attack. Environ. Model. Softw. 49, 98–102.

Page, W.G., Alexander, M.E., Jenkins, M.J., 2013b. Wildfire's resistance to control in mountain pine beetle-attacked lodgepole pine forests. For. Chron. 89, 783–794.

Page, W.G., Jenkins, M.J., Runyon, J.B., 2014a. Spruce beetle-induced changes to Engelmann spruce foliage flammability. For. Sci. 60, 691–702.

Page, W.G., Jenkins, M.J., Alexander, M.E., 2014b. Crown fire potential in lodgepole pine forests during the red stage of mountain pine beetle attack. Forestry 87, 347–361.

Parker, T.J., Clancy, K.M., Mathiasen, R.L., 2006. Interactions among fire, insects and pathogens in coniferous forests of the interior western United States and Canada. Agric. For. Entomol. 8, 167–189.

Pfeifer, E.M., Hicke, J.A., Meddens, A.J.H., 2011. Observations and modeling of 336 aboveground tree carbon stocks and fluxes following a bark beetle outbreak in the western United States. Glob. Chang. Biol. 17, 339–350.

Raffa, K.F., Aukema, B.H., Bentz, B.J., Carroll, A.L., Hicke, J.A., Turner, M.G., Romme, W.H., 2008. Cross-scale drivers of natural disturbances prone to anthropogenic amplification: the dynamics of bark beetle eruptions. Bioscience 58, 501–517.

Rehfeldt, G.E., Ferguson, D.E., Crookston, N.L., 2009. Aspen, climate, and sudden decline in western USA. For. Ecol. Manag. 258, 2353–2364.

Reid, M., 1989. The response of understory vegetation to major canopy disturbance in the subalpine forests of Colorado. Masters Thesis, Geography, University of Colorado, Boulder.

Renkin, R.A., Despain, D.G., 1992. Fuel moisture, forest type, and lightning-caused fire in Yellowstone National Park. Can. J. For. Res. 22, 37–45.

Romme, W.H., Despain, D.G., 1989. Historical perspective on the Yellowstone Fires of 1988. Bioscience 39, 695–699.

Romme, W.H., Knight, D.H., Yavitt, J.B., 1986. Mountain pine beetle outbreaks in the Rocky Mountains: regulators of primary productivity? Am. Nat. 127, 484–494.

Romme, W.H., Clement, J., Hicke, J., Kulakowski, D., MacDonald, L.H., Schoennagel, T.L., Veblen, T.T., 2006. Recent Forest Insect Outbreaks and Fire Risk in Colorado Forests: A Brief Synthesis of Relevant Research, Colorado Forest Restoration Institute, Fort Collins, CO.

Rothermel, R.C., 1972. A Mathematical Model for Predicting Fire Spread in Wildland Fuels. Res. Pap. INT-115. U.S. Department of Agriculture, Intermountain Forest and Range Experiment Station, Ogden, UT, 40 p.

Saab, V.A., Latif, Q.S., Rowland, M.M., Johnson, T.N., Chalfoun, A.D., Buskirk, S.W., Heyward, J.E., Dresser, M.A., 2014. Ecological consequences of mountain pine beetle outbreaks for wildlife in western North American forests. For. Sci. 60, 539–559.

Safranyik, L., 2004. Mountain pine beetle epidemiology in lodgepole pine. In: Ke lowna B.C., Shore, T.L., Brooks, J.E., Stone, J.E. (Eds.), Mountain pine beetle symposium: Challenges and solutions, October 30–31, 2003. Natural Resources Canada, Canadian Forest Service, Pacific Forestry Centre, Victoria, B.C. Information Report BC-X-399, pp. 33–40.

Schmid, J.M., Amman, G.D., 1992. *Dendroctonus* beetles and old-growth forests in the Rockies. In: Kaufmann, M.R., Moir, W.H., Bassett, R.L. (technical coordinators), Old-Growth Forests of the Southwest and Rocky Mountain Regions: Proceedings of a Workshop, pages 51–59. USDA Forest Service General Technical Report RM-213.

Schmid, J.M., Frye, R.H., 1977. Spruce Beetle in the Rockies. General Technical Report RM-49. USDA Forest Service, Fort Collins, CO, USA.

Schoennagel, T., Veblen, T.T., Kulakowski, D., Holz, A., 2007. Multidecadal climate variability and climate interactions affect subalpine fire occurrence, western Colorado (USA). Ecology 88, 2891–2902.

Schoennagel, T., Veblen, T.T., Negron, J.F., Smith, J.M., 2012. Effects of mountain pine beetle on fuels and expected fire behavior in lodgepole pine forests, Colorado, USA. PLoS One 7, e30002.

Schroeder, D., Mooney, C., 2012. Fire behavior in simulated mountain pine beetle-killed stands. Final Report, Wildfire Operations Research, FP Innovations. Hinton, Alberta.

Scott, J.H., Reinhardt, E.D., 2001. Assessing crown fire potential by linking models of surface and crown fire behavior. Res. Pap. RMRS-RP-29. U.S. Department of Agriculture, Forest Service, Rocky Mountain Research Station, Fort Collins, CO, 59 p.

Sherriff, R.L., Platt, R.V., Veblen, T.T., Schoennagel, T.L., Gartner, M.H., 2014. Historical, observed, and modeled wildfire severity in montane forests of the Colorado Front Range. PLoS One 9, e106971.

Sibold, J.S., Veblen, T.T., Gonzalez, M.E., 2006. Spatial and temporal variation in historic fire regimes in subalpine forests across the Colorado Front Range in Rocky Mountain National Park. J. Biogeogr. 32, 631–647.

Simard, M., Powell, E.N., Griffin, J.W., Raffa, K.F., Turner, M.G., 2008. Annotated Bibliography for Forest Managers on Fire-bark Beetle Interactions. USFS Western Wildlands Environmental Threats Assessment Center, Prineville, Ore.

Simard, M., Romme, W.H., Griffin, J.M., Turner, M.G., 2011. Do bark beetle outbreaks change the probability of active crown fire in lodgepole pine forests? Ecol. Monogr. 81, 3–24.

Simard, M., Romme, W.H., Griffin, J.M., Turner, M.G., 2012. Do mountain pine beetle outbreaks change the probability of active crown fire in lodgepole pine forests? Reply. Ecology 93, 946–950.

Smith, J.M., Hart, S.J., Chapman, T.B., Veblen, T.T., Schoennagel, T., 2012. Dendroecological reconstruction of 1980s mountain pine beetle outbreak in lodgepole pine forests in northwestern Colorado. Ecoscience 19, 113–126.

Smith, J.M., Paritsis, J., Veblen, T.T., Chapman, T.B., 2015. Permanent forest plots show accelerating tree mortality in subalpine forests of the Colorado Front Range from 1982 to 2013. For. Ecol. Manag. 341, 8–17.

Stinson, G., Kurz, W.A., Smyth, C.E., Neilson, E.T., Dymond, C.C., Metsaranta, J.M., Boisvenue, C., Rampley, G.J., Li, Q., White, T.M., Blain, D., 2011. An inventory-based analysis of Canada's managed forest carbon dynamics, 1990 to 2008. Glob. Chang. Biol. 17, 2227–2244.

Taylor, S.W., Carroll, A.L., 2004. Disturbance, forest age dynamics, and mountain pine beetle outbreaks in BC: a historical perspective. In: Shore, T.L., Brooks, J.E., Stone, J.E. (Eds.), Challenges and Solutions. Proc. of the Mountain Pine Beetle Symp., Kelowna, BC, Canada, October 30–31, 2003, Canadian Forest Service, Pacific Forestry Centre, Information Report BC-X-399. NRC Research Press, pp. 41–51.

Temperli, C., Veblen, T.T., Hart, S.J., Kulakowski, D., Tepley, A.J. Interactions among spruce beetle disturbance, climate change and forest dynamics captured by a forest landscape model, in press.

Turner, M.G., 2010. Disturbance and landscape dynamics in a changing world. Ecology 91, 2833–2849.

Turner, M.G., Gardner, R.H., Romme, W.H., 1999. Prefire heterogeneity, fire severity, and early postfire plant reestablishment in subalpine forests of Yellowstone National Park, Wyoming. Int. J. Wildland Fire 9, 21–36.

van Mantgem, P.J., Stephenson, N.L., Byrne, J.C., Daniels, L.D., Franklin, J.F., Fulé, P.Z., Harmon, M.E., Larson, A.J., Smith, J.M., Taylor, A.H., Veblen, T.T., 2009. Widespread increase of tree mortality rates in the western United States. Science 323, 521–524.

Veblen, T.T., 1986. Age and size structure of subalpine forests in the Colorado Front Range. Bull. Torrey Bot. Club 113, 225–240.

Veblen, T.T., Hadley, K.S., Reid, M.S., Rebertus, A.J., 1991. The response of subalpine forests to spruce beetle outbreak in Colorado. Ecology 72, 213–231.

Veblen, T.T., Hadley, K.S., Nel, E.M., Kitzberger, T., Reid, M., Villalba, R., 1994. Disturbance regime and disturbance interactions in a Rocky Mountain subalpine forest. J. Ecol. 82, 125–135.

Veblen, T.T., Donnegan, J.A., 2006. Historical Range of Variability of Forest Vegetation of the National Forests of the Colorado Front Range. USDA Forest Service, Rocky Mountain Region and the Colorado Forest Restoration Institute, Fort Collins, 151 pages.

White, P.S., Pickett, S.T.A., 1985. Natural disturbance and patch dynamics: an introduction. In: Pickett, S.T.A., White, P.S. (Eds.), The Ecology of Natural Disturbance and Patch Dynamics. Academic, New York, pp. 3–13.

Williams, A.P., Allen, C.D., Macalady, A.K., Griffin, D., Woodhouse, C.A., Meko, D.M., Swetnam, T.W., Rauscher, S.A., Seager, R., Grissino-Mayer, H.D., Dean, J.S., Cook, E.R., Gangodagamage, C., Cai, M., McDowell, N.G., 2013. Temperature as a potent driver of regional forest drought stress and tree mortality. Nat. Clim. Chang. 3, 292–297.

Worrall, J.J., Marchetti, S.B., Egeland, L., Mask, R.A., Eager, T., Howell, B., 2010. Effects and etiology of sudden aspen decline in southwestern Colorado, USA. For. Ecol. Manag. 260, 638–648.

Section 2

Global and Regional Perspectives on Mixed- and High-Severity Fires

Chapter 7

High-Severity Fire in Chaparral: Cognitive Dissonance in the Shrublands

Richard W. Halsey[1] **and Alexandra D. Syphard**[2]
[1]*Chaparral Institute, Escondido, CA, USA,* [2]*Senior Research Ecologist, Conservation Biology Institute, Corvallis, OR, USA*

7.1 CHAPARRAL AND THE FIRE SUPPRESSION PARADIGM

The conflict between facts and beliefs concerning fire in California's native shrublands is an example of cognitive dissonance—the psychological discomfort caused when an individual is confronted with new facts or ideas that are in conflict with currently held opinions (Festinger, 1957).

The most characteristic native shrubland in California is chaparral, a drought-hardy plant community composed of such iconic species as manzanita (*Arctostaphylos* sp.) and ceanothus (*Ceanothus* sp.) (Figure 7.1). Once the preferred habitat of the California grizzly bear (*Ursus arctos californicus*), chaparral covers many of the state's hills and mountains with rich biodiversity that reaches its peak on the central coast. Chaparral is also the most extensive vegetation characterizing the California Floristic Province and extends north to southern Oregon, south into Baja California, and as disjunct patches in central and southeastern Arizona and northern Mexico (Keeley, 2000). Although often portrayed as a fire-adapted ecosystem, a more accurate description is one adapted to a particular fire pattern or regime that is characterized by large, infrequent, and high-intensity fires (Keeley et al., 2012). Increase the frequency, reduce the intensity, or change the seasonality of fire and chaparral species can be eliminated, often replaced by nonnative weeds and grasses. As ignitions have increased as a result of human activity, chaparral is being threatened by too much fire in much of its range, particularly in southern California.

The role fire plays in chaparral is often misunderstood by policymakers, land and fire managers, forest scientists, and the public (Keeley et al., 2012). The primary cause of this misunderstanding is a powerful belief system that has formed around what can be characterized as the *fire suppression paradigm*,

FIGURE 7.1 Chaparral is a unique plant community characterized by large, contiguous stands of drought-hardy shrubs, a Mediterranean-type climate, and infrequent, high-intensity/high-severity crown fires (photo: R.W. Halsey).

resulting in cognitive dissonance as new scientific information has emerged. The fire suppression paradigm asserts that a century of fire suppression policy has eliminated fires and allowed vegetation (fuels) to accumulate to unnatural levels so that today when wildfires begin they burn uncontrollably, often producing catastrophic effects (Keeley et al., 1999). For many plant communities, especially chaparral, little could be further from the truth. Data for the past 100 years show that despite a policy of fire suppression, wildfires have not been able to be eliminated in most southern California landscapes; in fact, fires are more common today than historically. Because of this misconception about fire suppression, managers have been trained to believe that wildfires are fuel-driven events and, as a consequence, can be controlled by modifying vegetation.

Deeply embedded in the paradigm is the preconception that small, low-intensity/low-severity surface fires are natural and large, high-intensity/high-severity crown fires are not. When a high-severity fire burns more than ~40 ha (100 acres), it is often considered a direct result of past fire suppression. The paradigm was originally developed to describe the surface fire regime found in lightning-saturated, dry ponderosa pine (*Pinus ponderosa*) forests of the Southwest, where it is a relatively accurate representation (Steel et al., 2015).

Because the fire suppression paradigm is forest-centric, understory shrubs and small trees are viewed as fuel rather than important components of habitat. This has led to a set of values, facilitated by lumber and ranching interests, that view chaparral as "worthless brush," an "invader" of forests and rangeland, and an "unsightly menace" (Halsey, 2011). The bias has led to other pejorative characterizations such as the erroneous claim that chaparral plants are so pyrogenic that they are literally "oozing combustible resins" (Shea, 2008). The paradigm has effectively demonized a native ecosystem that supports significant biodiversity.

The key point is that chaparral fires are unlike forest fires, yet forest fire ecology has been misapplied to explain how fire should burn in chaparral.

Clearing up this confusion is one of the reasons this chapter was written. The basic facts about chaparral fires can be summarized as follows:

- Fire suppression has not caused excessive amounts of chaparral to accumulate (Keeley et al., 1999).
- Fire suppression has played a critical role in protecting many chaparral stands from ecological damage resulting from excessive fire.
- Infrequent, large, high-intensity crown fires are natural in chaparral (Keeley and Zedler 2009).
- There are few, if any, justifiable ecological/resource benefits in conducting prescribed burning or other vegetation treatments in chaparral (Keeley et al., 2009a).

Research over the past two decades has rejected the fire suppression paradigm when applied to ecosystems subject to crown fires, especially shrublands like chaparral. Not surprisingly, the cognitive dissonance caused by this research (e.g., Conard, S.G. and Weise, D.R., 1998, Mensing et al., 1999, Keeley et al., 1999, Keeley and Zedler, 2009, Lombardo et al., 2009) has fostered resistance by the supporters of the challenged paradigm (e.g., Minnich 2001). Consequently, it continues to influence public policy and opinion about chaparral specifically and wildfire in general. But, as the evidence has accumulated, the fire suppression paradigm is slowly shifting to a new one that acknowledges that infrequent, large, high-intensity crown fires do in fact represent the natural fire regime for chaparral and that weather, not fuel type, is the most important variable controlling fire intensity, spread, and size (e.g., Moritz et al., 2004, Keeley and Zedler, 2009).

7.2 THE FACTS ABOUT CHAPARRAL FIRES: THEY BURN INTENSELY AND SEVERELY

The natural, physical structure of chaparral shrubs (contiguous cover, dense accumulation of fine leaves and stems, and retention of dead wood) and the seasonal pattern of drought that includes low humidity, high temperatures, and low live fuel moistures create conditions favoring high-intensity crown fires (Figure 7.2).

Crown fires are those that burn into the canopies of the dominant vegetation. These are opposed to surface or understory fires that burn vegetation close to the ground. *Surface fires* are common in certain forested ecosystems where there is a distinct separation between understory vegetation and the tree canopy. Chaparral creates a contiguous fuel bed from the ground up that makes high-intensity crown fires inevitable.

Fire intensity represents the energy released during various phases of a fire. High-intensity fires typically consume most of the living, aboveground plant material, leaving behind only charred stems and branches.

Fire-severity is also used to describe wildland fire but in relation to how fire intensity affects ecosystems. It is typically measured by the amount of organic material consumed by the flames (above- and belowground), or plant mortality. The manner in which fire intensity and severity are used interchangeably by

FIGURE 7.2 The natural, physical structure of chaparral shrubs (contiguous fuel from the ground to the crown) and the seasonal pattern of drought create conditions favoring high-intensity/high-severity crown fires (photo: R.W. Halsey).

different authors sometimes leads to considerable confusion (Keeley, 2009). For chaparral, however, severity measures may not be particularly helpful because high-intensity chaparral fires typically burn all the aboveground living material, leaving behind only dead, charred shrub skeletons. Fire severity has been measured by the twig diameter remaining on the terminal branches of shrub skeletons and has been shown to correlate with one measure of fire intensity (Moreno and Oechel, 1989). Even though the mature, aboveground forms of some plant species are killed, the belowground portions remain alive as lignotubers that resprout vigorously within a few weeks after the fire. In the first year after fire, massive numbers of seeds from fire-killed obligate seeding shrubs and fire-following annuals are stimulated by fire cues to germinate in the postfire environment (Keeley, 1987, Keeley and Keeley, 1987). Obligate seeding shrubs are nonsprouting species, like many Ceanothus and manzanita species, that require a fire cue for seed germination. As long as fire arrives above the lower limit of the natural fire return interval of 30-40 years, the severely burned postfire chaparral ecosystem is extraordinarily resilient and vibrant (Figure 7.3).

The size of chaparral fires varies, but the seasonal occurrence of high winds, usually from September through December at the end of California's drought period, nearly guarantees periodic large, high-intensity fire events across the shrubland landscape. The historical, natural occurrence of such large crown fires two to three times per century has been confirmed by multiple investigators studying charcoal sediments (Mensing et al., 1999), tree rings of big-cone Douglas-fir (*Pseudotsuga macrocarpa*) that occur in small populations on steep slopes within the chaparral (Lombardo et al., 2009), and historic records (Keeley and Zedler, 2009).

Large crown fires that have historically burned with high intensity characterize all Mediterranean-type climate shrublands around the world (California,

FIGURE 7.3 A large variety of chaparral plant species quickly resprout from underground lignotubers after a fire. In addition, the germination of seeds of obligate seeding (non-respouting) shrubs is stimulated by heat, charred wood, or smoke. Resprouting species shown include chamise (*Adenostoma fasciculatum*; front right), two laurel sumac (*Malosma laurina*; center left), and three mission manzanita (*Xylococcus bicolor*). Obligate seeding *Ceanothus tomentosus* seedlings are pictured in the middle of the photo. Note the diameter of the burned stems. The lack of small twigs indicates a high-severity fire (photo: R.W. Halsey).

central Chile, South Africa, southwestern Australia, and the Mediterranean Basin) (Keeley, 2012). In particular, the likely scenario for chaparral-dominated wildfires in California before human settlement was one of large, infrequent fires (once or twice per century) that were ignited by lightning in forested areas at higher elevations during the moderate summer monsoon period between August and September. Remnants of the fire, such as smoldering logs, persisted into the fall. When extreme weather variables coincided, for example, several years of drought, low humidity, high temperature, and strong winds, the fire would have reignited and rapidly spread. Fires stopped when they reached the coast or when the weather changed. Today fires ignited at higher elevations during monsoonal storms are extinguished, and at lower elevations fires are vastly more frequent as a result of human-caused ignitions (Keeley, 2001).

Although counterintuitive, chaparral plant communities are much more resilient to infrequent, high-intensity fires than they are to cooler, more frequent, lower-intensity fires (Keeley et al., 2008). If chaparral does not have sufficient time to replenish the soil seed bank, accumulate the biomass necessary to produce fires hot enough to successfully germinate fire-cued seeds, or allow resprouting species time to restore starch supplies in underground lignotubers, a cascading series of events begins that can significantly change or completely eliminate the plant community. If the fire return interval is less than 10 to 20 years, biodiversity is reduced and nonnative weeds and grasses typically invade, ultimately type-converting native shrubland to nonnative grassland (Brooks et al., 2004).

Today the average fire rotation interval (time between fires) for wildlands in southern California is 36 years, but this varies widely among different locations.

Fire return intervals can vary from fires every few years in some locations to fires every 100 years or more at others (Keeley et al., 1999).

7.3 FIRE MISCONCEPTIONS ARE PERVASIVE

In conflict with ecological facts is the presumption of the fire suppression paradigm that large, high-intensity fires in chaparral are unnatural. Popularized versions of the paradigm as characterized by public opinion, the press, and Congressional testimony claim these fires are so hot that they destroy plant communities and leave behind lifeless moonscapes that are prone to mudslides that occur because of cooked soils. It is concluded that this is the direct result of twentieth-century fire suppression that allowed the chaparral to become overgrown with dense shrubs, creating massive amounts of fuel. Also, the fact that postfire recovery is so dramatic has likely reinforced the false notion of a fire-adapted community that "needed" fire to "rejuvenate" itself. These conclusions are clearly not supported by a plethora of studies (reviewed in Keeley et al., 2012).

Following the logic of the fire suppression paradigm is that chaparral fires should be allowed to burn without efforts to suppress them. In fact, some have used the paradigm to support artificially igniting fires to the landscape. The reality of the situation, however, makes such an approach both dangerous and ecologically damaging.

First, fire is suppressed for a reason: When near human communities it can destroy property and kill people. No responsible fire manager is going to allow a wildland fire to burn anywhere near a community. The much maligned US Forest Service's "10 a.m. policy," whereby all possible resources are thrown at the fire with the intention of suppressing it by 10 a.m. the next day, or the California Department of Forestry and Fire Protection's goal of keeping all fires confined to less than 4 ha, are critical public safety policy objectives near homes. "We're protecting private lands and public lands where there's many lives at stake and homes at stake, [and] infrastructure," Duane Shintaku, California Department of Forestry and Fire Protection's Deputy Director for Resource Protection said. "... [A]nd you can't tell someone 'You know what? We're just going to see what would happen if we wait to see if it gets big.'" (Goldenstein, 2015).

Second, as we discuss later, too much fire—rather than not enough—is threatening many native shrubland ecosystems. The overgeneralization and misapplication of the fire suppression paradigm is the underlying cause of many of the misconceptions about wildland fire in chaparral. Ironically, fire suppression often is criticized by the very agencies responsible for doing it and by citizens who have been misled by the publicity supporting the fire suppression paradigm, yet whose lives and property are being protected.

Confusing Fire Regimes

In forests, the idealized behavior of frequent, low-intensity fire caused by lightning has been characterized as the "good" kind of fire because it is considered controllable, typically burning 40 ha or less and only "pruning plants" rather than "consuming" them (Sneed, 2008, Kaufmann et al., 2005; also see Chapter 13). However, such a fire is physically impossible in vegetation with the characteristics of chaparral (Figure 7.4).

Emblematic of the impact caused by the misapplication of the forest-centric fire suppression paradigm is a statement made by the chair of the Santa Barbara County Fish and Game Commission, who criticized a proposal to designate chaparral as a protected, environmentally sensitive habitat (Giorgi, 2014):

> *Fire in our local ecosystems is one of the best ways to achieve the goal of good biodiversity. The local Native Americans burned almost every year. Early Spanish explorer records prove this to be true. There are many lightning-caused fires in our area, but we routinely put them out, creating an unnatural condition of heavy, dense fuel loading and harming our ecosystem in the process.*

The chair's statement would have been supportable if it had only referred to the region's few higher-elevation pine forests or the mixed-conifer forests on the western slopes of the Sierra Nevada (see Chapters 1 and 2). Extending it to the chaparral ecosystem that dominates the surrounding Los Padres National Forest, however, is inappropriate. In addition, unlike high-elevation forests where lightning is common, the south coastal region of southern California does not experience sufficient lightning frequency to sustain the kind of fire imagined by the board's chair. In fact, the region has one of the lowest lightning frequencies in North America (Keeley, 2002).

This information was provided to the Fish and Game Commission through testimony before and during the hearing. The commission voted to reject the proposal that chaparral be designated as a sensitive habitat.

Native American Burning

The burning of landscapes by Native Americans has become an integral part of the fire suppression paradigm because it supports the practice of prescribed burning to reduce fuel loads. While it is true Native Americans burned the landscape along the central coast of California, there is strong evidence that such burning led to the elimination of shrublands near population centers, rather than maintaining them in a healthy condition (Keeley, 2002). The assumption that anthropogenic burning is important to maintain healthy vegetation communities in North America is in conflict with the fact that these communities existed as functioning ecosystems for millions of years before human settlement.

The important point is that Native American burning practices were performed to modify selected parts of the landscape in an artificial manner to support a hunter-gatherer existence. We cannot afford to emulate this pattern today. Most shrubland ecosystems already experience more fire than they can tolerate (e.g., Keeley et al., 1999). In addition, Native Americans did not

FIGURE 7.4 The 2007 Zaca Fire burned more than 97,200 ha in the Los Padres National Forest, the third-largest recorded fire in California after the 1889 Santiago Canyon Fire and the 2003 Cedar Fire. Although there are unburned patches within the perimeter (note vegetation strips at the lower right, along the central ridge, and the unburned area to the left), wherever the flames burned, they did so at high intensity/high severity. The fire burned over the entire scene shown in the photo (photo: R.W. Halsey).

need to be concerned with the spread of combustible, non-native weeds and increased ignitions caused by millions of additional people on the landscape.

Some have also speculated that Native Americans used "controlled burning" to prevent large wildfires (Anderson, 2006, SBCFWC, 2008). Evidence of Native American burning shows it was for localized management within a half-day's walk from villages (Keeley, 2002), not that they were able to reduce the severity and frequency of uncontrolled wildfires. There is little reason to believe Native Americans could prevent the occurrence of large wildfires on the broader landscape. Indeed, one ethnographic report describes a massive wildfire in San Diego County before European contact that resulted in a significant migration of Native American residents to the desert (Odens, 1971).

Succession Rather Than Destruction

The notion that high-intensity fires "destroy" the natural environment is a common theme in media stories after nearly every wildland fire (see Chapter 13). The concept is so pervasive it makes its way from public media to professional reports for decision makers. For example, Los Angeles City Council staff reported that the 2007 Griffith Park Fire "... caused significant damage to the vegetation, destroying the majority of the mixed chaparral and mixed shrub plant communities" (LACC, 2007).

As long a fire is within the parameters of the natural fire regime, a more accurate view is that large, high-intensity fires are part of a natural successional process for chaparral. Interestingly, chaparral is "autosuccessional," meaning that after chaparral burns, chaparral returns (Hanes 1971). The first year or two after a fire, ephemeral fire-following annuals and short-lived perennials dominate but then begin to be replaced by shrub seedlings and resprouts. The shrubs continue to grow and eventually re-form the chaparral canopy within 10 to 15 years. This confounded early ecologists and foresters who were trained in traditional ecology to value trees over shrubs. Their response during the 1920s was to plant over a million conifers, a substantial share of which were nonnative in the San Gabriel Mountains in Los Angeles County. Most were soon killed by drought or eventually by fire, convincing most foresters that chaparral, not forest, was the most sustainable plant community in the area (Halsey, 2011).

Although postfire ecological succession stories do sometimes make the news, they are generally overwhelmed by sensationalized reports of flames, destruction, and blackened landscapes. As remarkable as the postfire chaparral environment is—with hills covered with colorful wildflowers, resprouting shrubs, and large clusters of seedlings emerging from the dark soil—the perception that the environment has been destroyed by fire remains a pervasive image.

Decadence, Productivity, and Old-Growth Chaparral

When discussing the impact of fire, one must take care not to fall into the trap of anthropomorphizing a wild ecosystem like chaparral and thinking fire is needed to "refresh" or "clean out" old, "decadent" or "senescent" growth (Hanes, 1971). These characterizations of older chaparral stands have not been supported by subsequent research (see, e.g., Moritz et al., 2004, Keeley, 1992).

Multiple studies have demonstrated the ability of old-growth chaparral, nearly a century old or more, to maintain productive growth and recover with high biodiversity after a fire (Hubbard, 1986, Keeley and Keeley, 1977, Larigauderie et al., 1990). In fact, long fire-free periods are required for many species to properly regenerate (Odion and Tyler, 2002, Odion and Davis, 2000, Keeley, 1992).

With legacy manzanitas having waist-sized trunks, a rich flora of lichens rarely found anywhere else (Lendemer et al., 2008), and a dense canopy forming a protective watershed, old-growth chaparral provides an important habitat for a wide array of species and valuable ecosystem services to surrounding human communities. As such, old-growth chaparral represents a crucial component in the preservation of California's biodiversity (Keeley, 2000) (Figure 7.5).

Sometimes, a trailside sign or textbook description of chaparral includes the specter of "undisturbed climax chaparral" eventually becoming so thick that it will either "choke itself," "die out," or be replaced by woodland (Ricciuti,

FIGURE 7.5 Old-growth chaparral in San Diego County, California. A big-berry manzanita (*Artctostaphylos glauca*) has wrapped itself around an Engelmann oak (*Quercus engelmannii*). The manzanita is estimated to be over a century old (photo: R.W. Halsey).

1996). While trees will overtop and shade out chaparral in areas with higher annual rainfall and richer soil conditions than exist in the vast majority of chaparral sites, the general belief that chaparral will eventually disappear because of age is not supported by data (Keeley, 1992).

The imagined fate of old-growth chaparral illustrates the common genesis of many misconceptions where anecdotal evidence has replaced scientific investigation—observations that may have merit in a limited, specific instance but have been broadly misapplied to support a binary, black-and-white paradigm. The remarkable nuances of nature as revealed by science are ignored.

Unfortunately, with increasing fire frequency, old-growth stands of chaparral (in excess of 75 years old) are becoming increasingly rare (Knudsen, 2006). And, while biodiversity does temporarily increase after a fire, because of the germination of ephemeral fire-following species, there is no danger that this biodiversity is threatened by long fire-return intervals. The soil seed bank can likely remain viable for a significant amount of time. Shrublands burned after approximately 150 years respond with a rich array of seedlings (Keeley et al., 2005b) (Figure 7.6). Considering the number of human-caused ignitions, there is no need to be concerned over the lack of fire. The flames will come.

Allelopathy

Another factor mentioned to support the notion that fire is "needed" in chaparral is allelopathy, the theorized phenomenon of plants releasing chemicals to suppress the growth or germination of neighboring competitors. It was suggested that such chemical inhibition explained the lack of plant growth under the canopy of mature chaparral stands in southern California (Muller et al., 1968). When the chaparral burned, the theory suggested, flames denatured the toxic

FIGURE 7.6 A large number of fire-following annuals and short-lived perennials emerge from the soil seed bank after a high-intensity chaparral fire. In addition, geophytes emerging from underground tubers, like this brodiaea (*Dichelostemma capitatum*), are likely stimulated to flower by additional sunlight provided by the removal of the chaparral canopy by fire (photo: R.W. Halsey). *(Tyler and Borchert (2007))*.

substances in the soil, thereby releasing the seeds from inhibition and suggesting the need for fire. One problem with this explanation is that the soil chemicals suspected of suppressing growth actually increase after a fire (Christensen and Muller, 1975).

The seeds of most chaparral plants are innately dormant before they make contact with the ground because of their dependency on fire cue-stimulated germination. In addition, the presence of herbivores has been demonstrated to be a major factor in eliminating seedlings that do germinate (Bartholomew, 1970). Therefore, the lack of seedlings under the canopy and the postfire seedling response in chaparral can be easily explained without considering chemical inhibition (Halsey 2004). Despite the research, however, allelopathy in chaparral is still presented as fact in college courses and texts (SBCC, 2002, George et al., 2014).

Fire Suppression Myth

Quickly to follow most fire stories are attempts to explain why the fire happened in the first place."Fuel build-up," as per the fire suppression paradigm, is invariably blamed despite the science that has demonstrated otherwise.

In analyzing the California Statewide Fire History Database since 1910, Keeley et al. (1999) concluded that for shrub-covered landscapes of southern and central coastal California, "there is no evidence that fire suppression has altered the natural stand replacing fire regime in the manner suggested by others." In fact, fire suppression in California's Pacific south coast has played an important role in *protecting* much of the chaparral from too much fire. The authors of a comprehensive summary of the literature about fires in the region concluded the following (Keeley et al., 2009a):

> *The fire regime in this region is dominated by human caused ignitions, and fire suppression has played a critical role in preventing the ever increasing anthropogenic ignitions from driving the system wildly outside the historical fire return interval. Because the net result has been relatively little change in overall fire regimes, there has not been fuel accumulation in excess of the historical range of variability, and as a result, fuel accumulation or changes in fuel continuity do not explain wildfire patterns.*

Unfortunately, fire suppression in shrublands has not been completely successful in protecting chaparral and sage scrub habitats from too much fire. Shrublands in areas surrounding the San Diego, Los Angeles, and Santa Barbara metropolitan areas have some of the most negative fire return interval departures in California, meaning they are experiencing more fire than they have historically, threatening the chaparral's resilience (Safford and Van de Water, 2014). The problem seems to be spreading north into the northern Santa Lucia Range and may likely continue to spread as climate change and population growth increase the potential for ignitions.

Too Much Fire Degrades Chaparral

Chaparral is highly resilient to periodic fire, within the natural range of variability, and postfire communities are remarkable in their capacity to return to prefire composition within a decade or so after fire, with the community assembly finely balanced with resprouting and seeding species. Nevertheless, given increases in fire frequency, this resiliency can be interrupted. "Type conversion" is the term given to changes in vegetation type caused by changes in the external environment, and one of the most common disturbances is accelerated fire frequency. When keystone, non-resprouting (obligate seeding) shrub species, like most *Ceanothus* species, experience closely spaced fires, their populations often are decimated and effect a type conversion to a less diverse, resprouting-dominated chaparral (Zedler et al., 1983). Such stands become more open and often are subsequently invaded by nonnative herbaceous species. Fire return intervals of less than 6 years have been shown to be highly detrimental to the persistence of non-resprouting chaparral species (Jacobsen et al., 2004); in fact, multiple fires within a 6-year interval have even reduced resprouting species, further opening the chaparral environment (Haidinger and Keeley, 1993).

That this type conversion has been an ongoing process since the arrival of humans in California is apparent (Wells, 1962). The process is complex, dependent on fire history, community composition, and site factors. The loss of shrub cover and the invasion of combustible grasses creates a positive feedback process (Keeley et al., 2005a) whereby the community assembly changes, further increasing fire frequency and causing further type conversion away from the original stand composition. The speed of the type conversion process can be increased dramatically by numerous variables such as drought, cool-season fires (Knapp et al., 2009), livestock grazing, soil type, soil disturbance, and mechanical clearance activities (Bentley, 1967).

During extended periods of drought, seedling success of obligate seeding shrubs, like many *Ceanothus* species, is reduced after fire. In fact, excessive soil temperatures resulting from drought-induced canopy reduction after adult die back between fires has been shown to cause the premature germination of *Ceanothus megacarpus* seedlings just before the seasonal drought period (Burns et al., 2014). Seedling survival under such conditions is questionable, and the process depletes the seed bank.

Record drought conditions after fire also increase the mortality of resprouting chaparral shrubs like chamise (*Adenostoma fasciculatum*) and greenbark (*Ceanothus spinosus*). Resprouting shrub species likely deplete their carbohydrate reserves during the resprouting process, making them particularly vulnerable to drought because of the need to transpire water to acquire carbon dioxide that is used to supply energy to a large, respiring root system (Pratt et al. 2014). An additional fire within a 10-year window adds even more stress to resprouting species.

That type conversions occur and that severe type conversion from evergreen chaparral to alien-dominated grasslands has significantly altered the Californian landscape in the past are beyond question (Wells, 1962, Keeley, 1990), but an important issue is the extent of this contemporary threat. Talluto and Suding (2008) found that, over a 76-year period, 49% of the sage scrub shrublands in one southern California county had been replaced by annual grasses and that a substantial amount of this could be attributed to fire frequency.

In recent years, southern California has experienced some rather extensive reburns at anomalously short intervals (Keeley et al., 2009b), potentially setting the stage for the disruption of natural ecosystem processes and type-converting these shrublands to a mosaic of exotic and native species. This has already been documented clearly for a number of sites (Keeley and Brennan, 2012), where short-interval fires have extirpated some native species and greatly enhanced alien species. As discussed above, within the four southern and central/coastal national forests in California, most of the shrublands—the dominant plant communities within these federal preserves—are threatened by excessive fire, whereas the mountain forests of southern California have an overall fire deficit (Figure 7.7).

Quantifying how much chaparral has been compromised or completely type converted is a challenging research question because much of the damage likely was accomplished before accurate records of plant cover were kept. Based on interesting relic patches of chamise and historical testimony, Cooper (1922) speculated that extensive areas of chaparral have been eliminated and converted to grasslands, including the floor of the Santa Clara Valley, large portions of the

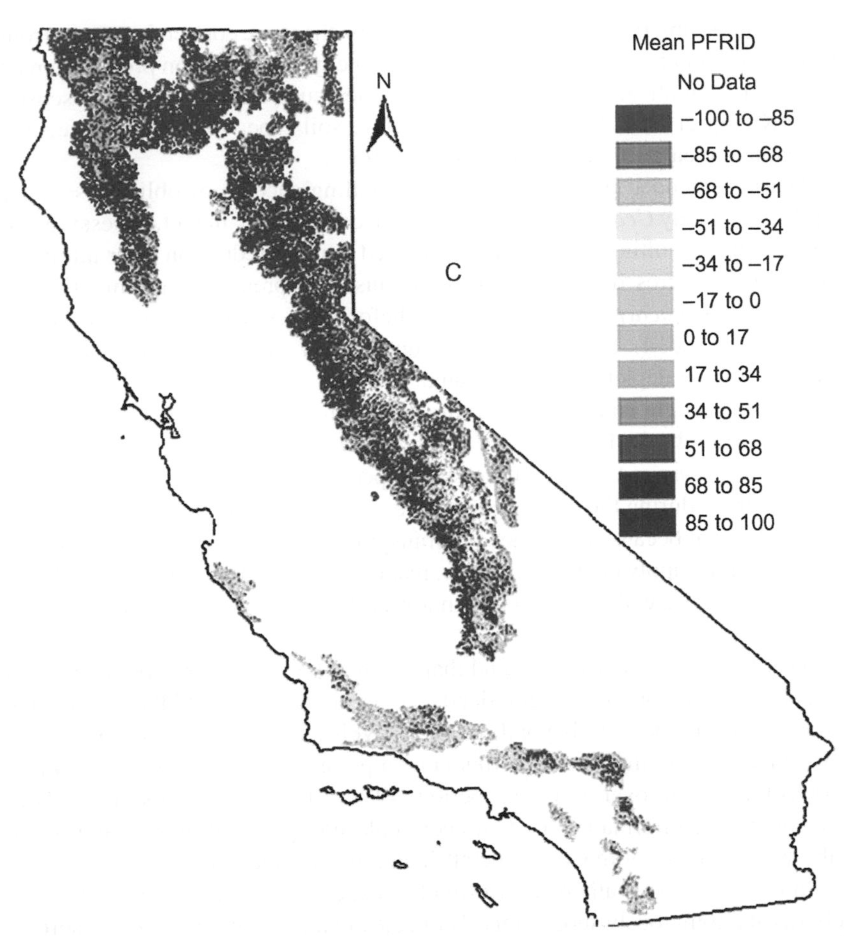

FIGURE 7.7 Most chaparral in California is threatened by too much fire, as shown by the map's color variations representing the fire return interval departure (PFRID) percentages for national forest lands in California. Note the color differences between the southern California national forests, which are dominated by chaparral (yellows), and the conifer-dominated forests in the Sierra Nevada (blues). The warm colors identify areas where the current fire return interval is shorter than that before European settlement (negative PFRID percentages). Cool colors represent current fire return intervals that are longer than those before European settlement (positive PFRID percentages) (photo: R.W. Halsey). *(From Safford and Van de Water (2014)).*

Sacramento and San Joaquin Valleys, and many of the grassy regions in the Coast Ranges and the western Sierra foothills. Large areas along Interstate 5 in the Cajon Pass region, the foothills above San Bernardino, and the Chino Hills south of Pomona also appear to be type-converted landscapes.

The focus on complete type conversion to grassland has led some to ignore the beginning stages of the process: the simplification of habitat by the loss of biodiversity (Keeley, 2005). For example, in a comment letter on the draft 2010

FIGURE 7.8 The impact of excessive fire on chaparral. The entire area shown was burned in 1970. The middle/left area burned again in 2001 and is returning with a full complement of native chaparral species. In the right portion, which burned again in 2003, obligate seeding species are absent, the number of resprouting species has been reduced, and nonnative weeds have invaded. The interval between the last two fires was too short, causing a dramatic reduction in biodiversity and leading to type conversion. The location pictured is near Alpine, San Diego County, California (photo: R.W. Halsey).

California Fire Plan, San Diego County claimed that chaparral burned in both the 2003 and 2007 wildland fires "remained chaparral and is recovering" (Steinhoff, 2010). In fact, much of the chaparral in question was not recovering well at all because of the loss of several keystone shrub species, and it was showing significant invasion by nonnative grasses (Keeley and Brennan, 2012) (Figure 7.8).

Meng et al. (2014) recently raised some skepticism about the ability of repeat fire to effect type conversion by pointing out the difficulty early twentieth-century range managers experienced when using fire to "improve" ranges that were supposedly plagued by chaparral. These managers typically relied on herbicides and mechanical destruction for thorough replacement of shrubs to create more useful grazing lands. As pointed out by Keeley and Brennan (2012), however, managers utilize fire only under narrow prescription conditions, which are generally not capable of carrying repeat fires at short fire return intervals—hence their difficulty in meeting their objective. By contrast, wildfires typically burn outside prescription, often with 100 km/h (about 60 mile/h) wind gusts and relative humidity less than 5%.

Using remote sensing, Meng et al. (2014) attempted to answer the question of how extensive type conversion is caused by repeat fires occurring in the past decade. While the technique cannot address changes in diversity and species composition that are known to occur with short-interval fires, it has some potential for viewing more gross changes in functional types such as shrubs and annual plants. Although these authors concluded that widespread type conversion is not an immediate threat in southern California, this conclusion deserves closer scrutiny because documenting fire-related vegetation change across large landscapes over just a 25-year period using remote sensing is fraught with potential errors and cannot serve as an effective proxy for field data.

One reason for error is that numerous spatially and temporally different human and biophysical factors can influence the process of postfire recovery; these factors should be controlled for before attribution can be determined. In the paper by Meng et al. (2014), the control and overlap areas were located on somewhat adjacent, but very different, parts of the landscape that varied by factors such as aspect, terrain, or soil type. The areas also could have experienced different landscape disturbance histories. This is especially possible given the topographic complexity of the region and researchers' use of the California's Fire Resource and Assessment Program's Fire History Database (FRAP) for discerning precise stand ages. This database is broadly useful for management planning but must be used carefully in a research context. For example, Keeley et al. (2008) found that across 250 sites the FRAP database did not accurately portray stand age (as determined by ring counts) for 47% of the sites, presumably because of the scale at which fires are mapped and by generally ignoring fires less than 40 ha.

Another concern is that the method of documenting vegetation change used by Meng et al. (2014) may not be sensitive enough to resolve gradual shifts in composition that would likely occur after only one repeat fire event. They used a vegetation index derived from imagery sensed remotely from a satellite as a way of assessing vegetation "cover," or the "greenness," of each 30-m image pixel. Because different pigments are stimulated by different parts of the light spectrum, this index essentially assesses chlorophyll content, which is correlated with biomass and assumed to represent the relative cover of evergreen shrubs. It does not, however, account for differences among chaparral species, whose composition in the plots was unknown. In addition, different species of chaparral have varying sensitivities to repeat fires, and thus multiple repeat fires of differing intervals might be required to discern enough vegetation change to be detected by this index.

Given that vegetation change is likely a gradual, cumulative process, the results reported by Meng et al. (2014) are actually consistent with a potential for widespread chaparral conversion—contrary to their conclusions. Over half of the area that burned twice in their study did have lower cover, as defined by the index, than the control area. Given enough fire on the landscape over a long enough period of time, gradual shifts may result in significant change and impact.

Type Conversion and Prescribed Fire

Unfortunately, the priorities of land management agencies have led some to deny the existence of chaparral type conversion. For example, in the same comment letter mentioned above, San Diego County wrote that it "strongly disagreed" with the draft 2010 California Fire Plan because it contained the following statement:

> *...fires have been too frequent in many shrublands, especially those of southern California, which are then at risk of type conversion from native species to invasives that can pose a fire threat every fire season.*

The county explained that recognizing the threat of chaparral type conversion in the Fire Plan would impact its ability to obtain funding to carry out vegetation clearance activities.

Prescribed burning—one of the clearance activities that San Diego County was hoping to conduct—has been shown to seriously compromise chaparral plant communities. In a study that simulated the effect of frequent fire on southern California coastal shrublands, Syphard et al. (2006) concluded that, "Due to this potential for vegetation change, caution is advised against the widespread use of prescribed fire in the region."

One of the problems with prescribed burning in chaparral is that there is a narrow window when such burns can occur: in the cool season (late spring). Plants have too much moisture in their tissues in the winter and early spring months to carry a fire. In the summer and fall, the risk of wildfire is too high because of low moisture levels and weather conditions. As a consequence, prescribed burns are conducted when the chaparral ecosystem is most vulnerable. The plants are growing, the soil is still moist, many animal species are breeding, and some birds are occupying the chaparral during their annual migrations. Thus significant ecological damage can occur as a result of a prescribed burn (Knapp et al., 2009).

The exact mechanisms are not clearly understood, but cool-season burns likely cause significant damage to plant growth tissues and destroy seeds in the soil as soil moisture turns into steam. A prescribed burn conducted in the 1990s in Pinnacles National Park, California, led to immediate type conversion of chaparral to nonnative grassland (Keeley, 2006). An escaped prescribed burn in 2013 consumed more than 1090 ha of fragile desert habitat in San Felipe Valley, California, much of which was chaparral that was recovering from a fire 11 years before. The fire seriously compromised one of the last old-growth desert chaparral stands in the region (CCI, 2013) (Figure 7.9).

Combustible Resins and Hydrophobia

There is no question that the loss of vegetation after a fire exposes more soil surface and increases the kinetic force of precipitation on the soil, which can increase the flow of water on the surface. The result can be significant erosion, flash flooding, and large debris flows. However, a factor that seems to get more attention than its proven influence justifies is water repellency, or "hydrophobic soils."

The observation that heat during a fire can change or intensify the water repellency of soil depending on temperature and other factors has been studied

FIGURE 7.9 Photo shows an escaped, 40 ha prescribed fire in the San Felipe Valley Wildlife Area, San Diego County, California, that ultimately burned more than 1000 ha, most of which was 11-year-old desert chaparral. Considering the ecological fragility of the area because of its age and the multiple fires that have burned much of the valley over the previous decade, there likely will be a significant reduction of biodiversity in the region (photo: R.W. Halsey).

extensively (DeBano, 1981, Hubbert et al., 2006) and was first identified after chaparral fires. The hydrophobic soils theory suggests that because of the gas released by burning plants and soil litter, hot fires create an impermeable "waxy layer" a few inches below the surface. According to popular accounts, this layer then prevents water from permeating the ground, causing large chunks of topsoil to break loose during rain storms and slide down the hill (LAT, 2014). Warnings about the hazards of such waterproof layers are commonly raised by the media after fires.

However, the actual impact hydrophobic soils have on erosion is questionable. Contrary to the impression often left by popular accounts, water repellency is not like a layer of plastic wrap under the surface; instead it is quite patchy and transient, abating once soils are wetted. Water repellency is also a natural condition of many unburned soils. In fact, high-severity fires have been found to destroy repellency (Doerr et al., 2006). In a review of the literature, Busse et al. (2014) concluded the following:

> *Most studies have only inferred a causal link between water repellency and erosion, and have failed to isolate the erosional impacts of water repellency from the confounding effects of losses in vegetation cover, litter cover, or soil aggregate stability.*

Unfortunately, the theorized role hydrophobic soils play in erosion has been repeated so many times that it has taken on the power of myth and is used to justify questionable, and sometimes expensive, land management decisions. The chaparral has been especially targeted for blame.

FIGURE 7.10 Postfire treatments in chaparral are costly and often of questionable value. Strips of mulch were dropped by aircraft on the side of the Viejas Mountain in San Diego County after the 2003 Cedar Fire (photo: R.W. Halsey).

To justify the clearance of native chaparral habitat, the Arizona Game and Fish Department claimed that "... catastrophic wildfire in the chaparral type can burn intensely enough to create hydrophobic soils, reducing soil productivity, increasing erosion, and causing severe downstream flooding" (AGFD, 2007). The City of Los Angeles spent $2 million to spread mulch after the 2007 Griffith Park Fire in part because "... chaparral vegetation has a natural tendency to develop water repellent or hydrophobic soils due to their natural high wax content. As a result, burned watersheds generally respond to runoff faster than unburned watersheds..." (LACC, 2007).

More than $1.25 million was spent laying down strips of mulch on Viejas Mountain in San Diego County after the 113,473 ha, high-intensity 2003 Cedar Fire, ostensibly to control erosion (Figure 7.10). However, Viejas Mountain is composed of gabbro-type soils that are not typically prone to extensive erosion (Halsey, 2008). Hydrophobic soils also have been used to justify postfire "salvage" logging after the 2013 Rim Fire in the Stanislaus National Forest (USFS, 2013).

7.4 REDUCING COGNITIVE DISSONANCE

Despite clear research that disproves many of the commonly held misconceptions about fire in chaparral that are fostered by the fire suppression paradigm, misconceptions persist. Many have found their way into land management plans that advocate landscape-scale "fuel treatments" or vegetation management projects for the stated purpose of "returning" California's chaparral ecosystem to a more "natural" and supposedly less dangerous fire regime. How the media, policymakers, and managers have responded to the cognitive dissonance that occurs when their assumptions about fire are challenged by science provides insight into the difficulties encountered when new ideas confront embedded paradigms.

Festinger (1957) suggested there are several ways an individual can reduce the tension caused by facts or ideas that conflict with their own opinions. Individuals can respond with cognitive competence by accepting the new data point or idea and change their opinions accordingly. Alternatively, individuals can respond incompetently by rejecting the new data point or idea either by ignoring or denying it or by justifying their opinion with new information or beliefs, occasionally using logical fallacies in the process.

For example, as mentioned earlier, when the chair of the Santa Barbara County Fish and Game Commission cited Native American burning as an argument for why we should not suppress fires in chaparral, he was using the common logical fallacy of appealing to antiquity. Such an appeal assumes older ideas or practices are better than newer ones because they have been around for a long time.

Local Agency

In an attempt to alter the natural fire regime, San Diego County tried to establish a chaparral clearance program that targeted more than 780 square kilometers of back country habitat with "prescribed fire, mechanical or biochemical fuel treatments" (SDCBS, 2009). This effort was based on a report issued earlier by the county. In misapplying the fire suppression paradigm to native shrublands, the report claimed that, "A fire regime of smaller, more frequent fires was being replaced by one of fewer, larger, and more intense fires" because of an unnatural density of "fuel" as a result of past fire suppression (SDCBS, 2003).

Despite volumes of data submitted by reviewing scientists over a period of more than 4 years indicating the county was basing its policies on incorrect assumptions, the county's Planning and Land Use Department repeatedly issued new drafts of its vegetation management plan without correcting the errors (Halsey, 2012).

In a comment letter by the Conservation Biology Institute, scientists wrote (Spencer, 2009):

> *Although this fourth draft is an improvement over previous drafts, it reflects partial and piece-meal updating based on various submitted comments and the workshop discussions rather than the comprehensive re-write that is necessary. This results in the report being internally inconsistent, confusing, and often self-contradictory. Moreover, despite scientific facts and logic presented to the county by numerous individuals, the report continues to perpetuate disproved myths about fires and fire management in southern California.*

In addition to ignoring information contrary to its position, the county misinterpreted the science in a manner that justified its viewpoint. One of the scientists whose work was the subject of the county's misinterpretations in the 2003 report wrote:

> *We were disturbed by the way our research findings were completely mischaracterized in this report on page 8. Not only are the specific statements about our findings completely false, but also, more generally, our research does not support the claims and recommendations of this section of the report.*
>
> Schoenberg and Peng (2004).

The San Diego County Board of Supervisors eventually adopted a final vegetation management plan in 2009, with most of the inaccurate information removed but with some of the questionable ecological assumptions about chaparral remaining. Within a month of the plan's adoption, the county attempted to implement the report's first clearance project without conducting the appropriate environmental review as required by the California Environmental Quality Act. The county claimed an "emergency exemption."

However, the California Chaparral Institute (CCI), an environmental nonprofit organization based in San Diego County, successfully challenged the project in court. The court rejected the county's position that a 3- to 4-year, $7 million vegetation management project was a "short-term project" addressing an immediate, emergency occurrence. In an attempt to influence the judge, county counsel used the logical fallacy of appealing to emotion by warning of death and destruction during future fires if the court ruled against the county. Since the hearing was considering a point of law, not evaluating emotional pleas, the judge was not swayed. The court ordered the county to follow the proper procedures under the law.

The county ultimately produced a full environmental impact report (EIR) on the project after being challenged again by CCI when it attempted to avoid the review process a final time through a negative declaration. The EIR was certified and the county completed the initial site-specific clearance project in 2012. The county later dropped the larger regional effort that had been so severely criticized.

State Agency

The California Board of Forestry and Fire Protection proposed in 2012 a statewide Vegetation Treatment Plan (VTP) that targeted more than one-third of the state for potential vegetation clearance operations. The VTP stated that large-scale wildland treatments should focus on areas "... up to the watershed scale, or even greater, that are treated to reduce highly flammable or dense fuels, including live brushy plants in some vegetation types (such as chaparral), a buildup of decadent herbaceous vegetation or, dead woody vegetation." One of the rationales for the VTP was that "[p]ast land and fire management practices (fire suppression) have had the effect of increasing the intensity, rate of spread, as well as the annual acreage burned on these lands" (CSBF, 2012).

As with the San Diego County example, there is no scientific support for this conclusion in chaparral, where most of California's largest wildland fires occur. Commenting on the VTP's stated intent to "reintroduce fire into (natural) communities where fire has been excluded through past fire suppression efforts," the California Department of Fish and Wildlife (CDFW, 2013) wrote the following:

> *There is substantial evidence that the frequency of fires continues to increase in coastal southern California (USDI NPS, 2004; Keeley et al., 1999). Fire management of California's shrublands has been heavily influenced by policies designed for coniferous forests; however, fire suppression has not effectively excluded fire from chaparral and coastal sage scrub landscapes and catastrophic wildfires are not the result of unnatural fuel accumulations (Keeley, 2002). There is also considerable evidence that high fire frequency is a very real threat to native shrublands in southern California, sometimes leading to loss of species when fire return intervals are shorter than the time required to reach reproductive maturity (Keeley, 2002).*

In contrast to San Diego County's reluctance to accept new scientific research, the state responded with cognitive competence. After the state board received criticism from fire scientists that the VTP did not reflect the most current research, the California State Legislature asked the California Fire Science Consortium, an independent network of fire scientists and managers, to review the proposal. The Consortium recommended that the VTP "undergo major revision if it is to be a contemporary, science-based document" (CFSC, 2014). The board then began the process of rewriting the document in 2014, with assurances they would be modifying their plan by incorporating the new information and offering opportunities for the original reviewers to provide input on the developing draft.

Media

The popular media poses a particular problem because reporters often do not specialize in one topic long enough to become familiar with contrary data that question prevailing paradigms. When confronted with new information, however, the media outlet has several options. It can provide time or space for an editorial response, publish another story on the subject, or make a concerted effort to incorporate the new information into its editing process for future stories.

For example, significant cognitive competence has been demonstrated by one of California's most influential newspapers, the *Los Angeles Times*. The paper has become familiar with the science and has helped its readers understand that too much fire is threatening the chaparral (LAT, 2009), recommended the California Board of Forestry withdraw its original vegetation management plan and produce a new one using the best available science (LAT, 2013), and commonly describes the state's characteristic ecosystem as chaparral rather than using the older, pejorative term "brush."

A San Francisco Bay-area publication, on the other hand, provides an example of how older, inaccurate information was allowed to persist. In an article about fires in the chaparral-dominated Ventana Wilderness area of the Los Padres National Forest, Rowntree (2009) wrote:

> *Because of fire suppression policies and strategies put into place in 1907, fires became relatively infrequent. But when fires happened, and Marble Cone is a prime example, the immense accumulated fuel led to hotter, more intense fires compared to those associated with a more natural fire regime.*

There is no scientific evidence to support the claim that chaparral fires are burning hotter or more intensely than they have historically. The Marble Cone Fire cited in the article burned approximately 72,000 ha in 1997. In 1906, however, before the fire suppression era was said to have begun, approximately 60,700 ha burned with equal intensity in the same area. Other large, intense fires in the region were recorded even earlier (J. Keeley, unpublished data).

In southern California, the 2003 Cedar Fire in San Diego County, which burned 113,473 ha, is often referred to as California's largest fire. But in 1889, the Santiago Canyon Fire burned an estimated 125,000 ha (and possibly as much as 200,000 ha, depending on the estimates used) in San Diego, Orange, and Riverside Counties (Keeley and Zedler, 2009). Although the capacity for large fires has not changed, the number of people and homes in the way of the flames certainly has increased. Over the past century, high-intensity chaparral-related wildfires have continued to be some of the largest and most devastating conflagrations in the United States in terms of property and lives lost (Halsey, 2008).

The author of the aforementioned article on the Marble Cone Fire reinforced the misconception that large, high-intensity fires are unnaturally destructive because they roar across the landscape, "destroying oak, madrone, chamise, manzanita, and all other shrubs and trees in its path." The impression made was that if the Marble Cone Fire had been natural, it would have been a low-intensity surface fire that "smolders as it slowly works its way through grasslands and chaparral." The presumed destructive nature of hydrophobic soil also was cited in the article as being responsible for creating "the slippery foundation for the mud-flows that caused havoc on Highway 1. . .."

After receiving a critique from the CCI citing the errors, the publisher decided, after consulting with the author, that the article was accurate and stood by its perspective. Although a website-based opportunity was offered for a short critique of the story, the publisher rejected publishing a follow-up article or comment letter because there was not enough room in the magazine for an additional discussion of an issue as complex as fire (D. Loeb, personal correspondence, 2010).

7.5 PARADIGM CHANGE REVISITED

In his seminal work on the structure of scientific revolutions, Thomas Kuhn (1962) wrote ". . .the proponents of competing paradigms practice their trades in different worlds." For the proponents of the fire suppression paradigm, wildfire is primarily a fuel-driven event. Thus controlling fuels controls fires, as the thinking goes, and native vegetation is viewed not as habitat but rather merely as unwanted fuel.

Alternatively, an increasingly common paradigm shift is framing wildfire in context of the entire environment, whereby other variables such as weather can play more important roles than fuel and whereby vegetation is viewed as wildlife habitat. The first paradigm is embedded in a controllable world where nature can be tamed, whereas in the second one, nature will ultimately defeat control. One sees nature as fuel; the other sees nature as providing important habitat in both its pre- and postburned conditions (also discussed in the Preface and Chapters 1–6 and 13). One focuses on manipulating wildlands to control wildfire, the other on community retrofits and planning to make them more fire-safe (Penman et al., 2014). As Kuhn explains, the two groups:

> *... see different things when they look from the same point in the same direction. Again, that is not to say that they can see anything they please. Both are looking at the world, and what they look at has not changed. But in some areas they see different things, and they see them in different relations one to the other. That is why a law that cannot even be demonstrated to one group of scientists may occasionally seem intuitively obvious to another.*

The feeling one may have during an argument that the other party is operating in another universe can in fact be an accurate description of what is happening.

Although other drivers of fire behavior are sometimes acknowledged, the practical implementation of policies resulting from the fire suppression paradigm is an exclusive focus on fuels (wildland vegetation). In this view, any fuel is too much fuel. Such a viewpoint was offered by a timber industry advocate during congressional testimony after the 2003 chaparral-dominated wildfires in southern California (Bonnicksen, 2003):

> *Some people believe that horrific brushland fires are wind-driven events. They are wrong. Science and nearly a century of professional experience shows that they are fuel driven events. Wind contributes to the intensity of a fire, but no fire can burn without adequate fuel, no matter how strong the wind.*

Besides the logical fallacy of appealing to unnamed authorities, this argument sets up the classic straw man fallacy. By misrepresenting the science that challenges the fuel-centric position and then refuting it, the congressional witness concludes that the science itself has been refuted. This is a fallacy because the science that is claimed to be refuted is actually being misrepresented.

Clearly, fire needs fuel to burn. Excepting extreme situations, all terrestrial environments have some kind of fuel, be it grass, shrubs, trees, or houses; all can provide adequate fuel for a fire under the right conditions. The science that challenges the fire suppression paradigm does not hold that fire can burn without fuel.

As the ~365,000 ha East Amarillo Complex grassland fire in Texas demonstrated in 2006, fine, grassy fuels also can also cause horrific fires. Twelve people died and 89 structures were destroyed in a fire that moved 72 km in just 9 h and had flame lengths >3.5 m (Zane et al., 2006).

While fuel reduction projects can help fire suppression efforts and reduce fire intensity, they have been shown to be ineffective when it matters most: during extreme fire weather. During such conditions, the fire is not controllable because it will burn through, over, or around fuel treatments (Keeley et al., 2004, Keeley et al., 2009b). Many fuel breaks never intersect fires, but those that do nearly always require the presence of a fire crew to be effective, demonstrating the importance of a fuel break's strategic location (Syphard et al., 2011). An extensive study of chaparral fires throughout central and southern California showed that there is not a strong relationship between fuel age and fire probabilities (Moritz et al., 2004). Even in fuels-reduced forests, burning under extreme weather conditions can produce large areas of high-severity fire (Lydersen et al., 2014). Extensive fuel treatments in a forest can also fail to prevent extensive damage to a community, such as Lake Arrowhead during the 2008 Grass Valley Fire, if the structures themselves are not fire-safe (Rogers et al., 2008).

Paradigms have a challenging intellectual duality because not only can they guide productive research, they can also blind. Proponents of an older paradigm can ignore overwhelming, contrary evidence or force it to fit their model. As Thomas Chamberlin (1890) wrote in his paper concerning the value of multiple working hypotheses, "There is an unconscious selection and magnifying of the phenomena that fall into harmony with the theory and support it, and an unconscious neglect of those that fail of coincidence."

In addition to the force of paradigm, financial pressure can be involved in propelling an idea beyond its proven effective value. When the 2003 Healthy Forests Restoration Act was passed by Congress, a significant source of money was made available for fuel treatments on public and private land. Shortly after the passing of the act, a US Forest Service supervisor summit was held in Nebraska, where forest supervisors were asked to sign a pledge to meet their forests' hazardous fuel targets. A clear signal was being sent from Washington, DC, that clearing vegetation was going to be a primary goal. The act codified the fire suppression paradigm and encouraged the perspective of habitat as hazardous fuel, regardless of the natural fire regime.

Don G. Despain, one of the original scientists who advocated allowing fire to perform its natural role in ecosystems, met with other wildland fire pioneers like Les Gunzel, Robert Mutch, and Bruce Kilgore in Missoula, Montana, in 1972 to discuss ways they could change how fire was viewed. "We were a pretty lonely bunch back then," Don explained in a 2006 interview (D. Despain, personal communication). But as time went on and attitudes about fire began to shift, Don began to notice that the impact of past fire suppression was being taken too far. Alternative variables that may have influenced fire behavior in the West were being ignored. "So many assumptions about fire were being made that had never been observed," Don said. "I came to think I was the only person to watch a fire actually burn. People need to get out and observe and apply natural curiosity with what is going on instead of running to the legislature."

7.6 CONCLUSION: MAKING THE PARADIGM SHIFT

In the 1990s the predominate view of chaparral within region 5 (California, mostly) of the US Forest Service was that the ecosystem represented primarily fuel, needed more fire, and that large chaparral wildfires were a direct product of twentieth-century fire suppression. Although there are Forest Service managers who still hold these views, the agency has demonstrated cognitive competence by accepting new information, rejecting the fire suppression paradigm as it had been applied to chaparral, and adjusting its official policies accordingly.

The shift began in 2000, after three papers that seriously questioned the prevailing views were published (Keeley et al., 1999, Mensing et al., 1999, Zedler and Seiger, 2000). These papers stimulated a significant volume of research, confirming that the fire suppression paradigm was not applicable to California's chaparral ecosystem.

John Tiszler (2000) wrote a white paper questioning the use of prescribed fire in the chaparral-dominated Santa Monica Mountains National Recreation Area (SMMNRA) within the National Park system. After the 2000 Cerro Grande Fire in Colorado, the National Park Service established a moratorium on prescribed fire and began a reexamination of its parks' fire management policies. New fire ecologists reexamined the SMMNRA's approach to fire and rejected the fire suppression paradigm. By 2005, a new fire management plan was formalized for the park (SMMNRA, 2005). The new approach is summarized on the park's website (SMMNRA, 2015):

> *In the last forty years fire managers have promoted the idea that prescribed fire is necessary to protect ecosystems and communities by restoring fire's natural role in the environment to thin forest stands and to reduce hazardous fuels. This is true for western forests where the natural fire regime was frequent, low intensity surface fires started by lightning ... However, this is not true for the shrubland dominated ecosystems of southern California and the Santa Monica Mountains.*

After the 2003 Cedar Fire in San Diego County, California, the California Chaparral Institute was established for the purpose of protecting and raising awareness about the value of native shrublands. Through publications, public outreach, and occasional legal challenges, the organization helped to communicate the new science to both the public and government agencies.

By 2013, the paradigm shift occurred and the US Forest Service published a guiding document that redefined their view of chaparral and recognized how excessive fires were threatening the ecosystem (USFS, 2013):

> *There is an additional crisis taking place in our Southern California Forests as an unprecedented number of human-caused fires have increased fire frequency to the extent that fire-adapted chaparral can no longer survive and is being replaced with non-native annual grasses at an alarming rate ... Only an environmental restoration program of unprecedented scale can alter the direction of current trends.*

On June 18, 2013, during an important US Forest Service symposium at the headquarters of the Angeles National Forest, Martin Dumpis, the coordinator for a new Forest Service initiative focusing on the protection and restoration of chaparral, summarized the new approach well. Standing at the podium and speaking with his characteristically disarming midwestern accent, he said, "Chaparral should be seen as a natural resource, rather than a fire hazard."

We still have a long way to go for a complete paradigm shift—from one that views mature chaparral as no more than an unnatural fuel load to one that recognizes expansive, contiguous stands of old-growth chaparral as natural and valuable. However, we are seeing the process of change accelerate. We believe that the recent forward progress with which the shift is occurring is not only the result of solid, compelling science but also constructive citizen involvement, persistence, and especially relationships based on trust and respect. While comment letters and lawsuits can speed up the process, we have found that relationships fuel change in the most productive, lasting way.

For example, when CCI won its lawsuit against San Diego County, its efforts to expand its educational programs were stymied by informal resistance from bureaucrats whose vegetation treatment programs were curtailed (Halsey, 2012). However, relationships developed through volunteer work, professional interactions, and sincere efforts to collaborate by environmental organizations like CCI and the Endangered Habitat League persisted and ultimately outlasted the resistance. The lawsuit was critical in protecting habitat, but it was relationships that implemented successful solutions. Such relationships also likely shaped legislative action on defensible space regulations and vegetation treatment programs, Forest Service policy shifts concerning chaparral, and the successful implementation of new public outreach efforts involving wildland preserves in San Diego County.

We have learned that, in the long run, science, involvement, and relationship building are all vital to ensure that the policies affecting our lives are based on the latest facts from paradigm shifts, rather than from unproductive responses to cognitive dissonance.

REFERENCES

AGFD, 2007. South Mowry Habitat Improvement Project. Arizona Game and Fish Department. Habitat Partnership Program. Habitat Enhancement and Wildlife Management Proposal. Project N. 07-521. Tucson, Arizona.

Anderson, M.K., 2006. Tending the Wild. University of California Press, Berkeley, 256 p.

Bartholomew, B., 1970. Bare zone between California shrub and grassland communities: the role of animals. Science 170, 1210–1212.

Bentley, J.R., 1967. Conversion of Chaparral Areas to Grassland: Techniques used in California. Agriculture Handbook No. 328. US Department of Agriculture, Washington, DC, 35 p.

Bonnicksen, T., 2003. Testimony to the Subcommittee on Forests and Forest Health of the Committee on Resources, U.S. House of Representatives. In: One Hundred Eighth Congress, First Session. December 5, 2003, Lake Arrowhead, California.

Brooks, M.L., D'Antonio, C.M., Richardson, D.M., DiTomaso, J.M., Grace, J.B., Hobbs, R.J., Keeley, J.E., Pellant, M., Pyke, D., 2004. Effects of invasive alien plants on fire regimes. Bioscience 54, 677–688.

Burns, A., Homlund, H.L., Lekson, V.M., Davis, S., 2014. Seedling survival after novel drought-induced germination in Ceanothus megacarpus. In: Abstract. Berea College Research Symposium, Kentucky.

Busse, M.D., Hubbert, K.R., Moghaddas, E.E.Y., 2014. Fuel reduction practices and their effects on soil quality. Gen. Tech. Rep. PSW-GTR-241, U.S. Department of Agriculture, Forest Service, Pacific Southwest Research Station, Albany, CA, 156 p.

CCI, 2013. Escaped Cal Fire Prescribed Burn, San Felipe Valley Wildlife Area in a letter from the California Chaparral Institute to the South Coast Region. Department of Fish and Wildlife, California, July 8, 2013.

CDFW, 2013. California Department of Fish and Wildlife Memorandum on the California Board of Forestry and Fire Protection Draft Vegetation Treatment Program Environmental Impact Report. Attachment A.

CFSC, 2014. Panel Review Report of Vegetation Treatment Program Environmental Impact Report by California Board of Forestry and Fire Protection in Association with CAL FIRE Agency. Coordinated by California Fire Science Consortium, August 2014.

Chamberlin, T.C., 1890. The method of multiple working hypotheses. Science Feb 7. Also reprinted in 1965. Science 148, 754–759.

Christensen, N.L., Muller, C.H., 1975. Effects of fire on factors controlling plant growth in Adenostoma chaparral. Ecol. Monogr. 45, 29–55.

Conard, S.G., Weise, D.R., 1998. Management of fire regimes, fuels, and fire effects in southern California chaparral: lessons from the past and thoughts for the future. Tall Timbers Ecology Conference Proceedings 20, 342–350.

Cooper, W.S., 1922. The Broad-Sclerophyll Vegetation of California. An Ecological Study of the Chaparral and its Related Communities. Carnegie Institution of Washington, Washington, DC, Publication No. 319. 124 p.

CSBF, 2012. Draft Programmatic Environmental Impact Report for the Vegetation Treatment Program of the California State Board of Forestry and Fire Protection. The California Department of Forestry, Sacramento, CA. October 30, 2012.

DeBano, L.F., 1981. Water repellent soils: a state-of-the-art. United States Department of Agriculture Forestry Service General Technical Report, PSW-46. Berkley, California.

Doerr, S.H., Shakesby, R.A., Blake, W.H., Chafer, C.J., Humphreys, G.S., Wallbrink, P.J., 2006. Effects of differing wildfire severities on soil wettability and implications for hydrological response. J. Hydrol. 319, 295–311.

Festinger, L., 1957. A Theory of Cognitive Dissonance. Stanford University Press, Stanford, CA.

George, M.R., Roche, L.M., Eastburn, D.J., 2014. Ecology. Annual Rangeland Handbook, Division of Agriculture and Natural Resources, University of California, Davis, CA.

Giorgi, W.T., 2014. Testimony at the Santa Barbara County Fish and Game Commission, November 20, 2014.

Goldenstein, T., 2015. Water Stress Takes Toll on California's Large Trees, Study Says. Los Angeles Times, Los Angeles, CA, January 19, 2015.

Haidinger, T.L., Keeley, J.E., 1993. Role of high fire frequency in destruction of mixed chaparral. Madrono 40, 141–147.

Halsey, R.W., 2004. In search of allelopathy: an eco-historical view of the investigation of chemical inhibition in California coastal sage scrub and chamise chaparral. J. Torrey Bot. Soc. 131, 343–367.

Halsey, R.W., 2008. Fire, Chaparral, and Survival in Southern California. Sunbelt Publications, San Diego, CA.

Halsey, R.W., 2011. Chaparral as a natural resource. In: Proceedings of the California Native Plant Society Conservation Conference, January 17-19, 2009, pp. 82–86.

Halsey, R.W., 2012. The politics of fire, the struggle between science & ideology in San Diego County. Serialized from The Chaparralian 40 (3/4), 6–20. Independent Voter Network, May 15, 2013. http://ivn.us/2012/05/15/the-politics-of-fire-the-struggle-between-science-ideology-in-san-diego-county/May 1, 2012.

Hanes, T.L., 1971. Succession after fire in the chaparral of southern California. Ecol. Monogr. 41, 27–52.

Hubbard, R.F., 1986. Stand Age and Growth Dynamics in Chamise Chaparral. Master's thesis, San Diego State University, San Diego, CA.

Hubbert, K.R., Preisler, H.K., Wohlgemuth, P.M., Graham, R.C., Narog, M.G., 2006. Prescribed burning effects on soil physical properties and soil water repellency in a steep chaparral watershed, southern California, U.S.A. Geoderma 130, 284–298.

Jacobsen, A.L., Davis, S.D., Fabritius, S.L., 2004. Fire frequency impacts non-sprouting chaparral shrubs in the Santa Monica Mountains of southern California. In: Arianoutsou, M., Papanastasis, V.P. (Eds.), Ecology, Conservation and Management of Mediterranean Climate Ecosystems. Millpress, Rotterdam, Netherlands.

Kaufmann, M.R., Shlisky, A., Marchand, P., 2005. Good fire, bad fire: how to think about forest land management and ecological processes. U.S. Department of Agriculture, Forest Service, Rocky Mountain Research Station, Fort Collins, CO.

Keeley, J.E., 1987. Role of fire in seed germination of woody taxa in California Chaparral. Ecology 68 (2), 434–443.

Keeley, J.E., 1990. The California valley grassland. In: Schoenherr, A.A. (Ed.), Endangered Plant Communities of Southern California. Southern California Botanists, Fullerton, California, pp. 2–23.

Keeley, J.E., 1992. Demographic structure of California chaparral in the long-term absence of fire. J. Veg. Sci. 3, 79–90.

Keeley, J.E., 2000. Chaparral. In: Barbour, M.G., Billings, W.D. (Eds.), North American Terrestrial Vegetation, second ed. Cambridge University Press, Cambridge, UK, pp. 203–253.

Keeley, J.E., 2001. We still need Smokey bear! Fire Manage. Today 61 (1), 21–22.

Keeley, J.E., 2002. American Indian influence on fire regimes in California's coastal ranges. J. Biogeogr. 29, 303–320.

Keeley, J.E., 2005. Fire as a threat to biodiversity in fire-type shrublands. Gen. Tech. Rep. PSW-GTR-195: 97-106, USDA Forest Service.

Keeley, J.E., 2006. Fire management impacts on invasive plant species in the western United States. Conserv. Biol. 20, 375–384.

Keeley, J.E., 2009. Fire intensity, fire severity and burn severity: a brief review and suggested usage. Int. J. Wildland Fire 18, 116–126.

Keeley, J.E., Brennan, T.J., 2012. Fire-driven alien invasion in a fire-adapted ecosystem. Oecologia 169, 1043–1052.

Keeley, J.E., Keeley, S.C., 1977. Energy allocation patterns of a sprouting and nonsprouting species of *Arctostaphylos* in the California chaparral. Am. Midl. Nat. 98, 1–10.

Keeley, J.E., Keeley, S.C., 1987. Role of fire in the germination of chaparral herbs and suffrutescents. Madrono 34, 240–249.

Keeley, J.E., Zedler, P.H., 2009. Large, high-intensity fire events in southern California shrublands: debunking the fine-grain age patch model. Ecol. Appl. 19, 69–94.

Keeley, J.E., Fotheringham, C.J., Morais, M., 1999. Reexamining fire suppression impacts on brushland fire regimes. Science 284, 1829–1832.

Keeley, J.E., Fotheringham, C.J., Moritz, M., 2004. Lessons from the 2003 wildfires in southern California. J. For. 102, 26–31.

Keeley, J.E., Keeley, M., Fotheringham, C.J., 2005a. Alien plant dynamics following fire in Mediterranean-climate California shrublands. Ecol. Appl. 15, 2109–2125.

Keeley, J.E., Pfaff, A.H., Safford, H.D., 2005b. Fire suppression impacts on postfire recovery of Sierra Nevada chaparral shrublands. Int. J. Wildland Fire 14, 255–265.

Keeley, J.E., Brennan, T.J., Pfaff, A.H., 2008. Fire Severity and ecosystem responses from crown fires in California shrublands. Ecol. Appl. 18 (6), 1530–1546.

Keeley, J.E., Aplet, G.H., Christensen, N.L., Conard, S.C., Johnson, E.A., Omi, P.N., Peterson, D.L., Swetnam, T.W., 2009a. Ecological foundations for fire management in North American forest and shrubland ecosystems. Gen. Tech. Report PNW-GTR-779. USDA, USFS PNW Research Station, Portland, OR, 92 p.

Keeley, J.E., Safford, H., Fotheringham, C.J., Franklin, J., Moritz, M., 2009b. The 2007 southern California wildfires: lessons in complexity. J. For. 107, 287–296.

Keeley, J.E., Bond, W.J., Bradstock, R.A., Pausas, J.G., Rundel, W., 2012. Fire in Mediterranean Climate Ecosystems: Ecology, Evolution and Management. Cambridge University Press, Cambridge, UK. 528 p.

Knapp, E.E., Estes, B.L., Skinner, C.N., 2009. Ecological effects of prescribed fire season: a literature review and synthesis for managers. Gen. Tech. Report PSW-GTR-224, USDA, Forest Service. PSW Research Station, 80 p.

Knudsen, K., 2006. Notes on the Lichen Flora of California # 2. Bull. Calif. Lichen Soc. 13 (1), 10–13.

Kuhn, T.S., 1962. The Structure of Scientific Revolutions. The University of Chicago Press, Chicago, IL.

LACC., 2007. Griffith park fire recovery plan/expenses. Council File Number 07-0600-S38, Los Angeles City Council, Los Angeles, CA, August 1, 2007.

LAT, 2009. A Burning Problem. Los Angeles Times editorial, Los Angeles, CA, April 22, 2009.

LAT, 2013. Cal Fire's Flawed Fire Plan. Critics Say It's Outdated, Contains Many Inaccuracies and Could Cause Major Environmental Damage. Los Angeles Times editorial, Los Angeles, CA, March 11, 2013.

LAT, 2014. Flash Floods after Fires. Los Angeles Times, Los Angeles, CA, December 18, 2014.

Larigauderie, A., Hubbard, T.W., Kummerow, J., 1990. Growth dynamics of two chaparral shrub species with time after fire. Madrono 37, 225–236.

Lendemer, J.C., Kocourkov, J., Knudsen, K., 2008. Studies in lichens and lichenicolous fungi: notes on some taxa from North America. Mycotaxon 105, 379–386.

Lombardo, K.J., Swetnam, T.W., Baisan, C.H., Borchert, M.I., 2009. Using bigcone Douglas-fir fire scars and tree rings to reconstruct interior chaparral fire history. Fire Ecol. 5, 32–53.

Lydersen, J.M., North, M.P., Collins, B.M., 2014. Severity of an uncharacteristically large wildfire, the Rim Fire, in forests with relatively restored frequent fire regimes. For. Ecol. Manag. 328, 326–334.

Meng, R., Dennison, P.E., D'Antonio, C.M., Moritz, M.A., 2014. Remote sensing analysis of vegetation recovery following short-interval fires in southern California shrublands. PLoS One 9, e110637.

Mensing, S.A., Michaelsen, J., Byrne, 1999. A 560 year record of Santa Ana fires reconstructed from charcoal deposited in the Santa Barbara Basin, California. Quat. Res. 51, 295–305.

Minnich, R.A., 2001. An integrated model of two fire regimes. Conservat. Biol. 15, 1549–1553.

Moreno, J.M., Oechel, W.C., 1989. A simple method for estimating fire intensity after a burn in California chaparral. Acta Oecol. 10, 57–68.

Moritz, M.A., Keeley, J.E., Johnson, E.A., Schaffner, A.A., 2004. Testing a basic assumption of shrubland fire management: how important is fuel age? Front. Ecol. Environ. 2, 67–72.

Muller, C.H., Hanawalt, R.B., McPherson, J.K., 1968. Allelopathic control of herb growth in the fire cycle of California chaparral. Bull. Torrey Bot. Soc. 95, 225–231.

Odens, P., 1971. The Indians and I. Imperial Printers, El Centro, CA, 80 p.

Odion, D.C., Davis, F.W., 2000. Fire, soil heating, and the formation of vegetation patterns in chaparral. Ecol. Monogr. 70, 149–169.

Odion, D., Tyler, C., 2002. Are long fire-free periods needed to maintain the endangered fire-recruiting shrub *Arctostaphylos morroensis* (Ericiaceae)? Conserv. Ecol. 6, 4.

Penman, T.D., Collins, L., Syphard, A.D., Keeley, J.E., Bradstock, R.A., 2014. Influence of fuels, weather and the built environment on the exposure of property to wildfire. PLoS One 9 (10), e111414.

Pratt, R.B., Jacobsen, A.L., Ramirez, A.R., Helms, A.M., Traugh, C.A., Tobin, M.F., Heffner, M.S., Davis, S.D., 2014. Mortality of resprouting chaparral shrubs after a fire and during a record drought: physiological mechanisms and demographic consequences. Global Change Biol. 20, 893–907.

Ricciuti, E.R., 1996. Chaparral. Biomes of the WorldBenchmark Books. Marshall Cavendish, New York, 64 p.

Rogers, G., Hann, W., Martin, C., Nicolet, T., Pence, M., 2008. Fuel Treatment Effects on Fire Behavior, Suppression Effectiveness, and Structure Ignition. US Department of Agriculture, Forest Service, Region 5, Vallejo, CA, R5-TP-026a. 35 p.

Rowntree, L., 2009. Forged by Fire. Bay Nat. 9 (4), 24–29.

Safford, H.D., Van de Water, K.M., 2014. Using fire return interval departure (FRID) analysis to map spatial and temporal changes in fire frequency on national forest lands in California.

Res. Pap. PSW-RP-266, U.S. Department of Agriculture, Forest Service, Pacific Southwest Research Station, Albany, CA, 59 p.

SBCC, 2002. Biology 100. Concepts in Biology. Introduction to Chaparral. Santa Barbara City College, Santa Barbara, CA. Summary on college website, Retrieved December 18, 2014, http://www.biosbcc.net/b100plant/.

SBCFWC, 2008. Draft resolution to the Santa Barbara County Board of Supervisors from the Santa Barbara County Fish and Wildlife Commission, February 26, 2008.

Schoenberg, F.P., Peng, R.D., 2004. Letter to the San Diego Board of Supervisors, January 26, 2004.

SDCBS, 2003. Mitigation Strategies for Reducing Wildland Fire Risks. San Diego County Wildland Fire Task Force Findings and Recommendations, San Diego County Board of Supervisors, San Diego, CA, August 13, 2003.

SDCBS, 2009. County of San Diego Vegetation Management Report. Final Draft, 2/11/09. San Diego County Board of Supervisors.

Shea, N., 2008. Under Fire. National Geographic Magazine, Washington DC. July, 2008.

SMMNRA, 2005. Final Environmental Impact Statement for a Fire Management Plan, Santa Monica Mountains National Recreation Area. US Department of the Interior, National Park Service, Washington, DC.

SMMNRA, 2015. Why This Park Does Not Use Prescribed Fire. US Department of the Interior, National Park Service, Washington, DC. Santa Monica Mountains National Recreation Area website: http://www.nps.gov/samo/parkmgmt/prescribedfires.htm.

Sneed, D., 2008. Fires in California Devastate Wildlife, Sensitive Habitats. The Tribune, San Luis Obispo, July 12, 2008.

Spencer, W.D., 2009. Letter to the County of San Diego Planning Commission. Conservation Biology Institute, San Diego, CA, January 5, 2009.

Steel, Z.L., Safford, H.D., Viers, J.H., 2015. The fire frequency-severity relationship and the legacy of fire suppression in California forests. Ecosphere 6 (1), 8.

Steinhoff, 2010. San Diego County comment letter on the draft California Fire Plan.

Syphard, A.D., Franklin, J., Keeley, J.E., 2006. Simulating the effects of frequent fire on southern California coastal shrublands. Ecol. Appl. 16 (5), 1744–1756.

Syphard, A.D., Keeley, J.E., Brennan, T.J., 2011. Comparing the role of fuel breaks across southern California national forests. For. Ecol. Manag. 261, 2038–2048.

Talluto, M.V., Suding, K.N., 2008. Historical change in coastal sage scrub in southern California, USA in relation to fire frequency and air pollution. Landsc. Ecol. 23, 803–815.

Tiszler, J., 2000. Fire Regime, Fire Management, and the Preservation of Biological Diversity in the Santa Monica Mountains NRA. Report for the Santa Monica National Recreation Area, Thousand Oaks, CA, May, 2000.

Tyler, C.M., Borchert, M.I., 2007. Chaparral geophytes: fire and flowers. Fremontia 35 (4), 22–24.

USFS, 2013. Ecological Restoration Implementation Plan. R5-MB-249. US Department of Agriculture. Forest Service. Pacific Southwest Region, Vallejo, CA, 154 p.

USDI NPS, 2004. Draft Environmental Impact Statement Fire Management Plan, Santa Monica Mountains National Recreation Area. United States Department of the Interior, National Park Service, Washington, DC.

Wells, P.V., 1962. Vegetation in relation to geological substratum and fire in the San Luis Obispo quadrangle, California. Ecol. Monogr. 32, 79–103.

Zane, D., Henry, J.H., Lindley, C., Pedergrass, P.W., Galloway, D., Spencer, T., Stanford, M., 2006. Surveillance of Mortality during the Texas Panhandle Wildfires (March 2006). Regional and Community Coordination Branch, Public Health Preparedness Unit, Texas Department of State Health Services.

Zedler, P.H., Seiger, L.A., 2000. Age mosaics and fire size in chaparral: a simulation study. In: 2nd Interface between Ecology and Land Development in California, pp. 9–18, USGS Open-File Report 00-02.

Zedler, P.H., Gautier, C.R., McMaster, G.S., 1983. Vegetation change in response to extreme events: the effect of a short interval between fires in California chaparral and coastal scrub. Ecology 64, 809–818.

Chapter 8

Regional Case Studies: Southeast Australia, Sub-Saharan Africa, Central Europe, and Boreal Canada

Case Study: The Ecology of Mixed-Severity Fire in Mountain Ash Forests

Laurence E. Berry[1] and Holly Sitters[2]

[1]Conservation and Landscape Ecology Group, Fenner School of Environment and Society, The Australian National University, Canberra, ACT, Australia, [2]Fire Ecology and Biodiversity Group, School of Ecosystem and Forest Sciences, University of Melbourne, Creswick, VIC, Australia

8.1 THE SETTING

The eucalypt forests of southeast Australia are among the most flammable ecosystems worldwide (Pyne, 1992). Most of the forest in the region is dominated by a single overstory eucalypt species. Here we describe stands of mountain ash (*Eucalyptus regnans*), which is the tallest flowering plant species in the world, at heights approaching 100 m (Beale, 2007). Mountain ash forest occurs in the states of Victoria and Tasmania, and our focus is on a 121,000 ha region in the state of Victoria's Central Highlands, where undulating landscapes merge with mountainous terrain at the foothills of the Great Dividing Range.

The region lies at altitudes of between 200 and 1100 m above sea level and experiences mild, humid winters and warm summers; mean annual rainfall increases from approximately 1200 mm in the west to 1800 mm in the east (Bureau of Meteorology, 2014). The productive wet forest comprises a layer

The Ecological Importance of Mixed-Severity Fires: Nature's Phoenix

of understory trees that reach heights of more than 20 m and include myrtle beach (*Nothofagus cunninghamii*), southern sassafras (*Atherosperma moschatum*) and silver wattle (*Acacia dealbata*). A shrub layer of 2-15 m supports tree ferns (*Dicksonia antarctica* and *Cyathea australis*), hazel pomaderris (*Pomaderris aspera*), musk daisy bush (*Olearia argophylla*), and blanket leaf (*Bedfordia salicina*). The well-developed understory and shrub layers of mountain ash forest accumulate high fuel loads that are too wet to burn except following periods of drought or during extremely hot and dry conditions (Ashton, 1981). Consequently, large fires in mountain ash forest are infrequent and intense (Jackson, 1968).

Unlike many eucalypt species, mountain ash is considered fire-sensitive because it does not reproduce vegetatively and is killed by severe fire (Lindenmayer, 2009). As an obligate seeder, however, its regeneration is dependent on high-intensity fire, which desiccates seed capsules and releases up to 14 million seeds per hectare (Attiwill, 1994). Seedlings are shade intolerant and rarely establish in mature forest; instead they thrive under the high light levels of the nitrogen-rich ash bed that characterizes the postfire landscape (Ashton and Martin, 1996). For several decades the prevailing paradigm was that regenerating stands were primarily even-aged (Ashton, 1976; Griffiths, 1992; Loyn, 1985).

It is increasingly recognized that large, intense fires create mosaics of fire severity and generate multi-aged forest (Simkin and Baker, 2008; Lindenmayer et al., 2000). Because of the substantial topographic reliefs (up to 1000 m) and resulting gradients in wetness, large fires in the Victorian Central Highlands (VCH) rarely burn homogeneously (Figures 8.1 and 8.2). Patterns of fire severity in the region are a function of interplay among topography, vegetation, and weather (Berry et al., 2015a,b). In general, high-severity crown fire occurs most predictably on ridgetops because of wind exposure and associated decreases in fuel moisture, whereas unburnt patches tend to occur in sheltered valleys where fuel moisture levels are elevated (Wood et al., 2011; Bradstock et al., 2010). Eighty-four unburnt patches of >1 ha were present within the perimeter (~250,000 ha) of the 2009 wildfire (average patch size was 27 ha; Leonard et al., 2014). Collectively, they covered only 1% of the fire-affected area, but they probably provided sufficient refuge habitat to allow animals and fire-intolerant plants to survive, persist, and recolonize following the fire (Robinson et al., 2013). Further, a spectrum of moderate- to high-severity fire causes only partial mortality of overstory trees (mountain ash survival averages 39% in areas of moderate crown scorch and is over 90% in areas affected by low-severity fire; Benyon and Lane (2013)), providing additional refuge habitat for fauna and giving rise to multi-aged stands (Simkin and Baker, 2008). We discuss the implications of mixed- and high-severity fire for the flora and fauna of mountain ash forest before considering the consequences of increased fire predicted under future climate scenarios (Bradstock et al., 2014).

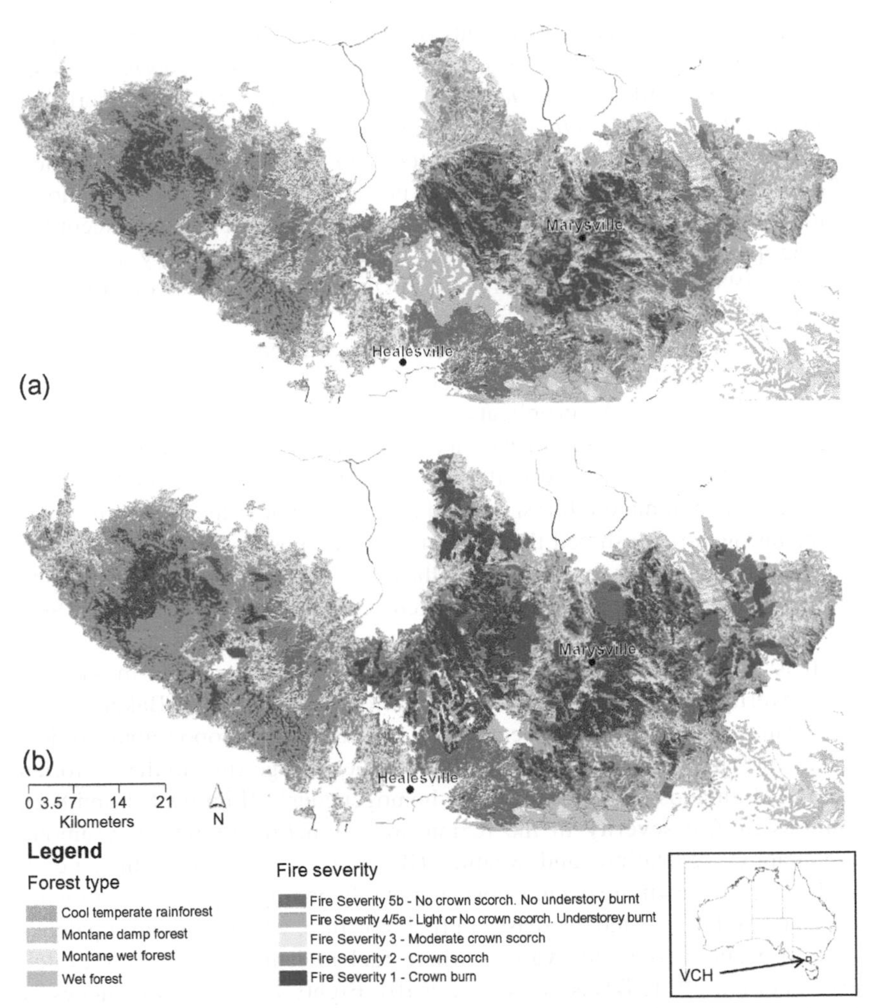

FIGURE 8.1 (a) Map of the 2009 Kilmore-Murrindindi fire complex in the Victorian Central Highlands (VCH), Australia. The map displays the extent of forest types containing mountain ash and major roads and towns. (b) The map is overlayed with the extent of logging per decade beginning in 1900 (light blue) to 2010 (dark blue).

8.2 MOUNTAIN ASH LIFE CYCLE

As an obligate seeder, mountain ash is susceptible to rapid decline under altered fire regimes (McCarthy et al., 1999) (Figure 8.3). Crucially, the fire return interval must be longer than the species' maturation age (15-20 years) and shorter than its life span (350-500 years) (Wood et al., 2010). If successive stand-replacing fires occur <20 years apart, mountain ash is succeeded by other species more tolerant of shorter fire intervals, such as silver wattle (Lindenmayer, 2009).

FIGURE 8.2 A topographically sheltered, unburnt gully embedded within a forest that burned at high severity. These areas may act as fire refuges within the burn extent, enabling the persistence of fauna otherwise vulnerable to the effects of fire on habitat. *(Photo credit: Laurence Berry.)*

FIGURE 8.3 Mountain ash trees displaying fine-scale variability in canopy consumption and a high density of new saplings. This photograph was taken approximately 5 years after fire. *(Photo credit: Laurence Berry.)*

The spatial and temporal attributes of fire regimes interact with climate, topography, and other disturbances to potentially alter species' distribution.

It is plausible, however, that the direct effects of climate change, such as increased temperature or rainfall, pose a greater threat to obligate seeders than altered fire regimes per se (Lawson et al., 2010). Smith et al. (2014) sought to disentangle the relative importance of fire-related factors (return interval and severity) and environmental factors (climate and topography) on seedling establishment in mountain ash. Regenerating mountain ash stands feature several million seedlings per hectare, and rapid self-thinning reduces sapling density to around 400 per hectare 40 years after fire. Only 40-80 mature trees per hectare remain after 150-200 years (Ashton and Attiwill, 1994). Competition causes the death of many growing seedlings, and the collapse of small, suppressed pole and sapling trees compounds mortality. Smith et al. focused their investigation on the critical stage of seedling establishment (within a year of fire) to identify factors that drive successful rejuvenation.

Climatic variables, topographic position, and fire return interval all were identified as important determinants of eucalypt seedling establishment (Smith et al., 2014). Seedling abundance was greater in flat, elevated areas of comparatively high rainfall and low temperature. Moreover, seedlings were more abundant at sites with a longer fire return interval, indicating that seed storage potential and rejuvenation capacity increase with forest age (Ashton, 1975). There was no detectable difference in seedling establishment between sites affected by moderate- and high-severity fire. Conceivably, opposing forces hid fire-severity effects; for example, light levels at high-severity sites may be optimal for seedling growth, but this positive effect could have been counteracted by greater seed mortality during fire (Smith et al., 2014). The authors suggest that both climate change and altered fire regimes influence the regeneration capacity of mountain ash, the distribution of which will potentially shift as the climate warms.

8.3 INFLUENCE OF STAND AGE ON FIRE SEVERITY

A warmer, drier climate is predicted to increase the frequency of large, intense fires in southeast Australia (Cary et al., 2012). Taylor et al. (2014) investigated the influence of stand age on fire severity in mountain ash forest across an array of growth stages and disturbance histories with a view toward better understanding the implications of increasingly frequent fire for species' capacity to regenerate. Mountain ash forests have been a key source of pulp wood and sawlogs since the 1930s (Lutze et al., 1999), so stand age classes were derived from multiple data sets relating to both fire and logging history, and fire severity data were sampled at 100 m intervals across a grid comprising nearly 10,000 sites.

Strong nonlinear relationships between forest age and fire severity were identified (Taylor et al., 2014). Severity was highest in stands that were 7-36 years old, which sustained canopy consumption and scorching. By contrast, canopy consumption rarely occurred in very young stands (<7 years old), and was uncommon in stands >40 years old. Regenerating stands support a highly flammable fuel layer of densely spaced seedlings, and self-thinning yields large volumes of fire-prone fine fuels. Fine fuel loads are thought to peak at 35 years after fire, when stand height and floristic composition can also compound flammability. A lack of dead fine fuel in stands <7 years old may explain the scarcity of crown fire in very young stands (Taylor et al., 2014). Correspondingly, the flammability of older forest is reduced by the establishment of rainforest species such as myrtle beach, which lessen light penetration to the ground layer and foster the development of a cool, moist microclimate (Wood et al., 2014). High-severity fire is intrinsic to mountain ash forest, but Taylor et al. (2014) emphasize the importance of old stands in which fire intensity and rates of spread are reduced by higher moisture levels and the probability of tree survival is greater.

8.4 DISTRIBUTION OF OLD-GROWTH FORESTS

Large, old trees in mountain ash forests provide biological legacies that perform myriad ecological roles; they store carbon, provide faunal habitat, and promote stand rejuvenation (Bowman et al., 2014). Sixty to 80% of mountain ash forest in VCH was historically considered old growth 100 to 150 years ago (Lindenmayer, 2009). Wildfire and logging during the past century have reduced old-growth forests to only 1.96% of the area within the extent of current mountain ash forests in the VCH (Lindenmayer et al., 2011). Old-growth stands naturally occur in flat plateaus and valley bottoms, where fire is less intense; clearcutting often targets such areas, precipitating the replacement of less flammable forest with dense regenerating stands of self-thinning trees and potentially altering patterns of spatial propagation of fire behavior (Lindenmayer et al., 2011). Rapid advances in fire severity mapping and modeling are benefitting the study of the spatial characteristics of large, intense fires and will inform management strategies that seek to promote the distribution and conservation of older forest stands (Berry et al., 2015a,b).

8.5 MIXED-SEVERITY FIRE AND FAUNA OF MOUNTAIN ASH FORESTS

Wildfires are a major form of disturbance in mountain ash forests. Extensive, high-intensity, stand-replacing fires dramatically alter habitat structure and resource distribution across large spatial scales (Bowman et al., 2009; Bradstock et al., 2005). How fauna respond to these fires is dependent on the interacting properties of the fire regime and the behavioral ecology of each species. The underlying history of fire in the landscape and the immediate effects of wildfire influence the availability and distribution of key wildlife habitat resources. Mountain ash forests often exhibit complex early seral mosaics resulting from the layering of historic and contemporary disturbances.

8.6 FAUNA AND FIRE-AFFECTED HABITAT STRUCTURES

Hollow-bearing trees are an essential habitat feature of tall mountain ash forests. Tree hollows begin to form in mature trees, known as "stags," after 120 years (Figure 8.4). Stags provide essential denning habitat for a diverse fauna of arboreal marsupials, birds, bats, and lizards. The density of hollow-bearing trees per hectare is the primary determinant of the presence of the globally endangered Leadbeater's possum (*Gymnobelideus leadbeateri*), a small, colonial, nocturnal, arboreal marsupial endemic to the VCH. Recent declines in the distribution and numbers of the possum are directly related to the fire- and logging-influenced decline in the number and distribution of hollow-bearing trees (Lindenmayer et al., 2013a,b).

Large, high-severity wildfires radically alter vegetation structure in mountain ash forests. Fires that burn the understory remove important habitat features such as silver wattle, which provides a major food resource (sap) for Leadbeater's

FIGURE 8.4 A large, hollow-bearing tree, known as a "stag," in an area of long unburnt, late successional forest. This particular example may have taken over 400 years to form. *(Photo credit: Laurence Berry.)*

possum and also provides a dense canopy that facilitates the movement of fauna through the forest. Leadbeater's possum has a complex social system, where many individuals form part of a large matriarchal colony. These colonies need to gather food resources over large areas. The impact of fire on food resource availability may reduce the total foraging area available to support these large colonies. Fires that consume the canopy lead to a decline in the abundance of foliage feeders such as the greater glider (*Petauroides volans*) (Lindenmayer et al., 2013b). Some birds that feed on open ground, such as the buff-rumped thornbill (*Acanthiza reguloides*), scarlet robin (*Petroica boodang*), and flame robin (*Petroica pheonicea*), become more abundant immediately after understory fires (Loyn, 1997).

Following the infamous 2009 "Black Saturday" wildfires that burned under extreme weather (high temperatures, drought, high winds) through ~250,000 ha of mountain ash forest in Victoria, Australia, the mountain brushtail possum (*Trichosurus cunninghami*) was the only arboreal marsupial occupying severely burnt sites 1 to 3 years after the fires (Lindenmayer et al., 2013b). This possum is a large (2-4 kg), nocturnal, arboreal marsupial with generalized foraging habits and diet (Seebeck et al., 1984). A study that fitted proximity measuring radio transmitters to possums during the 2009 fires found no evidence of fire-associated possum mortality (Banks et al., 2011). However, 80% of the dead, hollow-bearing trees within the range of the collared animals were consumed during the fire. Individuals adapted to this loss in critical resources by displaying flexibility in den site selection. Banks et al. (2011) concluded that although the adaptability in resource use displayed by the mountain brushtail possum may buffer populations against decline in the short term, increased frequency of stand-replacing fires is likely to cause major limitations on shelter availability and pose significant conservation challenges for hollow-dependent fauna.

8.7 FAUNAL RESPONSE TO THE SPATIAL OUTCOMES OF FIRE

Recent research in the mountain ash forests of Australia has highlighted the importance of the spatial outcomes of fire on biodiversity responses (Lindenmayer et al., 2013b; Robinson et al., 2014; Berry et al., 2015a). Species able to persist within recently burned forest may benefit from mixed-severity fire. For example, Sitters et al. (2014) tested whether the diversity and configuration of vegetation types and forest age classes influenced bird diversity in tall, open eucalypt forests in the Otways in southwest Victoria, Australia. They found that spatial variability in fire regimes can increase the diversity of bird assemblages in eucalypt forests.

Unburnt patches of forest embedded within large burns may provide essential habitat for species whose life cycles are dependent on resources reduced by fire. The relevance of these fire refuges to species conservation following large fires may be dependent on their size and the amount of additional unburnt resources available in the surrounding landscape. For example, Robinson et al. (2014) compared bird response to fire severity in patches of mountain ash forest with varying fire intervals: short (<3 years) and long (>20 years). They found that unburnt patches that persisted within the extent of large fires may act as refuges for birds associated with late-successional conditions. In particular, unburnt patches of forest with a long time since fire contained more individuals (56%) and higher species richness (20-40) than severely burned forest <3 years after fire. The rapid increase in some bird species to near, or above, prefire levels by the third year after severe fire (Lindenmayer et al., 2014a,b (Appendix S1)), however, suggests a need for additional study. Berry et al. (2015a) found that unburnt patches of forest located in mesic gullies at the edge of extensively burned landscapes may facilitate the presence of the mountain brushtail possum in fire-affected forest. The greater glider, however, occurred more often in large, intact, unburnt areas of forest. Lindenmayer et al. (2013b) found that the greater glider, sugar glider, and Leadbeater's possum declined as the amount of burned forest increased in the surrounding landscape. The landscape context effects of fires on fauna are expected to be transitory as rapid postfire recovery of vegetation occurs (Lindenmayer et al., 2013b). However, the temporal period in which landscape context effects should be considered is related to the rate of recovery of limiting resources, the scale and extent of the fire, and the time since previous fire.

8.8 CONSERVATION CHALLENGES AND FUTURE FIRE

Recent large fires in the mountain ash forests of Victoria coupled with short fire intervals and extensive clearcutting and postfire "salvage" logging operations have reduced the total area of available habitat for a range of late-successional specialists. Clearcutting removes keystone habitat structures before fire and may alter the spread of fire throughout the landscape by providing homogenous fuels across large areas. Postfire logging removes important biological legacies that may remain after fire and further fragments habitat after fire (see Chapter 11). These multilayered disturbances may interact to cause irreparable shifts in ecosystem state. For nearly a century, extensive postfire logging operations have followed large fires in the mountain ash forests of Victoria (Lindenmayer and Ough, 2006). The abundance of hollow-bearing trees is greatly reduced in such logged forests. Lindenmayer and Noss (2006) outlined key measures to improve biodiversity retention following large wildfires, including the preservation of unburnt or partially burned forest patches and excluding postfire logging from nature reserves and water catchments, extensive areas of old-growth forest, and areas with few roads.

A recent risk assessment of the mountain ash forests of the VCH ranked the ecosystem as critically endangered, with a predicted 92% chance of ecosystem collapse by 2067 (Burns et al., 2015). The authors of the study recommended immediate protection of undisturbed areas of mountain ash, substantial restorative action in degraded stands, and the cessation of widespread industrial logging. Spies et al. (2012) outlined a list of common conservation objectives in fire-prone, temperate, tall forests in southeast Australia and the Pacific Northwest of the United States, including identifying the socioeconomic context of appropriate fire management, identifying desirable disturbance for biodiversity needs, moving beyond fuel treatment as an end point in fire management by basing management goals on species ecology, and planning for the spatial outcomes of future large fire events.

The future integrity of mountain ash ecosystems is in doubt. The increasing frequency of large, unplanned fires coupled with ongoing extensive logging operations is decreasing the availability of suitable habitat for a range of late-successional specialist species (Lindenmayer et al., 2012). In some areas the interaction of fire and logging disturbances are creating irreversible shifts in ecosystem state (Lindenmayer et al., 2011). Studies show that recently logged forest burns at higher severity than old-growth forests (Taylor et al., 2014). Large areas of homogenously aged young forest regenerating from logging will likely increase the severity of a fire passing through the landscape. This process may catalyze an increase in fire severity in adjacent areas of forest that, because of topography, stand age, and forest type, may otherwise have burned at lower severity. Contemporary land uses may be mitigating the influence of landscape-level mediators of fire severity. These changes in the spatiotemporal mosaic of fire regimes will contribute to the continued decline of already endangered species such as Leadbeater's possum and the yellow-bellied glider and to making other common species, such as the greater glider, rarer. The authors of a major long-term research project in the VCH concluded that the only way to counter downward species trajectories is to remove the principal cause of the decline in keystone habitat structures and ecologically important biological legacies: widespread industrial logging (Lindenmayer et al., 2013a). At present, the majority of logged forest areas are regrowth from the massive 1939 "Black Friday" wildfires. Within these areas, most trees are felled approximately 40 years from developing the key ecological features required by fauna associated with late-successional conditions. The lack of accurate fire severity maps from the 1939 fires also leads to the felling of old trees embedded within regrowth forest that survived the fire. A Great Forest National Park has been proposed to encompass the extent of mountain ash forest in the VCH, from Kinglake north of Melbourne to Mount Baw-Baw in the southeast (Lindenmayer et al., 2013a). The protection of large areas of ash forest increases the likelihood that a greater area of old-growth forest will remain unburnt following future large wildfires. It would also ensure that regrowth forests across large spatial extents will reach the necessary level of maturity needed to provide essential habitat structures for endangered native, late-successional fauna in the region. Further, it would allow for ecologically appropriate levels of mixed- and high-severity fire to provide unmanaged complex early successional conditions for wildlife that benefits from higher-severity fire at some postfire time period.

REFERENCES

Ashton, D.H., 1975. Studies of flowering behaviour in *Eucalyptus regnans* F. Muell. in central Victoria. Aust. J. Bot. 23, 399–411.

Ashton, D.H., 1976. Development of even-aged stands of *Eucalyptus regnans* F. Muell. in Central Victoria. Aust. J. Bot. 24, 397–414.

Ashton, D.H., 1981. Fire in tall open forests (wet sclerophyll forests). In: Gill, A.M., Groves, R.H., Noble, I.R. (Eds.), Fire and the Australian Biota. Australian Academy of Science, Canberra.

Ashton, D.H., Attiwill, P.M., 1994. Tall open forests. In: Groves, R.H. (Ed.), Australian Vegetation. Cambridge University Press, Melbourne, pp. 157–196.

Ashton, D.H., Martin, D.G., 1996. Regeneration in a pole-stage forest of Eucalyptus regnans subjected to different fire intensities in 1982. Aust. J. Bot. 44, 393–410.

Attiwill, P.M., 1994. Ecological disturbance and the conservative management of eucalypt forests in Australia. For. Ecol. Manag. 63, 301–346.

Banks, S.C., Knight, E.J., Mcburney, L., Blair, D., Lindenmayer, D.B., 2011. The effects of wildfire on mortality and resources for an arboreal marsupial: resilience to fire events but susceptibility to fire regime change. PLoS One 6, e22952.

Beale, R., 2007. If Trees Could Speak. Stories of Australia's Greatest Trees. Allen & Unwin, Sydney.

Benyon, R.G., Lane, P.N.G., 2013. Ground and satellite-based assessments of wet eucalypt forest survival and regeneration for predicting long-term hydrological responses to a large wildfire. For. Ecol. Manag. 294, 187–207.

Berry, Laurence E., Driscoll, Don A., Banks, Samuel C., Lindenmayer, David B., 2015a. The use of topographic fire refuges by the greater glider (Petauroides volans) and the mountain brushtail possum (Trichosurus cunninghami) following a landscape-scale fire. Australian Mammalogy 37, 39–45.

Berry, L.E., Driscoll, D.A., Stein, J., Blanchard, W., Banks, S., Bradstock, R., Lindenmayer, D.B., 2015b. Identifying the location of fire refuges in wet forest ecosystems. Ecological Applications (in press).

Bowman, D., Balch, J.K., Artaxo, P., Bond, W.J., Carlson, J.M., Cochrane, M.A., D'antonio, C.M., Defries, R.S., Doyle, J.C., Harrison, S.P., Johnston, F.H., Keeley, J.E., Krawchuk, M.A., Kull, C.A., Marston, J.B., Moritz, M.A., Prentice, I.C., Roos, C.I., Scott, A.C., Swetnam, T.W., Van Der Werf, G.R., Pyne, S.J., 2009. Fire in the earth system. Science 324, 481–484.

Bowman, D., Murphy, B.P., Neyland, D.L.J., Williamson, G.J., Prior, L.D., 2014. Abrupt fire regime change may cause landscape-wide loss of mature obligate seeder forests. Glob. Chang. Biol. 20, 1008–1015.

Bradstock, R.A., Bedward, M., Gill, A.M., Cohn, J.S., 2005. Which mosaic? A landscape ecological approach for evaluating interactions between fire regimes, habitat and animals. Wildl. Res. 32, 409–423.

Bradstock, R.A., Hammill, K.A., Collins, L., Price, O., 2010. Effects of weather, fuel and terrain on fire severity in topographically diverse landscapes of south-eastern Australia. Landsc. Ecol. 25, 607–619.

Bradstock, R., Penman, T., Boer, M., Price, O., Clarke, H., 2014. Divergent responses of fire to recent warming and drying across south-eastern Australia. Glob. Chang. Biol. 20, 1412–1428.

Bureau of Meteorology, 2014. Climate Data Online [Online]. Bureau of Meteorology. Available: http://www.bom.gov.au/climate/data (accessed November 2014).

Burns, E.L., Lindenmayer, D.B., Stein, J., Blanchard, W., Mcburney, L., Blair, D., Banks, S.C., 2015. Ecosystem assessment of mountain ash forest in the Central Highlands of Victoria, south-eastern Australia. Austral Ecology (in press).

Cary, G.J., Bradstock, R.A., Gill, A.M., Williams, R.J., 2012. Global change and fire regimes in Australia. In: Bradstock, R.A., Gill, A.M., Williams, R.J. (Eds.), Flammable Australia: Fire Regimes, Biodiversity and Ecosystems in a Changing World. CSIRO Publishing, Melbourne.

Griffiths, T., 1992. Secrets of the Forest. Allen & Unwin, Melbourne.

Jackson, W., 1968. Fire, air, water and earth—an elemental ecology of Tasmania. Proc. Ecol. Soc. Aust. 3, 9–16.

Lawson, D.M., Regan, H.M., Zedler, P.H., Franklin, A., 2010. Cumulative effects of land use, altered fire regime and climate change on persistence of Ceanothus verrucosus, a rare, fire-dependent plant species. Glob. Chang. Biol. 16, 2518–2529.

Leonard, S.W.J., Bennett, A.F., Clarke, M.F., 2014. Determinants of the occurrence of unburnt forest patches: potential biotic refuges within a large, intense wildfire in south-eastern Australia. For. Ecol. Manag. 314, 85–93.

Lindenmayer, D., 2009. Forest Pattern and Ecological Process: A Synthesis of 25 Years of Research. CSIRO Melbourne, Melbourne.

Lindenmayer, D., Noss, R., 2006. Salvage logging, ecosystem processes, and biodiversity conservation. Conserv. Biol. 20, 949–958.

Lindenmayer, D., Ough, K., 2006. Salvage logging in the montane ash eucalypt forests of the Central Highlands of Victoria and its potential impacts on biodiversity. Conserv. Biol. 20, 1005–1015.

Lindenmayer, D.B., Cunningham, R.B., Donnelly, C.F., Franklin, J.F., 2000. Structural features of old-growth Australian montane ash forests. For. Ecol. Manag. 134, 189–204.

Lindenmayer, D.B., Hobbs, R.J., Likens, G.E., Krebs, C.J., Banks, S.C., 2011. Newly discovered landscape traps produce regime shifts in wet forests. Proc. Natl. Acad. Sci. U. S. A. 108, 15887–15891.

Lindenmayer, D.B., Blanchard, W., Mcburney, L., Blair, D., Banks, S., Likens, G.E., Franklin, J.F., Laurance, W.F., Stein, J.A., Gibbons, P., 2012. Interacting factors driving a major loss of large trees with cavities in a forest ecosystem. PLoS One 7, e41864.

Lindenmayer, D.B., Blair, D., Mcburney, L., Banks, S.C., Stein, J.A., Hobbs, R.J., Likens, G.E., Franklin, J.F., 2013a. Principles and practices for biodiversity conservation and restoration forestry: a 30 year case study on the Victorian montane ash forests and the critically endangered Leadbeater's possum. Aust. Zoo. 36, 441–460.

Lindenmayer, D.B., Blanchard, W., Mcburney, L., Blair, D., Banks, S.C., Driscoll, D., Smith, A.L., Gill, A.M., 2013b. Fire severity and landscape context effects on arboreal marsupials. Biol. Conserv. 167, 137–148.

Lindenmayer, B., Blair, B., Mcburney, L., Banks, S., 2014a. Preventing the extinction of an iconic globally endangered species—Leadbeater's possum (*Gymnobelideus leadbeateri*). J. Biodiv. Endanger. Species 2, 2.

Lindenmayer, D.B., Blanchard, W., McBurney, L., Blair, D., Banks, S.C., Driscoll, D.A., Smith, A.L., Gill, A.M., 2014b. Complex responses of birds to landscape-level fire extent, fire severity and environmental drivers. Divers. Distrib. 20, 467–477.

Loyn, R.H., 1985. Bird populations in successional forests of mountain ash *Eucalyptus regnans* in Central Victoria. EMU 85, 213–230.

Loyn, R.H., 1997. Effects of an extensive wildfire on birds in far eastern Victoria. Pac. Conserv. Biol. 3, 221.

Lutze, M.T., Campbell, R.G., Fagg, P.C., 1999. Development of silviculture in the native state forests of Victoria. Aust. For. 62, 236–244.

Mccarthy, M.A., Gill, A.M., Lindenmayer, D.B., 1999. Fire regimes in mountain ash forest: evidence from forest age structure, extinction models and wildlife habitat. For. Ecol. Manag. 124, 193–203.

Pyne, S.J., 1992. Burning Bush: A Fire History of Australia. Allen & Unwin, Sydney.

Robinson, N., Leonard, S., Ritchie, E., Bassett, M., Chia, E., Buckingham, S., Gibb, H., Bennett, A.F., Clarke, M.F., 2013. Refuges for fauna in fire-prone landscapes: their ecological function and importance. J. Appl. Ecol. 50, 1321–1329.

Robinson, N.M., Leonard, S.W.J., Bennett, A.F., Clarke, M.F., 2014. Refuges for birds in fire-prone landscapes: the influence of fire severity and fire history on the distribution of forest birds. For. Ecol. Manag. 318, 110–121.

Seebeck, J., Warneke, R., Baxter, B., 1984. Diet of the bobuck, Trichosurus caninus (Ogilby) (Marsupialia: Phalangeridae) in a mountain forest in Victoria. Possums and Gliders. In: Smith, A.P., Hume, I.D. (Eds.), Possums and Gliders. Surrey Beatty and Sons, Sydney, pp. 145–154.

Simkin, R., Baker, P.J., 2008. Disturbance history and stand dynamics in tall open forest and riparian rainforest in the Central Highlands of Victoria. Aust. Ecol. 33, 747–760.

Sitters, H., Christie, F.J., Di Stefano, J., Swan, M., Penman, T., Collins, P.C., York, A., 2014. Avian responses to the diversity and configuration of fire age classes and vegetation types across a rainfall gradient. For. Ecol. Manag. 318, 13–20.

Smith, A.L., Blair, D., Mcburney, L., Banks, S.C., Barton, P.S., Blanchard, W., Driscoll, D.A., Gill, A.M., Lindenmayer, D.B., 2014. Dominant drivers of seedling establishment in a fire-dependent obligate seeder: climate or fire regimes? Ecosystems 17, 258–270.

Spies, T.A., Lindenmayer, D.B., Gill, A.M., Stephens, S.L., Agee, J.K., 2012. Challenges and a checklist for biodiversity conservation in fire-prone forests: perspectives from the Pacific Northwest of USA and Southeastern Australia. Biol. Conserv. 145, 5–14.

Taylor, C., McCarthy, M.A., Lindenmayer, D.B., 2014. Nonlinear effects of stand age on fire severity. Conserv. Lett. 7, 355–370.

Wood, S.W., Hua, Q., Allen, K.J., Bowman, D., 2010. Age and growth of a fire prone Tasmanian temperate old-growth forest stand dominated by Eucalyptus regnans, the world's tallest angiosperm. For. Ecol. Manag. 260, 438–447.

Wood, S.W., Murphy, B.P., Bowman, D., 2011. Firescape ecology: how topography determines the contrasting distribution of fire and rain forest in the south-west of the Tasmanian Wilderness World Heritage Area. J. Biogeogr. 38, 1807–1820.

Wood, S., Bowman, D., Prior, L.D., Lindenmayer, D.B., Wardlaw, T., Robinson, R., 2014. Tall eucalypt forests. In: Lindenmayer, D.B., Burns, E., Thurgate, N.Y., Lowe, A.J. (Eds.), Biodiversity and Environmental Change: Monitoring, Challenges and Direction. CSIRO Publishing, Melbourne.

Case Study: The Importance of Mixed- and High-Severity Fires in sub-Saharan Africa

Ronald W. Abrams
Dru Associates, Inc., Glen Cove, New York, USA

8.9 THE BIG PICTURE

Biodiversity in sub-Saharan Africa evolved and persists in this complex region in part because of the various mechanisms of habitat change initiated by mixed- and high-severity fires. The greatest expanse of vegetated habitat in Africa is savannah, where large, severe fires are uncommon and anthropogenic burns are widespread (Figure 8.5). Lehmann et al. (2014) indicate that savannah ecology sustains as much as one-fifth of the global human population and—in Africa more than anywhere—hosts "most of the remaining megafauna." Across Africa where conservation is practiced, resource managers have embraced the importance of mixed- and high-severity fire, particularly where national parks and preserves serve ecotourism. A reigning leader in fire management is South

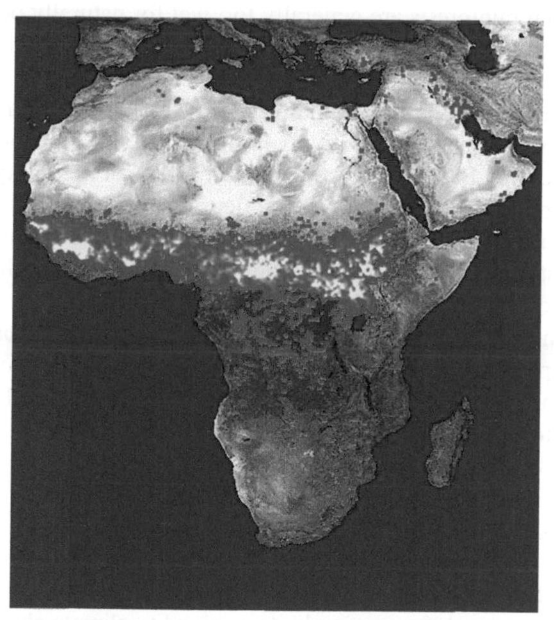

FIGURE 8.5 MODIS Rapid Response System Global Fire Maps showing regions burning (red) as of November 29, 2008. The image has been cropped to show sub-Saharan Africa. *(http://modis.gsfc.nasa.gov/gallery/individual.php?db_date=2008-11-29; accessed January 8, 2015.)*

Africa's Kruger National Park (see various articles in Du Toit et al., 2003; Biggs and Rogers, 2003). In sub-Saharan African savannah key drivers of the balance between grasses and woody vegetation include both natural wildfire (rare) and controlled burns (Du Toit et al., 2003; Govender et al., 2006). Controlled burns in the grasslands of African savannah burn fast and hot, spread quickly, and create habitat mosaics that in turn are associated with physiographic diversity. In this discussion, all fires that have biodiversity implications are mixed with portions that are high severity (termed "mixed severity").

In contrast to the story told by most of this book, Africa is comparatively closer to a "historic" condition for both ecological and anthropological reasons. The preface to this book sets a framework that seeks to overturn unreasonable "fear" of fire, as well as a suite of misunderstandings about the importance of mixed- and high-severity fire to various biomes. Pyne (2004) called for a new attitude that embraces the reality of wildfire, changes in land use, and prevention of continuing habitat loss, and optimistically expected such progress in 5-6 years. Over a decade later, we tackle the issue, and this section discusses how people and natural resource managers have perceived mixed-severity fire in sub-Saharan Africa—a very different perspective than that of the developed regions of the world.

8.10 WHERE IS FIRE IMPORTANT IN SUB-SAHARAN AFRICA?

Large tracts of Congo, Lakes Tanganyika and Victoria, tropical montane forests, and coastal rainforests are generally too wet for naturally occurring fires, and while anthropogenic fires occur, they are largely associated with pastoral agriculture (Cochrane, 2003; Thonicke et al., 2001; Brncic et al., 2007; Krawchuk and Moritz, 2011). Bowman et al. (2009) showed from a global perspective that fire frequency is relatively high in central Africa, but because the region of the Congo Basin is too wet for natural fires at high frequency, this is clearly anthropogenically induced fire frequency. The Congo Basin and its margins of tropical forests have lightning frequencies that are among the world's highest (Christian et al., 2003), but the high degree of moisture limits not only spread but also the intensity of fire, and thus severity is not high. Yet Laurance (2003) explained that the subtle nature of fire effects on tropical forests and their biota is widespread but poorly inventoried. Brncic et al. (2007) noted that lightning strikes in tropical forests tend to burn one spot (individual trees), which may contribute on a very small scale to mosaics of habitat but do not create the effects of large or intense fires.

In rain forest and other habitats, however, the potential for fire to create ubiquitous ecological change does exist; this is underscored by Bardgett and van der Putten (2014) in their discussion of the high sensitivity of microbial biota to microhabitat changes: fire can influence such changes, especially with new evidence of subtle range shifts as microbiota respond to climate change. Govender et al. (2006) described how variable fire intensity differentially heats the ground, which would influence desiccation and soil exposure to both wind and water erosion and thus can substantially alter the substrate and fauna. Microbiota may be the elephant in the room with respect to the future of biodiversity in Africa.

In sub-Saharan Africa fire-induced habitat change promotes biodiversity in both tropical forest and savannah, although most of the research in Africa is focused on savannah, the most widespread habitat on the subcontinent (Thonicke et al., 2001; van Wilgen et al., 2003, 2004; Brncic et al., 2007). Some research on montane habitat also confirms the contribution of mixed- and high-severity fire to biodiversity, but at smaller scales that are bounded by physical habitat barriers (Geldenhuys, 2004), a different scenario than the wildfires of interest in the northern hemisphere. These are often intense (severe), but measures of this factor are sparse for Africa. As explained in the preface of this book, wildfire has an important role in shaping vegetation communities, and we need to help managers to understand where the "free fire" should occur. This book is focused on a conservation-oriented understanding of wildfire, but much of what is discussed here about fire's contribution to biodiversity in Africa comes from studies of controlled burns or long-term observations that infer fire patterns from landscape history.

8.11 WHAT ABOUT PEOPLE AND FIRE?

Many people in Africa have a daily relationship with fire (cooking, heating). Though occasionally situations become dire because of uncontrolled, human-induced burns or the rare, naturally occurring wildfire, for many rural Africans fire is not something to be feared as a threat. This is in contrast to the perception of fire in developed nations, where the fear of wildfire limits its potential to perform a fundamental role in biodiversity maintenance (Bowman et al., 2009). Some people in Africa respond with suppression when fires threaten sacred woodlands or specimen trees, so there is an influence of religion or culture on habitat protection that may or may not support biodiversity (Hens, 2006; Ormsby, 2012). Criminal fires in Africa from war (Dudley et al., 2002) or civil conflicts (Hamilton et al., 2000) are clearly political and beyond the scope of this book. Nevertheless, criminal disturbance of habitats is a high priority in Africa, with a potentially positive effect on biodiversity as part of the overall relationship between the people and fire (as opposed to poaching or irresponsible waste disposal). Successful conservation strategies in Africa must consider the benefits that fire provides to a local human population (heat and fuel), just as conservation research now needs to consider the agenda of people living on the land (Abrams et al., 2009).

But is it possible for rural people to combine judicious use of fire that is both productive for them and coincidentally for the ecosystem? In South America studies of cycads (plants that are abundant in Africa) show that fires are used to encourage seed production for human use, and at the same time cycad reproduction increases in postfire recovery (Norstog and Nichols, 1997). The cycad's coralloid root adaptation seems to enable the plants to persist and even proliferate after either brush fires or anthropogenic burns (Norstog and Nichols, 1997). The only negative effect cited is when a fire is intense (hot) enough to kill the invertebrate pollinators residing in the surrounding leaf litter (which can result in shifting of the reproductive periodicity while the litter is recolonized). Of course there is the perspective that the extra cycads (and other species encouraged by human-set fire) lead to ecological imbalance, but for a group of plants as resilient as cycads, their dominance in the ecosystem may occur in a wide range of conditions. In fact, this group of plants is among the most resilient of organisms, with species that change their phenotype when transplanted between latitudes (Heenan, 1977, David Heenan, personal communication).

8.12 COEVOLUTION OF SAVANNAH, HERBIVORES, AND FIRE

Fire has influenced African ecological systems over evolutionary timescales, for example, in the historical shift of southwestern African grasslands from C3 (3-carbon molecules) to C4 (4-carbon molecules) dependent photosynthesis over millennia (Hoetzl et al., 2013; Lehmann et al., 2014). A similar historical trend has been debated for North America, however, with a more specific look at the mechanisms at work (see Chapters 1, 9 and 13 in this book). Heisler et al. (2003) described how fire may not always control domination by woody cover; shrub cover increased despite annual fire in a long-term study of North American grasslands. Johnson (2009) explained the decline of North American megafauna as a cause of the loss of prehistoric grasslands, an illustration from history that concurs with observations about Africa's herbivores as integral to the maintenance of Africa's grasslands.

In contrast to North America, sub-Saharan Africa is still dominated by savannah, where heterogeneity is maintained by two chief mechanisms. First, mixed-severity fires and herbivores (moderated by substrates that contain limited moisture for at least part of the year) directly suppress woody vegetation and favor grass-dominated habitat (Westbroek et al., 1993; Sankaran et al., 2005; Archibald, 2008). Second, and perhaps more important, is the way fire influences structural features that effect biodiversity, such as heating/desiccation of soils that alters chemistry and microbiota, which in turn influences the germination of species through control of nutrient availability in soils (Venter et al., 2003; Bardgett and van der Putten, 2014). In tropical peatlands (perhaps found in marshlands of the Congo Basin) fire poses a threat to biodiversity when apparently low-intensity smoldering burns occur because of the long burnout times (again, Laurance's (2003) insidious fire characterization) and the deep and wide reach of smoldering peat layers (subject to desiccation by climate change) that result in adverse effects to soils reaching beyond the surfaces affected (Turetsky et al., 2015). In other situations fire in marshland may create habitat mosaics that can enhance structural habitat diversity (Turetsky et al., 2015), but the subject needs new research in Africa.

The mid-twentieth century reasoning for biodiversity conservation through controlled burns had support because of the rising fear that Africa, through the loss of naturally balanced herbivore populations (i.e., the keystone herbivore hypothesis), would follow the path of other continents that lost grasslands and native forest diversity (Owen-Smith, 1989). As early as the 1980s, Owen-Smith (1989) confirmed that in Africa the natural cycling of fire effects was disrupted by fire suppression and elephant (*Loxodonta Africana*) population declines, which together have led to more wooded savannah habitat, and a loss of grassland, in the Tsavo area of Kenya (Owen-Smith, 1989).

8.13 HERBIVORES AND FIRE

Herbivore consumption and excretion are critical processes supporting biodiversity in African savannahs. Similarly, Bond and Keeley (2005) described fire as a global herbivore with significant influence on biodiversity through its capacity to alter habitats. Habitat alteration by grazing, browsing, and burning release plant species held in check by other plants that have a competitive advantage without disturbance, altering species assemblages and ecosystem dynamics, even if only for a short time. The resulting vegetative diversity, in height, form, and percentage of cover at different layers of woodland and savannah habitat, combine to support food diversity and thus faunal diversity, making the role of fire integral to the ecology and sociology of sub-Saharan Africa.

Sankaran et al. (2005) modeled the tree–grass balance in African savannah, showing that variations in rainfall, soils, and fire/herbivory drive and maintain diversity. Infrequent or low-intensity fire (>10.5-year return cycle) promotes woody cover, and conversely, frequent mixed-severity fires deter woody cover (except on sandy soils) (Sankaran et al., 2005). Of course, a savannah region that is drier accumulates greater fuel loads, leading to increased fire frequency and intensity, further suppressing woody habitat in favor of grasslands (Higgins et al., 2000) upon which many megafaunal species depend. Sankaran et al. (2005) found that most savannahs maintain woody cover well below the resource-limited upper bound, not generally reaching their climatic bounds, suggesting that drivers other than simply climate prevail (e.g., mixed severity fire, herbivores, soils), leading to the conclusion "… that water limits the maximum cover of woody species in many African savannah systems, but that disturbance dynamics control savannah structure below the maximum."

The long-term research in the Serengeti ecosystem of East Africa led by Sinclair et al. (2007) gives a useful picture of the relationship between fire and savannah biodiversity. Population size and dispersion of wildebeest (*Connochaetes taurinus*) and African buffalo (*Syncerus caffer*) have a negative correlation with fire return frequency and intensity (more or less accumulated fuel), and standing elephant population levels have a negative correlation with woodland recovery (i.e., a positive correlation with mixed-severity fire that causes direct tree mortality or weakening, which makes individuals more susceptible to lower-intensity fires); both mechanisms show strong biofeedback with the frequency and intensity (areal coverage = severity) of fires. As reduced fire activity continued over the years since the high fire activity of the 1960s, however, wildebeest populations later declined as woodlands expanded into grasslands, providing lions (*Panthero leo*) with better hiding cover and hunting success (Sinclair et al., 2007). In the Serengeti-Mara system there have been no known naturally occurring fires in recent decades—all were anthropogenic (Sinclair et al., 2007). Records of the extent of burns during the dry season in East Africa, documented by aerial reconnaissance, show a steady decline in the area burned from the early 1960s to the 1980s. More recent records confirm this trend (unpublished data cited by Sinclair et al. (2007)). These trends reflect an increase in herbivore reduction of grass biomass caused by grazing, meaning fires cannot progress as mixed-high severity burns, which can affect the grassland-woodland balance, leading to further shifts from grassland savannah to woodland.

In contrast to East Africa, in the Kruger Park of South Africa, where up to 20% of savannah fires are caused by lightning (van Wilgen et al., 2003), and in deference to the protection of villages and homes, most fires are subject to some form of control because the majority of them are set by humans. South African National Parks, however, have a long history in Kruger Park of maintaining grasslands through a combination of mechanisms (i.e., manipulation of vegetation by elephants, herbivore grazing, and frequent fire return). In 1992 Kruger Park managers decided to use a "lightning fire" approach, but within 10 years of monitoring they recognized that almost 80% of fires in the park were anthropogenic, so the experiment could not be continued, and they reverted to patch-mosaic burns and allowed natural fires to reach high intensity and to burn out, using controlled burns where needed and monitoring biodiversity with fire as a parameter (Biggs and Rogers, 2003; van Wilgen et al., 2004).

In modeling the triggers that switch habitats between grass- or tree-dominated savannahs, Higgins et al. (2000) confirmed Warner and Chesson's (1985) view that the "storage effect" hypothesis explains observations that varying fire intensities produce different species releases and thus coexistence of potentially competitive species (Warner and Chesson, 1985). Howe (2014) modernized this concept, especially for tropical tree diversity maintenance, terming it the diversity storage hypothesis. Higgins et al. (2000) hypothesized that grass–tree habitat-sharing is strongly influenced by the fortune of tree seedlings in surviving drought or avoiding the flame zone of fires intense enough to kill young trees, wherein tree recruitment is controlled by rainfall, which limits seed establishment, and by mixed- to high-severity fire that prevents recruitment to adult sizes. Thus variations in fire intensity are a factor controlling the ability of trees to escape the flame zone, where mortality is greater. Variable fire intensity also influences species "selection" as the more fire-resistant size classes are adapted to recolonizing more quickly or, where enough individuals escape severe fires, to maintaining the local population (van Wilgen et al., 2003). In other words, variance in fire intensity produces the variance in recruitment rates that is necessary for the storage effect to operate (Higgins et al., 2000).

A pair of studies reviewed by Mayer and Khalyani (2011; also see Figure 8.6) showed fire frequency drives shifts between forest, savannah, and grassland biomes. Staver et al. (2011) found that, globally, fire is a strong indicator of savannah distribution, where the absence of fire in sub-Saharan Africa leads to bimodal tree cover in terms of frequency when rainfall is intermediate (1000-2500 mm Mean Annual Rainfall (MAR)). Hirota et al. (2011) similarly found that fire (interacting with rainfall regimes) has the potential to create three modes: forest, savannah, and grasslands. Mayer and Khalyani (2011), concurring with Sinclair et al. (2007), summarized their interpretation of these data by writing: "Both reports identify an unstable state at 50-60 per cent tree cover; either trees take hold and promote their own growth hydrologically (and suppress fire), or grasses take hold and promote their expansion through fire (Figure 8.6)."

Staver et al. (2011) and Hirota et al. (2011) identified the transitions among biomes using global data sets showing that areas remain forested (>60% tree cover) with rainfall routinely >2500 mm/year. Habitats that receive a middle range of precipitation (1000-2500 mm/year) were forest or savannah depending on the strength of the fire–grass feedback. Habitats are unstable when tree cover is between 50% and 60%, a condition of transition between forest and savannah. When habitats are subject to strong seasonal rainfall (i.e., between ~750 and 1500 mm/year) they were either savannah or grassland, depending on fire frequency.

8.14 BEYOND AFRICA'S SAVANNAH HABITAT

Geldenhuys (2004) described how forests in South Africa once might have covered 7% of the terrain based on suitable microclimate, but now cover only 0.1% of their potential range. Geldenhuys attributed this to anthropogenic activities, with fire setting the boundaries of many habitats. In the coastal and certain inland regions of southern Africa, mountain range physiography creates wind patterns that drive the dispersion of fire (and intensity), and these locations actually serve as "fire refugia" that, based on Geldenhuys' description, also provide biodiversity storage depots (again, the storage effect or Howe's diversity storage hypothesis). Latimer et al. (2005) presented data showing that speciation in South Africa's fynbos biome (a Mediterranean-type habitat) is more rapid than in other comparable regimes, and release by fire is a leading management option.

Du Toit et al. (2003), Sinclair, and Geldenhuys interpret the research on fire ecology in sub-Saharan Africa such that land use managers must seek a balance between fire suppression and preserving its role in biodiversity maintenance. But they face a daunting challenge. Conservationists are wary of any non-

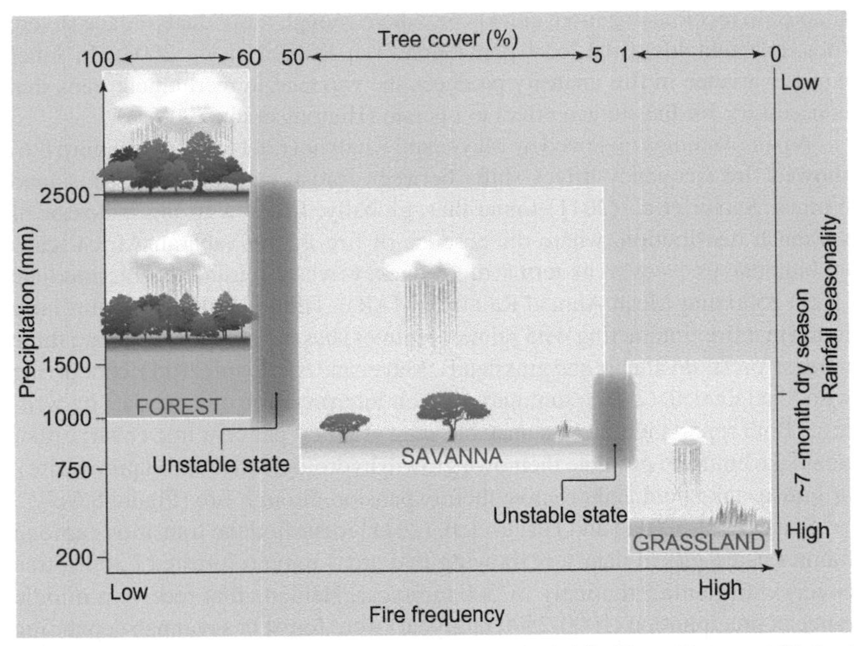

FIGURE 8.6 Vegetation dynamics mediated by fire and rainfall. *(From Mayer and Khalyani 2011 with permission.)*

natural alteration of natural cycles (e.g., climate-related and anthropogenic loss of the natural fire cycles) as the potential downfall of entire food webs, including in sub-Saharan Africa, where the bird and mammal populations have a prevailing role in biodiversity maintenance (Laurance and Useche, 2009; Mayer and Khalyani, 2011; Howe, 2014). Nimmo et al. (2013) showed how fire-created habitat mosaics contribute to faunal diversity in Australia and echoed the "patch-mosaic burning" strategy used globally. Gregory et al. (2010) confirmed, by studies in East African savannahs, that avifaunal diversity and species persistence in a given area are dependent on both herbivore grazing and fire return cycles. The adaptive management strategy used in Kruger Park in South Africa has derived the concept of "thresholds of potential concern" (TPCs), which takes a holistic approach to habitat, and thus wildlife management, by following ecological responses to TPCs and adapting their controlled burns schedule accordingly (Biggs and Rogers, 2003).

8.15 HABITAT MANAGEMENT THROUGH CONTROLLED BURNS

Venter et al. (2003) described how the interaction of fire with patch dynamics and soil biogeochemistry are important drivers of physiographic and, consequently, overall ecological heterogeneity, including the wildlife of Kruger Park.

In considering the size or intensity of two historic episodes of severe and intense fires in Kruger (>20% of the park's area), Venter et al. (2003) concluded that, while these fires resulted in temporary shifts in vegetation in any given location, the broader ecosystem remained stable, and they argued that prescribed burns can serve as scale-independent disturbances. In addition, while Latimer et al. (2005) agree that wildlife assemblages are structured in response to fire-driven cycles, they also found that species release follows, to some extent, Hubbell's neutral theory of ecology in the fynbos habitat because all of the many species are present and their appearance follows the patterns of disturbance (i.e., drought and fire) that cause microhabitat variations. Over evolutionary time scales, the system remains stable, with similar biodiversity metrics, but results in dynamic, shifting patches to which wildlife respond as the plants respond to mixed- and high-severity fire (Gregory et al., 2010). Thus, what would sub-Saharan Africa look like without anthropogenic fire?

But there are additional questions about our collective knowledge of the results of controlled burns for biodiversity conservation. Ironically, Romme et al. (1998) demonstrated that, at least in certain systems, adaptation to long-standing natural fire return cycles can actually suppress species release. The practitioners who manage Kruger National Park manipulate fire to prevent such suppression, using patch-mosaic strategies as a basis to ensure varying stages of succession (Biggs and Rogers, 2003). Parr and Andersen (2006) question whether sufficient monitoring of patch-mosaic burning has been used to parlay results in the field into adaptive management strategies, although they credit South African National Parks with a comprehensive approach to the issue (http://celtis.sanparks.org/parks/kruger/conservation/scientific/key_issues/6.Fire.pdf accessed January 30, 2015), and they do not doubt that pyrodiversity contributes to biodiversity. It seems that when one considers the combined effects on savannah of large herbivores and fire, they amount to biodiversity enhancement because herbivores maintain long-term, consistent pressure on grasses that is comparable to that of large, intense wildfire (widespread in its vegetative effects), which also mediates the interactions between forest, woodland, and grassland. In Kruger Park the herbivore contribution to grassland maintenance is supplemented with rotated patch-mosaic burns, making the combined outcome comparable to a circumstance in which large, mixed-severity fires were common and allowed to burn out, all adaptively managed by the park's TPC approach (Biggs and Rogers, 2003).

The TPC approach has achieved the overall objective of using an understanding of mixed- and high-severity fire dynamics to optimize support of the natural food web. Govender et al. (2006) provided some of the limited measures of fire intensity and severity for sub-Saharan Africa, looking at the savannah of Kruger Park. They reported results of a 21-year experiment designed at the time to test ideas about eradication and recovery, and post hoc analysis using fire intensity measurements and an array of technological and evidentiary methods showed how the occurrence and role of mixed-severity fire relate to a set of variables. Van Wilgen et al. (2004) described fire cycling in the Kruger Park as dependent on grass biomass (influenced by wildlife and rainfall regimes) but also observed that it could be managed by season and fire intensity.

The practices used now, for example, at Tygerberg Preserve in the Western Cape and at Kruger integrate the effects of variable fire intensity in selection of time, size, and target species. In the Western Cape there is less grass and more woody material to burn than in Kruger's savannah. So, in Kruger, wildfire spread rates (which control the ultimate intensity of a controlled burn) are a management choice (Govender et al., 2006). In the Western Cape (see Box 8.1 below), and in areas where uncontrolled spread is less of a concern, like the self-bounded mountain areas of South Africa discussed by Geldenhuys (2004), mixed- and high- severity burns are used to remove woody species or alien species such as black wattle (*Acacia mearnsii*).

However, Govender et al. (2006) showed that in the Kruger Park mixed- and high-severity fire serves to suppress the recruitment of large trees because they must outgrow the flame zone, so such fires are a tool that are being used by managers to maintain the tree islands amid the grasslands. Govender et al.

BOX 8.1 Southwestern Cape Renosterveld Management

Habitat type: Swartland Shale Renosterveld (critically endangered) is restricted to the Cape floristic kingdom within a Mediterranean climate. Renosterveld vegetation occurs in nutrient-rich (clay-based) soils. Structurally, Renosterveld is much less complex than fynbos, consisting mostly of a single layer of shrubs. The most dominant species is *Elytropappus rhinocerotis* (Renosterbos); hence the name of the vegetation type. The grassy understory can be extremely rich in geophytes and annual plant species, which are particularly conspicuous after fire. Renosterveld is a late-successional vegetation type. If the vegetation is not burned, however, it is invaded (encroached upon) by forest precursor species such as *Olea europaea* (wild olive) and *Kiggelaria africana* (wild peach) (adapted from Trinder-Smith et al., 2006).

Species being suppressed:
Olea europaea (wild olive)—tree
Kiggelaria africana (wild peach)—tree
Searsia angustifolia (Wilgerkorentebos)—large shrub
Post-fire recovery species:
Drimia capensis (Maerman)—perennial geophyte
Watsonia marginata—perennial geophyte
Haemanthus sanguineus (April fool)—perennial geophyte
Oxalis species (e.g., *Oxalis ecklonia*na and *Oxalis livida*)—perennial herb
Otholobium hirtum—shrub (pioneer species)
Helichrysum cymosum—woody shrub
Felicia fruticosa (wild aster)—shrub
Chrysanthemoides monilifera (tick berry/Bietou)—shrub

(2006) used a 3000 kW m^{-1} threshold to define high intensity (and thus potential severity) from van Wilgen et al. (1990), citing the example that saplings shorter than 1 m would be top-killed, whereas those taller than 2 m could persist.

An interesting finding by Govender et al. (2006) is that for 4-5 years after a fire, grass growth is robust and fuel accumulates, but into the sixth year and beyond, growth levels off and grazing pressure reduces the vitality of the stand. Does this mean that an optimal, natural fire periodicity would be 4-5 years? The current periodicity in Kruger is 3-6 years, and this seems to be the management target for South Africa's savannah. These fires are often large in area (>5000 ha) and burn at a range of intensities, with the front or back fire lines being of very high intensity. As reported in Du Toit et al. (2003), Govender et al. (2006) found that such fires progress to high severity effects through an array of burn scenarios, including variations in intensity associated with time of day or night, slopes, and wind direction/force, so that by the time the fire burns out the result is a "fine-scale mosaic of varying fire intensities" that drive postfire habitat conditions.

8.16 SOUTHWESTERN CAPE RENOSTERVELD MANAGEMENT

For Africa, there are not as many studies specific to fire regimes as for other regions. I close with an example of habitat management that does parallel other parts of the world. Syphard et al. (2009) included South Africa's Cape floristic region as one of the Mediterranean ecotypes, where fire periodicity is critical to the maintenance of biodiversity through seed release mechanisms. Latimer et al. (2005) provided evidence that this region is among the most biodiverse on earth. As an example of current biodiversity management with fire in this region, a case study using the Renosterveld habitat of South Africa's Western Cape is presented. Here, mixed-severity fire (Figure 8.7) is used to remove three invasive plants, and the patches are allowed to burn out as if they were natural (but with a professional crew on site to protect against the fire running to an unwanted area); this results in an immediate release of eight native species essential to conserving this critically endangered Mediterranean type habitat (Box 8.1).

At Tygerberg Preserve, the pattern of controlled burn implementation follows the patch-mosaic concept, aimed at creating a pattern of a succession that mirrors that of a series of natural fires to release the "stored species" (Figure 8.8). Although this burn (Figure 8.9) was done at the end of summer (for fire protection reasons), the fast recovery of the native vegetation is evident even as winter begins in June/July.

The controlled burns at Tygerberg are supplemented by onsite inventories and monitoring of soil and hydrology, maintaining a flexibility that is responsive to local residents, season, and stages of succession, much as described by Biggs and Rogers (2003) for Kruger Park. Although they are no longer present, the Western Cape Renosterveld once supported rhinoceros (*Diceros bicornis* or

FIGURE 8.7 The ignition of a patch burn in March 2012 at Tygerberg Preserve in South Africa's Cape Floristic Region, a World Heritage region. *(Photo: Penelope Glanville.)*

FIGURE 8.8 The patch burn is allowed to burn hot and expend itself in a prescribed area at Tygerberg Preserve, attended by a professional fire ecology management crew. Different stages of succession can be seen in distant mosaic areas. Note the proximity of residential areas. *(Photo: Penelope Glanville.)*

FIGURE 8.9 Patch mosaic recovery at Tygerberg Preserve, July 2012, where the burn occurred merely 3 months prior (i.e., fall to winter period). *(Photo: Neal Schachat.)*

Ceratotherium simum simum) and still supports bontebok (*Damaliscus pygargus pygarus*) and many smaller mammals, as well as some herpetofauna (N. Schachat, personal communication). The Cape floristic region is extremely sensitive to disturbance but is so valued that even municipalities expend considerable effort at biodiversity management.

8.17 CONCLUSION

An evaluation of the benefits of fires in Africa and other emerging nations is complicated because people living off the land have different priorities than do conservation scientists seeking an academic understanding of the issues surrounding wildfire. In this context, is considering the values of agri-environmental approaches to biodiversity conservation necessary (Laurance and Useche, 2009; Gregory et al., 2010)? Depending on the circumstance, anthropogenic fires can be considered by some communities to be damaging to local ecosystems. Accordingly, the term benefits must be understood as the degree to which the ecologically "desired outcome" is achieved and collateral damage is recognized as unintended. Therefore a simple consensus model in sub-Saharan Africa cannot properly measure or evaluate pyrogenic benefits, so I prefer to discuss "outcomes" of fire management policy and research. In the literature and media coverage of fire in sub-Saharan Africa, there is no parallel to the "panic and fear" that occur in, for example, the western region of the United States. There is, of course, less political focus on subtle issues of environmental policy: much of the continent is justifiably absorbed in basic human survival.

The outcomes of either wild fire or anthropogenic fires are a concern shared by those planning the protection of an ecosystem's biodiversity. For a simplified example, if the return periodicity of fire in a savannah ecosystem is suppressed and too much tinder allowed to accumulate, then the result would be an excess of high-intensity fire that could go beyond reducing the ground and shrub layers (a desired outcome) to damaging the canopy and disrupting the balance of shade and nutrient use that maintains the grass-tree mosaic (van Wilgen et al., 2004; Staver et al., 2011; Lehmann et al., 2014). This suggests an ongoing and integral role for anthropogenic burning in African savannah.

The nexus between people and habitat management makes it difficult to get a clear broad picture of the benefits of mixed- and high-severity fires in sub-Saharan Africa. Controlled burns in Kruger Park's rigorous long-term "experimentation" have taught us much. Many people making their living from the land would consider fire suppression as a benefit that protects their holdings from unwanted damage, possibly including outcomes that are protective of crops or forage for livestock. Even in regions where wildlife are raised for game farming, ecotourism, or even hunting, fire suppression is sometimes seen as a benefit to those endeavors, whereas the natural return cycle of wildfires would produce a different, and alternatively beneficial, outcome.

Nevertheless, the history of fire and habitat management in sub-Saharan Africa teaches us two main lessons: (1) mixed- and high-severity fire is a critical mechanism in the maintenance of biodiversity; and (2) people can and do live with severe fires as part of their local ecosystem. There is a serious need to inventory and assess the role of mixed- and high-severity wildfire in Africa, but between politics, war, and basic human needs, this sort of research has not been done at the same levels at which it is occurring in, for example, North America. For sub-Saharan Africa, where savannah ecology has gripped researchers for five decades or more, results show that the combined effects of fire on savannah, as well as on elephants, rhinos, wildebeest, buffalo, and many more ungulates, is to maintain a balance of woody versus grassy vegetation. When current fire management practices are combined with herbivore population monitoring (as in Kruger Park), the outcome is similar to what must have been the long-term evolution of that balance as influenced by natural, mixed-severity fire. In those parts of Africa where conservation of biodiversity is a priority, and where there is a commitment of human resources in this regard (skilled labor led by science), fire is an important tool and is given its deserved respect.

REFERENCES

Abrams, R.W., Anwana, E.D., Ormsby, A., Dovie, D.B.K., Ajagbe, A., Abrams, A., 2009. Integrating top-down with bottom-up conservation policy in Africa. Conserv. Biol. 23, 799–804.

Archibald, S., 2008. African grazing lawns: How fire, rainfall, and grazer numbers interact to affect grass community states. J. Wildl. Manag. 72, 492–501.

Bardgett, R.D., van der Putten, W.H., 2014. Belowground biodiversity and ecosystem functioning. Nature 515, 505–511. http://www.nature.com/doifinder/10.1038/nature13855.

Biggs, H.C., Rogers, K.H., 2003. An adaptive system to link science, monitoring and management in practice. In: Toit, D. et al., (Ed.), The Kruger Experience: Ecology and Management of Savannah Heterogeneity. Island Press, Washington, DC.

Bond, W.J., Keeley, J.E., 2005. Fire as a global 'herbivore': The ecology and evolution of flammable ecosystems. Trends Ecol. Evol. 20, 387–394.

Bowman, D.M.J.S., Balch, J.K., Artaxo, P., Bond, W.J., Carlson, J.M., Cochrane, M.A., D'Antonio, C.M., Defries, R.S., Doyle, J.C., Harrison, S.P., Johnston, F.H., Keeley, J.E., Krawchuk, M.A., Kull, C.A., Marston, J.B., Moritz, M.A., Prentice, I.C., Roos, C.I., Scott, A.C., Swetnam, T.W., Van Der Werf, G.R., Pyne, S.J., 2009. Fire in the Earth system. Science 324, 481–484.

Brncic, T.M., Willis, K.J., Harris, D.J., Washington, R., 2007. Culture or climate? The relative influences of past processes on the composition of the Lowland Congo Rainforest. Philos. Trans. R. Soc. Lond. Ser. B Biol. Sci. 362, 229–242.

Christian, H.J., Blakeslee, R.J., Boccippio, D.J., Boeck, W.L., Buechler, D.E., Driscoll, K.T., Goodman, S.J., Hall, J.M., Koshak, W.J., Mach, D.M., Stewart, M.F., 2003. Global frequency and distribution of lightning as observed from space by the Optical Transient Detector. J. Geophys. Res. 108, 4-1–4-15.

Cochrane, M.A., 2003. Fire science for rainforests. Nature 421, 913–919.

Du Toit, J.T., Rogers, K.H., Biggs, H.C. (Eds.), 2003. The Kruger Experience: Ecology and Management of Savannah Heterogeneity. Island Press, Washington, DC.

Dudley, J.P., Ginsberg, J.R., Plumptre, A.J., Hart, J.A., Campos, L.C., 2002. Effects of war and civil strife on wildlife and wildlife habitats. Conserv. Biol. 16, 319–329.

Geldenhuys, C.J., 2004. Concepts and process to control invader plants in and around natural evergreen forest in South Africa. Weed Technol. 18, 1386–1391.

Govender, N., Trollope, W.S.W., van Wilgen, B.W., 2006. The effect of fire season, fire frequency, rainfall and management on fire intensity in savannah vegetation in South Africa. J. Appl. Ecol. 43, 748–758.

Gregory, N.C., Sensenig, R.L., Wilcove, D.S., 2010. Effects of controlled fire and livestock grazing on bird communities in East African Savannahs. Conserv. Biol. 24, 1606–1616.

Hamilton, A., Cunningham, A., Byarugaba, D., Kayanja, F., 2000. Conservation in a region of political instability: Bwindi Impenetrable Forest, Uganda. Conserv. Biol. 14, 1722–1725.

Heenan, D., 1977. Some observations on the cycads of Central Africa. Bot. J. Linn. Soc. 74, 279–288.

Heisler, J.L., Briggs, J.M., Knapp, A.K., 2003. Long-term patterns of shrub expansion in a C4-dominated grassland: fire frequency and the dynamics of shrub cover and abundance. Am. J. Bot. 90, 423–428.

Hens, L., 2006. Indigenous knowledge and biodiversity conservation and management in Ghana. J. Hum. Ecol. 20, 21–30.

Higgins, S.I., Bond, W.J., Trollope, W.S.W., 2000. Fire, resprouting and variability: a recipe for grass-tree coexistence in Savannah. J. Ecol. 88, 213–229.

Hirota, M., Holmgren, M., Van Nes, E.H., Scheffer, M., 2011. Global resilience of tropical forest and savannah to critical transitions. Science 334, 232–235.

Hoetzl, S., Dupont, L., SchefuB, E., Rommershirchen, F., Weber, G., 2013. The role of fire in Miocene to Pliocene C_4 grassland and ecosystem evolution. Nat. Geosci. 6, 1027–1030.

Howe, H.F., 2014. Diversity storage: implications for tropical conservation and restoration. Glob. Ecol. Conserv. 2, 349–358. http://dx.doi.org/10.1016/j.gecco.2014.10.004.

Johnson, C., 2009. Paleontology, megafaunal decline and fall. Science 326, 1072–1073.

Krawchuk, M.A., Moritz, M.A., 2011. Constraints on global fire activity vary across a resource gradient. Ecology 92, 121–132.

Latimer, A.M., Silander, J.A., Cowling, R.M., 2005. Neutral ecological theory reveals isolation and rapid speciation in a biodiversity hot spot. Science 309, 1722–1725.

Laurance, W.F., 2003. Slow burn: the insidious effects of surface fires on tropical forests. Trends Ecol. Evol. 18, 209–212.

Laurance, W.F., Useche, D.C., 2009. Environmental synergisms and extinctions of tropical species. Conserv. Biol. 23, 1427–1437.

Lehmann, C.E.R., Anderson, T.M., Sankaran, M., Higgins, S.I., Archibald, S., Hoffmann, W.A., Hanan, N.P., Williams, R.J., Fensham, R.J., Felfili, J., Hutley, L.B., Ratnam, J., San Jose, J., Montes, R., Franklin, D., Russell-Smith, J., Ryan, C.M., Durigan, G., Hiernaux, P., Haidar, R., Bowman, D.M.J.S., Bond, W.J., 2014. Savannah vegetation-fire-climate relationships differ among continents. Science 343, 548–552.

Mayer, A.L., Khalyani, A.H., 2011. Grass trumps trees with fire. Science 334, 188–189.

Nimmo, D.G., Kelly, L.T., Spence-Bailey, L.M., Watson, S.J., Taylor, R.S., Clarke, M.F., Bennett, A.F., 2013. Fire mosaics and reptile conservation in a fire-prone region. Conserv. Biol. 27, 345–353.

Norstog, K.J., Nichols, T.J., 1997. The Biology of Cycads. Cornell University Press, Ithaca.

Ormsby, A., 2012. Cultural and conservation values of sacred forests in Ghana. In: Pungetti, G., Ovideo, G., Hooke, D. (Eds.), Sacred Species and Sites: Advances in Biocultural Conservation. Cambridge University Press, Cambridge.

Owen-Smith, N., 1989. Megafaunal extinctions: the conservation message from 11,000 Years BP. Conserv. Biol. 3, 405–412.

Parr, C.L., Andersen, A.N., 2006. Patch mosaic burning for biodiversity conservation: a critique of the pyrodiversity paradigm. Conserv. Biol. 20, 1610–1619.

Pyne, S.J., 2004. Pyromancy: reading stories in the flames. Conserv. Biol. 18, 874–877.

Romme, W.H., Everham, E.H., Frelich, L.E., Moritz, M.A., Sparks, R.E., 1998. Are large, infrequent disturbances qualitatively different from small, frequent disturbances? Ecosystems 1, 524–534.

Sankaran, M., Hanan, N.P., Scholes, R.J., Ratnam, J., Augustine, D.J., Cade, B.S., Gignoux, J., Higgins, S.I., Le Roux, X., Ludwig, F., Ardo, J., Banyikwa, F., Bronn, A., Bucini, G., Caylor, K.K., Coughenour, M.B., Diouf, A., Ekaya, W., Feral, C.J., February, E.C., Frost, P.G.H., Hiernaux, P., Hrabar, H., Metzger, K.L., Prins, H.H.T., Ringrose, S., Sea, W., Tews, J., Worden, J., Zambatis, N., 2005. Determinants of woody cover in African savannahs. Nature 438, 846–849.

Sinclair, A.R.E., Mduma, S.A.R., Hopcraft, J.G.C., Fryxell, J.M., Hilborn, R., Thirgood, S., 2007. Long-term ecosystem dynamics in the Serengeti: lessons for conservation. Conserv. Biol. 21, 580–590.

Staver, A.C., Archibald, S., Levin, S.A., 2011. The global extent and determinants of savannah and forest as alternative biome states. Science 334, 230–232.

Syphard, A.D., Radeloff, V.C., Hawbaker, T.J., Stewart, S.I., 2009. Conservation threats due to human-caused increases in fire frequency in Mediterranean-climate ecosystems. Conserv. Biol. 23, 758–769.

Thonicke, K., Venevsky, S., Sitch, S., Cramer, W., 2001. The role of fire disturbance for global vegetation dynamics: coupling fire into a dynamic global vegetation model. Glob. Ecol. Biogeogr. 10, 661–677.

Trinder-Smith, T.H., Kidd, M.M., Anderson, F., 2006. Wild flowers of the Table Mountain National Park. Botanical Society of South Africa, Cape Town, South Africa.

Turetsky, M.R., Benscoter, B., Page, S., Rein, G., van der Werf, G.R., Watts, A., 2015. Global vulnerability of peatlands to fire and carbon loss. Nat. Geosci. 8, 11–14.

Van Wilgen, B.W., Everson, C.S., Trollope, W.S.W., 1990. Fire management in southern Africa: some examples of current objectives, practices and problems. In: Goldammer, G.J. (Ed.), Fire in the Tropical Biota: Ecosystem Processes and Global Challenges. Springer Verlag, Berlin, Germany.

Van Wilgen, B.W., Trollope, W.S.W., Biggs, H.C., Potgieter, A.L.F., Brockett, B.H., 2003. Fire as a Driver of Ecosystem Variability. In: Du Toit, J.T., Rogers, K.H., Biggs, H.C. (Eds.), The Kruger Experience: Ecology and Management of Savannah Heterogeneity. Island Press, Washington, DC.

Van Wilgen, B.W., Govender, N., Biggs, H.C., Ntsala, D., Funda, X.N., 2004. Response of savannah fire regimes to changing fire-management policies in a large African national park. Conserv. Biol. 18, 1533–1540.

Venter, F.J., Scholes, R.J., Eckhardt, H.C., 2003. The abiotic template and its associated vegetation pattern. In: Du Toit, J.T., Rogers, K.H., Biggs, H.C. (Eds.), The Kruger Experience: Ecology and Management of Savannah Heterogeneity. Island Press, Washington, DC.

Warner, R.R., Chesson, P.L., 1985. Coexistence mediated by recruitment fluctuations: a field guide to the storage effect. Am. Nat. 125, 769–787.

Westbroek, P., Collins, M.J., Jansen, J.H.F., Talbot, L.M., 1993. World archaeology and global change: Did our ancestors ignite the Ice Age? World Archaeol. 25, 122–133.

Case Study: Response of Invertebrates to Mixed- and High-Severity Fires in Central Europe

Petr Heneberg
Charles University in Prague, Prague, Czech Republic

8.18 THE SETTING

Forest fire is considered a marginal phenomenon in traditional central European forestry. According to the European Forest Fire Information System, the Czech Republic (78,866 km^2, forest cover over 33.9%) experienced forest fire over 296 ± 136 ha annually in 2004-2008.[1] The situation in other central European countries is similar. Forest fires are considered socially and economically unwelcome, and the burned areas are mandatorily managed within just 2 years of their formation.[2] Management consists of removing partially or fully burned trees and nearly immediate replanting of new tree seedlings (Kunt, 1967). Forest fires concentrate prevalently in regions with pine stands growing on gravel/sand sediments or sandstone rocks. Data from 2013 suggest that large-scale (>40 ha) fires affected 288,169 ha across Europe (data available for 24 countries excluding Russia), with forest fires representing over half of the entire burned area recorded during the year (ranging from 6% in Ireland and 8% in the United Kingdom to 96% in Kosovo and 97% in Sweden). Interestingly, 72,008 ha (29% of the affected area) were protected within the "Natura 2000" network containing habitats of high interest for nature conservation (Evans, 2012). The percentage of the Natura 2000 area affected by fire ranged between 0.002% in Germany (a single fire affecting 133 ha) and 2.137% in Portugal (100 fires affecting 40,837 ha).[3] However, Natura 2000 sites cover not only forested areas but also meadows, fields, and wetlands as long as they are of conservation interest; thus the percentage of Natura 2000-protected forests affected annually by fires is much higher and could be of conservation interest as sites where postfire succession occurs in a relatively unmanaged state.

1. http://forest.jrc.ec.europa.eu/effis/applications/fire-history/; cited February 18, 2015.
2. Czech forestry law no. 289/2005 Coll., §31.
3. http://forest.jrc.ec.europa.eu/media/cms_page_media/9/FireReport2013_final2pdf_2.pdf; cited February 19, 2015.

8.19 AEOLIAN SANDS SPECIALISTS ALONGSIDE THE RAILWAY TRACK NEAR BZENEC-PŘÍVOZ

Public awareness of the importance of mixed- and high-severity fires for biodiversity in central European forests is very limited. In 1990, however, the nature reserve "Váté písky" (the name is literally translated as "Aeolian sands") was established; it consists of a narrow (up to 60 m-wide and 5.5 km-long) deforested strip alongside the railway track between Rohatec and Bzenec-Přívoz in the southeastern Czech Republic (48.91°N, 17.24°E). In the eighteenth century native acidophilus and pannonian oak forests with admixed pine (*Festuco ovinae-Quercetum*, *Carici fritschii-Quercetum roboris*, locally in wet depressions also *Carpinion* and *Carici elongace-Alnetum*) were removed, and the whole region was subject to desertification; numerous active sand dunes were formed, and associated biota occurred. At the beginning of the nineteenth century, nearly the whole area was planted with pines, but in 1841 a railway track between Vienna (Austria) and Cracow (Poland) was built across the newly planted forest. Along the railway, the deforested strip was formed and maintained both by manual tree cutting and by frequent fires induced by sparks generated by the steam engines used by the railway company. The strip was maintained in a strictly deforested state from the 1840s to 1970, when the operation of steam engines terminated. Since the 1970s, the self-seeded Scots pine (*Pinus sylvestris*) and black locust (*Robinia pseudoacacia*) have been occasionally removed; however, solitary trees have remained present since then, and fires still occur, although less often than in the times of steam engine operations. Pine forests surrounding the railway strip still belong to the two Czech regions with the highest frequency of forest fires; however, all the burned areas are quickly replanted. This reduces the occurrence of numerous rare invertebrates, only part of which can survive at sites subject to other types of disturbances, such as military training ranges and sand quarries, which provide patches of bare, sandy ground (Cizek et al., 2013; Heneberg et al., 2013) but are devoid of standing fire-killed trees and downed logs.

The area is characterized by the presence of a 10- to 30 m-thick stratum of Aeolian sands, which is overlaid by arsenic campisols (inceptisol) or arenic regosols (entisol, orthent). Although the strip is limited in its extent, it hosts a rich assemblage of fungi, vascular plants, and animals. Typical plant species are represented by *Corynephorus canescens*, *Festuca vaginata* subsp. *dominii*, *Stipa borysthenica*, *Verbascum phoeniceum*, *Spergula morisonii*, *Helichrysum arenarium*, *Gypsophila paniculata*, and *Hierochloë repens*. Typical invertebrates are represented by *Melolontha hippocastani* and *Polyphylla fullo*; hundreds of aculeate hymenopteran species including *Bembix rostrata* and *Bombus cryptarum*; butterflies such as *Zerynthia polyxena*, *Hipparchia statilinus*, *Chamaesphecia leocopsiformis*, and *Synansphecia muscaeformis*; and spiders such as *Eresus kollari*.

Lowered frequency of fires in the past four decades decreased the availability of bare soil and reduced the diversity and abundance of bare soil specialists. For many of these species, the above-mentioned narrow strip and similar nearby habitats represent the only known sites of their occurrence in the Czech Republic. There are several other similar areas in the surrounding pine forest and its vicinity, which include two military training ranges and two large sandpits. When the military ceased operations at one of its two local training ranges, however, the particular area was quickly overgrown by dense vegetation, and most of the bare soil specialists decreased in abundance by over one order of magnitude. Some of them, such as *Bembix rostrata*, are now associated only with the illegal off-road motorcycle tracks present at some parts of the former military range, which is now protected as a nature reserve, and numerous more or less unsuccessful attempts to mimic the disturbance-induced removal of vegetation cover and to prevent sod formation have been made. In this area, allowing more fires to occur, and bare ground to persist, through the period of natural succession (i.e., without artificially planting trees) would increase the habitat for now-rare invertebrates that depend on sandy openings created by fire.

8.20 POSTFIRE SUCCESSION NEAR JETŘICHOVICE: A CHANCE FOR DEAD WOOD SPECIALISTS

Following 2 months of drought, on June 22, 2006, a week-long mixed-severity forest fire affected 18 ha of Scots pine and white pine (*Pinus strobus*) forest mixed with *Quercus petraea*, *Fagus sylvatica*, *Picea abies*, and *Betula pendula* in rugged terrain consisting of sandstone rocks (mean slope inclination of 35°; range, 0-90°) near Jetřichovice (50.86°N, 14.40°E, northern Czech Republic). Because the whole affected area was located within a national park's borders, the administration of the park was authorized to demand the burned forest to remain intact, allowing no logging or replanting efforts (Marková et al., 2011; Trochta et al., 2012). Extensive additional mortality of the remaining live trees occurred over the first 2 years after fire. The coverage of green tree crowns decreased from 83% before the fire to 39% within 2 months after the fire, and to 15% 1 year later (Trochta et al., 2012).

Within 3 weeks after the fire, the fire-affected area was occupied by antracophilous fungi such as *Pyronema omphalodes*. Four months after the fire, 13 fungal species were recorded; the most common ones were *Rhizina undulata*, *Pyronema omphalodes*, *Geopyxis carbonaria*, and *Pholiota highlandensis*. Some rare fungal species occurred as well, represented by, for example, *Rutstroemia carbonicola*. Several bryophytes and ferns (*Pteridium aquilinum*) also appeared during this very early succession phase.

By the first spring after the fire, the burned area was covered by the common liverwort *Marchantia polymorpha*, common mosses *Ceratodon purpureus* and *Funaria hygrometrica*, and seedlings of *Betula pendula*, *Pinus* spp., *Fagus sylviatica*, and *Acer* spp. In autumn most of the burned area was covered by

B. pendula, *Populus tremula*, *Salix caprea*, and young *Pinus* spp. There were recorded 37 species of fungi, with antracophilous fungi representing one quarter of the species found.

In the next 5 years the burned area was dominated by ferns (*Pteridium aquilinum* and *Dryopteris carhusiana agg.*) and vascular plants such as *Epilobium angustifolium* (attracting particularly abundant assemblages of aculeate hymenopterans), *Erythronium montanum*, *Exacum tetragonum agg.*, *Spergula morisonii*, *Cerastium holosteoides* subsp. *triviale*, *Digitalis purpurea*; invasive grasses *Calamagrostis epigejos* and *Agrostis capillaris*; and numerous ragworts and groundsels, including *Senecio sylvaticus*, *Senecio vulgaris*, *Senecio viscosus*, *Senecio vernalis*, and *Senecio ovatus* (Hadinec and Lustyk, 2011; Marková et al., 2011). The family Asteraceae of flowering plants was particularly diverse at the burned site and included *Taraxacum sect. Ruderalia*, *Cirsium vulgare*, *Conyza canadensis*, *Hypochaeris radicata*, *Mycelis muralis*, *Tussilago farfara*, and *Crepis capillaries*. Mosses such as *Polytrichastrum formosum*, *Polytrichum juniperinum*, *Polytrichum piliferum*, *Ceratodon purpureus*, and *Pohlia nutans* also occurred. About 50 fungal species were found annually in the burned area, with a markedly decreasing share of antracophilous fungi, which decreased to 11% in the third, 7% in the fourth, and <2% (one species) in the fifth year after fire (Marková et al., 2011).

The burned area was thoroughly examined for changes in aculeate hymenopteran assemblages. Of the 12 red-listed species (threatened according to the Czech Red List of invertebrate species (Farkač et al., 2005) and according to the International Union for the Conservation of Nature) occurring in the surrounding intact pine forest that served as a control (unburned) area, 10 also were present in the burned forest 1-7 years following the forest fire. More important, the burned forest stands also attracted another 30 red-listed species of bees and wasps, which were absent in the surrounding unburned pine forest. Among them were two species that were considered regionally extinct (*Dipogon vechti* and *Chrysis iris*), two critically endangered species (*Miscophus niger* and *Passaloecus monilicornis*), and numerous endangered and vulnerable species. In total, the burned forest stands hosted 252 bee and wasp species 1-7 years following the forest fire, which represented 19% of the total 1343 species reported so far from the Czech Republic (Bogusch et al., 2007; Bogusch et al., 2015). Importantly, the species spectrum associated with the burned forest stands differed not only from those of the unburned forest but also from the species associated with bare sand and heather patches with retained solitary pine trees at a site 12 km away (Blažej and Straka, 2010; Bogusch et al., 2015). The species absent at the unburned site with bare sand and heather patches consisted mostly of those requiring dead wood, such as standing fire-killed trees (snags), which is rarely available in the current commercially exploited forests of the Czech Republic as well as surrounding countries. "Cavity adopters" represented 43% of the red-listed species recorded in the burned forest studied. The forest fire caused a rapid but very temporary decrease in both the abundance and diversity of bees and wasps. By the first year after fire, however, several species were found in the burned forest that were absent in the surrounding unburned forest.

These were mainly the polylectic species of the open countryside and several broadly distributed cavity nesters (*Mellinus arvensis*, *Lasioglossum pauxillum*, *Trypoxylon minus*). After 2-3 years, however, the burned forest attracted a highly diverse and abundant assemblage of very specialized bee and wasp species, the abundance and diversity of which was several times higher than in the surrounding forest. The species occurring temporarily at this stage of succession were represented by, for example, *Lasioglossum nitidiusculum* and *Andrena lapponica*. At 5-7 years following the fire, both the diversity and abundance of aculeate hymenopterans decreased back to the levels experienced in the surrounding unburned forest, and the species spectrum changed. Even in these later phases of forest succession, the burned site hosted numerous species that were absent or rare in the unburned forest and absent or rare during the earlier phases of forest succession (*Ammophila sabulosa*, *Lasioglossum punctatissimum*, *Arachnospila hedickei*, and *Crossocerus exiguus*) (Bogusch et al., 2015).

Though this section is focused on invertebrates and postfire habitat, it is worth noting that avian and mammal research also was conducted in this fire area. While birds associated with dense, mature forest declined after the mixed-severity fire, numerous bird species increased or appeared on the site for the first time (Marková et al., 2011). Regarding small mammals, the unburned pine forests adjacent to the study area are characterized by species-poor assemblages, represented by only *Clethrionomys glareolus* (bank vole) and *Apodemus flavicollis* (yellow-necked mouse), both of which occur at very low densities (Bárta, 1986; Marková et al., 2011). During the first year after fire, the burned site also hosted only these two species; however, their abundance was relatively high. The onset of natural postfire succession in the follow-up years was associated with newly appearing species *Sorex araneus* (Eurasian shrew) and *Sorex minutus* (Eurasian pygmy shrew), which otherwise occur in this region only in the river floodplains and in deciduous forests at basaltic rocks, as well as *Crocidura suaveolens* (lesser white-toothed shrew) and *Apodemus sylvaticus* (wood mouse) (Marková et al., 2011).

8.21 CONCLUSIONS

Combined data support the benefits of a mosaic of postfire successional habitats of various ages. Conservation management that focuses on the formation of bare soil patches by fire (as in the first of the two above-presented case studies) supports a very diverse spectrum of organisms but is not sufficient to provide habitat for hymenopteran cavity adopters, which are associated with dead wood and would benefit from greater public acceptance of mixed-severity fire in forests and greater retention of snags after fire (Bogusch et al., 2015). Results similar to those described above have been found for various invertebrate taxa after mixed-severity fire, including increases in diversity of saproxylic (dependent on decaying wood) beetles in the Swiss Alps (Moretti et al., 2010) and enhanced species richness of ground beetles, hoverflies, bees, wasps, and spiders in the forests of the southern Alps (Moretti et al., 2004). Further support for these conclusions regarding benefits to arthropods from mixed-severity fire is also found in studies regarding bark beetles. Similar to postfire forest stands, forest gaps (snag patches) formed by spruce bark beetles are associated with the increased density and diversity of many red-listed species across numerous taxa in this region (Beudert et al., 2015), including, for example, dead wood aculeate hymenopteran specialists (Müller et al., 2008; Lehnert et al., 2013). These emerging data indicate a need to allow for more natural disturbance and natural succession, and an increase in protected reserves, in this region to benefit arthropods and numerous vertebrate species, including many International Union for the Conservation of Nature red-listed species that seem to thrive in unmanaged postfire habitat (Beudert et al., 2015).

REFERENCES

Bárta, Z. 1986. Drobní zemní savci skalních plošin Jetřichovických stěn (Děčínské mezihoří, CHKO Labské pískovce). Závěrečná zpráva úkolu č.: IU-2/Bá-1985-86. Správa CHKO Labské pískovce, Děčín.

Beudert, B., C. Bässler, S. Thorn, R. Noss, B. Schröder, H. Dieffenbach-Fries, N. Foullois, and J. Müller. 2015. Bark beetles increase biodiversity while maintaining drinking water quality. Conservation Letters (in press). DOI: 10.1111/conl.12153.

Blažej, L., Straka, J., 2010. Výsledky monitoring vybraných skupin hmyzu (Coleoptera: Carabidae, Hymenoptera: Aculeata) v bývalé lesní školce u Bynovce (CHKO Labské pískovce). Sborník Oblastního muzea v Mostě, řada přírodovědná 32, 23–42.

Bogusch, P., Straka, J., Kment, P. (Eds.), 2007. Annotated checklist of the Aculeata (Hymenoptera) of the Czech Republic and Slovakia. Acta Entomologica Musei Nationalis Pragae, 11 (supplementum), 1–300.

Bogusch, P., Blažej, L., Trýzna, M., Heneberg, P., 2015. Forgotten role of fires in Central European forests: critical importance of early post-fire successional stages for bees and wasps (Hymenoptera: Aculeata). Eur. J. For. Res. 134, 153–166.

Cizek, O., Vrba, P., Benes, J., Hrazsky, Z., Koptik, J., Kucera, T., Marhoul, P., Zamecnik, J., Konvicka, J., 2013. Conservation potential of abandoned military areas matches that of established reserves: plants and butterflies in the Czech Republic. PLoS One 8, e53124.

Evans, D., 2012. Building the European Union's Natura 2000 network. Nat. Conserv. 1, 11–26.

Farkač, J., Král, D., Škorpík, M. (Eds.), 2005. Red List of Threatened Species in the Czech Republic—Invertebrates. AOPK ČR, Prague.

Hadinec, J., Lustyk, P. (Eds.), 2011. In: Additamenta ad floram Reipublicae Bohemicae. IX, vol. 46. Zprávy České botanické společnosti, Praha, pp. 51–160.

Heneberg, P., Bogusch, P., Řehounek, J., 2013. Sandpits provide critical refuge for bees and wasps (Hymenoptera: Apocrita). J. Insect Conserv. 17, 473–490.

Kunt, A., 1967. Lesní požáry. Knižnice požární ochrany No. 28. Československý svaz požární ochrany, Praha.

Lehnert, L.W., Bässler, C., Brandl, R., Burton, P.J., Müller, J., 2013. Conservation value of forests attached by bark beetles: highest number of indicator species is found in early successional stages. J. Nat. Conserv. 21, 97–104.

Marková, I., Adámek, M., Antonín, V., Benda, P., Jurek, V., Trochta, J., Švejnohová, A., Šteflová, D., 2011. Havraní skála u Jetřichovic v národním parku České Švýcarsko: vývoj flóry a fauny na ploše zasažené požárem. Ochr. přír. 66, 18–21.

Moretti, M., Obrist, M.K., Duelli, P., 2004. Arthropod biodiversity after forest fires: winners and losers in the winter fire regime of the southern Alps. Ecography 27, 173–186.

Moretti, M., De Cáceres, M., Pradella, C., Obrist, M.K., Wermelinger, B., Legendre, P., Duelli, P., 2010. Fire-induced taxonomic and functional changes in saproxylic beetle communities in fire sensitive regions. Ecography 33, 760–771.

Müller, J., Bußler, H., Goßner, M., Rettelbach, T., Duelli, P., 2008. The European spruce bark beetle *Ips typographus* in a national park: from pest to keystone species. Biodivers. Conserv. 17, 2979–3001.

Trochta, J., Král, K., Šamonil, P., 2012. Effects of wildfire on a pine stand in the Bohemian Switzerland National Park. J. For. Sci. 58, 299–307.

The Role of Large Fires in the Canadian Boreal Ecosystem*

André Arsenault

Natural Resources Canada, Canadian Forest Service–Atlantic Forestry Centre, Corner Brook, NL, Canada; Thompson Rivers University, Kamloops, BC, Canada; Memorial University at Grenfell, Corner Brook, NL, Canada

8.22 THE GREEN HALO

The green halo, visible from space and forming a distinct ring around the northern hemisphere (Figure 8.10), is one of the most striking ecological features on earth (Gawthrop, 1999; Talbot and Meades, 2011). The circumboreal forest biome, encompassing 11% of the earth's land surface and one-fourth of the world's forests, is one of the largest floristic regions of the world, wrapping around the northern parts of the globe including North America, Europe, and

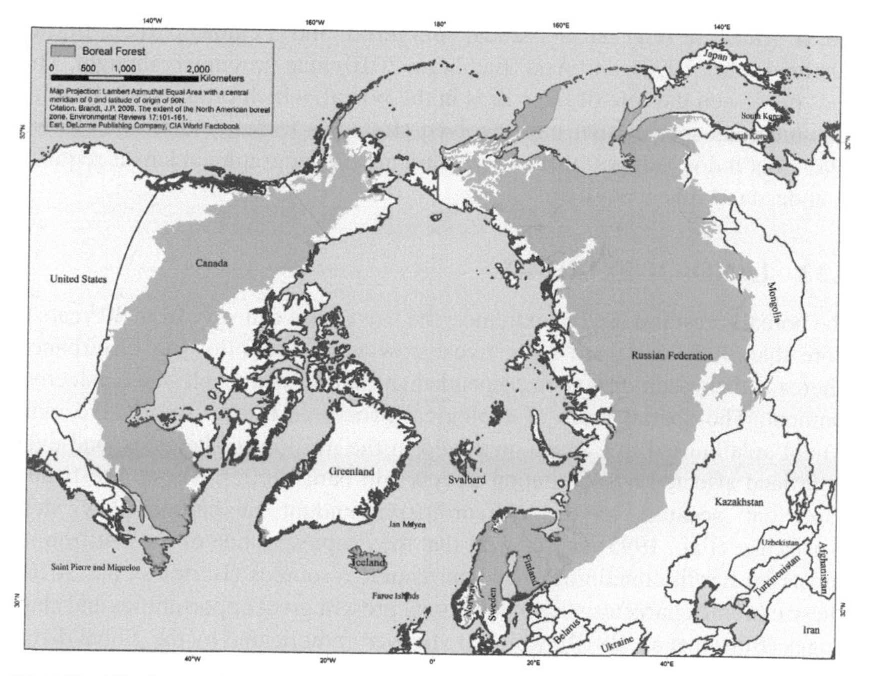

FIGURE 8.10 Extent of boreal forest and woodlands around the globe. *(Adapted from Brandt et al. 2013 with permission from James Brandt and NRC Press.)*

Asia. The boreal forest, named after the Greek god Boreas (north wind), is characterized by a cold, snowy climate, a short growing season, and relatively cool summers. Large wildfires tend to be more abundant in western Quebec in eastern Canada, whereas Newfoundland and Labrador, and Quebec's North shore, have very long fire cycles. Fires tend to increase westward from central Canada. Although fire activity is increasing in Canada as a result of climate change, predicted rates of increase are believed to be within the upper limits of the natural range of variation. Large fires are important in transforming forests into habitat for specialized ecological communities that require charred wood, by enabling a pulse of dead wood, which is important for many species and ecological processes over the course of natural succession, and by leaving refuges that allow species to persist within burns and recolonize burned areas.

Although some have described the boreal forest as a relatively simple ecosystem with low species diversity and frequent large crown fires, a new paradigm is emerging, that reveals a much more complex and interesting natural history. The vast range in fire cycles implies that old-growth and late seral features were historically abundant in certain parts of the boreal forests, more than was appreciated in the past. In fact, new evidence suggests that old-growth forests were a much more predominant feature of the preindustrial landscape in North America (Cyr et al., 2009; Bergeron and Fenton, 2012), Europe (Östlund et al., 1997), and Asia (Eichhorn, 2010) than previously thought. This does not lessen the role of large fires in the boreal, which produce vital habitat for numerous species that have evolved strategies to respond to fire (Rowe, 1983), but it does suggest that a more intricate biogeographical lens is required to understand this ecosystem.

8.23 LAND OF EXTREMES

The boreal forest is a land of extremes; the fire cycles can vary from 40 years to more than 900 years, and some trees grow very fast following disturbance, whereas others seem to live in suspended animation as a result of a harsh environment. The spatial scale of ecological processes is also impressive, with annual area burned in Canada measured in the millions of hectares, and even more land affected by defoliating insects and bark beetles. These forests also stand out because of the extraordinary amount of carbon they store (Kasischke et al., 1995) and the fact that they represent one of the last frontier forests on Earth, containing abundant natural resources (Burton et al., 2010). These extreme characteristics of the boreal present great opportunities and challenges (Burton et al., 2010), which are further complicated by the global distribution of this ecosystem across many jurisdictions and emerging issues of conservation of biodiversity and carbon in a changing climate (Bradshaw et al., 2009; Moen et al., 2014). Understanding the natural disturbance regime of the boreal forest and its effect on the environment is critical to ensure that the integrity and resilience of this ecosystem are maintained for future generations. This essay focuses on the role of large wildfires in the Canadian boreal forest ecosystem.

8.24 VEGETATION

Long, cold winters and short, cool summers are key features influencing the vegetation of the boreal. The distribution of the boreal forest in North America is vast, extending from one end of the continent to another (Figure 8.11). It crosses the state of Alaska and all the Canadian provinces and territories with the exception of the Maritime Provinces, making up 53% of Canada's forests (Brandt, 2009). Although the boreal forests of Canada share similar floristic and ecological characteristics across their distribution, they also exhibit significant differences. At a coarse scale, they are made up of seven different ecozones (Figure 8.12), representing formidable ecological gradients of elevation, longitude, latitude, and continentality. From the productive white spruce (*Picea glauca*) stands in the south, to the krumholtz forests shaped by the harsh climate of the northwestern mountains of the Yukon, to the fierce winds off the coast of Newfoundland to the east, this is a land of beauty and surprising ecological diversity.

Extensive coniferous forests, sometimes mixed with broadleaved trees, interspersed by bogs, barrens, fens, rivers, and lakes cover the landscape (Figure 8.13). Black spruce (*Picea marina*), white spruce, tamarack (*Larix laricina*), white

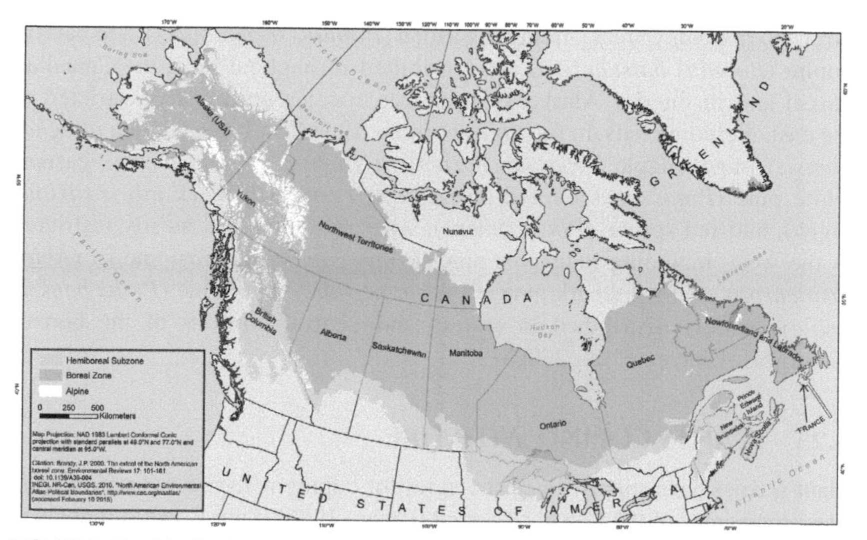

FIGURE 8.11 Distribution of the Canadian boreal forest and Alaska boreal interior. *(Adapted from Brandt 2009 with permission from James Brandt and NRC Press.)*

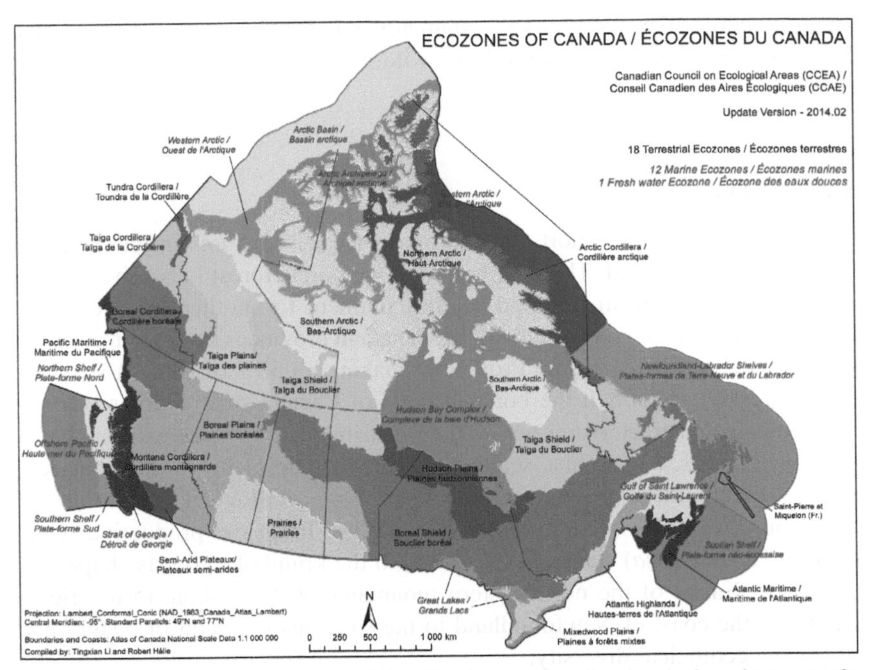

FIGURE 8.12 Map of the ecozones of Canada. The ecozones forming the boreal zone are the boreal shield, boreal plains, taiga plains, taiga shield, Hudson plains, taiga cordillera, and boreal cordillera. *(Source: Canadian Council on Ecological Areas (CCEA)).*

birch (*Betula papyrifera*), trembling aspen (*Populus tremuloides*), and balsam poplar (*Populus balsamifera*) are distributed throughout most the Canadian boreal and in interior Alaska. Several tree species are mostly restricted to the east, including balsam fir (*Abies balsamea*), yellow birch (*Betula alleghaniensis*), sugar maple (*Acer saccharum*), red maple (*Acer rubrum*), eastern white pine (*Pinus strobus*), red pine (*Pinus resinosa*), black ash (*Fraxinus nigra*), and red spruce (*Pinus rubens*), whereas others are mostly restricted to the west, including lodgepole pine (*Pinus contorta*), subalpine fir (*Abies lasiocarpa*), and whitebark pine (*Pinus albicaulis*). Jack pine (*Pinus banksiana*) is widely distributed in eastern and central portions of the boreal. (Bourgeau-Chavez et al., 2000a)

8.25 PLANTS COPING WITH FIRE

Plant species have sophisticated strategies for coping with fire and surviving in fire-prone areas of the Canadian boreal (Rowe, 1983). Stan Rowe, a renowned Canadian forest ecologist, developed a clever classification of plant functional groups according to how they cope with fire, inspired by the work on vital attributes and the plant regeneration niche developed by Noble and Slayter (1980). These functional groups are:

FIGURE 8.13 A boreal landscape in Terra Nova National Park, Newfoundland and Labrador. *(Picture courtesy of André Arsenault, Canadian Forest Service.)*

1. Invaders: examples include white birch, fire moss (*Ceratodon purpureus*), and fireweed (*Chamerion angustifolium*), which depend on copious amounts of wind-dispersed propagules.
2. Evaders: species that store seed in the canopy (e.g., lodgepole and jack pine), in the humus, and in mineral soil.
3. Avoiders: these species, such as balsam fir, have little adaption to fire.
4. Resisters: thick-bark old pine species.
5. Endurers: plants able to resprout after the passage of fire (i.e., the "phoenix" species), such as the trembling aspen (*Populus tremuloides*) and saskatoon (*Amelanchier alnifolia*).

Notably, some species such as fireweed can behave like both invaders and endurers. In general, this visionary framework links well with contemporary integrated approaches aimed toward understanding the role of plant functional traits in response to disturbance.

8.26 FIRE REGIME OF THE CANADIAN BOREAL FOREST

Large fires are considered a key controlling process in boreal forests (Payette 1992) (Figure 8.14). Although the growing season is relatively short, persistent high-pressure systems in mid-summer can lead to drought conditions, thereby drying out forest fuels and favoring the occurrence of large wildfires (Johnson, 1992). Unlike mixed-severity fires (Odion et al., 2014) or low-severity fires (Agee, 1993), crown fires in the boreal forest usually kill most trees within

FIGURE 8.14 Examples of large fires in the boreal forest of Canada. *(Pictures courtesy of the Canadian Forest Service.)*

the burned areas (Johnson, 1992), with the exception of fire skips, also known as remnants or refuges (Perera and Buse, 2014). A classic description of the fire regime for the boreal is perhaps best illustrated by Cogbill's (1985) fire history study of the Laurentian Highlands of Quebec, in which he described frequent catastrophic crown fires producing even-aged stands with fire cycles varying between 70 to 140 years. An important implication of such a fire regime is that a minor component of old stands would be embedded in a matrix of young forests. Bergeron and Fenton (2012) have argued that this view of the dynamics of the boreal forest has been used by some to justify clear-cut harvesting, which also results in a matrix of young stands under a sustained yield regime.

This classic interpretation of the fire regime has recently been challenged, suggesting that the fire cycle was variable and included intervals long enough to create an abundance of old-growth forests in northern Quebec (Cyr et al., 2009; Bergeron and Fenton, 2012). The fire regime also varies significantly across the spatial extent of the Canadian boreal. An updated map of large fires (Figure 8.15) from the Canadian large-fire database (Stocks et al., 2003; Canadian Forest Service 2015), reveals interesting patterns. The spatial extent of wildfires increases toward western Canada and decreases significantly in eastern Canada. This pattern is closely linked to fire cycles estimated from various sources (Table 8.1). Ironically, the largest burned areas are in ecozones least affected by anthropogenic influences but where fire weather is considered to be severe (Krezek-Hanes et al., 2011).

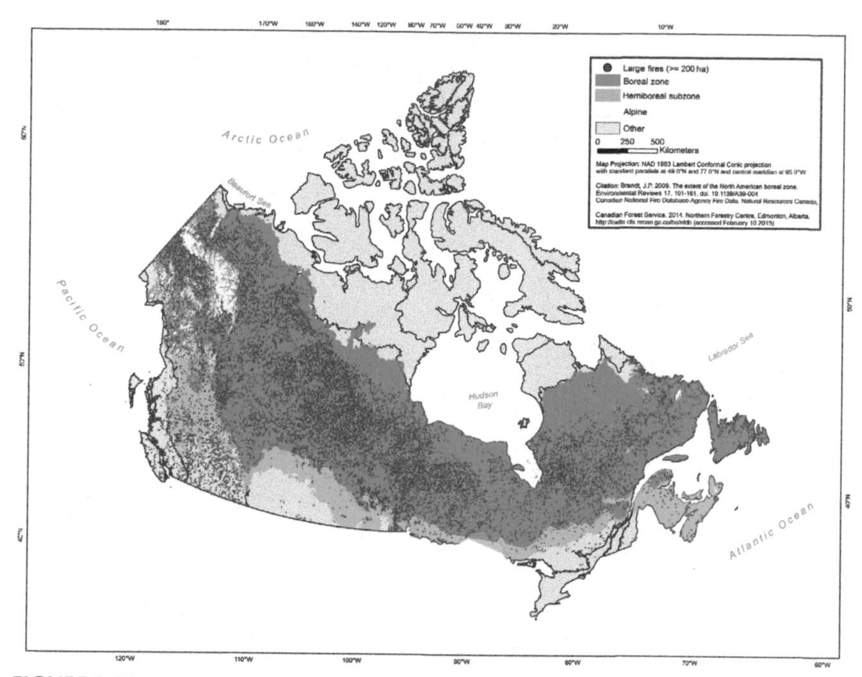

FIGURE 8.15 Distribution of large fires (>200 ha) in Canada and the distribution of the boreal region of Canada from 1959-2014. *(Source: Natural Resources Canada, Canadian Forest Service (2015)).*

TABLE 8.1 Fire Cycle Estimates for the Canadian Boreal Ecozones and Interior Alaska from Three Studies[a]

	Fire Cycle (Mean Forest Age) Years			
Ecozone	*Study 1*	*Study 2*	*Study 3*	*RC*
Boreal shield	204	498	122	36.9
Boreal shield east		905	166	
Boreal shield west		91	78	
Newfoundland boreal	769			0.8
Boreal plains	213	181	82	11.4
Taiga plains	141	179	142	13.8
Taiga shield	130	242	122	17.7
Taiga shield east		324	166	
Taiga shield west		160	78	

Continued

TABLE 8.1 Fire Cycle Estimates for the Canadian Boreal Ecozones and Interior Alaska from Three Studies—Cont'd

	Fire Cycle (Mean Forest Age) Years			
Ecozone	*Study 1*	*Study 2*	*Study 3*	*RC*
Hudson plains	588	506	813	3.3
Taiga cordillera	213		202	4.5
Boreal cordillera	263	396	255	7.8
Alaska boreal interior		184		

RC, relative contribution of the ecozone to the total area burned in Canada (percentage), per Krezek-Hanes et al., (2011).
[a]*Study 1: Krezek-Hanes et al. (2011); Study 2: Bourgeau-Chavez et al. (2000b); Study 3: Bergeron and Fenton (2012).*

The estimates of fire cycles among the three studies provided should not be compared because they used different methods. They do, however, provide a useful comparison of trends per ecozone. The first two studies, Krezek-Hanes et al. (2011) and Bourgeau-Chavez et al. (2000b), are based on an analysis of the Canadian large-fire database, whereas Bergeron and Fenton (2012) uses a combination of field-based reconstruction of time-since-fire maps and the large-fire database. Unfortunately, fire history reconstruction studies do not yet cover the entire range of the Canadian boreal, and more work is needed in developing methods for interpretations across spatial and temporal scales. The combination of tree ring studies with paleoecological reconstruction of fire history is a promising area to explore, especially in areas where fire intervals are very long, as in Newfoundland and Labrador (Foster, 1983; Arsenault, unpublished data). The temporal variability of the fire cycle over centuries seems to be strongly linked with climate drivers such as the Little Ice Age (Bergeron, 1991) and has changed over the past centuries. This suggests that no single fire cycle can be used as a reference for natural variability (Johnson et al., 1998). Furthermore, most fire cycle calculations using the negative exponential assume that the rate of burning was constant over time, an assumption that is rarely completely met (Huggard and Arsenault 1999; Bergeron and Fenton, 2012).

The spatial pattern of fire occurrence operates at multiple scales. At a coarse scale, fire is less common in Newfoundland and Labrador, and Quebec's north shore, compared with western Quebec. A similar pattern occurs in the western boreal shield, where fire occurrence increases from western Ontario westward, with very little fire activity in the Hudson plains, which is a very wet ecozone. At a finer scale, fire in Newfoundland is more prevalent in the central ecoregion and virtually absent in western Newfoundland (Arsenault, unpublished data).

Interestingly, this pattern is associated with a higher proportion of black spruce in the central ecoregion and a higher proportion of balsam fir in the western ecoregion of Newfoundland. This could reflect at least in part a feedback mechanism between species composition and fire occurrence because black spruce stands are considered more flammable (Furyaev et al., 1983). At an even finer scale, Bergeron (1991) observed that differences in fire severity and frequency were influenced by location on an island and lakeshore landscape. He concluded that the boreal fire regime was controlled by long-term changes in climate at a regional scale and by strong interaction with landscape features at a local scale. The spatial pattern of ignition type also varies greatly in Canada; not surprisingly, most anthropogenic ignitions occur in the southern portion of the boreal (Stocks et al., 2003). All ecozones have larger area burned by lightning fires as opposed to fires of anthropogenic origin, with a notable exception in Newfoundland and Labrador. Interestingly, the ratio of lightning fires to human-set fires is much higher in Labrador than it is on the island of Newfoundland. Virtually all of the large fires on the island since 1959 were of anthropogenic origin. Lightning ignition is relatively rare on the island, and many large fires resulted from railway and forestry operations.

8.27 TEMPORAL PATTERNS OF FIRE AND OTHER CHANGES IN THE BOREAL

The temporal pattern of annual area burned by large fires over the past 55 years is shown in Figure 8.16. A major increase in the 1980s and 1990s was followed by a significant decrease in the past decade and a recent increase between 2010 and 2013. The total area burned by decade shows this general pattern quite well

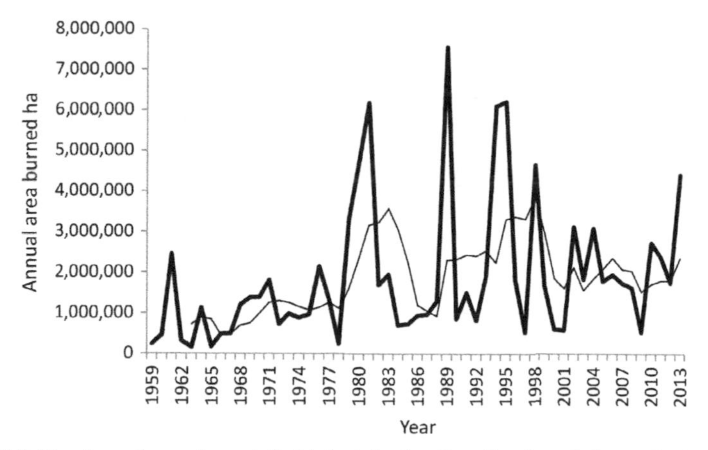

FIGURE 8.16 Annual area burned (bold line) in the Canadian boreal forests by large fires (>200 ha) from 1959 to 2013. A 5-year moving average trend line (lighter line) is included. *(Source: Natural Resources Canada, Canadian Forest Service (2015)).*

(Figure 8.16a). The drop in area burned in the last decade is clear when using the total area or the mean area but less clear with the median illustrating large annual variation. The pattern also varies among ecozones (Figure 8.16b). Initially, the large increase in the 1980s was attributed to a possible change in detection accuracy and more usage of the forests (Stocks et al., 2003). However, Krezek-Hanes et al. (2011) argue that several studies have shown a clear link between the increase in fire activity and temperature over the past 40 years (i.e., Girardin, 2007). This increase in fire activity, likely linked to the lengthening of the fire season, is expected to increase not only in the Canadian boreal but also globally (Flannigan et al., 2013). In addition, droughts during the fire season also are expected to increase, further increasing fire occurrence despite an overall increase in precipitation (Christensen et al., 2007). A comprehensive

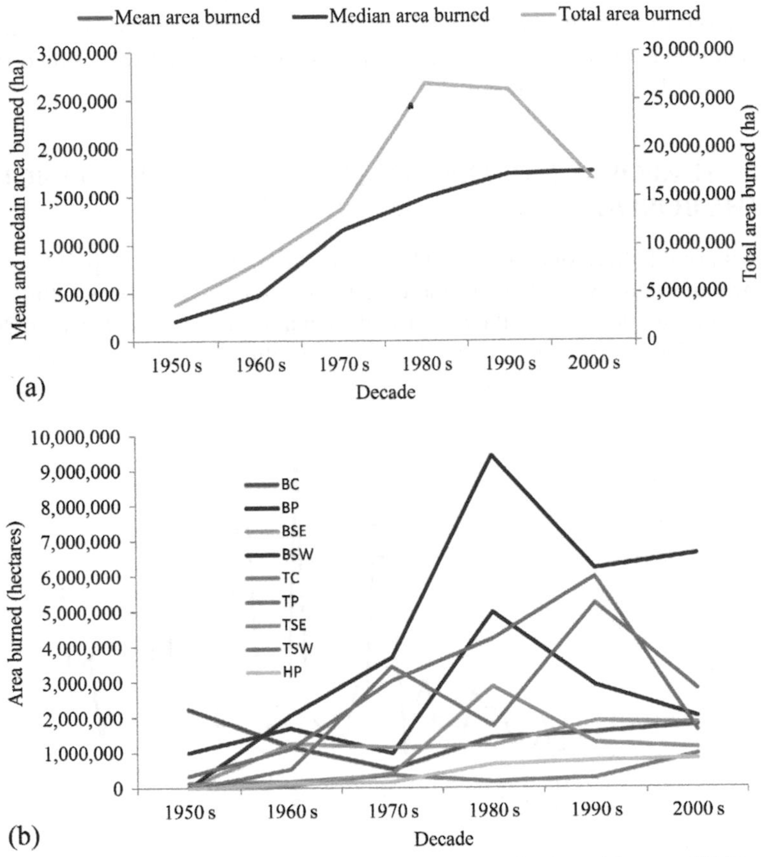

FIGURE 8.17 a) Metrics of area burned by decade in the Canadian boreal forest, and b) Total area burned by decade in Canadian ecozones. *(Source: Natural Resources Canada, Canadian Forest Service (2015)).*

meta-analysis of fire history studies in Canada clearly show a global decrease in fire activity from the preindustrial to the current era—a decrease linked to long-term changes in climatic pattern (Girardin et al., 2013), confirming that predicted increases in fire activity are still expected to be within the natural range of variability (Bergeron et al., 2004a, 2004b). However, it should be noted that fire frequency of recent decades, in the Yukon flats of interior Alaska, is estimated to have surpassed the natural range of variation estimated over a 10,000 year period (Kelly et al. 2013). The predicted proportion of area burned in the Eastern boreal forest of Canada is forecasted to be at the upper end of its historical range, going back 6000 years, by the end of this century (Bergeron et al. 2010; Girardin et al. 2013). Other factors could, however, complicate the forecast of forest changes resulting from rising temperature and increased fire activity (Hessl, 2011). Additional sources of uncertainty include the interaction of disturbances such as drought, fire, pathogens, and insects and their effect on vegetation, especially tree regeneration. An example of this is an interaction between fire severity and climate in Interior Alaska that triggers a change in successional trajectory favouring broadleaf species over black spruce (Johnstone et al. 2010; Beck et al. 2011).

8.28 BIODIVERSITY

The biodiversity of the Canadian boreal forest is not fully understood. However, it is clear that the diversity of physical habitat and the many ecotones among forests, bogs, fens, barrens, lakes, stream, and rivers are obviously key. The diversity of forest habitat resulting from a wide range of disturbance types and frequencies is also intimately linked to the distribution and abundance of organisms. For these reasons it is imperative that the boreal forest not be treated as a simple species-poor, fire-maintained ecosystem. It is much more complex. Large fires are important drivers of biodiversity in three key respects.

Firstly, they result in the arrival of an ephemeral but key pyro-community of organisms that are tightly connected to freshly burned trees (Figure 8.18A). Saproxylic insects, such as the spruce sawyer (*Monochamus scutellatus*) (Figure 8.18B), are usually first on the scene after fire (Boucher et al., 2012), soon followed by primary cavity-nesting birds such as the black-backed woodpecker (*Picoides arcticus*) (Hutto, 1995; Drapeau et al., 2002). This phenomenon is repeated over many different coniferous forests in North America and is a clear indication of the important role of large fires in biodiversity (Hutto, 2008). Recent studies have shown that this food web is linked to nutrient cycling in the boreal forest (Cobb et al., 2010). Several studies suggest the conservation of a mosaic of different burn severity patches is necessary to satisfy requirements of dead wood-associated species in burned forests (Nappi et al., 2010; Azeria et al., 2011, 2012).

FIGURE 8.18 Freshly killed trees (a) are prime habitat for the spruce sawyer beetle (b). *(Pictures courtesy of Sébastien Bélanger.)*

Secondly, large fires generate important pulses of dead wood that will be used over time by a variety of communities, including many fire-dependent species and other species associated with rotten wood such as saproxylic insects, lichens, bryophytes, cavity nesters, and fungi (Arsenault and Chapman, 2014).

Thirdly, large fires are often heterogeneous, creating patches of unburned forests and residual structures (Figure 8.19). These refuges can be of different sizes, ages, and permanence (DeLong and Kessler, 2000). They are instrumental in the survival of biota during a fire event, persistence of biota within large

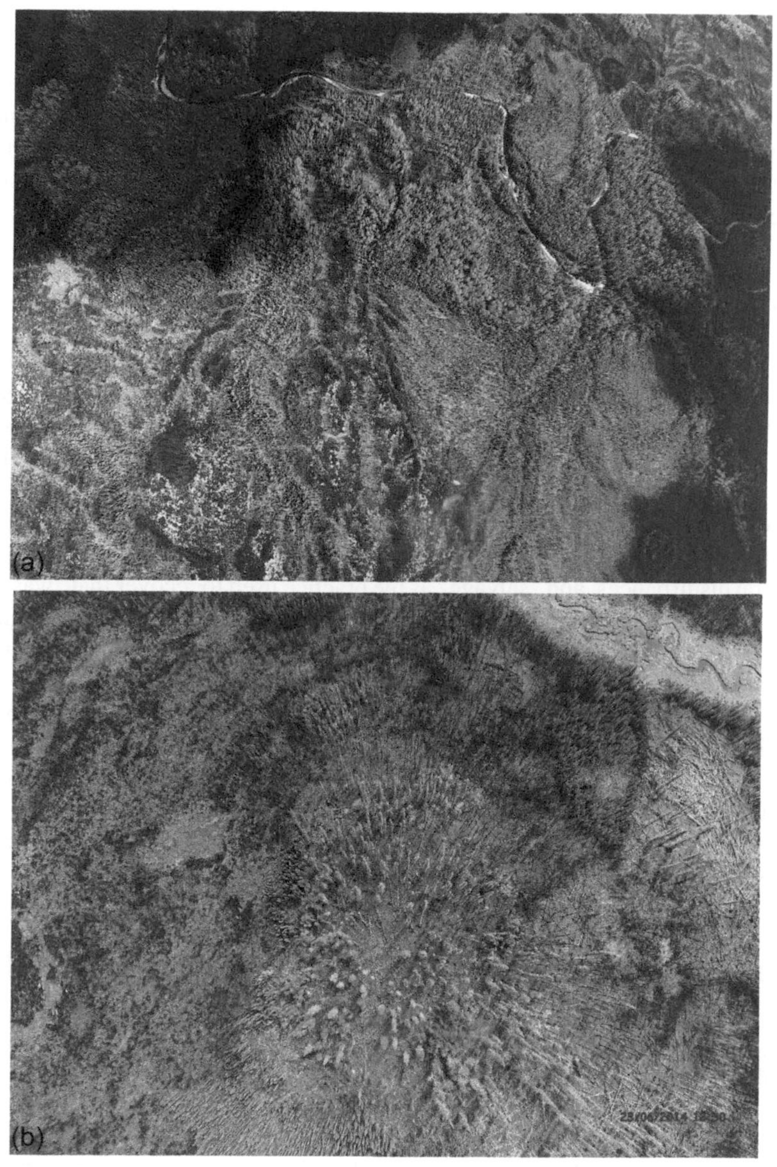

FIGURE 8.19 Examples of residual structure and remnants left after a large fire in Wood Buffalo National Park. *(Pictures courtesy of Brian Simpson, Canadian Forest Service.)*

burns, and as habitat for source populations for the recolonization of burned areas over time (Robinson et al., 2013; Perera and Buse, 2014).

The large range of fire intervals in the boreal also translates into a range of forest age classes, including old-growth forests (McCarthy and Weetman, 2006; Bergeron and Fenton, 2012). The conservation of old-growth forests and late seral conditions is essential for maintaining biodiversity in boreal forests (Desponts et al., 2004; Harper et al., 2003; Thompson et al., 2003; Bergeron and Fenton, 2012). This is especially true for areas with very long fire intervals, or where insects are the primary drivers of natural disturbance, creating gaps of various sizes and heterogeneous forest structure (McCarthy, 2001) that lead toward the development of some late seral characteristics (McCarthy and Weetman, 2007).

8.29 CONCLUSION

Notwithstanding such excellent research and writing on the boreal forest, new paradigms are still emerging. It is a sober reminder that critical thinking is an essential ingredient for keeping bias and assumptions in check. This is especially important when science is applied for managing forests. For a long time forest management was driven by a strong focus on timber extraction and developed a jargon that infiltrated the dialect of forestry with words such as "decadent" for old-growth forests, "waste wood" for trees that had been killed by natural disturbances, and "salvage" as the practice used to recover that "wasted" timber. Today, forest management in the boreal is strongly driven by themes like ecosystem-based management and sustainable development. The new era will require conservation of boreal forests at different ends of the disturbance spectrum, from newly created burn habitat to multicentury old-growth forests. The new information uncovered regarding the Canadian boreal is also an important reminder that investments in long-term ecological research pay off for the scientific community and for the people who live in, use, play in, and make decisions about this wonderful ecosystem, who surely has many more treasures of knowledge under her mossy carpet.

ACKNOWLEDGMENTS

Candice Power helped with preparation of figures and fire data; Ken Baldwin assisted with obtaining the map of the boreal distribution and the map of the Canadian ecozones; James Brandt and Roger Brett helped obtaining the distribution of the North American map. John Little assisted us in obtaining access to the large fire data base. Christian Hébert, Brad Hawkes, John Parminter, Brian Simpson kindly provided excellent pictures. Brian Hearn, Christian Hébert, and Caroline Simpson conducted official internal review of the manuscript and provided valuable comments. Additional comments were provided by Candice Power and Patricia Baines. NRC Press provided permissions to use figures published in two papers published in Environmental Reviews.

REFERENCES

Agee, J.K., 1993. Fire ecology of Pacific Northwest forests. Island Press, Washington, DC.

Arsenault, A., Chapman, B., 2014. How much dead wood is enough in Canadian forests? A review of the science and policy. Page 181, In: Parrota, J.A., Moser, C.F., Scherzer, A.J., Koerth, N.E., Lederle, D.R. (Eds.), Sustaining Forests, Sustaining People: The Role of Research. XXIV

IUFRO World Congress, 5-11 October, 2014, Salt Lake City, USAIn: The International Forestry Review, vol. 16(5). p. 578.

Azeria, E.T., Ibarzabal, J., Boucher, J., Hébert, C., 2011. Towards a better understanding of beta diversity: deconstructing composition patterns of saproxylic beetles breeding in recently burnt boreal forests. In: Pavlinov, I.Y. (Ed.), Research in Biodiversity: Models and Applications. InTech Open Access Publisher, Rijeka, pp. 75–94.

Azeria, Ermias T., et al., 2012. Differential effects of post-fire habitat legacies on beta diversity patterns of saproxylic beetles in the boreal forest. Ecoscience 19, 316–327.

Beck, P.S.A., Goetz, S.J., Mack, M.C., Alexander, H.D., Jin, Y., Randerson, J.T., Loranty, M.M., 2011. The impacts and implications of an intensifying fire regime on Alaskan boreal forest composition and albedo. Global Change Biol. 17, 2853–2866.

Bergeron, Y., 1991. The influence of island and mainland lakeshore landscapes on boreal forest fire regimes. Ecology 72, 1980–1992.

Bergeron, Y., Fenton, N., 2012. Boreal forests of eastern Canada revisited: old growth, non-fire disturbances, forest succession, and biodiversity. Botany 90, 509–523.

Bergeron, Y., Cyr, D., Girardin, M.P., Carcaillet, C., 2010. Will climate change drive 21st century burn rates in Canadian boreal forest outside of its natural variability: collating global climate model experiments with sedimentary charcoal data. Int. J. Wildland Fire 19, 1127–1139.

Bergeron, Y., Gauthier, S., Flannigan, M., Kafka, V., 2004a. Fire regimes at the transition between mixedwood and coniferous boreal forest in northwestern Quebec. Ecology 85, 1916–1932.

Bergeron, Y., Flannigan, M., Gauthier, S., Leduc, A., Lefort, P., 2004b. Past, current and future fire frequency in the canadian boreal forest: implications for sustainable forest management. Ambio 33, 356–360.

Boucher, J., E.T., Azeria, J. Ibarzabal, Hébert, C., 2012. Saproxylic beetles in disturbed boreal forests: temporal dynamics, habitat associations, and community structure. Ecoscience 19, 328–343.

Bourgeau-Chavez, L.L., Alexander, M.E., Stocks, B.J., Kasischke, E.S., 2000a. Distribution of forest ecosystems and the role of fire in the North American boreal region. In: Kasischke, E.S., Stocks, B.J. (Eds.), Fire, Climate Change, and Carbon Cycling in the Boreal Forest. In: Ecological Studies, vol. 138. pp. 111–131.

Bourgeau-Chavez, L.L., Kasischke, E.S., Mudd, J.P., French, N.H.F., 2000b. Characteristics of Forest Ecozones in the North American Boreal Region. In: Kasischke, E.S., Stocks, B.J. (Eds.), Fire, Climate Change, and Carbon Cycling in the Boreal Forest.In: Ecological StudiesVolume 138. pp. 258–273.

Bradshaw, C.J.A., Warkentin, I.G., Sodhi, N.S., 2009. Urgent preservation of boreal carbon stocks and biodiversity. Trends Ecol. Evol. 24, 541–548.

Brandt, J.P., 2009. The extent of the North American boreal zone. Environ. Rev. 17, 101–161.

Brandt, J.P., Flannigan, M.D., Maynard, D.G., Thompson, I.D., Volney, W.J.A., 2013. An introduction to Canada's boreal zone: ecosystem processes, health, sustainability, and environmental issues. Environ. Rev. 21, 207–226.

Burton, P.J., Bergeron, Y., Bogdanski, B.E.C., Juday, G.P., Kuuluvainen, T., McAfee, B.J., Ogden, A., Teplyakov, V.K., Alfaro, R.I., Francis, D.A., Gauthier, S., Hantula, J., 2010. Sustainability of boreal forests and forestry in a changing environment. In: Mery, G., Katila, P., Galloway, G., Alfaro, R.I., Kanninen, M., Lobovikov, M., Varjo, J. (Eds.), Forests and Society—Responding to Global Drivers of Change.In: IUFRO World Series, vol. 25. pp. 247–282, International Union of Forest Research Organizations (IUFRO) Vienna.

Canadian Forest Service, 2015. National Fire Database – Agency Fire Data. Natural Resources Canada, Canadian Forest Service, Northern Forestry Centre, Edmonton, Alberta. http://cwfis.cfs.nrcan.gc.ca/en_CA/nfdb.

Christensen, J.H., Hewitson, B., Busuioc, A., Chen, A., Gao, X., Held, I., Jones, R., Kolli, R.K., Kwon, W.T., Laprise, R., Magana Rueda, W., Mearns, L., Menéndez, C.G., Räisänen, J., Rinke, A., Sarr, A., Whetton, P., 2007. Regional climate projections. In: Solomon, S., Qin, D., Hanning, M., Chen, Z., Marquis, M., Averyt, K.B., Tignor, H., Miller, H.L. (Eds.), Climate Change 2007: The Physical Science Basis. Contribution of Working Group I to the Fourth Assessment Report of the Intergovernmental Panel on Climate Change. Cambridge University Press, Cambridge, UK, pp. 847–940.

Cobb, T.P., Hannam, K.D., Kischuk, B.E., Langor, D.W., Quideau, S.A., Spence, J.R., 2010. Wood-feeding beetles and soil nutrient cycling in burned forests: implications of post-fire salvage logging. Agric. For. Entomol. 12, 9–18.

Cogbill, C.V., 1985. Dynamics of the boreal forests of the Laurentian Highlands, Canada. Can. J. For. Res. 15, 252–261.

Cyr, D., Gauthier, S., Bergeron, Y., Carcaillet, C., 2009. Forest management is driving the eastern North American boreal forest outside of its natural range of variability. Front. Ecol. Environ. 10, 519–534.

DeLong, S.C., Kessler, W.B., 2000. Ecological characteristics of mature forest remnants left by wildfire. For. Ecol. Manag. 131, 93–106.

Desponts, M., Brunet, G., Bélanger, L., Bouchard, M., 2004. The eastern boreal old-growth balsam fir forest: a distinct ecosystem. Can. J. Bot. 82, 830–849.

Drapeau, P., Nappi, A., Giroux, J.F., Leduc, A., Savard, J.P., 2002. Distribution patterns of birds associated with snags in natural and managed eastern boreal forests. In: Laudenslayer, B., Valentine, B. (Eds.), Ecology and Management of Dead Wood in Western forests. USDA Forest Service Pacific Southwest Research Station, Albany, Calif, USDA Forest Service General Technical Report PSW-GTR 181.

Eichhorn, M.P., 2010. Boreal forests of Kamchatka: structure and composition. Forests 1, 154–176.

Flannigan, M., Cantin, A.S., de Groot, W.J., Wotton, M., Newbery, A., Gowman, L.M., 2013. Global wildland fire season severity in the 21st century. For. Ecol. Manag. 294, 54–61.

Foster, D.R., 1983. The history and pattern of fire in the boreal forest of southeastern Labrador. Can. J. Bot. 61, 2459–2471.

Furyaev, V.V., Wein, R.W., MacLean, D.A., 1983. Fire influences in Abies-dominated forests. In: Wein, R.W., MacLean, D.A. (Eds.), The Role of Fire in Northern Circumpolar Ecosystems. John Wiley & Sons Ltd, New York, pp. 221–234.

Gawthrop, D., 1999. Vanishing halo: saving the boreal forest. Greystone Books, Vancouver, BC.

Girardin, M.P., 2007. Interannual to decadal changes in area burned in Canada from 1781 to 1982 and the relationship to Northern Hemisphere land temperatures. Glob. Ecol. Biogeogr. 16, 557–566.

Girardin, M.P., Ali, A.A., Carcaillet, C., Gauthier, S., Hely, C., Le Goff, H., Terrier, A., Bergeron, Y., 2013. Fire in managed forests of eastern Canada: risks and options. For. Ecol. Manag. 294, 238–249.

Harper, K., Boudreault, C., De Grandpré, L., Drapeau, P., Gauthier, S., Bergeron, Y., 2003. Structure, composition and diversity of old-growth black spruce boreal forest of the Clay Belt region in Québec and Ontario. Environ. Rev. 11 (Suppl. S1), S79–S98.

Hessl, A.E., 2011. Pathways for climate change effects on fire: models, data, and uncertainties. Prog. Phys. Geogr. 35, 393–407.

Huggard, D.J., Arsenault, A., 1999. Reverse cumulative standing age distribution in fire Frequency analysis. Can. J. For. Res. 29, 1449–1456.

Hutto, R.L., 1995. Composition of bird communities following stand-replacement fires in Northern Rocky Mountain (U.S.A.) Conifer forests. Conserv. Biol. 9, 1041–1058.

Hutto, R.L., 2008. The ecological importance of severe wildfires: some like it hot. Ecol. Appl. 18, 1827–1834.

Johnson, E.A., 1992. Fire and vegetation dynamics: studies from the North American boreal forest. Cambridge University Press, Cambridge, UK, 129 p.

Johnson, E.A., Miyanishi, K., Weir, J.M.H., 1998. Wildfires in the Western Canadian boreal forest: landscape patterns and ecosystem management. J. Veg. Sci. 9, 603–610.

Johnstone, J.F., Hollingsworth, T.N., Chapin III, F.S., Mack, M.C., 2010. Changes in fire regime break the legacy lock on successional trajectories in Alaskan boreal forest. Global Change Biol. 16, 1281–1295.

Kasischke, E.S., Christensen, N.L., Stocks, B.J., 1995. Fire, global warming, and the carbon balance of boreal forests. Ecol. Appl. 5, 437–451.

Kelly, R., Chipman, M.L., Higuera, P.E., Stefanova, I., Brubaker, L.B., Hu, F.S., 2013. Recent burning of boreal forests exceeds fire regime limits of the past 10,000 years. Proc. Natl. Acad. Sci. U. S. A. 110, 13055–13060.

Krezek-Hanes, C.C., Ahern, F., Cantin, A., Flannigan, M.D., 2011. Trends in large fires in Canada, 1959-2007. Canadian Biodiversity: Ecosystem Status and Trends 2010, Technical Thematic Report No. 6. Canadian Councils of Resource Ministers. Ottawa, ON. v+48 p. http://www.biodivcanada.ca/default.asp?lang=En&n=137E1147-0.

McCarthy, J., 2001. Gap dynamics of forest trees: a review with particular attention to boreal forests. Environ. Rev. 9, 1–59.

McCarthy, J.W., Weetman, G., 2006. Age and size structure of gap-dynamic, old-growth boreal forest stands in Newfoundland. Silva Fennica 40, 209.

McCarthy, J.W., Weetman, G., 2007. Self-thinning dynamics in a balsam fir (Abies balsamea (L.) Mill.) insect-mediated boreal forest chronosequence. For. Ecol. Manag. 241, 295–309.

Moen, J., Rist, L., Bishop, K., Chapin III, F.S., Ellison, D., Kuuluvainen, T., Petersson, H., Puettmann, K.J., Rayner, J., Warkentin, I.G., Bradshaw, C.J.A., 2014. Eye on the taiga: removing global policy impediments to safeguard the boreal forest. Conserv. Lett. 7, 408–418.

Nappi, A., Drapeau, P., Saint-Germain, M., Angers, V.A., 2010. Effect of fire severity on long-term occupancy of burned boreal conifer forests by saproxylic insects and wood-foraging birds. Int. J. Wildland Fire 19, 500–511.

Noble, I.R., Slatyer, R.O., 1980. The use of vital attributes to predict successional changes in plant communities subject to recurrent disturbances. Vegetatio 43, 5–21.

Odion, D.C., Hanson, C.T., Arsenault, A., Baker, W.L., DellaSala, D.A., Hutto, R.L., Klenner, W., Moritz, M.A., Sherriff, R.L., Veblen, T.T., Williams, M.A., 2014. Examining historical and current mixed-severity fire regimes in ponderosa pine and mixed-conifer forests of western North America. PLoS One 9, e87852.

Östlund, L., Zackrisson, O., Axelsson, A.L., 1997. The history and transformation of a Scandinavian boreal forest landscape since the 19th century. Can. J. For. Res. 27, 1198–1206.

Payette, S., 1992. Fire as a controlling process in the North American boreal forest. In: Shugart, H.H., Leemans, R., Bonan, G.B. (Eds.), A Systems Analysis of the Global Boreal Forest. Cambridge University Press, New York, pp. 144–169.

Perera, A., Buse, L., 2014. Ecology of Wildfire Residuals in Boreal Forests. John Wiley & Sons, Chichester, West Sussex, UK.

Robinson, N.M., Leonard, S.W.J., Ritchie, E.G., Bassett, M., Chia, E.K., Buckingham, S., Gibb, H., Bennett, A.F., Clarke, M.F., 2013. Refuges for fauna in fire-prone landscapes: their ecological function and importance. J. Appl. Ecol. 50, 1321–1329.

Rowe, J.S., 1983. Concepts of fire effects on plant individuals and species. In: Wein, R.W., MacLean, D.A. (Eds.), In: The Role of Fire in Northern Circumpolar Ecosystems18, J. Wiley, New York, pp. 135–154, SCOPE.

Stocks, B.J., Mason, J.A., Todd, J.B., Bosch, E.M., Wotton, B.M., Amiro, B.D., Flannigan, M.D., Hirsch, K.G., Logan, K.A., Martell, D.L., Skinner, W.R., 2003. Large forest fires in Canada. J. Geophys. Res. 108, 1959–1997, FFR 5-1 through 5-12..

Talbot, S.S., Meades, W.J., 2011. Circumboreal Vegetation Map (CBVM): Mapping the Concept Paper. CAFF Strategy Series Report No. 3. CAFF Flora Group (CFG), CAFF International Secretariat, Akureyri, Iceland. ISBN 978-9935-431-05-9.

Thompson, I., Larson, D., Montevecchi, W., 2003. Characterization of old "wet boreal" forests, with an example from balsam fir forests of western Newfoundland. Environ. Rev. 11, S23–S46.

Chapter 9

Climate Change: Uncertainties, Shifting Baselines, and Fire Management

Cathy Whitlock[1], Dominick A. DellaSala[2], Shaye Wolf[3] and Chad T. Hanson[4]

[1]Department of Earth Sciences and Montana Institute on Ecosystems, Montana State University, Bozeman, MT, USA, [2]Geos Institute, Ashland, OR, USA, [3]Center for Biological Diversity, San Francisco, CA, USA, [4]John Muir Project of Earth Island Institute, Berkeley, CA, USA

9.1 TOP-DOWN CLIMATE FORCING FIRE BEHAVIOR

There is no doubt that today's climate is changing, primarily from increased greenhouse gases from fossil fuel emissions (Romero-Lankao et al., 2014). The combination of rising temperatures and changes in seasonal and annual precipitation affects the size, severity, and occurrence of fires around the world (e.g., Krawchuk et al., 2009; Bowman et al., 2009; Flannigan et al., 2009). Because climate will increasingly dominate fire behavior in the future (Figure 9.1), it is important to draw on as broad a base of knowledge as possible to understand fire-climate interactions and identify appropriate management strategies.

In this chapter, we argue that the period chosen for comparison to current or future conditions is critical for understanding fire trends. Too short a period can overlook the influence of legacy conditions, the importance of extreme fire weather conditions, and the long-term climate conditions that have shaped fire activity in particular biomes. A suitable historical baseline or reference period must thus capture a long-enough span of time (reviewed by Papworth et al. (2008) and DellaSala et al. (2013)) to adequately reflect the dynamics of the disturbance and postfire recovery, as well as fire-climate variability. Selecting the wrong baseline, or one that is too short, can actually lead to poor management decisions and novel ecosystems (see DellaSala et al., 2013).

The Ecological Importance of Mixed-Severity Fires: Nature's Phoenix

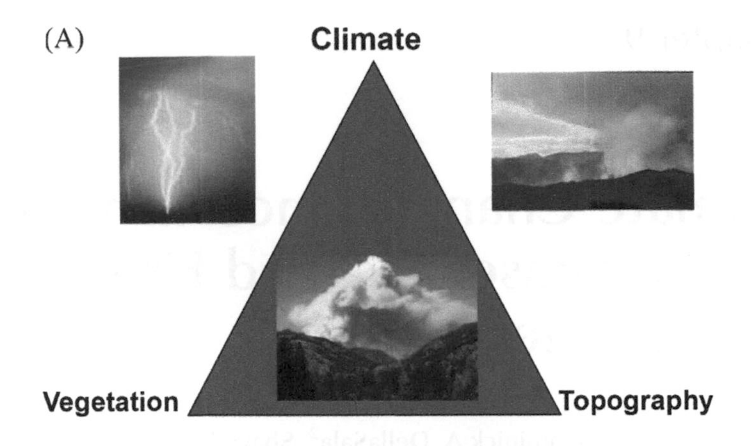

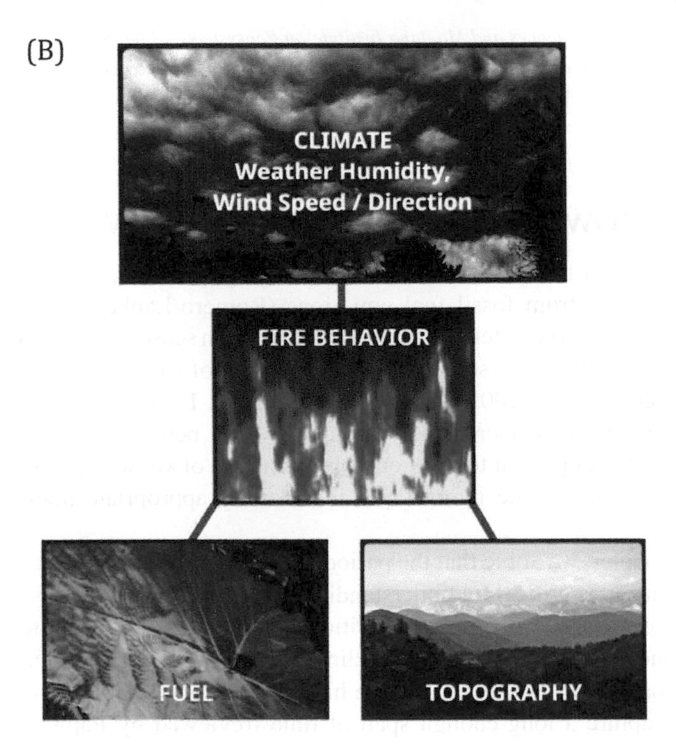

FIGURE 9.1 (A) Fuel-limited fire regimes depicting the interaction of climate, vegetation/fuels, and topography as generally equivalent influences of fire behavior. (B) Climate-limited fire regime depicting the top-down influence of climate on fire behavior. Many fire regimes are shifting from A to B as climate increasingly becomes the limiting factor of fire behavior. *(Also see Littell et al. (2009)).*

9.2 USING THE PALEO-RECORD TO CONSTRUCT A FIRE ENVELOPE

Fire history for any given location is a unique body of knowledge for establishing fire baselines because it describes fire causes and consequences over a wide range of climate conditions, land-use activities, and vegetation types. By providing a long-term perspective on fire regimes, historical data make us mindful of the short time span that serves as a reference condition for many forest management decisions, as well as the potential role of fire ahead with future climate and land-use changes. To effectively utilize historical fire information requires some level of understanding of the data sets that are available, as well as the time domains at which they describe fire. It also requires an appreciation of human influences on fire, including the degree to which people have altered past fire regimes through deliberate burning, land-use change, and the introduction of new species. Finally, fire history should be viewed not as irrelevant storytelling, but rather as vital information that describes the range of possible fire conditions under a broader array of spatial and temporal scales than we can observe at present.

9.3 RECONSTRUCTING PAST FIRE REGIMES

Multiple data sets are available to describe fire activity at different spatial and temporal scales (Gavin et al., 2007; Kehrwald et al., 2013) (Figure 9.2). On time scales of days to decades, remotely sensed data and historical documents register fire occurrence and are used to estimate global area burned (also see the Preface). On longer time scales of decades to centuries, tree-ring records, both fire scars on living trees and forest stand structures, provide information on prehistoric fire occurrence, fire frequency, and fire severity. Studies of tree rings in the western United States have been instrumental in describing low- and mixed-severity fire regimes (e.g., Brown et al., 1999), the character of postfire vegetation development following high-severity fires (Romme, 1982; Sherriff et al., 2001; Odion et al., 2014), and modes of climate variability that lead to years and decades of large fires (Swetnam and Betancourt, 1990; Heyerdahl et al., 2008; Trouet et al., 2010). Fire-scar tree-ring records can produce a reconstruction of fire history with yearly and sometimes seasonal precision and extend our knowledge of past fires back centuries and in some cases millennia, but they are less useful in understanding the history of forests that experience high-severity stand-replacing fires. In these settings, analysis of stand ages and postfire age structure provides information on past fire events as well as postfire vegetation development. In mixed fire regimes, a combination of stand age and fire scars has been effectively used to reveal the mosaic of burned and unburned vegetation patterns (Taylor and Skinner, 2003; Schoennagel et al., 2011; Odion et al., 2014).

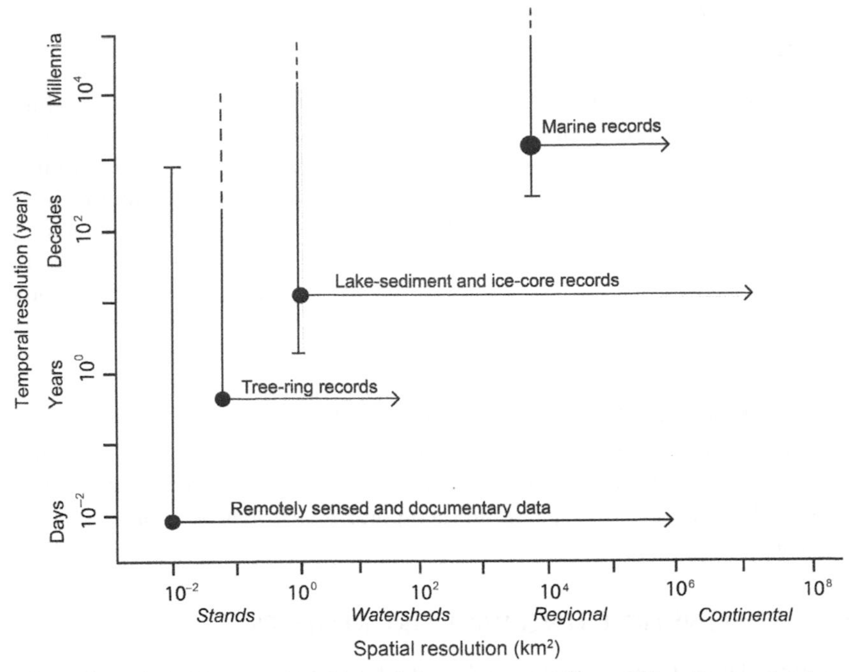

FIGURE 9.2 Types of data and models used to reconstruct past fire on different temporal and spatial scales. *(After Gavin et al. (2007).)*

Sedimentary Charcoal Analysis

On time scales of centuries to millennia, sedimentary records from lakes and natural wetlands provide information about fire history, and particulate charcoal is often a primary proxy. Charcoal records are less spatially resolved and temporally less precise in comparison to tree-ring data, but they have the advantage of examining fire response to a broader range of climate conditions and vegetation types than exist in the recent past. Evidence of fire in the form of black carbon, charcoal particles, and chemical signatures also is available in marine and ice cores that span several millennia (Daniau et al., 2013; Kehrwald et al., 2013).

Fire-history research based on sedimentary charcoal records has undergone a renaissance in recent decades, in part motivated by interest in understanding recent large, severe fires that seem to have little precedence in historic time. Whether such conflagrations occurred in the more distant past and the extent to which they were caused by unusual climate conditions or human activities are topics of both scientific and public concern. Charcoal analysis is based on the premise that charred particles are carried aloft during the fire and travel some distance in the atmosphere before settling on the ground and lake surface. The charcoal particles that fall on lakes and wetlands eventually become sequestered in sediment, and changes in particle abundance at different depths in sediment cores provide a proxy of past fire activity (Whitlock and Larsen, 2001). A suite of radiocarbon dates or other chronologic markers establishes an independent chronology in most fire-history studies. The primary data are presented as charcoal accumulation rates (CHAR particles per cm^2 per year), although several metrics have been used (Conedera et al., 2009). High-resolution charcoal investigations from lake sediments refer to the examination of large charcoal particles (>125 microns in diameter) in contiguous thin slices of the core. Because large particles are transported relatively short distances, they provide a record of local fires, and continuous sampling allows reconstructions with decadal precision (Figure 9.3).

The CHAR time-series data from a particular site are decomposed statistically to reconstruct the fire history (Marlon et al., 2006; Higuera et al., 2009). The long-term trend in the data is attributed to slowly varying changes in biomass burning, which is a function of both fuel composition and distribution

FIGURE 9.3 Fire-history analysis from lake sediments involves collecting a suite of cores from small lakes from an anchored platform (A); extruding and describing each sediment core in the field (B); slicing the core into contiguous 0.5-cm intervals and washing the material through sieves (C); and tallying black charcoal particles under the microscope for each sample (D). *(Photo 9.3D courtesy of Janet Wilmshurst, Landcare Research, Lincoln NZ.)*

(vegetation) and fire severity. Calibration studies and process-based models of charcoal transport suggest that CHAR trends are a good proxy of area burned within a <30 km radius of small lakes being studied, and the source area often matches that of pollen records from the same core used to reconstruct the vegetation history (Higuera et al., 2011; Kelly et al., 2013). Statistically significant CHAR peaks above a prescribed threshold are attributed to individual fire episodes (i.e., one or more fires occurring in the time span of the sample). Peak detection is used to identify fire episodes and describe variations in the frequency as well as the magnitude of fire episodes. Some studies make efforts to identify and separate grass and wood charcoal at each stratigraphic level as a tool to discriminate between surface and crown fires (Whitlock et al., 2006; Walsh et al., 2008). Additional precision also comes when the charcoal particles are themselves identified, a technique that comes from archeology (Carcaillet and Thinon, 1996; Marguerie and Hunot, 2007).

9.4 FIRE HISTORY ACROSS A MOISTURE GRADIENT

The goals of fire-history research are to distinguish the drivers of fire activity, be they climate, fuel, or anthropogenic factors; understand the extent and nature of past fire activity; and assess fire's long-term ecological effects. These objectives require (1) examining multiple charcoal records to separate local from regional patterns; (2) modeling studies to examine fire-ecosystem feedbacks; and (3) data-model comparisons (Henne et al., 2011; Marlon et al., 2013; Pfeiffer et al., 2013). One way to understand fire's role in different ecosystems is to examine its importance across a moisture gradient (Figure 9.4). At the dry end, deserts experience frequent ignitions and low fuel and soil moisture, but discontinuous fuel often prohibits fire spread, and fires of any significant size are infrequent. At the wet end, fuels are abundant in rainforests, but the dry

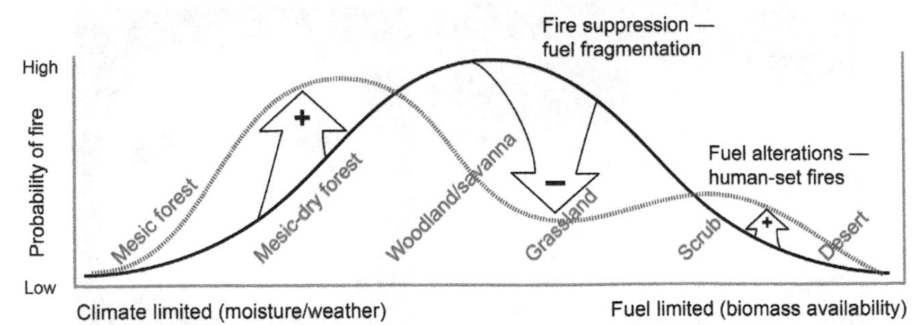

FIGURE 9.4 The magnitude of human influences on natural fire regimes varies along a broad moisture gradient of vegetation types. Climate exerts strong control over fire activity at the extreme wet and dry ends of the moisture gradient as a result of low combustion potential of fuels in mesic settings and the scarcity and disconnected arrangement of fuels in arid regions. Humans have the potential to alter fire regimes (shown by positive and negative arrows) by changing ignition frequency, fuel composition, and pattern as well as by suppressing fires (dashed line). *(After Whitlock et al. (2014)).*

season is short and natural ignitions are infrequent or do not coincide with the period of dry fuels. In such wet settings fires are also infrequent, although they can be severe when ignition and drought coincide. At the intermediate scale, temperate dry forests and savanna meet both requirements of sufficient amount and dryness of fuel and frequency of ignition, and these vegetation types support frequent low- and mixed-severity fires (Williams and Baker, 2012; Odion et al., 2014). Thus fires are infrequent at wet and dry ends of the moisture spectrum, and severe fires tend to characterize ignition-limited systems (Whitlock et al., 2010; Archibald et al., 2013; McWethy et al., 2013). Of course, these relationships are compromised by human activities, including where people alter the natural ignition frequency (e.g., the arrival of people on the Pacific Islands; McWethy et al., 2010; Chapter 7), introduce nonnative species that affect flammability (Brooks et al., 2004), or fragment natural landscape patterns (e.g., through logging and fire suppression; Odion et al., 2014).

9.5 CASE STUDIES OF LONG-TERM FIRE HISTORY IN THE WESTERN UNITED STATES

Tree-ring and charcoal data from middle- and high-elevation forests in the western United States indicate that past variations in fire activity are strongly linked to a changing climate. On long time scales, a primary driver of past fire activity has been slow variations in the seasonal cycle of insolation. In the early Holocene (~12,000-6000 calendar years before present (cal year BP), with "present" set at 1950 AD), summer insolation (generally, the degree of sun exposure) was 8% higher than at present, and winter insolation was lower by the same amount. Higher summer insolation led directly to higher-than-present summer temperatures and effectively decreased moisture; it indirectly produced a strengthened northeastern Pacific subtropical high-pressure system, which further suppressed summer moisture in the northwestern United States. Most parts of the northwestern United States show higher fire activity in the early Holocene compared with the late Holocene (Whitlock et al., 2008). At the same time, stronger-than-present monsoonal circulation, also driven by the summer insolation maximum, may have led to wet summer conditions and fewer fires in the southwestern United States (Bartlein et al., 1998; Anderson et al., 2008). On decadal to century time scales, ocean-atmosphere interactions (El Niño Southern Oscillation, Pacific Decadal Oscillation, American Multidecadal Oscillation) may contribute to fire occurrence and severity through atmospheric configurations that create persistent drought (Kitzberger et al., 2007; Trouet et al., 2010), although the strength of the these short-term relationships varies greatly from region to region.

Greater Yellowstone Region

In the greater Yellowstone ecosystem, regional analysis of charcoal records describe broad trends in climate, fire, and vegetation change over the past

15,000 years (Iglesias et al., 2015). These data indicate that highest fire activity in the region occurred between 12,000 and 10,000 cal year BP, when summers were warmer than today, winters were colder, and winter precipitation was generally high. The high-fire period was associated with decline in fire-vulnerable Engelmann spruce (*Picea engelmannii*) and an increase in whitebark pine (*Pinus albicaulis*) at all elevations.

On the rhyolite (a type of silica-rich volcanic rock) plateaus of central Yellowstone, charcoal data highlight the direct connections between fire and climate through time (Millspaugh et al., 2000). This area has supported lodgepole pine (*Pinus contorta*) forest for the past 11,000 years because of the strong edaphic (relating to the soil) controls on vegetation composition. By contrast, past fire activity was more dynamic than the vegetation history, showing the highest occurrence between 11,000 and 7,000 cal year BP during the summer insolation maximum and decreasing frequencies to the present day. Most prehistoric fires were likely mixed- or high-severity events, given the persistence of lodgepole pine. Other studies of Yellowstone show the occurrence of infrequent large fires during the Little Ice Age (1600-1900 AD), and fewer and likely small fire events during the Medieval Climate Anomaly (800-1200 AD) (Meyer et al., 1995; Pierce et al., 2004; Whitlock et al., 2012). By contrast, an analysis of post-fire sediment deposits in alluvial fans in ponderosa pine (*Pinus ponderosa*) forests in southern Idaho revealed large, severe-fire events well above recent levels during a warm period from 1050 to 650 cal year BP (Pierce et al., 2004).

Pacific Northwest

The fire history of the Pacific Northwest region also was strongly influenced by shifts in the duration and severity of summer drought and the composition of the forest. Between 9500 and 5000 cal year BP, drier-than-present summers supported forests with abundant Douglas-fir (*Pseudotsuga menziesii*), red alder (*Alnus rubra*), and bracken fern (*Pteridium*). This forest composition resembled current early seral forest stages, and—not surprisingly—fires were more frequent than today. In valley floors, woodland, prairie, and savanna habitats were expanded in the early Holocene compared with their present distribution, again in association with more fires. As summer insolation decreased in the late Holocene, summers became cooler and wetter than before, and forests of mesophytic (referring to plants adapted to moderate levels of moisture) conifers (e.g., western hemlock (*Tsuga heterophylla*), western red cedar (*Thuja plicata*), fir (*Abies* spp.), and Sitka spruce (*Picea sitchensis*)) prevailed. In association with this cooling trend, fires were less frequent, but, given the vegetation composition, they were likely more severe than in earlier times (Walsh et al., 2008; Whitlock et al., 2008; Gavin et al., 2013).

The temperate wet forests of the Pacific Northwest do not seem to have been particularly vulnerable to prehistoric human activities, even though people lived in the region throughout the Holocene and the population density at the time of European arrival was relatively high (Boyd, 1990). Several ecological and cultural factors may account for the limited influence of people in shaping Pacific Northwest fire regimes, but among them is the pyrogenicity of the dominant tree species in wet temperate forests: Douglas-fir. This conifer has evolved with fire and displays several life-history traits that allow it to persist across a wide range of fire frequencies and severities (Tepley et al., 2013). Its rapid establishment and growth of seedlings after mixed-severity fires and its ability to establish beneath and above competing shrubs promote rapid recovery of Douglas-fir canopy, often within decades after fire (Tepley et al., 2014). The presence of partially intact forest within most burned areas also enables Douglas-fir to rapidly colonize adjacent high-severity patches. Given these factors, it seems highly unlikely that a targeted ignition strategy by prehistoric peoples in the Pacific Northwest would have resulted in large-scale forest conversion, as occurred, for example, in the temperate wet forests of New Zealand (Whitlock et al., 2015).

Further south in the Pacific Northwest, the fire history is more complex in terms of spatial and temporal variability, particularly in the Klamath-Siskiyou region of southwestern Oregon and northern California (Taylor and Skinner, 2003; Colombaroli and Gavin, 2010; Odion et al., 2010; Briles et al., 2011). A study of Bolan Lake showed infrequent fires in the early postglacial period (17,000-14,500 cal year BP), when the climate was cooler than present and subalpine parklands of lodgepole pine, spruce, and mountain hemlock (*Tsuga mertensiana*) were present (Briles et al., 2005). Warming after 14,500 cal year BP was associated with forest closure and increased fire activity. After 11,000 cal year BP, open xerothermic (pertaining to plants adapted to relatively hotter, drier conditions) forests of pine, oak (*Quercus*), incense cedar (*Calocedrus decurrens*), and *Ceanothus* developed, and fires became more frequent than during the late-glacial period. During the middle Holocene (7000-4500 cal year BP), a closed forest of fir, Douglas-fir, red alder, and oak became established, and the frequency of fire episodes reached its highest levels. In the past 4000 years, fir-dominated forests have developed at middle elevations, and mountain hemlock has expanded at high elevations. At most sites, fire frequency has declined in the late Holocene, with the exception of elevated fire activity during the Medieval Climate Anomaly (Briles et al., 2011).

Colorado Rocky Mountains

The fire history of subalpine forests in the Colorado Rocky Mountains shows the importance of changes in forest composition and density on fire behavior (Higuera et al., 2014). Tree-ring records indicate that subalpine forests of Engelmann spruce, subalpine fir (*Abies lasiocarpa*), and lodgepole pine have supported low-frequency, stand-replacing fires in recent centuries (Buechling and Baker, 2004; Sibold and Veblen, 2006). The vegetation and fire frequency have shown little variation over the past 6000 years, despite long-term trends toward lower summer temperatures and less effective moisture (where effective moisture = precipitation − evaporation) (see Figure 9.5). Mean fire return intervals have ranged between 150 and 250 years during the past 6000 years, although the variability around the long-term fire return interval mean correlated well with shifts in summer moisture (i.e., more fires during drier summers). Levels of biomass

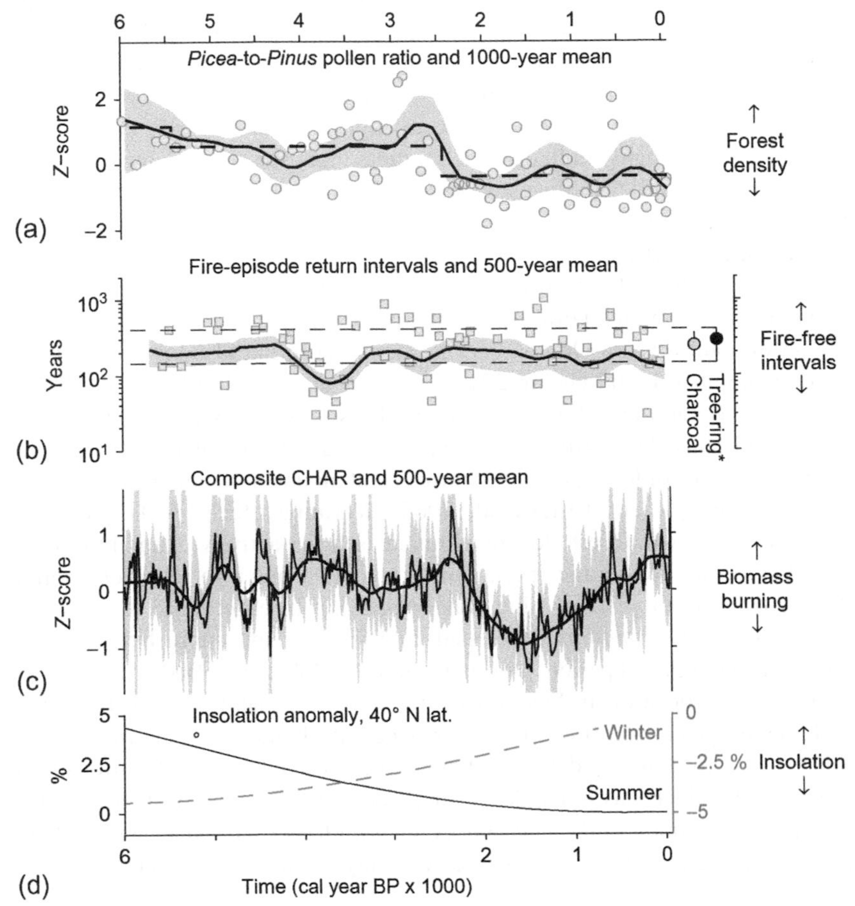

FIGURE 9.5 Millennial-scale vegetation, fire, and climate history from subalpine forests in Rocky Mountain National Park. (A) The *Picea*-to-*Pinus* pollen ratio (points) is a composite from three pollen records (black curves and grey envelopes represent the 95% confidence intervals). The ratio shifts to lower values (more pine) at 5400 and 2400 cal yr BP (calendar years before present). (B) Individual fire return intervals (FRIs) from each site (squares) and the composite mean FRI averaged and smoothed over 500-year periods (gray shading). Mean FRI from the charcoal record for the past 300 years (circa 1650-1950) compares well with estimated fire rotations from tree-ring records over the same period (Buechling and Baker, 2004; Sibold and Veblen, 2006) (grey and black circles on right). (C) Composite CHAR record at 15-year intervals (with 95% confidence intervals) and smoothed to 500 years. (D) Holocene insolation for the summer and winter solstice at 40° N latitude. *(From Higuera et al. (2014))*.

burned (inferred from CHAR trends) decreased significantly at 2400 cal year BP, despite little change in vegetation or fire frequency. This shift is interpreted as evidence of less biomass burned per fire and a decrease in crown fire severity. In the past 1500 years fire severity has steadily increased in these forests. Higuera et al. (2014) suggest that in Rocky Mountain subalpine forests (1) fire severity is likely more responsive to climate change than is fire frequency, and (2) the indirect influence of climate on vegetation and fuels is as important as the direct effects of climate on fire activity.

9.6 HISTORICAL RECORD AND THE FIRE ENVELOPE

Most studies describing current and projected trends in fire activity draw on a short baseline of historical data for comparison. In the conifer forests of North America a variety of methodological approaches have been used to establish historical reference conditions, including analysis of spatially explicit historical records of high-severity fire, reconstructions of past fire severity using aerial photographs to determine the number of emergent (surviving) trees in a particular study area, age analyses of stands in current unlogged forests, and analyses of stand structure based on historical field plot data (Baker, 2012; Williams and Baker, 2012; Baker, 2014; Odion et al., 2014; Hanson and Odion, 2015a; see also Chapter 1 for more detailed descriptions). An examination of historical data relative to current estimates of high-severity fire rates highlights the problem of using a baseline that is too short (and recent) to fully capture the historical variability of particular fire regimes. We focus, in particular, on areas where there is a tendency for recent trends in high-severity fire—whether increases or decreases—to be erroneously considered outside of the context of historical baseline variability. In such cases, land managers and policymakers assume that current fires exceed natural levels, despite historical data (pertaining, in this section, to the past few centuries) that indicate a substantial decline in fire activity (Baker, 2014; Odion et al., 2014; Hanson and Odion, 2015a). As a result, current conditions are misinterpreted in the development of management prescriptions.

For example, while area burned has increased in the boreal forests of Canada in recent decades (1959-1999) (Kasischke and Turetsky, 2006; Chapter 8), the pattern is not uniform when placed in a longer-term context and does not hold for high-severity fire ("stand-initiating" fire). Overall, high-severity fire and rotation intervals have decreased (i.e., less high-severity fire, on average, per year or decade) by two- to threefold in the boreal forests of eastern and western Canada. Based on stand-age analyses, high-severity fire rotations are much longer (i.e., less high-severity fire) currently than they were before the mid-1800s (Bergeron et al., 2001, 2004a,b). Studies analyzing longer time periods also report decreases in fire activity in many regions of western Canada. Wallenius et al. (2011) indicate a significant decrease in burned area in northwestern Canada (northeast British Columbia, northwest Alberta, southeast Yukon Territory, and southwest Northwest Territories) from 1800 to 2000, and the findings are consistent with other long-term studies that describe decreases in boreal fire activity since 1850 (see Girardin et al., 2009). This decline has been attributed mainly to long-term shifts toward warmer, wetter conditions (with some fluctuations in the opposite direction in the early twentieth century) since the mid-1800s (Larsen, 1996; Bergeron et al., 2001, 2004a, b). For example, Meyn et al. (2013) report a significant decrease in burned area in British Columbia from 1920 to 2000 that was linked to increased precipitation, which outweighed the effects of rising temperatures and drought severity.

In mixed-conifer and ponderosa pine forests of western North America, historical high-severity fire rotation intervals typically ranged from about 150 to 400 years before the effects of fire suppression and logging (Leiberg, 1900, 1902; Bekker and Taylor, 2001; Baker, 2012, 2014; Williams and Baker, 2012; Hanson and Odion, 2015a). Since the early twentieth century, high-severity fire has declined by approximately two- to fourfold in most forests, and current high-severity fire rotations are generally 600-1000 years across broad regions (Odion and Hanson, 2013; Odion et al., 2014). This decline in fire is likely a result of fire suppression, long-term climate change, or both. Because many species in these forests benefit from and depend on the unique habitat created by high-severity fires (Chapters 3-6), alteration of fire frequency or severity threatens biodiversity and ecosystem dynamics. Increases or decreases in fire occurrence in the past few decades, and those projected in the future, must be understood within an ecological context informed by long-term fire regimes and high-severity fire rotation intervals.

9.7 UNDERSTANDING THE INFLUENCE OF ANTHROPOGENIC CLIMATE CHANGE ON FIRE

Given the strong influence of climate on fire activity, anthropogenic climate change is likely to alter fire activity around the globe (Bowman et al., 2009; Flannigan et al., 2009; Krawchuk et al., 2009). Because climate change is placing stress on ecosystems (Parmesan 2006), a common assumption is that anthropogenic climate change will increase fire activity to levels that will be deleterious to forests. As illustrated in previous chapters, however, wildfire, including high-severity fire, provides important ecological benefits to forest ecosystems, and these types of fires have become uncommon in many regions of western North America (Odion et al., 2014). Therefore, analyses linking fire and climate change should also consider fire's ecological benefits and the degree to which fire has been removed from the ecosystem (the fire deficit). Assessment of fire-climate trends should be based on a sufficiently long-term baseline to capture the historical range of fire variability of the ecosystem and should also account for the role of other anthropogenic factors, such as changes in wildfire management policies. Most current studies of fire and climate change do not consider all of these components.

Using a historical baseline for detecting and interpreting the effects of climate change on fire activity is particularly important in western US forests where fire activity trends of the past century have been altered by land-use and management practices. Stringent policies on fire suppression on US federal lands throughout most of the twentieth century profoundly and abruptly decreased the area burned in many western forests (Mouillot and Field, 2005; Stephens et al., 2007; Marlon et al., 2012; Odion et al., 2014). These studies indicate that a baseline that at least considers fire variability before 1900 is needed to understand fire variability under a range of climate and fuel conditions. A long-term baseline also clarifies the relative influences of nonclimate drivers in shaping current fire conditions. The invasion of nonnative plants, introduction of nonnative grazers, land-use change, and changes in forest management practices, for example, have caused abrupt changes in fire regimes globally, independent of climate change (Pausas and Keeley, 2014).

Using the appropriate spatial scale is also important for understanding the relationships between climate change and fire activity. Many studies have documented spatial variability in fire-climate relationships among western ecoregions (Westerling et al., 2006; Littell et al., 2009; Parisien et al., 2011, 2012) and in the ways that climate change will affect temperature, precipitation timing and extent, drought severity, and other key drivers of fire activity (Hartmann et al., 2013; Melillo et al., 2014). Depending on the interplay between rising temperature and changing precipitation timing and amounts, climate change will affect fire activity differentially across regions and vegetation types (Krawchuk and Moritz, 2011). For example, in many northern and mountainous regions of the western United States, low precipitation and warmer temperatures in the seasons leading up to and including the fire season are strongly associated with increased burned area (Littell et al., 2009), whereas increased precipitation in summer suppresses fire (Moritz et al., 2012). By contrast, in the more fuel-limited arid ecosystems of the southwestern United States, increased precipitation before the fire season is strongly associated with increased burned area (Littell et al., 2009), but lower precipitation before the fire season suppresses fire activity by decreasing fuel biomass (Moritz et al., 2012).

9.8 OBSERVED TRENDS IN FIRE ACTIVITY LINKED TO CLIMATE CHANGE

Studies of fire trends in western North America in relation to recent climate change report a range of patterns depending on the fire activity metric (e.g., burned area, occurrence, severity), regional scale, and time period analyzed. Most studies have examined trends only over recent decades (e.g., 1970s/1980s to 2000s) rather than longer periods that would encompass a greater range of variability. Although some studies report increases in burned area linked to increased temperature and precipitation change in recent decades (e.g., Westerling et al., 2006), others indicate patterns of decrease (e.g., Meyn et al., 2013) and areas of relative fire stability (e.g., Dennison et al., 2014). Most current research has not detected a trend in fire severity in recent decades.

Westerling et al. (2006) is the most highly cited study linking wildfire activity with recent climate change in western North America. Using a study period from 1970-2003 and averaging across forested regions in the western United States, the study reported a marked shift during the mid-1980s toward a higher frequency of large fires, a greater average annual area burned, and a longer fire season, which the authors associated with increased spring and summer temperatures and an earlier spring snowmelt. However, trends since the mid-1980s are less clear (Westerling et al., 2006).

Most subsequent studies have examined fire-activity trends on an ecoregional level and have found differing geographic patterns over short time periods. Dillon et al. (2011) analyzed trends across six ecoregions in the southwestern and northwestern United States from 1984 to 2006 and detected no trends in annual area burned or proportion burned severely in the northwestern ecoregions (Pacific, Inland Northwest, and Northern Rockies). The study did report a significant increase in burned area and high-severity burned area in the three southwestern ecoregions (Southern Rockies, Colorado Plateau, and Mogollon Rim) and a significant upward trend in fire severity (proportion of high-severity fire) in one southwestern ecoregion (Southern Rockies). Topography (i.e., elevation, aspect/slope) was identified as the most important variable in determining severe fire occurrence, followed by climate conditions.

Dennison et al. (2014) examined trends in fire activity from 1984 to 2011 in nine ecoregions in the western United States. This study detected significant increases in annual fire area in three of nine ecoregions (Southern Plains, warm deserts, and Arizona-New Mexico Mountains) and significant increases in the number of large fires in four of nine ecoregions (Southern Plains, Arizona-New Mexico Mountains, Rocky Mountains, Sierra Nevada/Cascades). In contrast to Westerling et al. (2006), this study did not detect a significant trend toward an earlier fire season in any ecoregion. Dennison et al. (2014) caution against directly attributing increases in fire activity to climate change but note that ecoregions with increasing trends in the number of large fires and total fire area also experienced increasing drought severity over that period.

The few studies that have examined trends in fire severity also use short time periods and indicate that fire severity has not increased in recent decades in most forested regions in the western United States: Pacific Northwest and California (Schwind, 2008), Pacific Northwest and Southwest except the Southern Rockies (Dillon et al., 2011), northwestern California (Miller et al., 2012), the Klamath/Siskiyou region and Eastern Cascades (Hanson et al., 2009), and Sierra Nevada and Southern Cascades (Collins et al., 2009; Hanson and Odion, 2014; however, see Miller et al., 2009; Miller and Safford, 2012). Hanson and Odion (2014) found that use of a vegetation data set that postdates the time series being analyzed tends to result in a statistically significant bias toward reporting an increasing trend in severity. For example, conifer forest that experiences high-severity fire in the earlier years of the time series is disproportionately reclassified later as nonconifer vegetation, thus creating the false appearance of increasing severity. Safford et al. (2015) hypothesized that an increasing trend would be found if analysis focused solely on wildland fires in mixed-conifer and ponderosa/Jeffrey pine forests on national forest lands. Hanson and Odion (2015b) tested this hypothesis and again found no trend in increasing fire severity.

9.9 PROJECTED CHANGES IN FIRE ACTIVITY IN RESPONSE TO CLIMATE CHANGE

Studies projecting how climate change will affect future fire activity typically use one of three modeling approaches, each with its own limitations: statistical models, changes in fire activity indices, and dynamic global vegetation models (DGVMs) (see Yue et al., 2013). Statistical models correlate empirical observations of fire activity (e.g., area burned, fire occurrence, fire probability) with environmental variables expected to affect fire. The models are used to project fire activity under future climate conditions derived from a global or regional climate model. This approach is similar to species distribution models that forecast shifts in species ranges under climate change, and they have similar limitations (e.g., Guisan and Thuiller, 2005). A second approach projects changes in a fire activity index, such as a drought index, severity rating, or energy release component, to estimate future fire potential as a result of climate change; a primary limitation is the accuracy of the index in representing fire activity. A third approach is to incorporate a fire module into a DGVM, which is a process-based biogeochemical model that simulates vegetation dynamics in response to climate change driven by climate data from global climate models (GCMs). Modeling fire in DGVMs can be challenging because it requires a mechanistic understanding of how climate and fire interact, and this approach is often limited by the accuracy of representing historical fire activity patterns.

Fire projection studies differ not only in their modeling approaches but also in the number and choice of GCMs, emissions scenarios, climate variables, spatial scale (i.e., global or regional), and the historic baseline for deriving fire-climate relationships and for comparing projected versus historic fire activity, all of which can create significant variation among study results and interpretations. One important source of uncertainty is the large differences across GCMs in the projected change in precipitation timing and amount in western North America (Roy et al., 2012; Peterson et al., 2013). The choice of GCMs has the potential to create divergent projections of future fire activity depending on whether selected models forecast wetter or drier futures.

Modeling studies have projected a range of responses in future fire activity across the globe and in western North America, including areas of decrease, increase, and relative stability in wildfire probability, occurrence, and biomass burned (Scholze et al., 2006; Krawchuk et al., 2009; Gonzalez et al., 2010; Liu et al., 2010; Pechony and Shindell, 2010; Moritz et al., 2012). These global studies show a general lack of spatial concordance in their projections, likely because of differences in modeling approaches, climate variables used, and the number and selection of GCMs (see Moritz et al., 2012). For example, using changes in drought index to measure fire potential, Liu et al. (2010) projected future global fire patterns nearly opposite those of Moritz et al. (2012) that employed a statistical modeling approach.

Analyses of the western United States and Canada have primarily projected increases in fire activity (e.g., area burned and fire potential) in response to climate change, although there is significant variability among studies and ecoregions, in particular forested ones. Using one GCM in a statistical modeling approach, Spracklen et al. (2009) projected an average increase in burned area of 54% across the western United States overall by midcentury, although significant increases occurred in only three of six western ecoregions (Pacific Northwest forests, desert Southwest, Rocky Mountains forests). Yue et al. (2013), using 15 GCMs, projected an average increase in burned area of 61% across the western United States by midcentury, but increases in ecoregions varied substantially depending on whether a statistical or process-based modeling approach was used. Fire projection studies at smaller regional scales have suggested increases in fire activity for some regions—the Pacific Northwest (Rogers et al., 2011) and Southern Rockies (Litschert et al., 2012)—and conflicting patterns of increases and decreases for others: California, Nevada, southern Oregon, southwestern Idaho, western Utah, and western Arizona (Westerling and Bryant, 2008, Krawchuk and Moritz, 2012). Projection studies typically have not examined changes in fire severity (but see Rogers et al., 2011), but focus on occurrence, probability, and area burned.

Most fire projection studies use a short historical baseline spanning the past few decades, which does not provide a useful context for determining whether projected changes fall within the range of historical variability. Illustrating important exceptions, Bergeron et al. (2010) projected a 125% increase in burn rate in the eastern Canadian boreal forest by the end of the century compared with the recent period from 1961 to 1999, but they determined that the increase fell well within the long-term variability for this region during the past 7000 years, as well as a shorter baseline of the past 300 years. By contrast, Westerling et al. (2011) suggest enormous increases in area burned in the forests of the greater Yellowstone ecosystem, projecting a nearly 10-fold (900%) increase by midcentury and a 1000-fold (100,000%) increase by the end of the century. If true, this level of burning would lie well outside the range of variability of the past 10,000 years. Some studies have projected increases in total annual area burned in California ranging from 9% and 11% to 15% by the end of the century compared with that in 1895-2003 (Lenihan et al., 2008), and increases in the number of large fires ranging from 12%, 23%, and 34% to 53% by the end of the century compared with that during 1961-1990 (Westerling and Bryant, 2008). Given that the average annual burned area in

California in the past several decades (1950-2009) was at least several times lower than the burned area before 1800 (Stephens et al., 2007; Odion et al., 2014), these projected increases in fire activity in California would likely remain within the historical range of the past several centuries.

9.10 CONCLUSIONS

Understanding the causes and effects of wildland fire in forest ecosystems depends on the temporal and spatial scale of interest. In this regard, fire triangles are a common starting point for conceptualizing the suite of biophysical factors operating at particular scales as well as cross-scale interactions (Figures 9.1 and 9.6). Taken together, the fire envelope is defined by a hierarchy of temporal

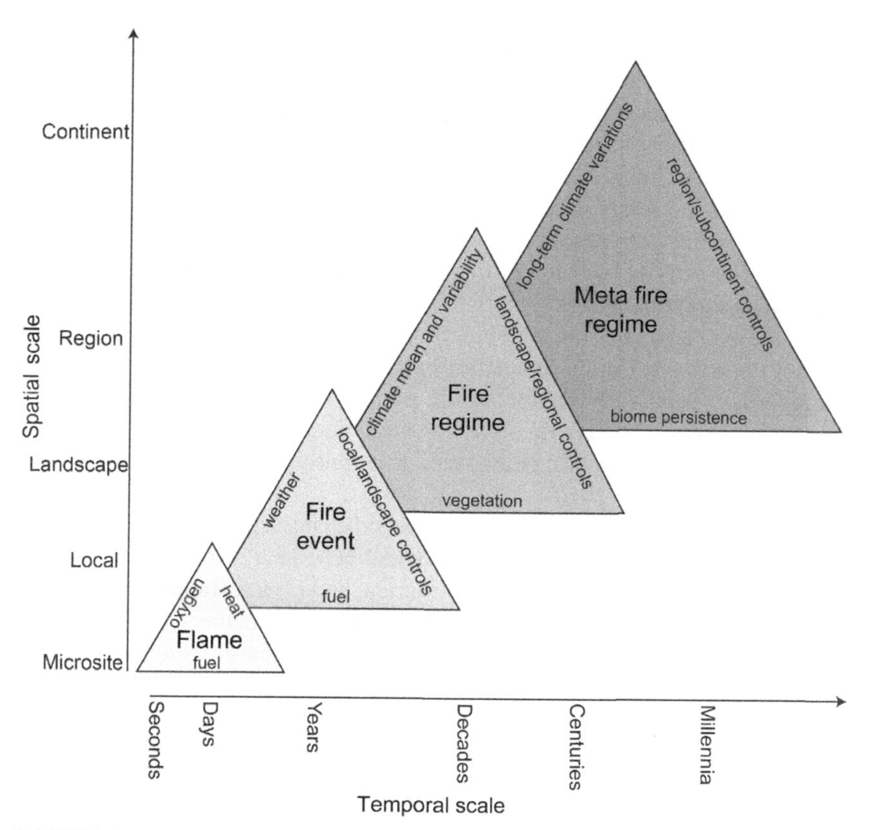

FIGURE 9.6 Controls of fire at multiple temporal and spatial scales conceptualized as fire triangles (modified from Parisien and Moritz, 2009). The side of each triangle indicates the dominant drivers at different temporal and spatial scales, and the overlap of triangles shows their nested nature. Paleoecological data suggest the need for a broader conceptualization of fire regimes that considers the variability of fire characteristics over the lifespan and spatial extent of a biome. *(From Whitlock et al. (2010)).*

and spatial conditions (or triangles) that shape biomass burning over time and space. At the smallest scale, the fire fundamentals triangle links oxygen, heat, and fuel at time scales of hours to years. At the next temporal and spatial scale, the fire event triangle links weather, fuels, and topography as factors that influence ignition probability, rate of fire spread, and fire intensity over seasons and years (Rothermel, 1972; Bowman et al., 2009). On decadal-to-millennial time scales, the fire regime triangle describes variables that determine the characteristic pattern, frequency, and intensity of fire at landscape and broader scales, reflecting the linkages between vegetation as a determinant of fuel, climate conditions as creators of fire weather, and ignition sources, be they human or natural (Parisien and Moritz, 2009; Krawchuk and Moritz, 2011). Our understanding of the paleofire record suggests that a larger and longer scale should also be considered in the fire envelope. A meta-fire regime triangle describes insights gained from the range of conditions that govern fire history over the duration of a vegetation type at time scales of centuries to millennia and fire variability at the scale of regions.

Understanding past human-vegetation-climate linkages of fire regimes has gained wider attention and appreciation in the face of projected future climate change. Although many definitions of a fire baseline implicitly consider time, historical data are rarely used to define a fire envelope. More often, baselines rest on recent fire statistics that are at best imprecise and at worst inaccurate in capturing fire activity over long time scales. What may seem like a stationary response on short time scales is often nonstationary when viewed on longer time scales and over a broader range of bioclimatic forces (Swetnam, 1993). In many parts of the western United States, for example, current levels of fire are considerably less than what climate would predict based on long-term linkages. This notion of a present-day fire deficit in many forest types implies that current fire management is decoupling the natural relationship between area burned and climate (Marlon et al., 2012).

We recommend that observed and projected changes in fire-climate linkages be understood in terms of (1) fire's ecological benefits, (2) the current fire deficit in most forested regions of North America, and (3) a sufficiently long baseline to capture the historical range of fire variability within the particular biome. Detecting and interpreting the significance of climate-driven fire patterns requires information on the magnitude and direction of change in comparison to the long-term fire occurrence within the ecosystem as well as the relative influences of climatic and nonclimatic drivers. Ideally, a fire regime should describe the size, severity, and frequency of fires at different stages of forest development and consider the climate, fuel properties, and human influences that have influenced fire history. This broad temporal and spatial context is essential if we are to accurately project and understand the consequences and benefits of fires in the future.

REFERENCES

Anderson, R.S., Allen, C.D., Toney, J.L., Jass, R.B., Bair, A.N., 2008. Holocene vegetation and fire regimes in subalpine and mixed conifer forests, southern Rocky Mountains, USA. Int. J. Wildland Fire 17, 96–114.

Archibald, S., Lehmann, C.E.R., Gómez-Dans, J.L., Bradstock, R.A., 2013. Defining pyromes and global syndromes of fire regimes. Proc. Natl. Acad. Sci. U. S. A. 110, 6442–6447.

Baker, W.L., 2012. Implications of spatially extensive historical data from surveys for restoring dry forests of Oregon's eastern Cascades. Ecosphere 3, Article 23.

Baker, W.L., 2014. Historical forest structure and fire in Sierran mixed-conifer forests reconstructed from General Land Office survey data. Ecosphere 5, Article 79.

Bartlein, P.J., Anderson, K.H., Anderson, P.M., Edwards, M.E., Mock, C.M., Thompson, R.S., Webb, R.S., Webb III, T., Whitlock, C., 1998. Paleoclimate simulations for North America over the past 21,000 years: features of the simulated climate and comparisons with paleoenvironmental data. Quat. Sci. Rev. 17, 549–585.

Bekker, M.F., Taylor, A.H., 2001. Gradient analysis of fire regimes in montane forests of the southern Cascade Range, Thousand Lakes Wilderness, California, USA. Plant Ecol. 155, 15–28.

Bergeron, Y., Gauthier, S., Kafka, V., Lefort, P., Lesieur, D., 2001. Natural fire frequency for the eastern Canadian Boreal forest: consequences for sustainable forestry. Can. J. For. Res. 31, 384–391.

Bergeron, Y., Flannigan, M., Gauthier, S., Leduc, A., Lefort, P., 2004a. Past, current, and future fire frequency in the Canadian boreal forest: implications for sustainable forest management. Ambio 33, 356–360.

Bergeron, Y., Gauthier, S., Flannigan, M., Kafka, V., 2004b. Fire regimes at the transition between mixedwood and coniferous boreal forest in northwestern Quebec. Ecology 85, 1916–1932.

Bergeron, Y., Cyr, D., Girardin, M.P., Carcaillet, C., 2010. Will climate change drive 21st century burn rates in Canadian boreal forest outside of its natural variability: collating global climate model experiments with sedimentary charcoal data. Int. J. Wildland Fire 19, 1127–1139.

Bowman, D.M.J.S., Balch, J.K., Artaxo, P., Bond, W.J., Carlson, J.M., Cochrane, M.A., D'Antonio, C.M., Defries, R.S., Doyle, J.C., Harrison, S.P., Johnston, F.H., Keeley, J.E., Krawchuk, M.A., Kull, C.A., Marston, J.B., Moritz, M.A., Prentice, I.C., Roos, C.I., Scott, A.C., Swetnam, T.W., van der Werf, G.R., Pyne, S.J., 2009. Fire in the Earth system. Science 324, 481–484.

Boyd, R., 1990. Demongraphic history, 1174-1874. In: Suttles, W. (Ed.), In: Handbook of North American Indians: Northwest Coast, vol. 7. Smithsonian Institution, Washington, D.C., pp. 135–148.

Briles, C.E., Whitlock, C., Bartlein, P.J., 2005. Postglacial vegetation, fire, and climate history of the Siskiyou Mountains, Oregon, USA. Quat. Res. 64, 44–56.

Briles, C.E., Whitlock, C., Skinner, C.N., Mohr, J., 2011. Holocene forest development and maintenance on different substrates in the Klamath Mountains, northern California, USA. Ecology 92, 590–601.

Brooks, M.L., D'Antonio, C.M., Richardson, D.M., Grace, J.B., Keeley, J.E., DiTomaso, J.M., Pyke, D., 2004. Effects of invasive alien plants on fire regimes. Bioscience 54, 677–688. http://dx.doi.org/10.1641/0006-3568(2004)054[0677:EOIAPO]2.0.CO;2.

Brown, P.M., Kaufmann, M.R., Shepperd, W.D., 1999. Long-term, landscape patterns of past fire events in a montane ponderosa pine forest of central Colorado. Landsc. Ecol. 14, 513–532.

Buechling, A., Baker, W.L., 2004. A fire history from tree rings in a high-elevation forest of Rocky Mountain National Park. Can. J. For. Res. 34, 1259–1273.

Carcaillet, C., Thinon, M., 1996. Pedoanthracological contribution to the study of the evolution of the upper treeline in the Maurienne valley (North French Alps): methodology and preliminary data. Rev. Palaeobot. Palynol. 91, 399–416.

Collins, B.M., Miller, J.D., Thode, A.E., Kelly, M., van Wagtendonk, J.W., Stephens, S.L., 2009. Interactions among wildland fires in a long-established Sierra Nevada natural fire area. Ecosystems 12, 114–128.

Colombaroli, D., Gavin, D.G., 2010. Highly episodic fire and erosion regime over the past 2,000 y in the Siskiyou Mountains, Oregon. Proc. Natl. Acad. Sci. U. S. A. 107, 18909–18914. http://dx.doi.org/10.1073/pnas.1007692107.

Conedera, M., Tinner, W., Neff, C., Meurer, M., Dickens, A.F., Krebs, P., 2009. Reconstructing past fire regimes: methods, applications, and relevance to fire management and conservation. Quat. Sci. Rev. 28, 555–576.

Daniau, A.-L., Sánchez Goñi, M.F., Martinez, P., Urrego, D.H., Bout-Roumazeilles, V., Desprat, S., Marlon, J.R., 2013. Orbital-scale climate forcing of grassland burning in southern Africa. Proc. Natl. Acad. Sci. U. S. A. 110, 5069–5073.

DellaSala, D.A., Anthony, R.G., Bond, M.L., Fernandez, E., Hanson, C.T., Hutto, R.L., Spivak, R., 2013. Alternative views of a restoration framework for federal forests in the Pacific Northwest. J. For. 111, 402–492.

Dennison, P.E., Brewer, S.C., Arnold, J.D., Moritz, M.A., 2014. Large wildfire trends in the western United States, 1984-2011. Geophys. Res. Lett. 41, 2928–2933.

Dillon, G.K., Holden, Z.A., Morgan, P., Crimmins, M.A., Heyerdahl, E.K., Luce, C.H., 2011. Both topography and climate affected forest and woodland burn severity in two regions of the western US, 1984 to 2006. Ecosphere 2 (12), 130.

Flannigan, M.D., Krawchuk, M.A., de Groot, W.J., Wotton, B.M., Gowman, L.M., 2009. Implications of changing climate for global wildland fire. Int. J. Wildland Fire 18, 483–507.

Gavin, D.G., Hallett, D.J., Feng, S.H., Lertzman, K.P., Prichard, S.J., Brown, K.J., Lynch, J.A., Bartlein, P., Peterson, D.L., 2007. Forest fire and climate change in western North America: insights from sediment charcoal records. Front. Ecol. Environ. 5, 499–506.

Gavin, D.G., Brubaker, L.B., Greenwald, D.N., 2013. Postglacial climate and fire- mediated vegetation change on the western Olympic Peninsula, Washington. Ecol. Monogr. 83, 471–489.

Girardin, M.P., Ali, A.A., Carcaillet, C., Mudelsee, M., Drobyshev, I., Hély, C., Bergeron, Y., 2009. Heterogeneous response of circumboreal wildfire risk to climate change since the early 1900s. Glob. Chang. Biol. 15, 2751–2769.

Gonzalez, P., Neilson, R.P., Lenihan, J.M., Drapek, R.J., 2010. Global patterns in the vulnerability of ecosystems to vegetation shifts due to climate change. Glob. Ecol. Biogeogr. 19, 755–768.

Guisan, A., Thuiller, W., 2005. Predicting species distribution: offering more than simple habitat models. Ecol. Lett. 8, 993–1009.

Hanson, C.T., Odion, D.C., 2014. Is fire severity increasing in the Sierra Nevada mountains, California, USA? Int. J. Wildland Fire 23, 1–8.

Hanson, C.T., Odion, D.C., 2015a. Historical forest conditions within the range of the Pacific Fisher and Spotted Owl in the central and southern Sierra Nevada, California. USA, Natural Areas Journal (in press).

Hanson, C.T., Odion, D.C., 2015b. Sierra Nevada fire severity conclusions are robust to further analysis: a reply to Safford et al. International Journal of Wildland Fire 24, 294–295.

Hanson, C.T., Odion, D.C., DellaSala, D.A., Baker, W.L., 2009. Overestimation of fire risk in the northern spotted owl recovery plan. Conserv. Biol. 23, 1314–1319.

Hartmann, D.L., Klein Tank, A.M.G., Rusticucci, M., Alexander, L.V., Brönnimann, S., Charabi, Y., Dentener, F.J., Dlugokencky, E.J., Easterling, D.R., Kaplan, A., Soden, B.J., Thorne, P.W., Wild, M., Zhai, P.M., 2013. Observations: atmosphere and surface. In: Stocker, T.F., Qin, D., Plattner, G.-K., Tignor, M., Allen, S.K., Boschung, J., Nauels, A., Xia, Y., Bex, V., Midgley, P.M. (Eds.), Climate Change 2013: The Physical Science Basis. Contribution of Working Group I to the Fifth Assessment Report of the Intergovernmental Panel on Climate Change. Cambridge University Press, Cambridge, United Kingdom and New York, NY, USA.

Henne, P.D., Elkin, C.M., Reineking, B., Bugmann, H., Tinner, W., 2011. Did soil development limit spruce (Picea abies) expansion in the Central Alps during the Holocene? Testing a palaeobotanical hypothesis with a dynamic landscape model. J. Biogeogr. 38, 933–949.

Heyerdahl, E.K., McKenzie, D., Daniels, L.D., Hessl, A.E., Littell, J.S., Mantua, N.J., 2008. Climate drivers of regionally synchronous fires in the inland Northwest (1650-1900). Int. J. Wildland Fire 17, 40–49.

Higuera, P.E., Brubaker, L.B., Anderson, P.M., Hu, F.S., Brown, T., 2009. Vegetation mediated the impacts of postglacial climate change on fire regimes in the south-central Brooks Range, Alaska. Ecol. Monogr. 79, 201–219.

Higuera, P.E., Whitlock, C., Gage, J., 2011. Linking tree-ring and sediment-charcoal records to reconstruct fire occurrence and area burned in subalpine forests of Yellowstone National Park, USA. The Holocene 21, 327–341.

Higuera, P.E., Briles, C.E., Whitlock, C., 2014. Fire-regime complacency and sensitivity to centennial-through millennial-scale climate change in Rocky Mountain subalpine forests, Colorado, USA. J. Ecol. 102, 1429–1441.

Iglesias, V., Krause, T.R., Whitlock, C., 2015. Complex response of subalpine conifers to past environmental variability increases understanding of future vulnerability. PLoS One.

Kasischke, E.S., Turetsky, M.R., 2006. Recent changes in the fire regime across the North American boreal region—spatial and temporal patterns of burning across Canada and Alaska. Geophys. Res. Lett. 33, L09703.

Kehrwald, N.M., Whitlock, C., Barbante, C., Brovkin, V., Daniau, a.-L, Kaplan, J.O., Marlon, J.R., Power, M.J., Thonicke, K., van der Werf, G.R., 2013. Fire research: linking past, present, and future data. EOS Trans. Am. Geophys. Union 94, 421–422.

Kelly, R., Chipman, M.L., Higuera, P.E., Stefanova, I., Brubaker, L.B., Hu, F.S., 2013. Recent burning of boreal forests exceeds fire regime limits of the past 10,000 years. Proc. Natl. Acad. Sci. U. S. A. 110, 13055–13060.

Kitzberger, T., Brown, P.M., Heyerdahl, E.K., Swetnam, T.W., Veblen, T.T., 2007. Contingent Pacific-Atlantic Ocean influence on multi-century wildfire synchrony over western North America. Proc. Natl. Acad. Sci. U. S. A. 104, 543–548.

Krawchuk, M.A., Moritz, M.A., 2011. Constraints on global fire activity vary across a resource gradient. Ecology 92, 121–132.

Krawchuk, M.A., Moritz, M.A., 2012. Fire and climate change in California. California Energy Commission. Publication number: CEC-500-2012-026.

Krawchuk, M.A., Moritz, M.A., Parisien, M., Van Dorn, J., Hayhoe, K., 2009. Global pyrogeography: the current and future distribution of wildfire. PLoS One 4, e5102.

Larsen, C.P.S., 1996. Fire and climate dynamics in the boreal forest of northern Alberta, Canada, from AD 1850 to 1989. The Holocene 6, 449–456.

Leiberg, J.B., 1900. Cascade Range Forest Reserve, Oregon, from township 28 south to township 37 south, inclusive; together with the Ashland Forest Reserve and adjacent forest regions from township 28 south to township 41 south, inclusive, and from range 2 west to range 14 east, Willamette Meridian, inclusive. In: 21st Annual Report of the U.S. Geological Survey, Part V. Government Printing Office, Washington, D.C.

Leiberg, J.B., 1902. Forest conditions in the northern Sierra Nevada, California. USDI Geological Survey, Professional Paper No. 8. U.S. Government Printing Office, Washington, D.C.

Lenihan, J.M., Bachelet, D., Neilson, R.P., Drapek, R., 2008. Response of vegetation distribution, ecosystem productivity, and fire to climate change scenarios for California. Climate Change 87 (Suppl. 1), S215–S230.

Litschert, S.E., Brown, T.C., Theobald, D.M., 2012. Historic and future extent of wildfires in the Southern Rockies Ecoregion, USA. For. Ecol. Manag. 269, 124–133.

Littell, J.S., McKenzie, D., Peterson, D.L., Westerling, A.L., 2009. Climate and wildfire area burned in western U.S. ecoprovinces, 1916-2003. Ecol. Appl. 19, 1003–1021.

Liu, Y., Stanturf, J.A., Goodrick, S.L., 2010. Trends in global wildfire potential in a changing climate. For. Ecol. Manag. 259, 685–697.

Marguerie, D., Hunot, J.Y., 2007. Charcoal analysis and dendrology: data from archaeological sites in north-western France. J. Archaeol. Sci. 34, 1417–1433.

Marlon, J.R., Bartlein, P.J., Whitlock, C., 2006. Fire-fuel-climate linkages in the northwestern USA during the Holocene. The Holocene 16, 1059–1071.

Marlon, J.R., Bartlein, P.J., Gavin, D.G., Long, C.J., Anderson, R.S., Briles, C.E., Brown, K.J., Colombaroli, D., Hallett, D.J., Power, M.J., Scharf, E.A., Walsh, M.K., 2012. Long-term perspective on wildfires in the western USA. Proc. Natl. Acad. Sci. U. S. A. 109, E535–E543.

Marlon, J.R., Barlein, P.J., Daniau, A.-L., Harrison, S.P., Power, M.J., Tinner, W., Maezumie, S., Vanniere, B., 2013. Global biomass burning: a synthesis and review of Holocene paleofire records and their controls. Quat. Sci. Rev. 65, 5–25.

McWethy, D.B., Whitlock, C., Wilmshurst, J.M., McGlone, M.S., Li, X., Fromont, M., Diffenbacher-Krall, A., Hobbs, W.O., Fritz, S., Cook, E.R., 2010. Rapid landscape tranformation in South Island, New Zealand following initial Polynesian settlement. Proc. Natl. Acad. Sci. U. S. A. 107, 21343–21348.

McWethy, D.B., Higuera, P.E., Whitlock, C., Veblen, T.T., Bowman, D.M.J.S., Cary, G.J., Haberle, S.G., Keane, R.E., Maxwell, B.D., Mcglone, M.S., Perry, G.L.W., Wilmshurst, J.M., Holz, A., Tepley, A.J., 2013. A conceptual framework for predicting temperate ecosystem sensitivity to human impacts on fire regimes. Glob. Ecol. Biogeogr. 22, 900–912.

Melillo, J.M., Richmond, T.C., Yohe, G.W. (Eds.), 2014. Climate Change Impacts in the United States: The Third National Climate Assessment. U.S. Global Change Research Program, Washington, D.C., p. 841.

Meyer, G.A., Wells, S.G., Jull, A.J.T., 1995. Fire and alluvial chronology in Yellowstone National Park: climatic and intrinsic controls on Holocene geomorphic processes. Geol. Soc. Am. Bull. 107, 1211–1230.

Meyn, A., Schmidtlein, S., Taylor, S.W., Giardin, M.P., Thonicke, K., Cramer, W., 2013. Precipitation-driven decreases in wildfires in British Columbia. Reg. Environ. Chang. 13, 165–177.

Miller, J.D., Safford, H., 2012. Trends in wildfire severity: 1984 to 2010 in the Sierra Nevada, Modoc Plateau, and southern Cascades, California, USA. Fire Ecol. 8, 41–57.

Miller, J.D., Safford, H.D., Crimmins, M.A., Thode, A.E., 2009. Quantitative evidence for increasing forest fire severity in the Sierra Nevada and southern Cascade Mountains, California and Nevada, USA. Ecosystems 12, 16–32.

Miller, J.D., Skinner, C.N., Safford, H.D., Knapp, E.E., Ramirez, C.M., 2012. Trends and causes of severity, size, and number of fires in northwestern California, USA. Ecol. Appl. 22, 184–203.

Millspaugh, S.H., Whitlock, C., Bartlein, P.J., 2000. Variations in fire frequency and climate over the past 17,000 yr in central Yellowstone National Park. Geology 28, 211–214.

Moritz, M.A., Parisien, M.A., Batllori, E., Krawchuk, M.A., Van Dorn, J., Ganz, D.J., Hayhoe, K., 2012. Climate change and disruptions to global fire activity. Ecosphere 3, Article 49.

Mouillot, F., Field, C., 2005. Fire history and the global carbon budget: a 1x1 fire history reconstruction for the 20th century. Glob. Chang. Biol. 11, 398–420.

Odion, D.C., Hanson, C.T., 2013. Projecting impacts of fire management on a biodiversity indicator in the Sierra Nevada and Cascades, USA: the Black-backed Woodpecker. Open For. Sci. J. 6, 14–23.

Odion, D.C., Moritz, M.A., DellaSala, D.A., 2010. Alternative community states maintained by fire in the Klamath Mountains, USA. J. Ecol. 98, 96–105.

Odion, D.C., Hanson, C.T., Arsenault, A., Baker, W.L., DellaSala, D.A., Hutto, R.L., Klenner, W., Moritz, M.A., Sherriff, R.L., Veblen, T.T., Williams, M.A., 2014. Examining historical and current mixed-severity fire regimes in ponderosa pine and mixed-conifer forests of Western North America. PLoS One 9, e87852.

Papworth, S.K., Rist, J., Coad L, L., Milner-Gulland, E.J., 2008. Evidence for shifting baseline syndrome in conservation. Conserv. Lett. 2, 93–100.

Parisien, M.-A., Moritz, M.A., 2009. Environmental controls on the distribution of wildfire at multiple spatical scales. Ecol. Monogr. 79, 127–154.

Parisien, M.-A., Parks, S.A., Krawchuk, M.A., Flannigan, M.D., Bowman, L.M., Moritz, M.A., 2011. Scale-dependent controls on the area burned in the boreal forest of Canada, 1980-2005. Ecol. Appl. 21, 789–805.

Parisien, M.-A., Snetsinger, S., Greenberg, J.A., Nelson, C.R., Schoennagel, T., Dobrowski, S., Moritz, M.A., 2012. Spatial variability in wildfire probability across the western United States. Int. J. Wildland Fire 21, 313–327.

Parmesan, C., 2006. Ecological and evolutionary responses to recent climate change. Annual Review of Ecology Evolution and Systematics 37, 637–669.

Pausas, J.G., Keeley, J.E., 2014. Abrupt climate-independent fire regime changes. Ecosystems 17, 1109–1120.

Pechony, O., Shindell, D.T., 2010. Driving forces of global wildfires over the past millennium and the forthcoming century. Proc. Natl. Acad. Sci. U. S. A. 107, 19167–19170.

Peterson, T.C., et al., 2013. Monitoring and understanding changes in heat waves, cold waves, floods, and droughts in the United States: State of knowledge. Bull. Am. Meteorol. Soc. 94, 821–834.

Pfeiffer, M., Spessa, A., Kaplan, J.O., 2013. A model for global biomass burning in preindustrial time: LPJ-LMfire (v1.0). Geosci. Model Dev. 6, 643–685.

Pierce, J.L., Meyer, G.A., Jull, A.J.T., 2004. Fire-induced erosion and millennial-scale climate change in northern ponderosa pine forests. Nature 432, 87–90. http://dx.doi.org/10.1038/nature03058.

Rogers, B.M., Neilson, R.P., Drapek, R., Lenihan, J.M., Wells, J.R., Bachelet, D., Law, B.E., 2011. Impacts of climate change on fire regimes and carbon stocks of the U.S. Pacific Northwest. J. Geophys. Res. 116, G03037.

Romero-Lankao, P., Smith, J.B., Davidson, D.J., Diffenbaugh, N.S., Kinney, P.L., Kirshen, P., Kovacs, P., Villers Ruiz, L., 2014. North America. In: Barros, V.R., Field, C.B., Dokken, D.J., Mastrandrea, M.D., Mach, K.J., Bilir, T.E., Chatterjee, M., Ebi, K.L., Estrada, Y.O., Genova, R.C., Girma, B., Kissel, E.S., Levy, A.N., MacCracken, S., Mastrandrea, P.R., White, L.L. (Eds.), Climate Change 2014: Impacts, Adaptation, and Vulnerability. Part B: Regional Aspects. Contribution of Working Group II to the Fifth Assessment Report of the Intergovernmental Panel on Climate Change. Cambridge University Press, Cambridge, United Kingdom and New York, NY, USA, pp. 1439–1498.

Romme, W.H., 1982. Fire and landscape diversity in subalpine forests of Yellowstone National Park. Ecol. Monogr. 52, 199–221.

Rothermel, R.C., 1972. A mathematical model for predicting fire spread in wildland fuels. USDA Forest Service Research Paper INT USA, 40 p.

Roy, S.B., Chen, L., Girvetz, E.H., Maurer, E.P., Mills, W.B., Grieb, T.M., 2012. Projecting water withdrawal and supply for future decades in the U.S. under climate change scenarios. Environ. Sci. Technol. 46, 2545–2556.

Safford, H.D., Miller, J.D., Collins, B.M., 2015. Differences in land ownership, fire management objectives, and source data matter: a reply to Hanson and Odion (2014). International Journal of Wildland Fire 24, 286–293.

Schoennagel, T., Sherriff, R.L., Veblen, T.T., 2011. Fire history and tree recruitment in the Colorado Front Range upper montane zone: Implications for forest restoration. Ecol. Appl. 21, 2210–2222.

Scholze, M., Knorr, W., Arnell, N.W., Prentice, I.C., 2006. A climate-change risk analysis for world ecosystems. Proc. Natl. Acad. Sci. U. S. A. 103, 13116–13120.

Schwind, B. (compiler). 2008. Monitoring trends in burn severity: report on the Pacific Northwest and Pacific Southwest Fires (1984 to 2005). U.S. Geological Survey, U.S. Forest Service, and Monitoring Trends in Burn Severity Project. Available online: http://mtbs.gov/.

Sherriff, R.L., Veblen, T.T., Sibold, J.S., 2001. Fire history in high elevation subalpine forests in the Colorado Front Range. Ecoscience 8, 369–380.

Sibold, J.S., Veblen, T.T., 2006. Relationships of subalpine forest fires in the Colorado Front Range with interannual and multidecadal-scale climatic variation. J. Biogeogr. 33, 833–842.

Spracklen, D.V., Mickley, L.J., Logan, J.A., Hudman, R.C., Yevich, R., Flannigan, M.D., Westerling, A.L., 2009. Impacts of climate change from 2000 to 2050 on wildfire activity and carbonaceous aerosol concentrations in the western United States. J. Geophys. Res. 114, D20301. http://www.altmetric.com/details.php?domain=onlinelibrary.wiley.com&doi 10.1111/j.1461-0248.2005.00792.x.

Stephens, S.L., Martin, R.E., Clinton, N.E., 2007. Prehistoric fire area and emissions from California's forests, woodlands, shrublands and grasslands. For. Ecol. Manag. 251, 205–216.

Swetnam, T.W., 1993. Fire history and climate change in giant sequoia groves. Science 262, 885–889.

Swetnam, T.W., Betancourt, J.L., 1990. Fire-southern oscillation relations in the southwestern United States. Science 249, 1017–1020.

Taylor, A.H., Skinner, C.N., 2003. Spatial patterns and controls on historical fire regimes and forest structure in the Klamath Mountains. Ecol. Appl. 13, 704–719.

Tepley, A.J., Swanson, F.J., Spies, T.A., 2013. Fire-mediated pathways of stand development in Douglas-fir/western hemlock forests of the Pacific Northwest, USA. Ecology 94, 1729–1743.

Tepley, A.J., Swanson, F.J., Spies, T.A., 2014. Post-fire tree establishment and early cohort development in conifer forests of the western Cascades of Oregon, USA. Ecosphere 5, 80, art.

Trouet, V., Taylor, A.H., Wahl, E.R., Skinner, C.N., Stephens, S.L., 2010. Fire-climate interactions in the American West since 1400 CE. Geophys. Res. Lett. 37, L04702. http://dx.doi.org/10.1029/2006GL027502.

Wallenius, T.H., Pennanen, J., Burton, P.J., 2011. Long-term decreasing trend in forest fires in northwestern Canada. Ecosphere 2 (5), Article 53.

Walsh, M.K., Whitlock, C., Bartlein, P.J., 2008. A 14,300-year-long record of fire-vegetation-climate linkages at Battle Ground Lake, southwestern Washington. Quat. Res. 70, 251–264.

Westerling, A., Bryant, B., 2008. Climate change and wildfire in California. Climate Change 87, S231–S249.

Westerling, A.L., Hidalgo, H.G., Cayan, D.R., Swetnam, T.W., 2006. Warming and earlier spring increases western US forest wildfire activity. Science 313, 940–943.

Westerling, A.L., Turner, M.G., Smithwick, E.A.H., Romme, W.H., Ryan, M.G., 2011. Continued warming could transform Greater Yellowstone fire regimes by mid-21st century. Proc. Natl. Acad. Sci. U. S. A. 108, 13165–13170.

Whitlock, C., Larsen, C., 2001. Charcoal as a fire proxy. In: Smol, J.P., Birks, H.J.B., Last, W.M. (Eds.), Tracking Environmental Change Using Lake Sediments. In: Terrestrial Algal and Siliceous Indicators, 3. Kluwer Academic Press, London, pp. 75–97.

Whitlock, C., Bianchi, M.M., Bartlein, P.J., Markgraf, V., Marlon, J., Walsh, M., Mccoy, N., 2006. Postglacial vegetation, climate, and fire history along the east side of the Andes (lat 41-42.5°S), Argentina. Quat. Res. 66, 187–201.

Whitlock, C., Marlon, J., Briles, C., Brunelle, A., Long, C., Bartlein, P., 2008. Long-term relations among fire, fuel, and climate in the north-western US based on lake-sediment studies. Int. J. Wildland Fire 17, 72–83.

Whitlock, C., Higuera, P.E., McWethy, D.B., Briles, C.E., 2010. Paleoecological perspectives on fire ecology: revisiting the fire-regime concept. Open Ecol. J. 3, 6–21.

Whitlock, C., Dean, W.E., Fritz, S.C., Stevens, L.R., Stone, J.R., Power, M.J., Rosenbaum, J.R., Pierce, K.L., Bracht-Flyr, B.B., 2012. Holocene seasonal variability inferred from multiple proxy records from Crevice Lake, Yellowstone National Park, USA. Palaeogeogr. Palaeoclimatol. Palaeoecol. 331-332, 90–103.

Whitlock, C., McWethy, D.B., Tepley, A.J., Veblen, T.T., Holz, A., McGlone, M.S., Perry, G.L.W., Wilmshurst, J.M., Wood, S.W., 2015. Past and present vulnerability of closed-canopy temperate forests to altered fire regimes: a comparison of the Pacific Northwest, New Zealand, and Patagonia. Bioscience 65, 151–163.

Williams, M.A., Baker, W.L., 2012. Spatially extensive reconstructions show variable-severity fire and heterogeneous structure in historical western United States dry forests. Glob. Ecol. Biogeogr. 21, 1042–1052.

Yue, X., Mickey, L.J., Logan, J.A., Kaplan, J.O., 2013. Ensemble projections of wildfire activity and carbonaceous aerosol concentrations over the western United States in the mid-21st century. Atmos. Environ. 77, 767–780.

Chapter 10

Carbon Dynamics of Mixed- and High-Severity Wildfires: Pyrogenic CO_2 Emissions, Postfire Carbon Balance, and Succession

Stephen Mitchell
Duke University, Durham, NC, USA

10.1 MIXED-SEVERITY FIRES: A DIVERSITY OF FUELS, ENVIRONMENTS, AND FIRE BEHAVIORS

Recent increases in global temperatures are projected by some research to increase the frequency and severity of wildfires in certain regions, particularly those experiencing warmer, drier summers (McKenzie et al., 2004; Flannigan et al., 2006). While the annual area burned in most forests of western North America remains well below historical levels (see Chapters 1, 9), many areas have experienced significant increases in annual burning, particularly from 1970 to 1986 (Westerling et al., 2006), prompting concerns about the additional release of carbon, primarily in the form of carbon dioxide. However, concerns over a positive feedback between wildfire-caused carbon emissions and temperature increase must be considered in the context of the physical magnitudes of pyrogenic carbon emissions and the respective constituents of forest carbon storage from which they are derived. Here I discuss the factors influencing the combustion of different constituents of forest carbon storage and how rates of fuel combustion vary among fires of low, medium, and high severity. This chapter also addresses the relationship of fuel reduction treatments with regard to reducing fire severity and carbon emissions at the potential expense of forest carbon storage. Finally, I discuss postfire carbon emissions from the decomposition of fire-killed biomass, postfire forest succession, and the eventual recovery of forest carbon storage.

The Ecological Importance of Mixed-Severity Fires: Nature's Phoenix

Rates of pyrogenic carbon emission from wildfires can be highly variable among mixed-severity wildfires. The consumption of each respective component of forest fuel is strongly determined by individual particle geometry, often expressed as the surface area-to-volume ratio for the purposes of quantifying the amount of fuel that is likely to be consumed. Combustion generally occurs at the surface of the fuel particle, and the size of each particle and its surface area-to-volume ratio control the amount of heat required for ignition and consumption. Fuels with large surface area-to-volume ratios, such as grasses and pine needles, require less heat for ignition and combustion. Conversely, large fuels with low surface area-to-volume ratios, such as standing trees, as well as snags, downed logs, and other forms of coarse woody debris, require considerably more energy for ignition and combustion. Fuel particle size also influences the rates of moisture absorption and release, as smaller fuel particles release moisture more rapidly than larger particles in response to increasing atmospheric vapor pressure deficits, as well as in response to the thermal energy brought about by an approaching flaming front. Consequently, large fuels are much more likely to burn during the smoldering stage, in which the emissions of combustible gases and vapors are too low to support flaming combustion (Lobert and Warnatz, 1993).

Fuel consumption also is influenced by the compactness of the fuel bed, in part because of the two-stage process of consumption through pyrolysis and combustion. While these processes are nearly simultaneous, pyrolysis occurs first and is the heat-absorbing reaction that converts fuel elements such as cellulose into char, carbon dioxide, carbon monoxide, water vapor, highly combustible vapors and gases, and particulate matter (DeBano et al., 1998; Ward, 2001; Ottmar, 2014). Pyrolysis is followed by combustion, in which escaping hydrocarbon vapors are released from the surface of the fuels and are oxidized. Thus fuel compaction presents a tradeoff between heat transfer and oxygen diffusion. Highly compacted fuels facilitate a more efficient transfer of heat between fuel particles while limiting the diffusion of oxygen and, by extension, limiting consumption. Conversely, low fuel compaction allows for high diffusion of oxygen, albeit with a low diffusion of heat between fuel particles (Hardy et al., 2001). Fuel consumption also is influenced by the spacing, or continuity, of fuels across the forest floor (Finney et al., 2010) (Figure 10.1).

While the amount of consumption that is to be expected can be strongly determined by the fuel's physical and chemical characteristics, it is also a function of climate and topography. Regional climate exerts a top-down influence on fire frequency through seasonal patterns of temperature and precipitation (Littell et al., 2010), whereas local factors such as topography, vegetative composition, and fuel loads exert a bottom-up influence on fire behavior (Perry et al., 2011; Miller et al., 2012). Topography can influence the species composition of a forest, the composition and accumulation of fuels from a forest, and the topographically mediated content of fuel moisture. Among landscapes at elevations dominated by ponderosa pine (*Pinus ponderosa*) in eastern Oregon and Washington, white fir (*Abies concolor*) and grand fir (*Abies grandis*) are more common on north-facing slopes

FIGURE 10.1 Aerial view of a smoke plume. *(Photo courtesy of M. Welling, Max Planck Institute for Chemistry).*

because of the cooler and moist conditions that result from less incoming solar radiation (Cowlin et al., 1942). Stand composition and structure interact with the edaphic (pertaining to soils) moisture gradients to determine patterns of fire severity (Hessburg et al., 2000; Miller, 2003; Hessburg et al., 2004). In areas north of the Klamath Mountains in northwestern California, north-facing slopes may burn with mixed severity, whereas south-facing slopes can burn with mixed or low fire severity. However, the opposite occurs in the more xeric (dry) forests of the Klamath Mountains, wherein mixed-severity fires have historically dominated on south- and west-facing aspects, whereas low-severity fires were dominant on north- and east-facing aspects (Taylor and Skinner, 1998). Extreme weather conditions can override these effects, however, as was the case in the Biscuit Fire of 2002 in southwest Oregon; hot, dry winds from the northeast drove the fire, thereby eclipsing much of the influence of topographic positions (Thompson and Spies, 2010). Other fires with severe conditions have shown a stronger response to topographic controls, such as the Megram Fire in northern California (Jimerson and Jones, 2000).

The expected fuel consumption for a given level of fire severity is often expressed as a combustion factor (CF). A CF is the proportion of a biomass constituent that is expected to be consumed in a wildfire. CFs vary with respect to different biomass components such as live foliage, litter, stem, branches, shrubs, and soil. CFs can also vary as a function of fire severity: lower levels of fire severity typically result in lower levels of combustion for each respective constituent of forest carbon storage. Note, however, that the use here of the term "fire severity," expressed as the proportion of mortality observed in overstory trees (Table 10.1), can be misleading when used as a determinant of fuel combustion. Fuel combustion is often determined by fire intensity, a measure of

TABLE 10.1 Mortality Factors (MF) for Different Fire Severities (Ghimire, 2012)

	Low-Severity	Medium-Severity	High-Severity
Dominant Vegetation	*MF*	*MF*	*MF*
Pinyon/juniper	0.24	0.53	0.94
Douglas-fir	0.17	0.48	0.96
Ponderosa pine	0.19	0.42	0.97
Fir/spruce/mountain hemlock	0.25	0.51	0.94
Lodgepole pine	0.29	0.59	0.88
Hemlock/sitka spruce	0.29	0.63	0.94
California mixed conifer	0.24	0.53	0.94
Elm/ash/cottonwood	0.27	0.57	0.99
Aspen/birch	0.26	0.56	0.98
Western oak	0.14	0.53	0.73
Tanoak/laurel	0.24	0.53	0.92

energy output from a fire (Keeley, 2009). A fire of relatively low intensity could conceivably result in a fire of medium or even high severity if it occurred among trees with relatively low tolerance to fire. Because this is a book concerned about forest ecosystems with mixed- and high-severity fire regimes, however, we are largely dealing with ecosystems that have evolved at least some adaptations to moderate- or high-severity fire.

An improper use of a CF in estimating the carbon emissions of a given fire can produce vastly different estimates of pyrogenic carbon emissions. Worldwide, forests store about 45% of terrestrial carbon (861 ± 66 pg carbon) in soils, ~42% in above- and belowground live biomass, ~8% in dead wood, and ~5% in litter (Bonan, 2008). Given the magnitude of carbon stored in, say, dead wood, a poorly derived CF for dead wood can have a considerable impact on the resulting estimates of carbon dioxide emissions. Estimates of average pyrogenic carbon emissions for a given time period can produce a considerable range of values, some of which can be over four times higher than those of others (Wiedinmyer and Neff, 2007; Ghimire et al., 2012), in part because of methodological differences in the approaches used to estimate biomass accumulation and area burned, as well as different approaches used by different studies to obtain CFs.

Here I discuss factors controlling the combustion of different constituents of carbon storage in forest ecosystems and how these constituents can influence, and can be influenced by, different levels of fire severity in forested landscapes with mixed- and high-severity fire regimes. I also discuss the indirect impacts of wildfire through the long-term carbon emissions of fire-killed biomass and how emissions after wildfire can influence the source-sink dynamics throughout a postfire landscape.

10.2 DUFF, LITTER, AND WOODY DEBRIS COMBUSTION

Duff carbon comprises the dead organic matter found in the O_a (almost complete decomposition) through the O_e (moderate composition) horizons, whereas litter comprises the dead materials found in the O_i horizon (undecomposed plant parts) and includes small, woody fragments <0.51 cm in diameter, also known as 1 h fuels. Small, woody debris consists of particles 0.51-2.54 cm in diameter, also known as 10-h fuels. While only a small fraction of total forest carbon storage, these components of carbon storage on the forest floor often constitute the majority of combusted fuel for fires of all severities. Campbell et al. (2007) estimated that duff, litter, and small, downed, woody debris consumption constituted about 60% of direct carbon emissions in the Biscuit Fire of 2002. High rates of combustion among these components are consistent with the principle that fuels with large surface area-to-volume ratios have higher CFs than fuels with lower surface area-to-volume ratios, much of which can be attributed to the short time periods required for woody materials (1- to 10 h fuels) to dry out. Seasonal variation in fuel moisture can thus have a considerable impact on carbon emissions. Knapp et al. (2005) found that early season burns, in which fuel moisture was higher, left approximately five times more litter and duff unconsumed in areas where fire passed over the forest floor than late season burns.

Noting that this pool of carbon storage is destined for biogenic emission to the atmosphere in the absence of wildfires is important. Pools of litter, foliage, and small, downed wood are thought to have a mean residence time of 10-20 years (Law et al., 2001), and while a portion of this eventually transitions into more stable forms of soil carbon storage, much of it is lost through decay. Furthermore, much of the carbon stored in a pool with such high turnover should equate to a subsequent reduction in heterotrophic (requiring organic matter for food) respiration until these pools become recharged by the addition of leaf litter and small, woody debris (Campbell et al., 2007).

Because additional energy is necessary to remove water before combustion is possible, more energy is required to propagate flaming combustion in moist fuels than dry fuels (Nelson, 2001). In theory (Finney et al., 2013), as well as in some modeling studies (Hargrove et al., 2000; Miller and Urban, 2000), the probability that fire will propagate to neighboring fuels is reduced at higher fuel moisture levels. Knapp et al. (2005) found that the amount of area within the fire perimeter burned, and greater patchiness of early season burns conducted under higher fuel moisture conditions, are consistent with these model predictions. Thus the combustion of large, woody debris (1000-h fuels) can be particularly sensitive to fuel moisture. Estimates of combustion of downed, coarse, woody debris suggest that the majority of carbon contained therein will remain after the fire, with CFs of 0.04 for low- and very-low-severity fires and up to 0.08 and 0.24 for medium- and high-severity fires (Table 10.2). CFs are even lower for standing coarse, woody debris, ranging from 0.02 for low- and very-low-severity fires to 0.04 and 0.12 for medium- and high-severity fires (Table 10.2).

TABLE 10.2 Constituents of Biomass Storage and Combustion Factors[a] for the 2002 Biscuit Fire in the Rogue River-Siskiyou National Forest in Southwestern Oregon

C Storage Constituent	C Storage (kg C ha^{-1})	High Severity CF (%)	Medium Severity CF (%)	Low Severity CF (%)	Very Low Severity CF (%)
Foliage					
Large conifers	3242	0.69	0.27	0.08	0.02
Large hardwoods	1698	0.58	0.29	0.12	0.03
Small conifers	1863	0.89	0.76	0.44	0.01
Small hardwoods	417	1.00	0.80	0.50	0.00
Grass and forbs	2	1.00	0.76	0.75	0.70
Branch					
Large conifers	9858	0.05	0.02	0.00	0.00
Large hardwoods	4350	0.05	0.02	0.01	0.00
Small conifers	609	0.64	0.69	0.41	0.00
Small hardwoods	579	0.79	0.63	0.40	0.00
Bark					
Large conifers	11,199	0.20	0.06	0.03	0.01
Large hardwoods	4523	0.22	0.11	0.03	0.01
Small conifers	597	0.70	0.70	0.42	0.01
Small hardwoods	69	0.79	0.63	0.40	0.00

Continued

TABLE 10.2 Constituents of Biomass Storage and Combustion Factors for the 2002 Biscuit Fire in the Rogue River-Siskiyou National Forest in Southwestern Oregon—Cont'd

C Storage Constituent	C Storage (kg C ha^{-1})	High Severity CF (%)	Medium Severity CF (%)	Low Severity CF (%)	Very Low Severity CF (%)
Bole					
Large conifers	57,419	0.00	0.00	0.00	0.00
Large hardwoods	30,748	0.00	0.00	0.00	0.00
Small conifers	288	0.61	0.68	0.40	0.00
Small hardwoods	700	0.79	0.63	0.40	0.00
Dead wood					
Large standing	5927	0.12	0.04	0.02	0.02
Small standing	1642	0.61	0.68	0.40	0.00
Large downed	9324	0.24	0.08	0.04	0.04
Medium downed	1798	0.79	0.73	0.67	0.62
Small downed	1543	0.78	0.58	0.61	0.62
Forest floor and soil					
Litter	9499	1.00	0.76	0.75	0.70
Duff	6335	0.99	0.51	0.54	0.44
Soil to 10 cm	45,500	0.08	0.04	0.04	0.02

Litter consists of materials in the O_i horizon, and duff is in the O_e and O_a horizon. Soil is all mineral soil to a depth of 10 cm, including fine roots. For live trees, small is a <7.62 cm diameter at breast height (DBH); large is a >7.62 cm DBH. For dead wood, small is 0.51-2.54 cm, medium is 2.54-7.62 cm, and large is a >7.62 cm diameter.
[a]*Data from Campbell et al. (2007).*

Interestingly, levels of fuel consumption for woody debris, duff, and litter exhibit a surprisingly high level of similarity at different levels of fire severity, even among different forest types (Table 10.3). CFs for woody debris (including all diameter classes) averaged 0.56, 0.63, and 0.79 for low-, medium-, and high-severity fires, respectively (Table 10.3). Average duff combustion (0.46) was lower than average woody debris combustion among stands burned by low-severity fires, but it was higher in stands burned by medium- and high-severity fires, with average CFs of 0.70 and 0.90, respectively (Table 10.3). The highest

TABLE 10.3 Combustion Factors (CFs) for Woody Debris (WD), Litter, and Duff Fuels for Different Forest Species Groups and Levels of Fire Severity[a]

	Low Severity			Medium Severity			High Severity		
Dominant Vegetation	*WD CF*	*Litter CF*	*Duff CF*	*WD CF*	*Litter CF*	*Duff CF*	*WD CF*	*Litter CF*	*Duff CF*
Pinyon/juniper	0.56	0.63	0.48	0.62	0.77	0.77	0.81	0.97	0.97
Douglas-fir	0.53	0.70	0.47	0.60	0.73	0.81	0.81	0.97	0.97
Ponderosa pine	0.52	0.65	0.54	0.65	0.75	0.84	0.82	0.96	0.97
Fir/spruce/mountain hemlock	0.53	0.60	0.44	0.63	0.76	0.69	0.77	0.92	0.83
Lodgepole pine	0.68	0.50	0.21	0.77	0.56	0.33	0.96	0.72	0.42
Hemlock/sitka spruce	0.59	0.75	0.54	0.58	0.76	0.51	0.77	1.00	0.99
California mixed conifer	0.56	0.64	0.48	0.62	0.77	0.77	0.80	0.97	0.97
Elm/ash/cottonwood	0.58	0.75	0.51	0.66	0.74	0.77	0.77	1.00	0.99
Aspen/birch	0.43	0.77	0.40	0.48	0.74	0.64	0.60	1.00	0.81
Western oak	0.56	0.76	0.50	0.67	0.80	0.79	0.81	0.98	0.95
Tanoak/laurel	0.59	0.75	0.54	0.68	0.66	0.79	0.77	1.00	0.99

[a]*Data from Ghimire et al. (2012).*

rates of combustion were observed in litter biomass, which had CFs of 0.68, 0.73, and 0.95 for low-, medium-, and high-severity fires, respectively (Table 10.3).

10.3 LIVE FOLIAGE COMBUSTION

Estimates of live, crown foliage combustion are difficult because few studies have attempted to distinguish between crown consumption and noncombustive mortality (Wyant et al., 1986; McHugh et al., 2003; Hull Sieg et al., 2006; Campbell et al., 2007; Keyser et al., 2008). While live foliage can be consumed by wildfires, foliage can also be scorched and damaged by direct contact with or indirect convective heating from flames, leaving a yellowing or browning of foliage. Once scorched, the foliage is usually killed and subsequently falls to the ground.

Understory and shrub-layer vegetation can have a significant impact on foliage consumption, but these effects depend on species composition. In the 2002 Biscuit Fire, open conifer forests with a predominantly sclerophyllous (trees and shrubs with hard, thick leaves) shrub understory experienced the most crown mortality (Thompson and Spies, 2009). Conversely, an assessment of the foliar moisture content of several grass and nonsclerophyllous shrub species suggested the possibility that the presence of a grass and/or shrub in the understory could reduce flame height throughout most of the fire season (Agee et al., 2002). If true under field conditions of fire ignition and development, such a finding would suggest a possible caveat to the common assumption that fuels with high surface area-to-volume ratios are among the most combustible and efficiently burning fuel types. The abundance of foliage fuel found throughout densely stocked, uniform forests, however, clearly has a high probability of combustion capable of propagating fires with high subsequent mortality. In a mixed conifer system in the Sierra Nevada range, North and Hurteau (2011) examined the effects of "thin from below" treatments, in which trees of a given diameter are removed to minimize the presence of ladder fuels that could propagate a crown fire. Following wildfire, differences in fire mortality between treated (53%) and untreated (97%) forest suggest that fuel reduction treatments can allow for a considerable reduction in the presence of foliage and ladder fuels throughout the stand, though this did not include the effects of direct mortality from the mechanical thinning itself, which would substantially increase overall mortality in the thinned areas.

The potential for fire to spread vertically to the forest canopy is highly dependent upon the successional stage of the forest stand. As densely stocked stands of shade-intolerant species mature, self-thinning raises the crown height, and the resulting shading discourages the development of ladder fuels, thereby reducing the probability of fire propagation from the ground fuels into the canopy (Odion et al., 2004; Perry et al., 2011). Collins and Stephens (2010) found that stands were most susceptible to high-severity reburn when they were between 17 and 30 years old (also see Chapter 1). Consequently, mature, closed conifer stands can be more resistant to foliage combustion and tree morality than their younger counterparts (Thompson and Spies, 2009, 2010). These findings bear relevance to the commonly held assumption that the probability of high severity fires tends to increase with stand age. Such assumption is often made on the premise that forests accumulate more biomass through time, and thus have more total fuel that could be burned, thereby resulting in fires of higher severity. However, the infrequent occurrence of high-severity wildfires is not necessarily the result of infrequently high amounts of forest fuel availability. For many ecosystems, it is the infrequent occurrence of extreme weather conditions that may lead to a high-severity, foliage-consuming crown fire (Perry et al., 2011).

Foliage combustion rates may thus be best thought of as a function of fire severity and the vertical strata of the foliage. CFs for grass and forbs range from 0.70 to 0.75 in very-low-/low-severity fires to 1.00 in high-severity fires, whereas the combustion of fuels of small (<7.62 cm diameter at breast height [dbh]) trees and shrubs at a slightly higher vertical strata is slightly less: CFs for low-, medium-, and high-severity fires are 0.44, 0.76, and 0.89 for conifers and 0.50, 0.80, and 1.00 for hardwoods, respectively. Estimated CFs for the foliage of large trees are, as expected, lower than the others because of the vertical distance between foliage and surface fuels, where the majority of combustion takes place. CFs for large (>7.62 cm dbh) foliage in low-, moderate- and high-severity fires are 0.09, 0.27, and 0.69 for conifers and 0.12, 0.29, and 0.58 for hardwoods, respectively (Table 10.2).

10.4 SOIL COMBUSTION

Soil represents a considerable fraction of forest carbon, comprising approximately 44% of total forest carbon storage worldwide (Bonan, 2008). Soil carbon storage is usually low among ecosystems with frequent, low-severity fire regimes, such as those found in semiarid ponderosa pine forests. Conversely, soil carbon storage can be very high in ecosystems with infrequent (i.e., a mean fire return interval of >200 years) fires. Fires of high intensity and severity typify many forests with infrequent fire regimes. Because of the high magnitude of soil carbon storage in stands with infrequent, high-severity fires, estimates of carbon emissions from wildland fires are highly sensitive to the CF used to estimate the proportion of soil carbon that is consumed. However, estimates of soil carbon combustion are difficult to obtain, particularly in high-severity wildland fires, because of the lack of prefire estimates of soil carbon content.

The process of soil carbon consumption is dominated by smoldering, as opposed to flaming, combustion. Smoldering combustion is a result of insufficient amounts of oxygen required to support flaming combustion and is most prevalent in organic soils and rotting logs. The combustion of forest soils is highly dependent on the magnitude of the temperatures they are exposed to and the duration of exposure. Agee (1993) suggested that soils can be combusted at temperatures as low as 100 °C, but laboratory-based experiments suggest that significant amounts of soil carbon volatilization require temperatures between 200 °C and 315 °C (Lide, 2004), with peak smoldering temperatures ranging from 300 °C to 600 °C (Rein et al., 2008). Work by Fernández et al. (1997) heated the top 10 cm of soil taken from a Scots pine (*Pinus sylvestris*) stand to 150° at a gradually increasing rate (+3 °C min^{-1}), at which point the soil was heated for 30 min thereafter, yet no significant amount of soil carbon combustion was observed. Upon applying the same heating regime at temperatures of 220°, 350°, and 490 °C, however, there were significant changes in the content of soil organic matter (i.e., soil carbon). Temperatures of 220°, 350°, and 490 °C resulted in losses of 37%, 90%, and nearly 100%, respectively. Others have noted that shorter heating times at 350 °C resulted in a 50% weight loss after only 180 s (Almendros et al., 2003), compared with 90% at 350 °C observed by Fernández et al. (1997). Consequently, exposure to increased temperatures is highly dependent on combustion times and rates of fire spread; the relatively high rates at which fire moves across western North American landscapes, combined with the relatively limited diffusion of oxygen into the relatively nonporous soil profile, limit soil carbon emissions. CFs for soils described by Campbell et al. (2007) were 0.04 for low- and medium-severity fires and 0.08 for high-severity fires (Table 10.2).

The combustion of soils in boreal forests represents an important exception to the relatively low rates of soil carbon emissions observed in most western US forests. Turetsky et al. (2011) and Kasischke and Hoy (2012) found that the combustion of soil carbon in Alaskan boreal forests can actually constitute the majority of carbon emissions during fires, representing 54-70% of total carbon emissions. Turetsky et al. found that three factors explained most of the variation in the depth of burning/carbon consumption in the surface organic layers of black spruce forests. First, topography was a significant control: Higher fractions of consumption were observed in upland sites compared with lowland sites. Second, season of the fire was also a factor: Seasonal thawing of permafrost resulted in drier ground layers as the growing season progressed. Finally, in upland sites, fires that exhibited higher consumption occurred in the early season in years where fires had a large spatial extent compared to those in years where fires had a smaller spatial extent because of drier conditions and more extreme fire behavior.

Large amounts of biomass with long-term smoldering potential also are found in pocosin shrublands (a type of wetland with deep, sandy, and acidic soils) in the southeastern United States. While pocosin systems can have substantial amounts of combustible fuel contained in deep peat layers, they differ most notably from boreal forests in their lack of both a freeze-thaw cycle and a strong, seasonally sensitive decline in moisture as the growing season progresses. Consumption of fuel beds in these systems is poorly understood, and additional research on moisture dynamics, biogeochemical processes, and combustion is needed (Reardon et al., 2007, 2009).

10.5 BOLE BIOMASS CONSUMPTION

While many studies report tree mortality rates, relatively little on the fraction of fire-killed trees that were combusted during wildfire has been reported. In estimates of pyrogenic carbon emissions taken from the Biscuit Fire in 2002, Campbell et al. (2007) found no combustion of bole biomass among large (>7.62 cm dbh) trees, regardless of fire severity (Table 10.2). The lack of combustion for the boles of large trees seems to have been effectively mediated by the combustion of bark, which had CFs of 0.03, 0.06, and 0.20 for conifers and 0.03, 0.11, and 0.22 for hardwoods in low-, medium-, and high-severity fires, respectively. Such a finding is consistent with what is expected of fuels with low surface area-to-volume ratios (Table 10.2).

Bark CFs were much higher for small trees; for low-, medium-, and high-severity fires there were CFs of 0.42, 0.70, and 0.70 for conifers and 0.40, 0.63, and 0.79 for hardwoods, respectively (Table 10.2). As expected, the thinner bark of smaller trees, much of which was combusted, was not effective in protecting the bole biomass from combustion. Estimates of the combustion of the boles of small trees for low-, medium-, and high-severity fires were 0.40, 0.68, and 0.61 for conifers and from 0.40, 0.63, and 0.79 for hardwoods, respectively (Table 10.2). Weighted CFs for all trees, adjusted for the abundance of small tree biomass versus large tree biomass, would be approximately 0.03, 0.07, and 0.08 for low-, medium-, and high-severity fires, respectively (Campbell et al., 2007). Others have used far higher CFs for high-severity fires in modeling studies. An estimated high-severity CF of 0.30 has been used for Siberian forests (Soja et al., 2004), which may be realistic, given the small diameters prevalent in boreal forest stands. Estimates of bole CFs, however, some of which are as high as 0.30 for North American forests (Wiedinmyer et al., 2006), seem to be at odds with those estimated by Campbell et al. (2007), given the majority of biomass is stored in boles of large trees, none of which is combusted by high-severity fires. Such estimates, if inaccurate, can result in substantial overestimates of pyrogenic carbon emissions because of the considerable stocks of carbon in bole biomass of large trees. Overall, the CFs for total forest biomass (i.e., trees, snags, shrubs, woody fuels, litter, duff, and soil), weighted according to their respective prefire biomass, were 0.13, 0.15, and 0.21 for low-, medium-, and high-severity fires, respectively, in the Biscuit Fire (Campbell et al., 2007) (Table 10.2).

10.6 FUEL REDUCTION TREATMENTS, CARBON EMISSIONS, AND LONG-TERM CARBON STORAGE

The application of fuel reduction treatments have become common in many fire-adapted forests throughout the western North America. Such treatments are intended to reduce the severity of fires, primarily out of concern over public safety in fire-prone regions, as well as to minimize widespread tree mortality. Fuel reduction treatments often include understory removal, whereby midstory and understory vegetation are removed through pruning or harvesting. Understory removal treatments are often followed by prescribed fire, which reduces surface fuels in order to limit the flame height of a wildfire that might enter the stand. This is done by removing fuel through prescribed fire or pile burning, both of which reduce the potential magnitude of a wildfire by making it more difficult for a surface fire to ignite the canopy. The timing of prescribed fire can be central to its effectiveness. If performed *after* an understory removal treatment, it may burn any additional residue created by the treatment. Additionally, performing prescribed fire under cooler and moisture conditions than those experienced during the fire season is also ideal to avoid the propagation of an unplanned fire. Other fuel reduction treatments involve a partial harvest of overstory trees to limit the potential of fire to spread from crown to crown.

While such treatments can sometimes be effective in reducing fire severity, if and when fires occur in thinned areas (Rhodes and Baker, 2008), they can come at the expense of carbon storage. The majority of carbon stored in leaves, leaf litter, and duff is typically consumed by high-severity wildfire and often constitutes the majority of the carbon emissions during the a given fire, yet most of the carbon stored in forest biomass (stem wood, branches, and coarse, woody debris) remains unconsumed even by high-severity wildfires. Consequently, fuel removal via forest thinning almost always reduces carbon storage more than the additional carbon that a stand is able to store when made more resistant to wildfire. For this reason, removing large amounts of biomass to reduce the fraction by which other biomass components are consumed via combustion is inefficient (Mitchell et al., 2009). Fuel reduction treatments that involve the removal of overstory biomass (i.e., intermediate-sized and large trees) are, perhaps unsurprisingly, the most inefficient methods of reducing wildfire-related carbon losses because they remove large amounts of carbon for only a marginal reduction in expected fire severity (Figure 10.2).

10.7 INDIRECT SOURCES OF CARBON EMISSIONS

Our discussion thus far has focused on the *direct* effects of wildfire on carbon emissions as a result of the combustion of live vegetation, dead biomass, and soil organic matter. *Indirect* effects, by contrast, are not the result of the active combustion of biomass or soil organic matter; instead, they result from the long-term decomposition of vegetation killed in wildfire. The magnitude of indirect emissions, and the temporal scales at which they affect the net ecosystem carbon balance, vary with different fire behaviors. Most of the mortality resulting from low-severity fires is limited to understory plants, shrubs, and small trees, which do not typically constitute a significant portion of total stand carbon storage and, by extension, do not represent a significant source of carbon emissions upon decomposition. High-severity fires, by contrast, result in the near-total death of all trees within a stand, including overstory dominants. While the addition of any unburned leaf litter and fine, woody debris from fire-killed trees

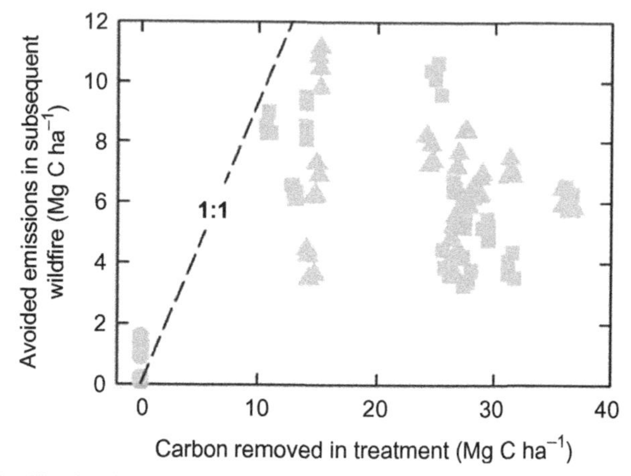

FIGURE 10.2 Simulated effectiveness of various fuel-reduction treatments in reducing future wildfire combustion in a ponderosa pine forest. In general, protecting one unit of carbon (C) from wildfire combustion came at the cost of removing approximately three units of carbon in treatment. At the very lowest (least biomass removed) treatment levels, more carbon was protected from combustion than removed during treatment; however, the absolute gains were extremely low. Circles show understory removal, squares show prescribed fire, and triangles show understory removal and prescribed fire. Simulations were run for 800 years with a treatment-return interval of 10 years and a mean fire-return interval of 16 years. Forest structure and growth were modeled to represent mature, semiarid ponderosa pine forest growing in Deschutes, Oregon. Further descriptions of these simulations are given by Mitchell et al. (2009).

represent pools with relatively high turnover (10-20 years), a large pool of coarse woody debris (e.g., logs, snags) can be a significant source of carbon emissions (Bond-Lamberty et al., 2003), one that can continue to release (and store) carbon for periods of up to, and even exceeding, 100 years (Kashian et al., 2006).

Fire severity has a significant impact on postdisturbance rates of net primary production and net ecosystem production (NEP). Net primary production is the difference between photosynthesis and autotrophic (i.e., plant) respiration, whereas NEP is a measure of net ecosystem carbon uptake, defined as the difference between photosynthesis and autotrophic respiration plus heterotrophic (i.e., decomposition) respiration. Following a high-severity disturbance, rates of heterotrophic respiration are, for a period of time, far higher than rates of photosynthesis, resulting in negative NEP (Harmon et al., 2011). While indirect sources of carbon emissions following fire can be substantial, particularly following high-severity fire, the postdisturbance regrowth of a new cohort of trees is also a significant contributor to total ecosystem carbon storage and the net ecosystem carbon balance (Figure 10.3).

The amount of time required for a recently disturbed forest to shift from a source to a sink depends on fire severity, forest type, and local climate. Following high-severity wildfires, forests with low rates of productivity, such as the ponderosa pine forests of the southwestern United States, take relatively longer to make

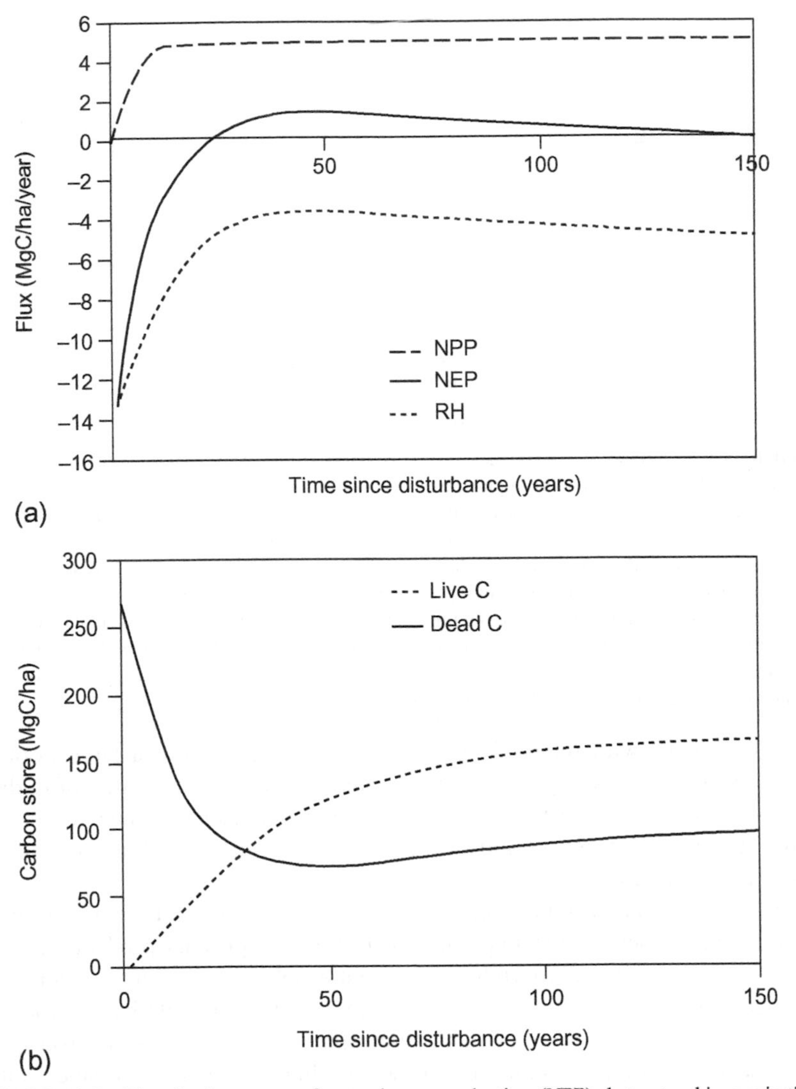

FIGURE 10.3 The classic pattern of net primary production (NPP), heterotrophic respiration (RH), and net ecosystem production (NEP) (A) and associated carbon stores (B) following a high-severity disturbance. *(From Harmon et al. (2011)).*

the postfire transition from carbon source to carbon sink (Ghimire et al., 2012). Dore et al. (2008) examined a ponderosa pine forest in northern Arizona 10 years after a stand-replacing fire and found it to be a moderate source of carbon (109 g carbon m^{-2} $year^{-1}$), but they observed a moderate carbon sink (164 g carbon m^{-2} $year^{-1}$) in an unburned stand nearby. The burned stand remained a source of carbon during all months of the year that were measured, even during the growing season in the summer months. Annual ecosystem respiration was 33% lower in the burned stand. The slow recovery of such stands is largely attributed to the climate, whereby cold winters combine with low springtime precipitation to limit gross primary production (GPP), whereas warm summers with periodic precipitation are conducive to respiration-driven losses of soil carbon (Dore et al., 2008). However, this analysis was based on only five plots with a 25 m radius; therefore, some caution regarding broader inferences is appropriate.

Differences in the postfire carbon balance of uptake were observed in semiarid, mixed-conifer forests of eastern Oregon. Meigs et al. (2009) found that 4-5 years after a mixed-severity fire, areas that burned at low severity were modest net carbon sinks. By contrast, ponderosa pine forests that also were affected by a low-severity fire were carbon neutral in low-severity fire areas. Differences in the recovery time to being a source of carbon once again may be because of differences in productivity; ponderosa pine forests are typically less productive than mixed-conifer forests (Franklin and Dyrness, 1973). Among areas affected by high severity fires, both ponderosa pine and mixed-conifer stands were sources of carbon emissions 4-5 years following fire. Modeled estimates of the postfire transition from carbon source to carbon sink suggest that ~40 years may necessary for low-productivity ponderosa pine forests to shift from being a carbon source to a carbon sink (Ghimire et al., 2012), though this analysis did not control for the potentially confounding effect of postfire logging, which is common after high-severity fire in ponderosa pine and mixed-conifer forests (see Chapter 11). Forests with higher rates of productivity, such as coastal range Sitka spruce (*Picea sitchensis*)/western hemlock (*Tsuga heterophylla*) forests in the Pacific Northwest, seem to make the postfire transition from carbon source to carbon sink in a shorter amount of time than any other coniferous western forest, potentially in <30 years. Harmon et al. (2011) reviewed the scientific literature on this question for various forest types and concluded that the transition from source to sink following fire sufficiently severe to reset the successional "clock" varied from 14-50 years in forests types characteristic of the Pacific northwestern United States and 5-15 years in boreal forests. High-severity fire rotation intervals are currently several hundred years to more than 1000 years in most mixed-conifer and ponderosa pine forest regions of the western United States, however, and these rates are generally substantially lower than historical rates (see Chapter 1). Thus a long-term spatio-temporal perspective is important to understand more fully the natural disturbance dynamics in these systems (see Chapter 9).

10.8 CONCLUSIONS

The majority of carbon stored in montane forest ecosystems of western North America remains unconsumed, even in high-severity wildfires. Large carbon stores in the bole biomass of large forest trees are not consumed, and the substantial proportion of carbon stored in forest soils is only slightly consumed. Most of the carbon emissions in a wildfire are from combustion of litter, duff, and woody debris. In the 2002 Biscuit Fire, CFs for total forest biomass (i.e., trees, snags, shrubs, woody fuels, litter, duff, and soil), weighted according to their respective prefire biomass, were 0.13, 0.15, and 0.21 for low-, medium-, and high-severity fires, respectively. Such factors can be even lower among stands with a higher proportion of carbon storage in bole biomass that likewise remains unconsumed in high-severity wildfires, such as Sitka spruce (*P. sitchensis*)/Western Hemlock (*T. heterophylla*) forests in the coast range of the Pacific Northwest (Smithwick et al., 2002; Mitchell et al., 2009). The application of fuel treatments can be effective in reducing fire severity and carbon emissions, but such treatments come at the cost of a net reduction in carbon storage relative to fire alone (Mitchell et al., 2009).

Postfire carbon emissions from fire-killed biomass can be substantial for decades following wildfires. Low- or even moderate-severity fires, however, do not necessarily result in a postfire source of carbon released to the atmosphere. High-severity fire temporarily creates a source of postfire carbon emissions as a result of the decomposition of fire-killed biomass, which lessens each year with natural postfire succession of vegetation, transitioning from a carbon source to a carbon sink within 5-50 years, depending on the ecosystem. Rates of postfire recovery are highest among systems with high productivity, whereas high-severity wildfires in forests with low productivity transition from source to sink over a relatively longer timeline, though there are important limitations in the amount and scope of existing studies of these systems. Additional research on the relationship between climatic change, disturbance regimes, and postdisturbance successional trajectories may prove to be a crucial step toward projecting the future of pyrogenic carbon emissions in mixed-severity fire regimes.

REFERENCES

Agee, J.K., 1993. Fire Ecology of Pacific Northwest Forests. Island Press, Washington, D.C., 494p.

Agee, J.K., Wright, C.S., Williamson, N., Huff, M.H., 2002. Foliar moisture content of Pacific Northwest vegetation and its relation to wildland fire behavior. For. Ecol. Manag. 167, 57–66.

Almendros, G., Knicker, H., González-Vila, F.J., 2003. Rearrangement of carbon and nitrogen forms in peat after progressive thermal oxidation as determined by solid-state ^{13}C-and ^{15}N-NMR spectroscopy. Org. Geochem. 34, 1559–1568.

Bonan, G.B., 2008. Forests and climate change: forcings, feedbacks, and the climate benefits of forests. Science 320, 1444–1449.

Bond-Lamberty, B., Wang, C., Gower, S.T., 2003. Annual carbon flux from woody debris for a boreal black spruce fire chronosequence. J. Geophys. Res. 108 (D3), 8220.

Bowman, D.M.J.S., Balch, J.K., Artaxo, P., Bond, W.J., Carlson, J.M., Cochrane, M.A., D'Antonio, C.M., DeFries, R.S., Doyle, J.C., Harrison, S.P., Johnston, F.H., Keeley, J.E., Krawchuk, M.A., Kull, C.A., Marston, J.B., Moritz, M.A., Prentice, I.C., Roos, C.I., Scott, A.C., Swetnam, T.W., Van der Werf, G.R., Pyne, S.J., 2009. Fire in the Earth system. Science 324, 481–484.

Campbell, J., Donato, D.C., Azuma, D., Law, B., 2007. Pyrogenic carbon emission from a large wildfire in Oregon, United States. J. Geophys. Res. Biogeosci. 112, 1–11

Collins, B., Stephens, S., 2010. Stand-replacing patches within a 'mixed severity' fire regime: quantitative characterization using recent fires in a long-established natural fire area. Landsc. Ecol. 25, 927–939.

Cowlin, R.W., Briegleb, P.A., Moravets, F.L., 1942. Forest Resources of the Ponderosa Pine Region of Washington and Oregon. US Department of Agriculture, Forest Service, Washington, D.C.

DeBano, L.F., Neary, D.G., Ffolliott, P.F., 1998. Fire Effects on Ecosystems. John Wiley & Sons, New York.

Dore, S., Kolb, T., Montes-Helu, M., Sullivan, B., Winslow, W., Hart, S., Kaye, J., Koch, G., Hungate, B., 2008. Long-term impact of a stand-replacing fire on ecosystem CO2 exchange of a Ponderosa pine forest. Glob. Chang. Biol. 14, 1801–1820.

Fernández, I., Cabaneiro, A., Carballas, T., 1997. Organic matter changes immediately after a wildfire in an Atlantic forest soil and comparison with laboratory soil heating. Soil Biol. Biochem. 29, 1–11.

Finney, M.A., Cohen, J.D., Grenfell, I.C., Yedinak, K.M., 2010. An examination of fire spread thresholds in discontinuous fuel beds. Int. J. Wildland Fire 19, 163–170.

Finney, M.A., Cohen, J.D., McAllister, S.S., Jolly, W.M., 2013. On the need for a theory of wildland fire spread. Int. J. Wildland Fire 22, 25–36.

Flannigan, M.D., Amiro, B.D., Logan, K.A., Stocks, B.J., Wotton, B.M., 2006. Forest Fires and Climate Change in the 21st Century. Mitig. Adapt. Strateg. Glob. Chang. 11, 847–859.

Franklin, J.F., Dyrness, C.T., 1973. Natural vegetation of Oregon and Washington. USDA Forest Service General Technical Report, Pacific Northwest Forest and Range Experiment Station, Portland.

Ghimire, B., Williams, C.A., Collatz, G.J., Vanderhoof, M., 2012. Fire-induced carbon emissions and regrowth uptake in western US forests: documenting variation across forest types, fire severity, and climate regions. J. Geophys. Res. Biogeosci. 2005-2012, 117.

Hardy, C.C., Ottmar, R.D., Peterson, J.L., Core, J.E., 2001. Smoke Management Guide for Prescribed and Wildland Fire: 2001 Edition. National Wildlife Coordinating Group, Ogden, UT.

Hargrove, W.W., Gardner, R., Turner, M., Romme, W., Despain, D., 2000. Simulating fire patterns in heterogeneous landscapes. Ecol. Model. 135, 243–263.

Harmon, M.E., Bond-Lamberty, B., Tang, J., Vargas, R., 2011. Heterotrophic respiration in disturbed forests: a review with examples from North America. J. Geophys. Res. Biogeosci. 116, 1–17.

Hessburg, P.F., Reynolds, K.M., Salter, R.B., Richmond, M.B., 2004. Using a decision support system to estimate departures of present forest landscape patterns from historical conditions: an example from the Inland Northwest region of the United States. In: Perera, A.H., Buse, L.J., Weber, M.G (Eds.), Emulating Natural Forest Landscape Disturbances: Concepts and Applications. Columbia University Press, New York, NY, United States, pp. 158–174 (Chapter 13).

Hessburg, P.F., Reynolds, K.M., Salter, R.B., Richmond, M.B., 2004. Using a decision support system to estimate departures of present forest landscape patterns from historical reference condition—an example from the inland Northwest region of the United States.

Hull Sieg, C., McMillin, J.D., Fowler, J.F., Allen, K.K., Negron, J.F., Wadleigh, L.L., Anhold, J.A., Gibson, K.E., 2006. Best predictors for postfire mortality of ponderosa pine trees in the Intermountain West. For. Sci. 52, 718–728.

IPCC, 2013. Working Group I Contribution to the IPCC Fifth Assessment Report Climate Change 2013: The Physical Science Basis, Draft Summary for Policymakers. 5.

Jimerson, T.M., Jones, D.W., 2000. Ecological and watershed implications of the Megram Fire. In: Proceedings of "Fire Conference2000: The First Congress on Fire Ecology, Prevention and Management".

Kashian, D.M., Romme, W.H., Tinker, D.B., Turner, M.G., Ryan, M.G., 2006. Carbon storage on landscapes with stand-replacing fires. Bioscience 56, 598–606.

Kasischke, E.S., Hoy, E.E., 2012. Controls on carbon consumption during Alaskan wildland fires. Glob. Chang. Biol. 18, 685–699.

Keeley, J.E., 2009. Fire intensity, fire severity and burn severity: a brief review and suggested usage. Int. J. Wildland Fire 18, 116–126.

Keyser, T.L., Lentile, L.B., Smith, F.W., Shepperd, W.D., 2008. Changes in forest structure after a large, mixed-severity wildfire in ponderosa pine forests of the Black Hills, South Dakota, USA. For. Sci. 54, 328–338.

Knapp, E., Keeley, J., Ballenger, E., Brennan, T., 2005. Fuel reduction and coarse woody debris dynamics with early season and late season prescribed fire in a Sierra Nevada mixed conifer forest. For. Ecol. Manag. 208, 383–397.

Law, B., Thornton, P., Irvine, J., Anthoni, P., Van Tuyl, S., 2001. Carbon storage and fluxes in ponderosa pine forests at different developmental stages. Glob. Chang. Biol. 7, 755–777.

Lide, D.R., 2004. CRC Handbook of Chemistry and Physics. CRC press, Boca Raton, FL.

Littell, J.S., Oneil, E.E., McKenzie, D., Hicke, J.A., Lutz, J.A., Norheim, R.A., Elsner, M.M., 2010. Forest ecosystems, disturbance, and climatic change in Washington State, USA. Clim. Chang. 102, 129–158.

Lobert, J.M., Warnatz, J., 1993. Emissions from the combustion process in vegetation. In: Crutzen, P.J., Goldammer, J.G. (Eds.), Fire in the Environment: The Ecological, Climatic, and Atmospheric Chemical Importance of Vegetation Fires. Wiley, New York, pp. 15–37.

McHugh, C.W., Kolb, T.E., Wilson, J.L., 2003. Bark beetle attacks on ponderosa pine following fire in northern Arizona. Environ. Entomol. 32, 510–522.

McKenzie, D., Gedalof, Z.E., Peterson, D.L., Mote, P., 2004. Climatic change, wildfire, and conservation (Cambio Climático, Incendios y Conservación). Conserv. Biol. 18, 890–902.

Meigs, G.W., Donato, D.C., Campbell, J.L., Martin, J.G., Law, B.E., 2009. Forest fire impacts on carbon uptake, storage, and emission: the role of burn severity in the eastern cascades, Oregon. Ecosystems 12, 1246–1267.

Miller, C., 2003. Simulation of effects of climatic change on fire regimes. In: Veblen, T.T., Baker, W.L., Montenegro, G., Swetnam, T.W. (Eds.), Fire and Climatic Change in Temperate Ecosystems of the Western Americas. Springer, New York, NY, pp. 69–94.

Miller, C., Urban, D.L., 2000. Connectivity of forest fuels and surface fire regimes. Landsc. Ecol. 15, 145–154.

Miller, J., Skinner, C., Safford, H., Knapp, E.E., Ramirez, C., 2012. Trends and causes of severity, size, and number of fires in northwestern California, USA. Ecol. Appl. 22, 184–203.

Mitchell, S.R., Harmon, M.E., O'Connell, K.E.B., 2009. Forest fuel reduction alters fire severity and long-term carbon storage in three Pacific Northwest ecosystems. Ecol. Appl. 19, 643–655.

Nelson Jr., R., 2001. Water relations of forest fuels. In: Johnson, E.A., Miyanishi, K. (Eds.), Forest Fires: Behavior and Ecological Effects. Academic Press, San Diego, CA, pp. 79–149.

North, M.P., Hurteau, M.D., 2011. High-severity wildfire effects on carbon stocks and emissions in fuels treated and untreated forest. For. Ecol. Manag. 261, 1115–1120.

Odion, D.C., Frost, E.J., Strittholt, J.R., Jiang, H., Dellasala, D.A., Moritz, M.A., 2004. Patterns of fire severity and forest conditions in the western Klamath Mountains, California. Conserv. Biol. 18, 927–936.

Ottmar, R.D., 2014. Wildland fire emissions, carbon, and climate: modeling fuel consumption. For. Ecol. Manag. 317, 41–50.

Perry, D.A., Hessburg, P.F., Skinner, C.N., Spies, T.A., Stephens, S.L., Taylor, A.H., Franklin, J.F., McComb, B., Riegel, G., 2011. The ecology of mixed severity fire regimes in Washington, Oregon, and Northern California. For. Ecol. Manag. 262, 703–717.

Reardon, J., Hungerford, R., Ryan, K., 2007. Factors affecting sustained smouldering in organic soils from pocosin and pond pine woodland wetlands. Int. J. Wildland Fire 16, 107–118.

Reardon, J., Curcio, G., Bartlette, R., 2009. Soil moisture dynamics and smoldering combustion limits of pocosin soils in North Carolina, USA. Int. J. Wildland Fire 18, 326–335.

Rein, G., Cleaver, N., Ashton, C., Pironi, P., Torero, J.L., 2008. The severity of smouldering peat fires and damage to the forest soil. Catena 74, 304–309.

Rhodes, J.J., Baker, W.L., 2008. Fire probability, fuel treatment effectiveness and ecological trade-offs in western U.S. public forests. Open For. Sci. J. 1, 1–7.

Smithwick, E.A.H., Harmon, M.E., Remillard, S.M., Acker, S.A., Franklin, J.F., 2002. Potential upper bounds of carbon stores in forests of the Pacific Northwest. Ecol. Appl. 12, 1303–1317.

Soja, A.J., Cofer, W.R., Shugart, H.H., Sukhinin, A.I., Stackhouse, P.W., McRae, D.J., Conard, S.G., 2004. Estimating fire emissions and disparities in boreal Siberia (1998-2002). J. Geophys. Res. Atmos. 109, 1–22.

Taylor, A.H., Skinner, C.N., 1998. Fire history and landscape dynamics in a late-successional reserve, Klamath Mountains, California, USA. For. Ecol. Manag. 111, 285–301.

Thompson, J.R., Spies, T.A., 2009. Vegetation and weather explain variation in crown damage within a large mixed-severity wildfire. For. Ecol. Manag. 258, 1684–1694.

Thompson, J.R., Spies, T.A., 2010. Factors associated with crown damage following recurring mixed-severity wildfires and post-fire management in southwestern Oregon. Landsc. Ecol. 25, 775–789.

Turetsky, M.R., Kane, E.S., Harden, J.W., Ottmar, R.D., Manies, K.L., Hoy, E., Kasischke, E.S., 2011. Recent acceleration of biomass burning and carbon losses in Alaskan forests and peatlands. Nat. Geosci. 4, 27–31.

van Mantgem, P.J., Stephenson, N.L., Byrne, J.C., Daniels, L.D., Franklin, J.F., Fule, P.Z., Harmon, M.E., Larson, A.J., Smith, J.M., Taylor, A.H., Veblen, T.T., 2009. Widespread increase of tree mortality rates in the western United States. Science 323, 521–524.

Ward, D., 2001. Combustion chemistry and smoke. In: Johnson, E.A., Miyanishi, K. (Eds.), Forest Fires: Behavior and Ecological Effects. Academic Press, New York, pp. 55–77.

Westerling, A.L., Hidalgo, H.G., Cayan, D.R., Swetnam, T.W., 2006. Warming and earlier spring increase western US forest wildfire activity. Science 313, 940–943.

Whitlock, C., Shafer, S.L., Marlon, J., 2003. The role of climate and vegetation change in shaping past and future fire regimes in the northwestern US and the implications for ecosystem management. For. Ecol. Manag. 178, 5–21.

Wiedinmyer, C., Neff, J.C., 2007. Estimates of CO2 from fires in the United States: implications for carbon management. Carbon Balance Manage. 2, 10.

Wiedinmyer, C., Quayle, B., Geron, C., Belote, A., McKenzie, D., Zhang, X., O'Neill, S., Wynne, K.K., 2006. Estimating emissions from fires in North America for air quality modeling. Atmos. Environ. 40, 3419–3432.

Wyant, J.G., Omi, P.N., Laven, R.D., 1986. Fire induced tree mortality in a Colorado ponderosa pine/Douglas-fir stand. For. Sci. 32, 49–59.

Section 3

Managing Mixed- and High-Severity Fires

Chapter 11

In the Aftermath of Fire: Logging and Related Actions Degrade Mixed- and High-Severity Burn Areas

Dominick A. DellaSala[1], David B. Lindenmayer[2], Chad T. Hanson[3] and Jim Furnish[4]

[1]Geos Institute, Ashland, OR, USA, [2]Fenner School of Environment and Society, The Australian National University, Canberra, Australia, [3]John Muir Project of Earth Island Institute, Berkeley, CA, USA, [4]Consulting Forester, Rockville, MD, USA

11.1 POSTFIRE LOGGING AS A KNEE-JERK RESPONSE

Three things are just about guaranteed every fire season: (1) forests will burn over large areas, occasionally reaching megafire proportions under extreme conditions (see Chapter 2); (2) land managers will proclaim burn areas to be disasters in need of "restoration," proposing logging and tree planting to speed up "recovery;" and (3) decision-makers (government officials, politicians) will want to optimize the economic value of fire-killed and live trees, often bypassing environmental safeguards to quickly cut the trees.

Shortly after fires have been extinguished, so-called salvage logging of dead and frequently live trees happens and, in intensively managed areas, most often includes road building, replanting with commercial trees with genomes selected for regional conditions, seeding with nonnative plants, use of straw bales for erosion abatement, and spraying herbicide or using mechanical methods to suppress native vegetation—especially shrubs—that land managers think might compete with commercially important trees. Such forestry activities make sense only if forests are viewed as commodities, but there are substantial tradeoffs given that they disproportionately target the most ecologically important areas where economic values are highest, thereby setting up conflicts with increasing regularity, as big fires become more frequent in a changing climate.

As discussed throughout this book, fear of fire is coupled to socioeconomic drivers that result in command-and-control actions during (see Chapter 12) and after fires (this chapter). Misperceptions about postfire landscapes begin with

The Ecological Importance of Mixed-Severity Fires: Nature's Phoenix.

the branding of the term "salvage" and postfire landscapes as "wastelands." Aside from maritime uses, salvage is defined as "the act of saving goods or property that were in danger of damage or destruction; save from ruin, destruction, or harm, and collect discarded or refused material" (WordNet Dictionary). It also refers to "an amount estimated as expected to be realized or actually realized on sale of a fixed asset at the end of its useful life—used in calculating depreciation" (Merriam-Webster online, http://www.merriam-webster.com/dictionary/salvage).

This largely sums it up regarding why postfire management activities are knee-jerk responses. That is, the prevailing view on fire is this: "disaster" (blackened forest) caused by the fire is bad and "recovery" (green forest) via active management is good because burned areas are "destroyed" by fire and therefore need to be "restored." The "discarded" materials in this case are fire-killed trees, "salvaged" before they "depreciate" in economic value or the end of the fixed asset's useful life, so to speak. But there is no ecological basis for salvage logging, and ecologists should refrain from using the term. In reality, a fire-dependent ecosystem is not being salvaged from a disaster but, rather, is being degraded by postfire logging and related actions compared with the ecologically beneficial role that fire just performed.

In this chapter we summarize how postfire logging and related activities can lead to compounded ecological disturbances (Paine et al., 1998) that, if implemented over large landscapes, exceed disturbance thresholds, flipping entire areas to altered ecosystem states that trigger type conversions (i.e., "landscape traps;" Lindenmayer et al., 2011). We also discuss how the aftermath of fire has been used as a driver for lifting environmental safeguards proposed by decision makers wanting to replace fire-dependent, high-quality, complex postfire forests with tree plantations (essentially tree crops, planted with a few commercially valued tree species, often grown in dense rows and treated with herbicides and fertilizers). In many areas tree plantations have, ironically, burned in uncharacteristically intense fires resulting from high fuel loads caused by densely stocked trees (e.g., Odion et al., 2004). Postfire logging proposals also tend to increase in proportion to the size of an individual fire (especially megafires), the accessibility of burned areas (e.g., high road densities), and the economic interests in expediting logging before trees diminish in economic value. To make matters worse, these activities are poised to scale up in intensity in places where climate change is expected to trigger more fires in the coming decades (see Chapter 9). Increases in postfire logging may combine with climate change threats that accumulate in space and time for rare and declining wildlife associated with high-quality habitat created by mixed- and high-severity fires (see Chapters 2–6).

Four case studies illustrate the kinds of ecosystem degradation that typically are associated with postfire management: (1) the Biscuit Fire of 2002 in southwest Oregon; (2) the Rim Fire of 2013 in the Sierra region of central California; (3) the Jasper Fire of 2000 in the Black Hills, South Dakota; and (4) fires in montane ash-eucalypt forests of Victoria, Australia. For the case studies, we provide exemplary methods for reducing the ecological footprint of postfire management where intervention occurs for economic reasons and recommendations for conserving ecologically valuable postfire landscapes where conservation is the priority. We stress that there is no ecological basis for postfire logging and, if forests are to be managed for ecological integrity, postfire logging—and its associated activities (chemical and mechanical removal of native shrubs and the establishment of artificial tree plantations)—is not a management practice that should continue. This particular chapter contains a mix of science, conservation, international postfire logging issues, and personal experiences in extreme postfire logging projects.

11.2 CUMULATIVE EFFECTS OF POSTFIRE LOGGING AND RELATED ACTIVITIES

Compared with the biologically diverse unlogged landscape created by fire, intensively managed postfire areas lack the pulse of legacy structures created by fire because most, if not all, of the ecologically valued dead and live trees (Box 11.1) are removed during logging operations (Appendix 11.1). These impacts occur when the postdisturbance landscape is especially vulnerable to soil compaction given its fragile state. Chronic management effects can inhibit the development of complex postfire seral stages for decades to centuries, given the slow rates of soil establishment in places (see McIver and Starr, 2000), and removes biological legacies over large areas (DellaSala et al., 2014). This affects a broad suite of postfire-dependent species, most notably, cavity-nesting (Figure 11.1) and shrub-nesting birds (Burnett et al., 2012, Hanson, 2014).

Notably, in congressional testimony to the House Subcommittee on Resources (November 10, 2005, hearing on HR4200), University of Washington Professor Jerry Franklin stated, "Timber salvage is most appropriately viewed as a 'tax' on ecological recovery. The tax can be very large or relatively small depending upon the amount of material removed and the logging techniques that are used."

Response of fire-adapted species and communities to postfire logging depends on the scale, intensity, degree of biological legacies removed (McIver and Starr, 2000, Lindenmayer and Noss, 2006), disturbance history of the site (Reeves et al., 2006, Hutto, 2006), and species-specific tolerance to logging. Documented effects span a broad range of taxa, ecosystem processes, and forest functions (see Karr et al., 2004, Lindenmayer et al., 2004, Hutto, 2006, DellaSala et al., 2006, Hanson and North, 2008, Lindenmayer et al., 2008, DellaSala et al., 2006, DellaSala et al., 2014; see also Appendix 11.1) that can be summarized as follows:

- Extensive degradation of stand structure and function
- Loss of soil nutrients
- Chronic sedimentation and erosion
- Reduction in carbon storage
- Increased fine fuel loads and potential reburn severity

BOX 11.1 Biological Legacies as the Building Blocks for Nature's Phoenix

Nothing in a forest is wasted, especially after a fire, as biological legacies link pre- and postdisturbance conditions, life and death in the forest, and aquatic and terrestrial ecosystems. Biological legacies such as large snags and downed logs typically have long "residence" times, persisting for decades to centuries and spanning successional stages. They include predisturbance elements (large live and dead trees, shrubs) that survive, persist, or regenerate in the burn area and are an important seed source for recolonization of plants in the new forest. They perform vital ecosystem functions such as anchoring soils (e.g., large root wads of live and dead trees); recycling nutrients (e.g., downed logs decomposed by detritovores); storing carbon long term (given slow rates of decomposition) and sequestering it, providing microsites for recolonizing plants and wildlife (e.g., so-called nurse logs that are substrate for conifer seedlings, large snags that provide shade for seedlings), and acting as refuges for numerous species (e.g., downed logs as moisture sites for salamanders, fungi, and invertebrates). Snags are used by hundreds of wildlife species for foraging (because they harbor numerous insects, particularly the larval stages), nesting, hiding, roosting, perching, and denning (examples include cavity-nesting birds, bats, and mammals, including many rare species). Many insectivorous species that use snags, in turn, perform vital trophic functions that help keep insects in check after fire. When large snags along streams eventually topple into the riverbed, they become hiding cover for fish, and pulses of postfire sedimentation (typically in the first winter after a fire) create spawning grounds for native fish, linking aquatic and terrestrial ecosystems. Despite their ecological importance, however, biological legacies are most often considered a "wasted resource" that will otherwise "rot" and need to be replaced by tree seedlings artificially grown in nurseries and planted in areas after burns, frequently in dense rows resembling corn fields, particularly in the western United States. The typical argument is that postfire logging and subsequent conifer plantings are needed to leap-frog over successional stages to a "forest," even though those actions degrade one of the most biologically diverse seral stages—complex early seral forest—and does not create a diverse forest ecosystem but, rather, creates a biologically diminished and simplified crop for lumber and wood fiber.

- Degradation of habitat for threatened, endangered, and sensitive species
- Reduced habitat and prey for apex predators and forest carnivores
- Greatly reduced snag densities for cavity-nesting birds and mammals
- Exotic species invasions
- Reduced resilience and resistance of postfire landscapes to future disturbances

Nearly unanimous results like those presented in Appendix 11.1 and illustrated in Figure 11.2 show a widespread and consistent pattern of postfire logging impacts across taxa and regions; that is, this type of logging has arguably more severe adverse impacts than logging in green forests. In addition, a feedback

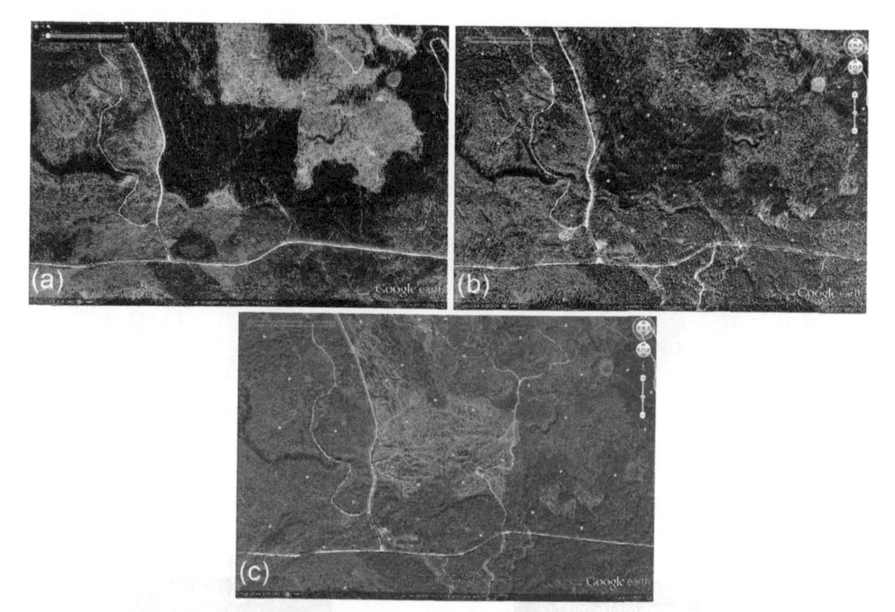

FIGURE 11.1 (a) Forest fragment in 1991 before the 2003 Wedge Fire, ~60 km north of Columbia Falls, Montana. (b) Subset of avian survey plots ($n=5000$ stations) distributed across >100 fire areas in western Montana since 1998. Exact center of the image had a nesting black-backed woodpecker (*Picoides arcticus*) after the Wedge Fire. (c) Postfire logging eliminated all biological legacies over a large landscape, including remaining nesting habitat for populations of black-backed woodpeckers not detected after the logging across the sample grids, including the center image area. *(Courtesy R. L. Hutto, University of Montana.)*

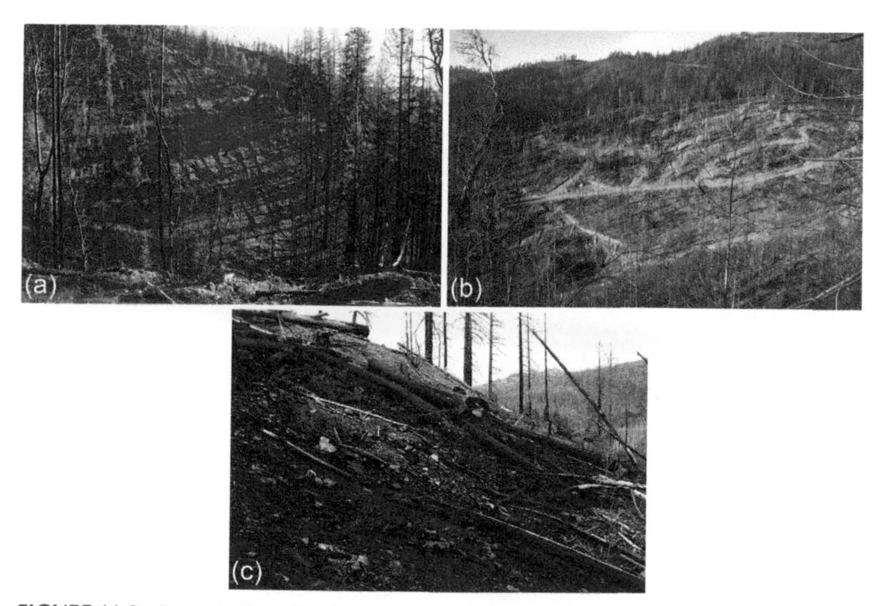

FIGURE 11.2 Impacts of postfire logging on soil in two areas in southwest Oregon: the Quartz Creek Fire area, showing extensive soil ruts from dragging logs upslope on private lands (a and b); and the Biscuit Fire area, showing soil damage from burning logging slash on public lands (c) (photos by D. DellaSala). Onset of productive soil horizons spans human generations, and thus soil degradation is a chronic postfire disturbance.

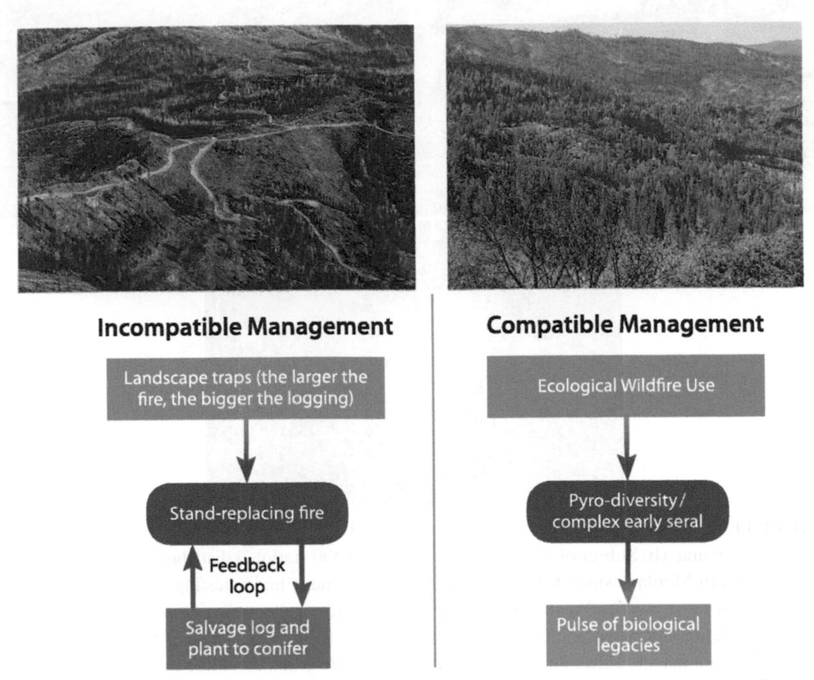

FIGURE 11.3 Incompatible vs. ecologically compatible post-fire management in large fire complexes. Most post-fire management in the western United States follows the flow chart on the left.

loop exists whereby areas burn in a fire, are logged and planted with commercial species, only to burn more intensely in the next fire, and then are logged again later (Figure 11.3). Exacerbating this cycle, the combination of postfire logging and removal of native shrubs through herbicides—as is commonly practiced in the western United States—dramatically increases the spread of invasive (and often highly combustible) weeds (McGinnis et al., 2010). The fire/postfire-logging/intense-fire feedback may accelerate in a changing climate in places where more fires are expected to trigger more logging, which already is occurring in the western United States.

11.3 POSTFIRE LOGGING LESSONS FROM CASE STUDIES

Biscuit Fire of 2002, Southwest Oregon

The Biscuit Fire of 2002 encompassed a fire perimeter of nearly 200,000 ha of southwest Oregon's Klamath Mountains, burning in a natural mosaic pattern of mixed severities (29% high-severity, 30% moderate-severity, and 41% low-severity fire [http://fsgeodata.net/MTBS_Uploads/data/2002/maps/OR4244112390420020713_map.pdf]; Figures 11.4 and 11.5).

At the time this was one of the nation's largest recorded fires in what is considered the most ecologically important (biodiverse) landscape in western

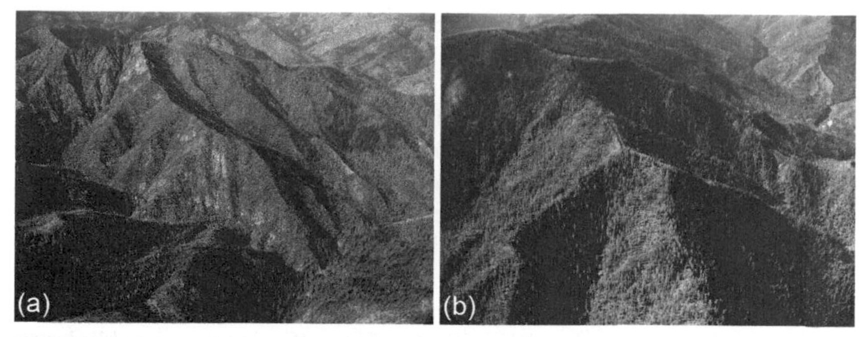

FIGURE 11.4 Two views of burn severities in the Biscuit fire area, southwest Oregon: (a) primarily high severity fire that burned through steep terrain and chaparral; and (b) more of a mixed-severity pattern that skipped around cooler, north-facing slopes (green areas). *(Photos courtesy of S. Whitney.)*

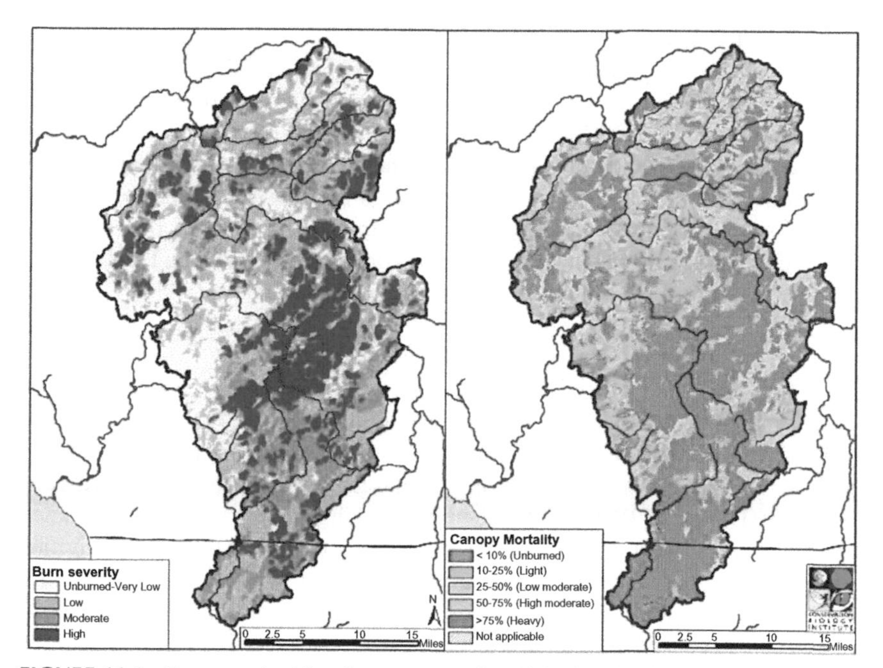

FIGURE 11.5 Burn severity (a) and canopy mortality (b) in the Biscuit Fire, as defined by the US Department of Agriculture postfire assessment team. It should be noted that, because the Forest Service's initial assessment was conducted very soon after the Biscuit fire, the canopy mortality map above is an overestimation (see http://fsgeodata.net/MTBS_Uploads/data/2002/maps/OR4244112390420020713_map.pdf), due to the fact that it was conducted too early to detect post-fire conifer responses, such as flushing (of pines) or epicormic branching (of Douglas-fir), as discussed in the next case study. *(Courtesy of the Conservation Biology Institute.)*

North America that is largely unprotected (DellaSala et al., 1999). Using the Biscuit Fire as an example, we show (1) how context and scale matter in postfire management decisions; (2) how decisions by federal land managers (US Department of Agriculture Forest Service and US Department of the Interior Bureau of Land Management) are often at odds with postfire maintenance of areas with high conservation value; and (3) a prioritization process for minimizing ecological damage in large postfire landscapes where the pressure to log outweighs conservation. This case study has broader implications in postfire management because the Biscuit logging project at the time was precedent setting (e.g., national legislation was proposed to expedite logging after fires in all the national forests). We reiterate that there is no ecological justification for postfire logging on public lands. Given that land managers and decision makers already slated this area for massive and controversial postfire logging, however, we present an approach that would have reduced some of the logging damage and perhaps some of the controversy.

To begin, context and scale matter in understanding patterns and processes in nature and are especially relevant in prioritization schemes. The use of "ecological screens" illustrates approaches that include recognition of context and scale in designating "go" and "no-go" zones for management that may be useful in minimizing post-fire management conflicts.

By "context," we mean knowledge of ecological condition, function, and management history that can be used to place a particular site or project area within its larger setting (Slosser et al., 2005). Along with context, planning at multiple scales is fundamental to understanding postfire processes and effects of management. Because ecological processes operate at multiple scales, the relative size of a management unit, the watersheds within which it lies, and the time frame over which natural processes operate all need to be factored into whether and how to treat landscapes following large fires.

Biscuit Project Scope

Unfortunately, in the planning stages for actions after the Biscuit Fire, the federal agencies downplayed one of the largest postfire logging proposals in history by focusing on the relatively small area logged over the very large burn perimeter (USFS and BLM, 2004), thereby masking the impacts and importance of context and scale. For instance, federal agencies claimed that their activities would minimally affect the burn area because only ~4% (~8000 ha and 877,920 cubic meters of timber) of the 200,000-ha burn perimeter was to be logged. An additional 12,600 ha was to be either seeded or planted with conifer seedling stock, construction of 480 km of fuel management zones (FMZs) was proposed to remove 5600 ha (2360 cubic meters) of timber with the stated purpose of lowering fuel hazards, and another 8000 ha project-wide would be mechanically "thinned" (managed for fuels, with 33,160 ha scheduled for prescribed burning even though the fire already lowered fuels). In actuality, the scale of postfire management was not 4% as claimed; rather ~51,360 ha (25%) would receive some form of postfire management activities. But this tells

BOX 11.2 Land Use and Postfire Logging

While the Biscuit Fire project area is governed by several resource management laws and forest planning documents, two particular policies stand out the Northwest Forest Plan (USFS and BLM, 1994), which governs the management of nearly 10 million hectares of federal lands, and the Roadless Conservation Rule (USFS, 2001), which protected over 25 million hectares of inventoried roadless areas (IRAs) across the nation. The Northwest Forest Plan resulted in dramatic reductions in logging levels on federal lands that included, in part, late-successional reserves (LSRs) managed for late-seral species; however, some logging is permitted only if it is "conservative" or "prudent" and consistent with the development of late-seral conditions (USFS and BLM, 1994). The Roadless Conservation Rule prohibited logging in IRAs (which lack roads and are at least 2000 ha), with the exception of "primarily small tree thinning" where fire is a concern. Further, several watersheds in the Biscuit Fire area are managed for their wild and scenic character under the National Wild and Scenic Rivers Act (1968). Depending on specific categories, this encompassed an adjoining corridor of approximately 400 m on either side of the designated river.

only part of the story because none of the planning involved issues of scale or context (see Box 11.2). Because postfire logging was heavily concentrated in high-severity fire areas, the effects on complex early seral forest from landscape fragmentation in particular were even higher (Figure 11.6).

Context and Scale Matter

Before the Biscuit Fire of 2002, the area had been nominated for national monument protections by conservation groups because its regional context lies within the globally important Klamath-Siskiyou ecoregion, which has the highest concentration of rare plants of any national forest group in the United States, the largest complex of IRAs along the Pacific Coast from the Mexican to Canadian borders, and one of the best wild salmon fisheries in the region (DellaSala et al., 1999). This is the last place that should be logged given its regional and global context and because fires have been an ongoing source of natural landscape heterogeneity

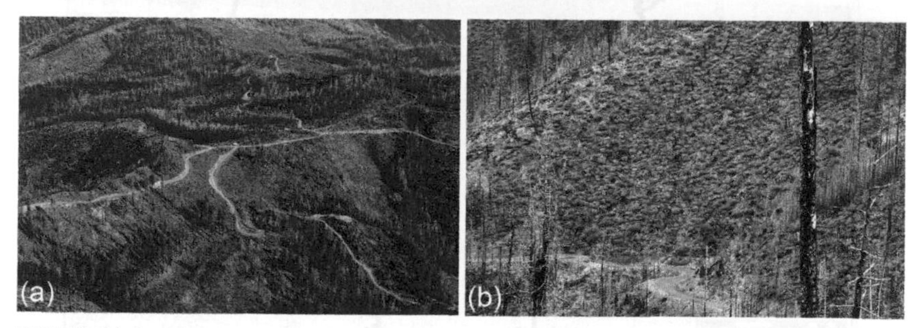

FIGURE 11.6 Biscuit post-fire logged area in southwest Oregon (2002) showing landscape fragmentation from roads and clearcuts (a) and near complete removal of biological legacies (b). *(Photos courtesy K. Schafer, D. DellaSala)*

associated with the region's extraordinary beta biodiversity (species turnover across environmental gradients) (Odion et al., 2010).

In the landscape where the Biscuit Fire occurred, the greatest proportion of larger trees was concentrated inside LSRs and IRAs—rather than distributed randomly throughout the project area—because of prior logging and natural vegetation patterns. Not only were the majority of proposed postfire management activities concentrated in those two conservation areas, but also a disproportionate amount of the total expected 877,920 cubic meters of logging volume (Figure 11.7) would occur there. Notably, the largest LSR in the project

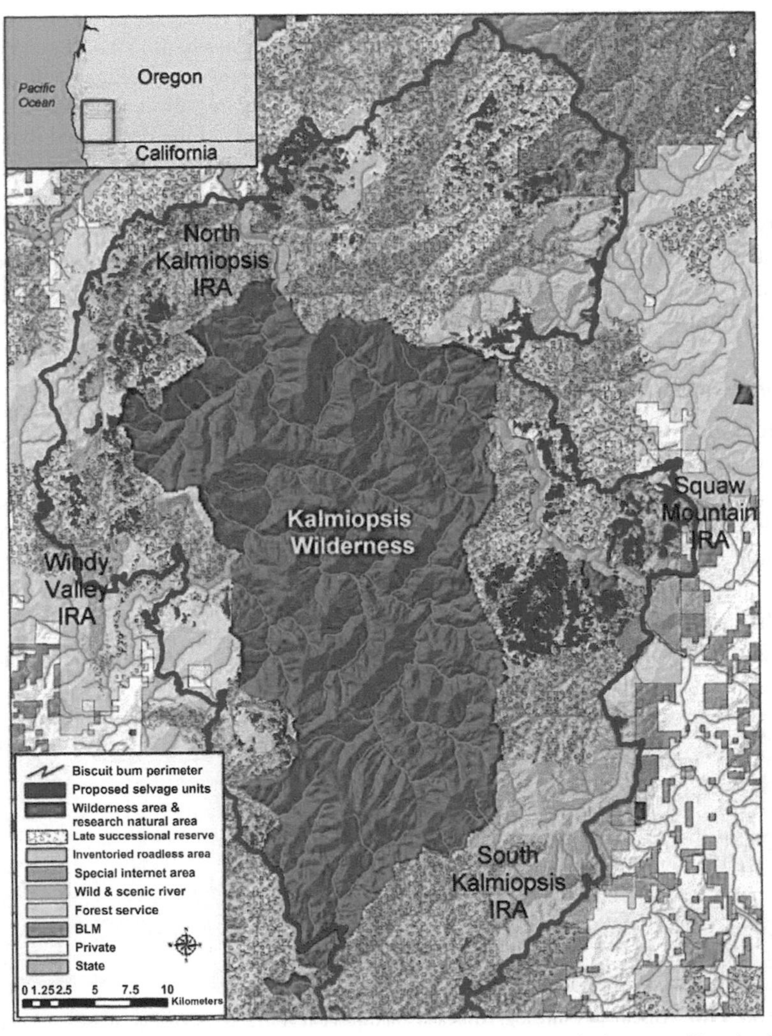

FIGURE 11.7 Map of the Biscuit Fire burn perimeter, showing agencies' proposed logging locations over major land management designations such as IRAs, LSRs, and other special interest areas.

area lies along the border of the Kalmiopsis Wilderness, and logging fragmented an otherwise intact ecosystem. Furthermore, 90% of the proposed logging units lie within watersheds whose streams flow directly into wild and scenic rivers located within the burn perimeter (scale and context matter).

When the appropriate scale is considered, rather than the percentage of the total burn perimeter to be logged, strikingly, 70% of the project-area volume would come from the collection of LSRs that represented only 42% of the total burn area. Similarly, a high percentage of proposed logging units (52%) would occur within IRAs. Finally, a total of 55 watersheds were proposed for logging to varying degrees, and therefore logging opportunities were not uniformly distributed but, rather, were clustered in areas of high ecological importance, making the project footprint much larger than claimed.

Integrating Context and Scale into Project Decisions

The lack of attention to context and scale in the Biscuit case study illustrates how land managers can grossly underestimate the postfire logging footprint. Thus land managers would benefit from incorporating context and scale in decision-making to truly assess project impacts.

As an example, researchers (Beschta et al., 2004, Karr et al., 2004) proposed the application of "ecological screens" for minimizing postfire logging damage in ecologically sensitive areas. In the case of the Biscuit Fire, three types of screens have been proffered: administrative, operational, and ecological prohibitions (Strittholt and Rustigian, 2004). Administrative screens are areas designated as off limits to logging by existing planning documents and environmental laws (e.g., congressionally and administratively withdrawn areas identified as no-logging areas in the Northwest Forest Plan, wilderness areas designated by Congress, IRAs). Operational screens are areas where steep terrain or lack of roads inhibits entry. Ecological screens are fine-scale filters related to specific retentions (e.g., large dead and live trees—biological legacies) to minimize impacts on site. Using multiple screens would yield a much different outcome.

After careful consideration of context and scale and applying the screens, a less ecologically damaging and more constrained response to the Biscuit area yields a much-reduced logging "footprint" while producing significant timber volume. For instance, a total of 3950 ha and an estimated volume of 177,000-224,200 cubic meters of timber would be available for logging using ecological screens (Figure 11.8), compared with the agencies' alternative of 7877 ha yielding an estimated 877,920 cubic meters. Under the ecological screens approach, postfire logging would be permitted only under strict guidelines such as those recommended by Beschta et al. (2004).

Biscuit Fire Case Study Conclusions

In the absence of context and scale, postfire landscapes are treated indifferently, leading to underestimates of the logging footprint. Given the high risk of doing

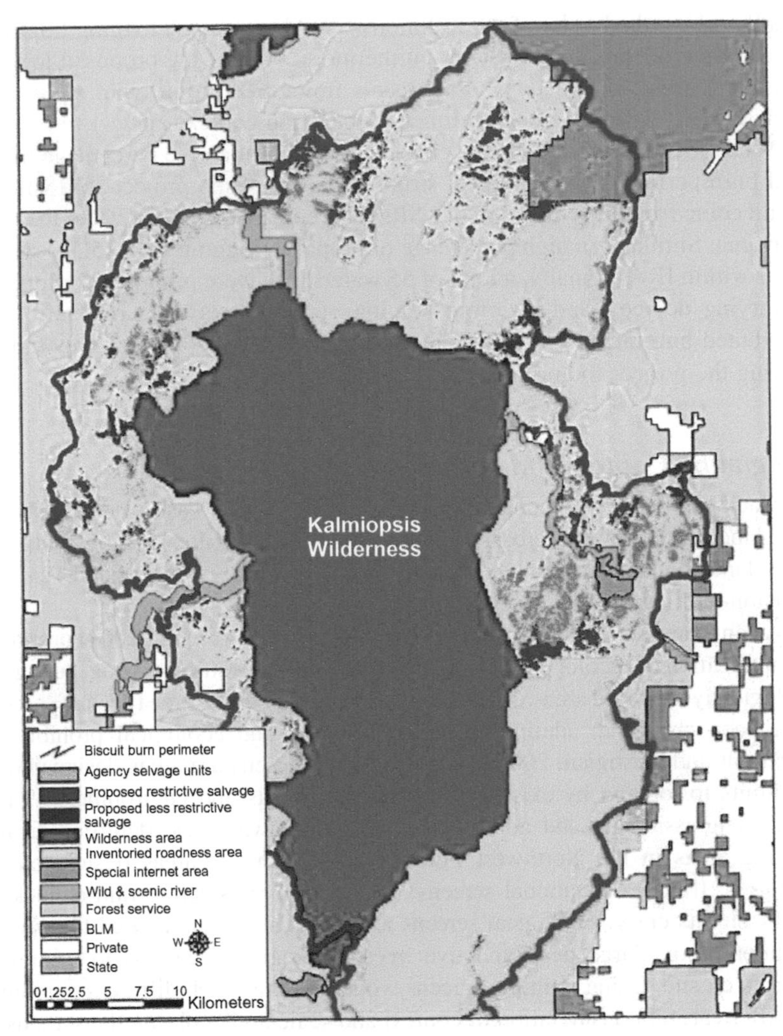

FIGURE 11.8 Comparison of agencies' logging units (gray) and units based on administrative, operational, and ecological screens highlighting two impact levels—restricted logging (red) and less restrictive (purple)—within the Biscuit Fire burn perimeter. *(Courtesy of the Conservation Biology Institute.)*

further damage to postfire landscapes through large-scale logging and associated conifer planting and shrub removal, proceeding cautiously is prudent, especially in areas of particular ecological significance where context matters most. Incorporating ecological screens into project-level decisions allows for proper attention to context and scale. Even in go zones (logging units), however, managers must proceed cautiously because there is a preponderance of evidence that postfire logging disrupts natural processes and harms the development of complex early seral conditions. Within LSRs, where a "conservative" amount of logging is permitted under the Northwest Forest Plan, managers should maintain all biological legacies. To do otherwise would place postfire landscapes with high conservation value at risk of significant ecological damage. Our findings are important for policy makers considering mandating logging, shrub removal, and tree planting following large-scale fire events, as is debated often by decision makers wanting to bypass environmental safeguards under the assumption that postfire landscapes are wastelands in need of recovery.

Rim Fire of 2013, Sierra Nevada, California

The Rim Fire was first detected on August 17, 2013, burning in a canyon in the Stanislaus National Forest west of Yosemite National Park in the central Sierra Nevada Mountains of California. Over the next several weeks, the fire would ultimately span 104,176 ha, mostly in the Stanislaus National Forest, but also including a portion of the western edge of Yosemite National Park, as well as several thousand hectares of private timberlands. Over a third of the fire area was comprised of non-conifer prefire vegetation, including grassland, foothill chaparral, oak woodlands, and numerous large rock outcroppings; the remainder comprised montane conifer forest (USFS, 2014a, b). Soon after the smoke cleared, the US Forest Service—which keeps 100% of the revenue from the sale of timber from postfire logging projects—was already proposing one of the largest national forest timber sales in history. Conservative members of Congress threatened to override environmental laws to mandate that such postfire logging occur across the Rim Fire area. In response, in the autumn of 2013, some 250 scientists sent a letter to Congress opposing postfire logging in the Rim Fire area, urging lawmakers to appreciate the high ecological value of postfire habitat and to not weaken or roll back federal environmental laws (DellaSala et al., 2013). The scientists concluded:

> *Though it may seem at first glance that a post-fire landscape is a catastrophe ecologically, numerous scientific studies tell us that even in patches where forest fires burned most intensely the resulting post-fire community is one of the most ecologically important and biodiverse habitat types in western conifer forest. Post-fire conditions serve as a refuge for rare and imperiled wildlife that depend upon the unique habitat features created by intense fire. These include an abundance of standing dead trees or 'snags' that provide nesting and foraging habitat for woodpeckers and many other wildlife species, as well as patches of native flowering shrubs that replenish soil nitrogen and attract a diverse bounty of beneficial insects that aid in pollination after fire.... This post-fire habitat, known as 'complex early seral forest', is quite simply some of the best wildlife habitat in forests and is an essential stage of natural forest processes. Moreover, it is the least protected of all forest habitat types and is often as rare, or rarer, than old-growth forest, due to damaging forest practices encouraged by post-fire logging policies.*

The scientists' letter carried the day with regard to the legislative threat, and the bill did not pass Congress, but the Forest Service continued to move forward with its plan to log the Rim Fire area, as discussed below.

Overestimation of Fire Severity

Based on the US Forest Service's initial "rapid assessment" using satellite imagery from just weeks after the fire, the agency reported approximately 40% high-severity fire effects (http://www.fs.fed.us/postfirevegcondition/index.shtml; accessed October 25, 2014), immediately released these results to the media, and called the Rim Fire area a "moonscape" that had been "nuked" by the fire (Cone, 2013), an exaggeration that was used to justify proposing postfire logging of many thousands of hectares (USFS, 2014a). Not only was this effort to deny the ecological value of postfire habitat inaccurate and misleading—as discussed in Chapters 2–5 and later in this chapter—the Forest Service's initial assessment also greatly exaggerated the fire severity by failing to account for postfire responses such as "flushing" in pine species (Hanson and North, 2009) and other rapid postfire vegetation regrowth. Through "flushing," several conifer species—including the most common species in the Rim Fire (ponderosa pine [*Pinus ponderosa*])—that initially seem to be dead because they have no remaining green needles after fire, produce new green needles from surviving terminal buds at the ends of branches 1 year after fire (Figure 11.9).

Through this natural adaptation to mixed-severity fire, numerous areas that initially seem to have very high, or complete, tree mortality ultimately have many or most trees survive, particularly larger overstory trees (Hanson and North, 2009). Within the conifer forests of the Rim Fire area, flushing was common and pervasive among ponderosa pines, and some other species, by the spring and summer of 2014, resulting in many forested areas that looked quite different than they did in the autumn of 2013 and winter of 2014 (Figure 11.10).

As a result, when using satellite imagery 1 year after the fire, accessed through the Monitoring Trends in Burn Severity (MTBS) system led by the

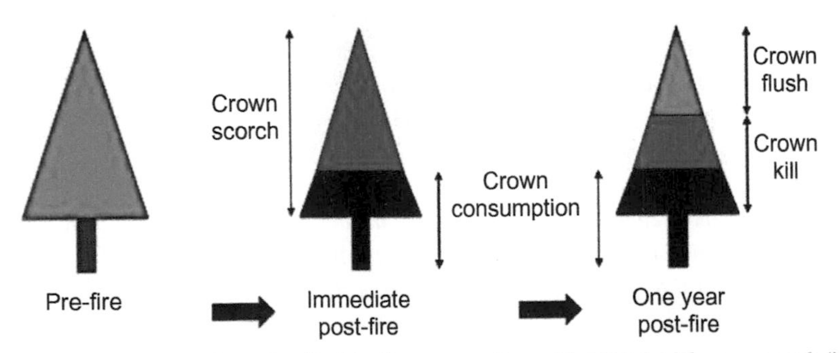

FIGURE 11.9 Process of postfire "flushing" among conifers with 100% initial crown scorch (i.e., no remaining green needles).

FIGURE 11.10 Early stages of postfire flushing of ponderosa pines, with 100% initial crown scorch, in May 2014. *(Photo by Chad Hanson.)*

US Geological Survey, high-severity fire comprised only 19.9% of the Rim Fire (www.mtbs.gov), not 40% as initially claimed by the Forest Service. The effect of this change can be seen in the difference between the Forest Service's preliminary assessment and the MTBS assessment 1 year after the fire; the latter shows much less high-severity fire, much smaller high-severity fire patches, and far more internal heterogeneity within large, high-severity fire patches (i.e., low-/moderate-severity inclusions within high-severity fire patches; Figure 11.11). None of these changes, however, were taken into account when the Forest Service issued the final decision to conduct postfire logging (including both "salvage" logging and roadside logging along nonpublic roads) over 14,000 ha of the Rim Fire area (USFS, 2014a). Moreover, as with the Biscuit Fire, the agency also minimized the overall effects of the planned logging by noting that the logging would comprise less than 14% of the total area within the Rim Fire's perimeter (USFS, 2014a). Because less than two-thirds of the fire area comprised conifer forest, however, and because only about one-fifth of this conifer forest experienced high-severity fire (www.mtbs.gov), the logging planned by the Forest Service actually represented the removal of the majority of the Rim Fire area's "complex early seral forest." Complex early seral forest is created by high-severity fire occurring in dense, mature/old conifer forest and is one of the rarest and most biodiverse habitat types in forests of western North America (DellaSala et al., 2014; see also Chapters 1–6).

Undisclosed Effects on California Spotted Owls (Strix occidentalis occidentalis)

California spotted owls use mostly forests burned by low-/moderate-severity fire or unburned forests for nesting and roosting, and they preferentially select unlogged, high-severity fire patches occurring in older conifer forest for foraging (hunting) (Bond et al., 2009), given the rich small-mammal prey base in complex early seral forest habitat (Bond et al., 2013). Thus, the species seems quite adapted to fires if sufficient postfire habitat is available in an unlogged condition and in a mixture of patch severities.

Based on the Forest Service's preliminary Rapid Assessment of Vegetation of fire severity in the Rim Fire, the agency concluded that the majority of spotted owl territories in the Rim Fire area would have little or no chance of being occupied by owls after the fire. This was based on the assumption that territories with substantial levels of high-severity fire would not likely retain occupancy (USFS, 2014a). This assessment was not based upon the MTBS fire severity data from 1 year after the fire, however, and it therefore substantially overestimated high-severity fire effects in these areas, as discussed above (Figure 11.11). Moreover, it included only effects to the ~120-ha cores of the territories (USFS, 2014b), rather than effects on the biological territory used for foraging by owls—a much larger area with a radius of at least 1.5 km, in general (Bond et al., 2009). In addition, when complex early seral forest is removed by postfire logging and other postfire management (e.g., removal of native shrubs via mastication or herbicides), owl occupancy in the affected territories is usually eliminated (Lee et al., 2012, Clark et al., 2011, Clark et al., 2013); mixed-severity fire alone has not been found to reduce occupancy

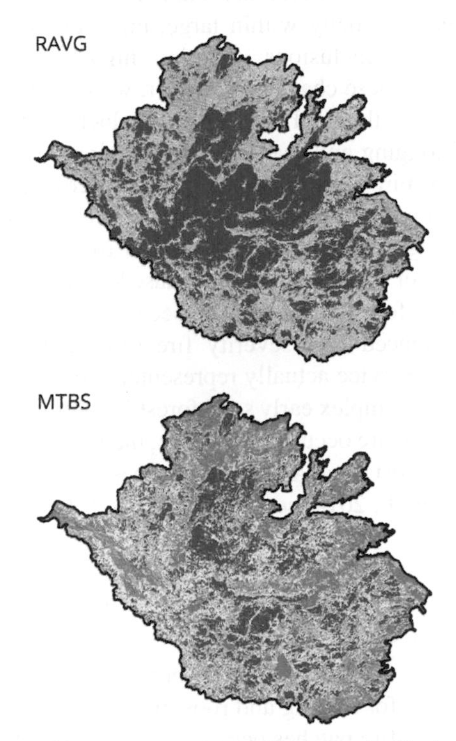

FIGURE 11.11 A large difference in the amount of high-severity fire is seen between the Forest Service's preliminary Rapid Assessment of Vegetation (RAVG) condition and the Monitoring Trends in Burn Severity (MTBS) assessment 1 year after the Rim fire.

(Lee et al., 2012). Scientists have recommended that postfire logging be completely avoided within at least 1.5 km of spotted owl nest/roost sites (Bond et al., 2009, Clark et al., 2013).

When the results of the Forest Service's own surveys of California spotted owl 1 year after the Rim Fire became available—weeks before final decision documents were issued—independent owl scientists analyzed the data for the territories fully surveyed and made some remarkable findings. First, they found 39 occupied spotted owl territories, and occupancy was 92% of prefire "historical" territories (territories occupied in one or more years in the past), which is substantially higher than the average annual occupancy in unburned mature forest, which generally ranges from 60% to 76% (in any given year, not all spotted owl territories are occupied; the owls occasionally leave a territory and return one or more years later) (Lee and Bond, 2015). Second, they found that increasing high-severity fire did not reduce the occupancy of spotted owl pairs in the Rim Fire area, and even the territories with mostly high-severity fire had over 90% occupancy; some of the occupied territories were entirely within the boundaries of large high-severity fire patches with significant low-/moderate-severity inclusions (Lee and Bond, 2015). These findings were submitted to the Forest Service over a week before the decision to log the Rim Fire area was signed, but the agency did not disclose this information in its decision. The independent analysis found that postfire logging units are located in every single occupied spotted owl territory in the Rim Fire area, and in numerous cases the majority of the entire territory would be postfire logged (Lee and Bond, 2015). In the Rim Fire area on national forest lands, this entails removal of all but 10 large snags per hectare where the prelogging snag patches typically have well over 125 large snags per hectare—generally over 90% removal (USFS, 2014a, b). Implementation of this logging project began in September 2014 (Figure 11.12).

Natural Postfire Conifer Regeneration

One of the key rationales used by the Forest Service to justify the Rim Fire logging project was the argument that little or no postfire conifer regeneration would occur within the large, high-severity fire patches except within several dozen meters of the patch edges (USFS, 2014a, b). The agency also argued that it needed to clearcut thousands of hectares of complex early seral forest in the Rim Fire area ostensibly to cover the costs of artificially planting conifers where the Forest Service claimed conifers would not naturally regrow (USFS, 2014a, b).

Site visits by several scientists in the spring and autumn of 2014, however, revealed abundant natural conifer regeneration (Figure 11.13), even deep into the interior of large, high-severity fire patches in the Rim Fire area (see also Chapter 2). Though these findings were conveyed to the Forest Service in the form of comments and photographs, the agency did not incorporate this information into the decision documents or provide any information on the amount of natural conifer regeneration in high-severity fire patches within

FIGURE 11.12 Postfire clearcutting in the Stanislaus National Forest in occupied California spotted owl territories in the autumn of 2014. *(Photo by Chad Hanson.)*

FIGURE 11.13 Natural postfire conifer regeneration, generally numbering hundreds of seedlings per hectare, within the interior of a large high-severity fire patch in the Rim Fire area in December 2014. *(Photos by Chad Hanson.)*

the fire area (USFS, 2014a, b). Troublingly, postfire logging—especially ground-based tractor logging, which comprises nearly all of the planned logging in the Rim Fire area (USFS, 2014a)—kills most of the existing natural postfire conifer regeneration, literally crushing it under the treads of heavy logging machinery and as logs are skidded to landings (Donato et al., 2006).

In reality, killing off the existing natural conifer regeneration in the Rim Fire area through logging will cost taxpayers millions. The Forest Service estimates that it will generate about $500 per hectare in revenue from the logging project (USFS, 2014a). However, the agency has received only about one-third of that amount in the timber sales after implementing the Rim Fire logging project decision. Further, Forest Service documents show that artificial planting (including site preparation and planting expenses) costs over $1700 per hectare, and sometimes more (USFS, 2014c). Therefore, on any given hectare, when postfire logging kills natural conifer regeneration, the net cost to taxpayers for replanting the areas is at least $1200 per hectare. The real cost could be even higher given that most of the timber sale receipts are required to be allocated to future postfire logging projects, not replanting, under the "Salvage Sale Fund."

Moreover, after soil damage from postfire tractor logging on fragile soils and artificial conifer planting with nursery-grown seedlings that are not naturally adapted to the microsites where they are planted, plantings commonly fail—often extensively. When this occurs, the US Forest Service tends to conduct a second project involving intensive herbicide application to eliminate nonconifer vegetation, followed by additional attempts to establish conifer tree plantations. In some cases, the Forest Service is on the third iteration of this practice in a single fire area following initial postfire logging (e.g., USFS, 2014e).

As the environmental assessment for another, much smaller, recent Forest Service postfire logging project—the Aspen postfire logging project in the Sierra National Forest—recently admitted, "Foregoing recovery and reforestation treatments would save taxpayers approximately $3,287,000 of appropriated funding needed to implement these activities" (USFS, 2014d). Because of the massive size of the Rim Fire, the net cost to taxpayers—just on this issue alone—could, by conservative estimates, be more than $15 million.

Rim Fire Case Study Conclusions

In the final decision documents for the Rim Fire area logging project, the Forest Service stated that the number 1 reason that the agency chose to propose and implement the project is that it would generate millions of dollars in revenue for the agency's budget (USFS, 2014a). The agency also noted that it was urgently interested in selling the timber to private logging companies and beginning logging as soon as possible to minimize natural postfire decay of merchantable timber and maximize the revenue to the Forest Service (USFS, 2014a). In this context of financial conflict of interest, the actual data regarding fire severity, flushing, spotted owl occupancy, and the rarity and ecological value of complex early seral forest were largely ignored or subordinated by an agency eager to begin logging and maximize financial returns.

Under the current regional forest plan that governs all national forests in the Sierra Nevada Mountains, the Southern Cascades in California, and the Modoc Plateau in northeastern California, there are no protections for complex early seral forest created by high-severity fire in mature/old conifer stands (USFS, 2004).

Postfire logging in the Sierra Nevada has been found to significantly reduce overall avian biodiversity (Burnett et al., 2012) and harm rare and imperiled wildlife species like the California spotted owl (Lee et al., 2012) and black-backed woodpecker (Hanson and North, 2008, Odion and Hanson, 2013, Siegel et al., 2013). Moreover, the avian species associated with the habitat created by high-severity fire, including several shrub-nesting species, are experiencing population declines, whereas birds associated with unburned forest are experiencing no such trend (Hanson, 2014). Ongoing fire suppression, postfire logging, subsequent eradication of native shrubs through mechanical or chemical means, establishment of artificial conifer plantations, and mechanical thinning have been identified as major threats to these declining species (DellaSala et al., 2014, Hanson, 2014). The ecological importance and rarity of complex early seral forest created by high-severity fire needs to be recognized and administrative and legislative protections put in place to maintain and recover this habitat, which is currently in substantial deficit relative to its extent before fire suppression (DellaSala et al., 2014, Odion et al., 2014).

Jasper Fire of 2000, Black Hills, South Dakota

On August 24, 2000, a woman dropped a match along a highway west of Custer City, South Dakota. Within 2 hours, a pyrocumulus cloud appeared over the southwestern part of the Black Hills National Forest (BHNF). Three days later nearly 33,600 ha had burned, including almost all of Jewel Cave National Monument and about 8% of BHNF lands.

Predictably, a battle over whether and how much fire-killed timber should be logged ensued immediately. Conditions in August 2000 were very dry and hot, making containment extremely difficult. Yet, as is often the case when the smoke clears, initial fears about the gravity of the fire were lessened by a postfire assessment revealing a mosaic of fire intensities (Figure 11.14). The burned area was managed primarily for timber production and was dominated by previously logged ponderosa pine (excluding Jewel Cave National Monument), where stems per acre were reduced and a stand of homogenous, evenly spaced mature trees were left.

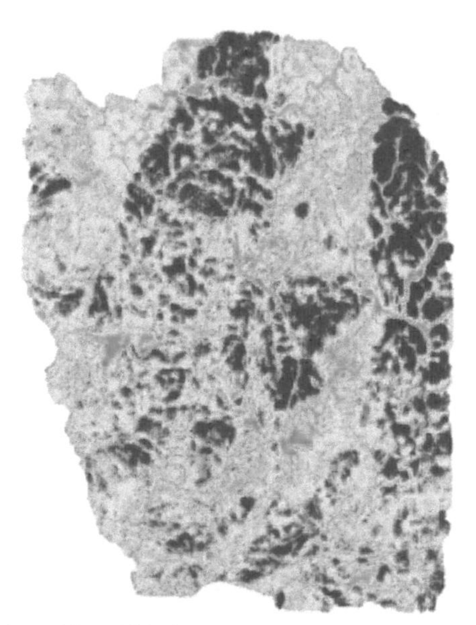

FIGURE 11.14 Burn intensities within the Jasper Fire burn perimeter, determined using ERDAS Imagery software to perform unsupervised classifications on the 3,4,7 band combination of Landsat 7 satellite imagery taken on September 3, 2000. Intensity included high (red areas, 12,718 ha, 38%), moderate (yellow areas, 10,421 ha, 31%), and low (green areas, 8047 ha, 24%) in Black Hills National Forest, as classified by the US Department of Agriculture Forest Service.

The Forest Service quickly produced a postfire assessment, estimating 542,740 cubic meters of tree mortality, and proposed to postfire log 141,584 cubic meters, concentrating on areas that had been severely burned with near total mortality.

Notably, the National Environmental Policy Act (NEPA) contains a provision to exempt certain emergency actions from environmental review, and the BHNF requested that the Jasper Fire post-fire logging project be exempted. As Forest Service Deputy Chief at the time, I (J.F.) reviewed such requests before they were submitted by Forest Service Chief Mike Dombeck to the Council for Environmental Quality (CEQ) for approval. The CEQ was known to be very stingy in granting exemptions, however.

We (J.F.) elected to limit our request to the more narrow circumstance where the Jasper Fire had burned through areas already under commercial timber sale contract (several timber sales totaling 103,828 cubic meters), rather than the BHNF's comprehensive request to log as much of the timber in severely burned areas as possible. We acknowledged the time constraint—generally agreed to be about 1 year—until fire mortality begins to quickly lose commercial value. Annual timber harvest on the BHNF averaged about 165,182-188,789 cubic meters; thus their plan to log 141,584 cubic meters equaled a "windfall substitution" of nearly a year's supply.

The chief's judgment was that the only postfire logging volume that merited emergency consideration from CEQ involved mortality in commercial timber sales where the government was involved in a contractual relationship with industry. Thus we stipulated that only dead trees within existing timber sales could be logged under the exemption; any other postfire logging would have to follow normal NEPA procedures. Surviving trees could not be cut, even if they had been previously designated for harvest. This resulted in far less postfire logging than the 141,584 cubic meters that the BHNF supervisor had requested.

In a political twist, this determination played out in the months subsequent to George W. Bush's election as president in November 2000 and before Chief Dombeck's resignation in April 2001, shortly after Anne Veneman was appointed the new Secretary of Agriculture. The election fundamentally altered the perspective of the administration from being cautious and environmentally sensitive to being pro-logging. The result is that the final decision, made after completing the environmental impact statement, opted for postfire logging of the entire 141,584 cubic meters originally sought by the BHNF.

Jasper Case Study Conclusions

Consistent with the prevailing dogma of the time, local Forest Service officials viewed the Jasper Fire as an opportunity for unprecedented postfire logging because timber production was the driving factor in forest management. The fire, the largest in recent history for the Black Hills, was termed a "disaster" even though most of the burn perimeter was low and moderate intensity. Tree mortality was considered "lost" and of no value unless it was converted to wood products. The primary limitation for logging interests was time and thus constituted an emergency because time impinged on the capacity to maximize logging. Even though the BHNF produced an 80-page fire report within a few weeks, they claimed to lack the necessary staff and budget to comply with ordinary NEPA procedures.

By contrast, the more enlightened Chief Dombeck confined the NEPA "emergency exemption" to burned areas within commercial timber sale areas in place before the fire and where the timber industry had a reasonable premise for economic loss. Beyond these contractual considerations, normal procedures would apply. Ultimately, concluding that more postfire logging occurred because of political considerations than would have been authorized had the fire erupted in 1999 when a more environmentally supportive administration was in place is reasonable.

Postfire conditions of the Jasper Fire area, particularly natural regeneration of pine seedlings, were affected by a severe drought from 2000 to 2008, resulting in ongoing reforestation efforts. There was widespread concern that fuel accumulations from snags would exacerbate ecosystem losses in the event of a reburn. To date there has been no reburn of the area, however, and the Jasper Fire was ecologically beneficial other than in areas logged after fire.

2009 Wildfires, Victoria, Australia

This case study describes some aspects of the postfire logging operations following the February 2009 "Black Friday" wildfires in the wet montane ash forests of the Central Highlands of Victoria (Figure 11.15). Mountain ash (*Eucalyptus regnans*) and alpine ash (*Eucalyptus delegatensis*) are dominant in these forests. The 2009 wildfires burned more than an estimated 78,300 ha of montane ash forest (Burns et al., 2014). Many people think of the 2009 wildfires as a single conflagration; however, it was actually a number of different kinds of fires that varied markedly in severity over a 2-week period (Cruz et al., 2012). The most intense fires occurred in the afternoon and evening of Saturday, February 7, 2009 (Taylor et al., 2014), with some fire-affected areas reputed to have experienced among the most intense fires ever recorded, reaching 88,000 kW/m (Cruz et al., 2012). Indeed, these were the most destructive fires in Australian history in terms of loss of human life and property

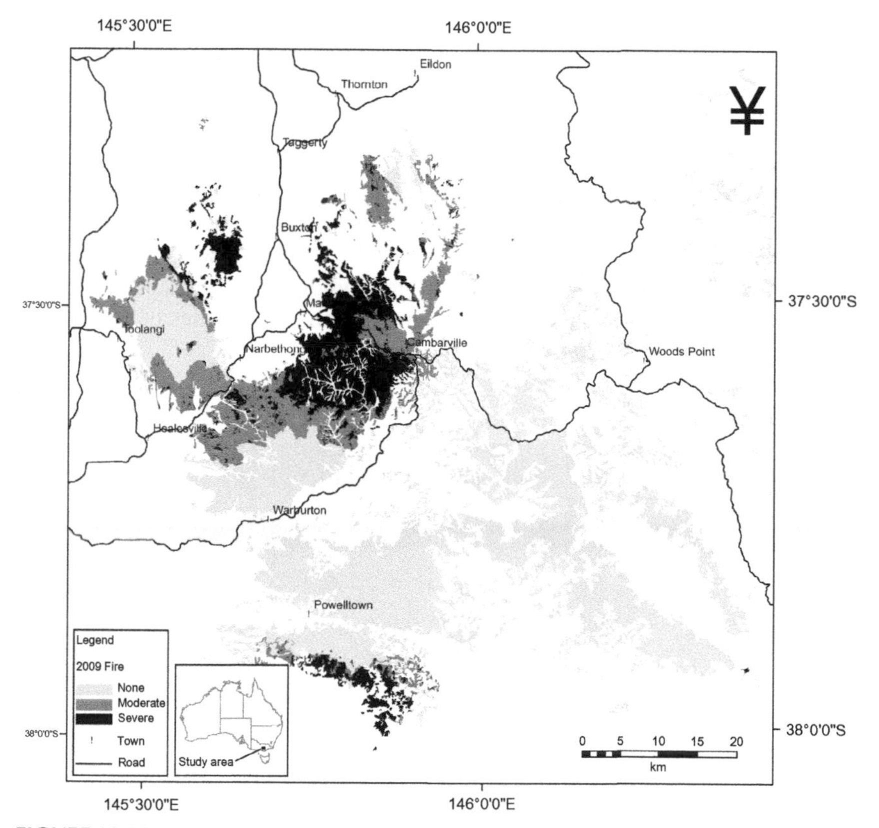

FIGURE 11.15 Mixed-severity fires of the 2009 "Black Friday" fires in wet montane ash forests of the Central Highlands of Victoria, Australia.

(Gibbons et al., 2012). By contrast, for 2 weeks after February 8, fires burned the forests around towns like Healesville at very low severity under a semisupervised "watching brief." Of course, such low-severity fires were markedly different in their effects on the forest and forest biodiversity relative to the fire in the afternoon and evening of February 7, 2009. These differences are critically important in terms of ecological and management understanding of montane ash forests, although some researchers have overlooked them and treated the 2009 fires as a single event, leading to flawed work.

Postfire logging is a prominent kind of logging in montane ash forests. It takes place after natural disturbances, especially wildfires, although it also occurs following windstorms in the region. The steps involved in postfire logging operations are akin to conventional clearfelling (clearcutting) except that the removal of all merchantable trees on a site occurs in burned forest rather than green (unburned) forest. Following the completion of harvesting operations, logging slash such as tree heads and lateral branches are burned to produce a bed of ashes into which seeds of eucalypts are dropped to regenerate a new stand of trees. Hence areas logged after fire are subject to three disturbances in rapid succession: wildfire, logging, and a regeneration burn.

As in western North America, postfire logging is conducted in an attempt to recover some of the economic value of the timber in burned stands. Significant areas of Central Highland ash forest were logged after the 1983 and 2009 wildfires. Of approximately 72,000 ha of montane ash forest that was burned in the fires that occurred in February 2009, about 3000 ha was logged, and logging was concentrated in areas of very high-severity fire following the 2009 fires (Figure 11.16). The area of forest logged after the 2009 fires was comparatively limited relative to the extensive and prolonged logging operations that followed the 1939 wildfires and that continued for more than two decades before finally being halted in the 1960s (Noble, 1977).

FIGURE 11.16 Two views of post-fire logging operations at Paradize Plains, near Marysville in the Central Highlands of Victoria, showing extensive soil damage (a) and removal of biological legacies (b). *(Photo by D. Lindenmayer.)*

Much has been written about the *potential* impacts of postfire logging in montane ash forests (Lindenmayer and Ough, 2006, Lindenmayer et al., 2008). They include accelerating the loss of large, old, hollow-bearing trees (Lindenmayer and Ough, 2006) and damaging the fire-triggered regeneration of understory and ground cover plants that are sensitive to mechanical disturbance by logging machinery. To quantify the actual impacts of postfire logging, we (D.L.) initiated formal experimental studies immediately following the 2009 wildfires. Two taxa—plants and birds—have been the focus of these studies, now in their fifth year. Results for birds in the postfire logging experiment indicate that very few species inhabit areas subject to postfire logging. One clear exception is the flame robin (*Petroica phoenicea*), which is closely associated with early succession (postfire and postlogging) environments in montane ash forests (Lindenmayer et al., 2014). No other species seem to be clear early successional specialists in this ecosystem, at least within the first 2 years after fire; however, several bird species returned to approximately prefire levels, or above prefire levels, by 3 or 4 years after fire (Lindenmayer et al., 2014: Appendix S3). Thus additional data may reveal relationships not apparent in the earliest years after fire.

Preliminary results from work to date make it clear that the effects of postfire logging on plants are more severe than the effects of traditional clearfelling of "green" (unburned) forest (Blair et al. unpublished data). Postfire logging effects on plants are also more severe than those of wildfire. Some groups of plants, especially resprouting species like tree ferns and the musk daisy bush (*Olearia argophylla*), are particularly vulnerable to postfire logging effects. As an example, preliminary data sets indicate that overall plant species richness is reduced by almost 30% at sites that have been subject to postfire logging relative to clearfelled sites and sites subjected to high-severity wildfire (Blair et al. unpublished data). In addition, our work to date has highlighted the potentially negative impacts of postfire logging resulting from the loss of, or damage to, key substrates for bryophytes, such as large logs and tree ferns.

The results of detailed experimental studies often become apparent only after many years of measurements, and this likely will be the case for our (D.L.) postfire logging work. Nevertheless, our work to date on plants in montane ash forests suggests the importance of conserving burned areas and exempting them from postfire logging operations. In addition, empirical evidence makes it clear that logging montane ash forests makes young stands regenerating after harvesting more prone to subsequent crown fires (Taylor et al., 2014). Moreover, recent analyses using International Union for Conservation of Nature ecosystem assessment criteria suggest that the mountain ash ecosystem should be classified as critically endangered because of its risk of collapse in the next 30-50 years, particularly as a result of additional wildfires and ongoing logging (Burns et al., 2014). Given relationships between logging and fire, together with the parlous state of this ecosystem, the amount of additional logging and possible future salvage logging needs to be carefully reviewed.

11.4 CONCLUSIONS

Based on a review of the literature and the case studies presented from four regions, postfire logging impedes natural postfire processes by removing some of the rarest and most biodiverse wildlife habitat in many forest ecosystems, compacting soils, causing chronic erosion, delaying natural succession, and introducing or spreading invasive species (effects are more severe for ground- and cable-based logging than helicopter logging), among other damage. Further, rather than jump-starting forests as claimed, postfire logging damages or removes complex early seral forests and inhibits the return of forest ecosystem conditions over time by removing the very components (large dead, dying, and downed trees) crucial to their development and by eradicating core components of forest biodiversity, such as native shrub patches. Postfire logging can also elevate fine fuels by removing the least combustible portion of trees (trunks) and leaving logging slash (in places where logging slash is treated with pile burning, damage to soils can have long-term consequences). Naturally regenerating landscapes following fire are biologically rich and need to be conserved for their unique ecological value.

Therefore, we recommend that (1) postdisturbance landscapes be allowed to regenerate on their own because evidence from several studies (e.g., see Turner and Dale, 1998, Donato et al., 2006, Lindenmayer et al., 2008) indicates postdisturbance processes can be surprisingly productive; (2) road building (including temporary roads) be avoided because it damages regenerative processes; (3) postfire logging in dense, mature/old forest stands that experience intense fire be avoided because such areas tend to provide the highest quality, and spatially rarest, complex early seral forest habitat (Hutto, 2006, DellaSala et al., 2014); (4) biological legacies (large dead and dying trees) be protected to aid in regenerative processes; (5) interventions be made only in ways that promote natural processes (i.e., do no harm); and (6) fragile areas be avoided using ecological screens to establish "go" and "no-go" zones if postfire logging occurs in some areas for economic reasons.

The case studies presented demonstrate that ecosystem-damaging feedback is created when a large fire strikes an area, which is logged over large landscapes after the fire, and especially where such logging occurs disproportionately in the most ecologically important areas. These degrading activities are likely to scale up in places where climate change produces more and larger fires.

APPENDIX 11.1 EFFECTS OF POSTFIRE MANAGEMENT ACROSS REGIONS WHERE MOST STUDIES HAVE BEEN CONDUCTED

Location	Attribute	Effects on Ecosystems	Sources
Broadly applicable	Process, structure, and function	Altered and diminished structural complexity of stands, ecosystem processes and functions, and populations of species and community composition	Karr et al. (2004), Lindenmayer and Noss (2006a), Burnett et al. (2012), DellaSala et al. (2014)
Mostly western United States but broadly applicable	Chronic soil erosion	Erosion is greatest when logging is associated with road building, conducted with ground-based log retrieval systems, or undertaken in areas with steep slopes and sensitive soils; erosion occurs when logs are dragged across steep slopes, and damage to soil horizons occurs from burning of slash piles	McIver and Starr (2000), Beschta et al. (2004), Karr et al. (2004)
	Aquatic and hydrological processes	Substantial disturbance of hydrological systems especially from chronic sediments from roads	Karr et al. (2004)
		Removal of burned trees that provide shade may hamper tree regeneration, especially in high-elevation or dry sites; increased frequency and magnitude of erosive high flows and raising of sediment loads cause changes that alter the character of river channels, harming aquatic species; construction and reconstruction of roads and landings accelerate runoff and chronic erosion harmful to aquatic systems	

Continued

Location	Attribute	Effects on Ecosystems	Sources
	Ecosystem restoration	Postfire logging inconsistent with comprehensive restoration goals	Beschta et al. (2004), Donato et al. (2006), Swanson et al. (2011), DellaSala et al. (2014), Hanson (2014)
Western United States	Riparian areas	Inhibit riparian functions	Reeves et al. (2006)
	Logging slash and fuels	Increased combustible fuels left on site	Weatherspoon and Skinner (1995), Duncan (2002), Donato et al. (2006)
	Herbicides	Used to kill shrubs viewed as competitors of commercially valuable trees, though such shrubs are important to nutrient cycling and mycorrhizae development often lacking in industrial settings (e.g., private lands), and data do not indicate that shrub cover precludes conifer regeneration; herbicide spraying strongly tends to increase invasive weeds	Beschta et al. (2004), Shatford et al. (2007), McGinnis et al. (2010)
Pacific Northwest and northern California, United States	Threatened, endangered, and sensitive species	Local extirpation of northern spotted owl (*Strix occidentalis caurina*) territories (Clark et al., 2011, Clark et al., 2013) and California spotted owl territories	Lee et al. (2012), Clark et al. (2013)
West-central Alberta, Canada	Acorn predators	Alters the guild of acorn predators and may reshape the pattern of seedling establishment	Puerta-Pinero et al. (2010)
Victoria, Australia		After a 1939 wildfire in Victoria, logging contributed to shortage of cavity trees for more than 40 vertebrate species, including some endangered ones.	Lindenmayer et al. (2004)

Continued

Location	Attribute	Effects on Ecosystems	Sources
Mostly western United States; Quebec, Canada	Plant richness and biomass, understory vegetation	Reduced vegetation biomass, increased graminoid (grass) cover, overall reduced plant species richness, and survival of planted seedlings relative to unlogged areas; reduced understory abundance, richness, and diversity	McIver and Starr (2000), Donato et al. (2006), Titus and Householder (2007), Purdon et al. (2004)
Western United States; Victoria, Australia	Biological legacies	Removal of a large percentage of large, dead, woody structure significantly alters postfire wildlife habitat (partial removal less so)	Lindenmayer et al. (2004), Beschta et al. (2004), Russell et al. (2006)
	Cavity-nesting mammals and birds	Removal of hollows reduced the persistence of an array of cavity-using species, including Leadbeater's possum (*Gymnobelideus leadbeateri*), an endangered arboreal marsupial	Lindenmayer and Ough (2006)
		Reduced multiaged montane ash forests that typically support the highest diversity of arboreal marsupials and forest birds	Lindenmayer and Ough (2006)
		Reduced the abundance and nesting density of cavity-nesting birds	Caton (1996), Hitchcox (1996), Hejl et al. (1995), Saab and Dudley (1998), Smucker et al. (2005), Hutto (2006), Hutto and Gallo (2006), Cahall and Hayes (2009) (some open-nesting birds increased), Hanson and North (2008), Hutto (2008), Burnett et al. (2012)

Continued

Location	Attribute	Effects on Ecosystems	Sources
New England, United States	Resistance/ resilience to disturbance	Postdisturbance logging and silvicultural attempts after hurricanes and insect outbreaks failed to improve resistance or resilience of forests and were degrading overall	Foster and Orwig (2006)
Mostly western United States	Exotic species	Increases in invasions related to soil disturbance, livestock, road pathways, greater human (vector of spread) site access, and herbicide spraying	McIver and Starr (2000), Beschta et al. (2004), Karr et al. (2004), McGinnis et al. (2010)
Southwest Oregon, United States	Conifer seedlings	Natural conifer regeneration 2 years after the 2002 Biscuit Fire, although variable, was abundant even in high-severity burn areas where conifer seedling densities (>120/ha) exceeded regional standards for fully stocked stands Postfire logging reduced median conifer regeneration density by 71%, affected conifer seedlings by damaging soils and by physically burying seedlings by woody material as a result of logging; significantly increased fine and coarse, woody fuel loads	Donato et al. (2006)
Canadian Rockies, Alberta, Canada; eastern Oregon	Apex predators and forest carnivores	Avoidance of logged areas in wolf-ungulate systems; postfire logging, thinning, and conversion from fir to pine adversely affects fishers	Bull et al. (2001), Hebblewhite et al. (2008)
Mediterranean and Sierra Nevada conifer forests	Carbon storage	Reduced from removal of woody biomass	Powers et al. (2013), Serrano-Ortiz et al. (2011)
Pacific Northwest United States	Burn severity	Increased between successive fire events in logged areas	Thompson et al. (2007)

Continued

Location	Attribute	Effects on Ecosystems	Sources
Northeastern Alberta, Canada	Byrophytes	Negative effect on species richness and species composition	Bradbury (2006)
Northwest Quebec, Canada	Soil nutrients	Loss of calcium, magnesium, and potassium for at least 110-year timber rotation	Brain et al. (2000)

REFERENCES

Beschta, R.L., Rhodes, J.J., Kauffman, J.B., Gresswell, R.E., Minshall, G.W., Karr, J.R., Perry, D.A., Hauer, F.R., Frissell, C.A., 2004. Postifre management on forested public lands of the western United States. Conserv. Biol. 18, 957–967.

Bond, M.L., Lee, D.E., Siegel, R.B., Ward Jr., J.P., 2009. Habitat use and selection by California Spotted Owls in a postfire landscape. J. Wildl. Manag. 73, 1116–1124.

Bond, M.L., Lee, D.E., Siegel, R.B., Tingley, M.W., 2013. Diet and home-range size of California spotted owls in a burned forest. Western Birds 44, 114–126.

Bradbury, S.M., 2006. Response of the post-fire bryophyte community to salvage logging in boreal mixedwood forests of northeastern Alberta, Canada. For. Ecol. Manag. 234, 313–322.

Brain, S., David, P., Ouimet, R., 2000. Impacts of wild fire severity and salvage harvesting on the nutrient balance of jack pine and black spruce boreal stands. For. Ecol. Manag. 137, 231–243.

Bull, E.L., Aubry, K.B., Wales, B.C., 2001. Effects of disturbance on forest carnivores of conservation concern in eastern Oregon and Washington. Northwest Sci. 75, 180–184.

Burnett, R.D., Preston, M., Seavy, N., 2012. Plumas Lassen Study 2011 Annual Report. U.S. Forest Service, Pacific Southwest Region, Vallejo, CA.

Burns, E.L., Lindenmayer, D.B., Stein, J.A., Blanchard, W., McBurney, L., Blair, D., Banks, S.C., 2014. Ecosystem assessment of mountain ash forest in the Central Highlands of Victoria, south-eastern Australia. Austral Ecol. 39, 1–14, http://dx.doi.org/10.1111/aec.12200.

Cahall, R.E., Hayes, J.P., 2009. Influences of postfire salvage logging on forest birds in the eastern Cascades. For. Ecol. Manag. 257, 1119–1128.

Caton, E.L., 1996. Effects of fire and salvage logging on the cavity-nesting bird community in north-western Montana. University of Montana, Missoula, Montana.

Clark, D.A., Anthony, R.G., Andrews, L.S., 2011. Survival rates of northern spotted owls in postfire landscapes of southwest Oregon. J. Raptor Res. 45, 38–47.

Clark, D.A., Anthony, R.G., Andrews, L.S., 2013. Relationship between wildfire salvage logging, and occupancy of nesting territories by Northern Spotted Owls. J. Wildl. Manag. 77, 672–688.

Cone, T., 2013. Nearly 40 Percent of Rim Fire Land a Moonscape. Associated Press news story, Fresno, California, USA, September 19, 2013.

Cruz, M.G., Sullivan, A.L., Gould, J.S., Sims, N.C., Bannister, A.J., Hollis, J.J., Hurley, R.J., 2012. Anatomy of a catastrophic wildfire: the Black Saturday Kilmore East fire in Victoria, Australia. For. Ecol. Manag. 284, 269–285.

DellaSala, D.A., Reid, S.B., Frest, T.J., Strittholt, J.R., Olson, D.M., 1999. A global perspective on the biodiversity of the Klamath-Siskiyou ecoregion. Nat. Area. J. 19, 300–319.

DellaSala, D.A., Karr, J.R., Schoennagel, T., Perry, D., Noss, R.F., Lindenmayer, D., Beschta, R., Hutto, R.L., Swanson, M.E., Evans, J., 2006. Postfire logging debate ignores many issues. Science 314, 51–52.

DellaSala, D.A., et al., 2013. Open letter to members of congress from 250 scientists concerned about post-fire logging. In: October 30, 2014, http://www.geosinstitute.org/images/stories/pdfs/Publications/Fire/Scientist_Letter_Postfire_2013.pdf.

DellaSala, D.A., Bond, M.L., Hanson, C.T., Hutto, R.L., Odion, D.C., 2014. Complex early seral forests of the Sierra Nevada: what are they and how can they be managed for ecological integrity? Nat. Area. J. 34, 310–324.

Donato, D.C., Fontaine, J.B., Campbell, J.L., Robinson, W.D., Kauffman, J.B., Law, B.E., 2006. Post-wildfire logging hinders regeneration and increases fire risk. Science 311, 352.

Duncan, S., 2002. Postfire logging: is it beneficial to a forest? Science Findings Issue 47, USDA Forest Service, Pacific Northwest Research Station.

Foster, D.R., Orwig, D.A., 2006. Preemptive and salvage harvesting of New England forests: when doing nothing is a viable alternative. Conserv. Biol. 20, 959–970.

Gibbons, P., van Bommel, L., Gill, M.A., Cary, G.J., Driscoll, D.A., Bradstock, R.A., Knight, E., Moritz, M.A., Stephens, S.L., Lindenmayer, D.B., 2012. Land management practices associated with house loss in wildfires. PLoS ONE 7, e29212.

Hanson, C.T., North, M.P., 2009. Post-fire survival and flushing in three Sierra Nevada conifers with high initial crown scorch. Int. J. Wildland Fire 18, 857–864.

Hanson, C.T., North, M.P., 2008. Postfire woodpecker foraging in salvage-logged and unlogged forests of the Sierra Nevada. Condor 110, 777–782.

Hanson, C.T., 2014. Conservation concerns for Sierra Nevada birds associated with high-severity fire. Western Birds 45, 204–212.

Hebblewhite, M., Munro, R.H., Merrill, E.H., 2008. Trophic consequences of postfire logging in a wolf–ungulate system. For. Ecol. Manag. 257, 1053–1062.

Hejl, S.J., Hutto, R.L., Preston, C.R., Finch, D.M., 1995. Effects of silvicultural treatments in the Rocky Mountains. In: Martin, T.E., Finch, D.M. (Eds.), Ecology and Management of Neotropical Migratory Birds. Oxford University Press, New York, pp. 220–244.

Hitchcox, S.M., 1996. Abundance and nesting success of cavity-nesting birds in unlogged and salvage-logged burned forest in northwestern Montana. M.S. Thesis, University of Montana, Missoula, Montana.

Hutto, R.L., 2006. Toward meaningful snag-management guidelines for postfire salvage logging in North America conifer forests. Conserv. Biol. 20, 984–993.

Hutto, R.L., 2008. The ecological importance of severe wildfires: some like it hot. Ecol. Appl. 18, 1827–1834.

Hutto, R.L., Gallo, S.M., 2006. The effects of postfire salvage logging on cavity-nesting birds. Condor 108, 817–831.

Karr, J.R., Rhodes, J.J., Minshall, G.W., Hauer, F.R., Beschta, R.L., Frissell, C.A., Perry, D.A., 2004. The effects of postifre salvage logging on aquatic ecosystems in the American West. Bioscience 54, 1029–1033.

Lee, D.E., Bond, M.L., 2015. Occupancy of California spotted owl sites following a large fire in the Sierra Nevada, California. The Condor 117, 228–236.

Lee, D.E., Bond, M.L., Siegel, R.B., 2012. Dynamics of breeding-season site occupancy of the California spotted owl in burned forests. Condor 114, 792–802.

Lindenmayer, D.B., Foster, D.R., Franklin, J.F., Hunter, M.L., Noss, R.F., Schmeigelow, F.A., Perry, D., 2004. Salvage harvesting policies after natural disturbance. Science 303, 1303.

Lindenmayer, D.B., Noss, R.F., 2006. Salvage logging, ecosystem processes, and biodiversity conservation. Conserv. Biol. 20, 949–958.

Lindenmayer, D.B., Ough, K., 2006. Salvage logging in the montane ash eucalypt forests of the central highlands of Victoria and its potential impacts on biodiversity. Conserv. Biol. 20, 1005–1015.

Lindenmayer, D.B., Burton, P.J., Franklin, J.F., 2008. Salvage Logging and Its Ecological Consequences. Island Press, Washington, D.C.

Lindenmayer, D.B., Hobbs, R.J., Likens, G.E., Krebs, C.J., Banks, S.C., 2011. Newly discovered landscape traps produce regime shifts in wet forests. PNAS 108 (38), 15887–15891. www.pnas.org/cgi/doi/10.1073/pnas.1110245108.

Lindenmayer, D.B., Blanchard, W., McBurney, L., Blair, D., Banks, S.C., Driscoll, D.A., Smith, A.L., Gill, A.M., 2014. Complex responses of birds to landscape-level fire extent, fire severity and environmental drivers. Divers. Distrib. 20, 467–477.

McGinnis, T.W., Keeley, J.E., Stephens, S.L., Roller, G.B., 2010. Fuel buildup and potential fire behavior after stand-replacing fires, logging fire-killed trees and herbicide shrub removal in Sierra Nevada forests. For. Ecol. Manag. 260, 22–35.

McIver, J.D., Starr, L., 2000. Environmental effects of postfire logging: literature review and annotated bibliography. USDA Forest Service Pacific Northwest Research Station, Portland, OR, Gen. Tech. Report PNW-GTR-486.

Noble, W.S., 1977. Ordeal by Fire. The Week a State Burned Up. Hawthorn Press, Melbourne.

Odion, D.C., Hanson, C.T., 2013. Projecting impacts of fire management on a biodiversity indicator in the Sierra Nevada and Cascades, USA: the Black-backed Woodpecker. Open For. Sci. J. 6, 14–23.

Odion, D.C., Strittholt, J.R., Jiang, H., Frost, E., DellaSala, D.A., Moritz, M., 2004. Fire severity patterns and forest management in the Klamath National Forest, northwest California, USA. Conserv. Biol. 18, 927–936.

Odion, D.C., Moritz, M.A., DellaSala, D.A., 2010. Alternative community states maintained by fire in the Klamath Mountains, USA. J. Ecol. 98, 96–105.

Odion, D.C., Hanson, C.T., Arsenault, A., Baker, W.L., DellaSala, D.A., Hutto, R.L., Klenner, W., Moritz, M.A., Sherriff, R.L., Veblen, T.T., Williams, M.A., 2014. Examining historical and current mixed-severity fire regimes in ponderosa pine and mixed-conifer forests of western North America. PLoS ONE 9, e87852.

Paine, R.T., Tegener, M.J., Johnson, E.A., 1998. Compounded perturbations yield ecological surprises. Ecosystems 1, 535–545.

Powers, E.M., Marshall, J.D., Zhang, J., Wei, L., 2013. Post-fire management regimes affect carbon sequestration and storage in a Sierra Nevada mixed conifer forest. For. Ecol. Manag. 291, 268–277.

Purdon, M., Brais, S., Bergeron, Y., 2004. Initial response of understory vegetation to fire severity and salvage logging in the southern boreal forest of Quebec. Appl. Veg. Sci. 7, 49–60.

Puerta-Pinero, C., Sanchez, A., Leverkus, M.A., Catro, J., 2010. Management of burnt wood after fire affects post-dispersal acorn predation. For. Ecol. Manag. 260, 345–352.

Reeves, G.H., Bisson, P.A., Riema, B.E., Benda, L.E., 2006. Postfire logging in riparian areas. Conserv. Biol. 20, 994–1004.

Russell, R.E., Saab, V.A., Dudley, J.G., Rotella, J.J., 2006. N to wildfire and postfire salvage logging. For. Ecol. Manag. 232, 179–187.

Saab, V.A., Dudley, J.G., 1998. Responses of cavity-nesting birds to stand-replacement fire and salvage logging in ponderosa pine/Douglas-fir forests of southwestern Idaho. USDA Forest Service, Ogden, Utah, Research paper RMRS-RP-11:1–17.

Serrano-Ortiz, P., Marañón-Jiménez, S., Reverter, B.R., Sánchez-Cañete, E.P., Castro, J., Zamora, R., Kowalski, A.S., 2011. Postfire salvage logging reduces carbon sequestration in Mediterranean coniferous forest. For. Ecol. Manage. 262 (2011), 2287–2296.

Shatford, J.P.A., Hibbs, D.E., Puettmann, K.J., 2007. Conifer regeneration after forest fire in the Klamath-Siskiyous: how much, how soon? J. For. April/May, 139–146.

Siegel, R.B., Tingley, M.W., Wilkerson, R.L., Bond, M.L., Howell, C.A., 2013. Assessing home range size and habitat needs of Black-backed Woodpeckers in California: Report for the 2011 and 2012 field seasons. Institute for Bird Populations, Point Reyes Station, CA.

Slosser, N.C., Strittholt, J.R., DellaSala, D.A., Wilson, J., 2005. The landscape context in forest conservation: integrating protection, restoration, and certification. Ecol. Restor. 23, 15–23.

Smucker, K.M., Hutto, R.L., Steele, B.M., 2005. Changes in bird abundance after wildfire: importance of fire severity and time since fire. Ecol. Appl. 15, 1535–1549.

Strittholt, J.R., Rustigian, H., 2004. Ecological issues underlying proposals to conduct salvage logging in areas burned by the Biscuit fire. Unpublished report, Conservation Biology Institute, Corvallis, OR.

Swanson, M.E., Franklin, J.F., Beschta, R.L., Crisafulli, C.M., DellaSala, D.A., Hutto, R.L., Lindenmayer, D., Swanson, F.J., 2011. The forgotten stage of forest succession: early-successional ecosystems on forest sites. Front. Ecol. Environ. 9, 117–125.

Taylor, C., McCarthy, M.A., Lindenmayer, D.B., 2014. Nonlinear effects of stand age on fire severity. Conserv. Lett. 7, 355–370.

Thompson, J.R., Spies, T.A., Ganio, L.M., 2007. Reburn severity in managed and unmanaged vegetation in a large wildfire. PNAS 104, 10743–10748. http://dx.doi.org/10.1073/pnas.0700229104.

Titus, J.H., Householder, E., 2007. Salvage logging and replanting reduce understory cover and richness compared to unsalvaged-unplanted sites at Mount St. Helens, Washington. Wes. N. Am. Naturalist 67, 219–231.

Turner, M.G., Dale, V.H., 1998. Comparing large, infrequent disturbances: what have we learned? Ecosystems 1, 493–496.

USFS, BLM, 1994. Record of Decision for Amendments to Forest Service and Bureau of Land Management Planning Documents Within the Range of the Northern Spotted Owl. USDA Forest Service and USDI Bureau of Land Management, Portland, OR, http://www.reo.gov/library/reports/newroda.pdf.

USFS, 2001. Forest Service Roadless Conservation Final Environmental Impact Statement. USDA Forest Service, Washington, D.C.

USFS, 2002. Burned area Emergency Rehabilitation Report for the Biscuit Fire. Rogue and Siskiyou National Forests, Medford, Oregon.

USFS, BLM, 2004. Final Environmental Impact Statement: The Biscuit Fire Recovery Project. Rogue River-Siskiyou National Forest, Josephine and Curry Counties, OR.

USFS, 2004. Sierra Nevada Forest Plan Amendment, Final Supplemental Environmental Impact Statement and Record of Decision. U.S. Forest Service, Pacific Southwest Region, Vallejo, California, USA.

USFS, 2014a. Rim Fire Recovery, Final Environmental Impact Statement and Record of Decision. U.S. Forest Service, Stanislaus National Forest, Sonora, California, USA.

USFS, 2014b. Rim Fire Recovery, Vegetation Report. U.S. Forest Service, Stanislaus National Forest, Sonora, California, USA.

USFS, 2014c. Big Hope Fire Salvage and Restoration Project, Environmental Assessment. U.S. Forest Service, Tahoe National Forest, Nevada City, California, USA.

USFS, 2014d. Aspen Recovery and Reforestation Project, Environmental Assessment. U.S. Forest Service, Hi Sierra Ranger District, Sierra National Forest, Prather, California, USA.

USFS, 2014e. Power Fire Reforestation Project, Proposed Action. U.S. Forest Service, Eldorado National Forest, Placerville, California, USA.

Weatherspoon, C.P., Skinner, C.N., 1995. An assessment of factors associated with damage to tree crowns from the 1987 wildfire in northern California. For. Sci. 41, 430–451.

Chapter 12

The Rising Costs of Wildfire Suppression and the Case for Ecological Fire Use

Timothy Ingalsbee[1] and Urooj Raja[2]
[1]*Firefighters United for Safety, Ethics, and Ecology, Eugene, OR, USA,* [2]*Department of Sociology, University of Colorado, Boulder, CO, USA*

12.1 BURNED AND BUSTED: THE RISING COST OF FIGHTING FIRES

In 1908, just 3 years after the creation of the US Forest Service (USFS), a US Congressional legislative rider established the Forest Fires Emergency Act that gave the fledging federal agency the authority to engage in deficit spending for wildfire suppression operations (Pyne, 1997). This extraordinary power freed up the agency's suppression program from the normal budgetary constraints that affect every other federal agency or program. It was first used during the infamous "Big Blowup" of the 1910 fires when the USFS spent over $1 million that year. A century later, a million dollars is the average daily cost of suppressing large wildfires, and the average annual expenditures exceed one billion tax dollars (Headwaters Economics, 2009). The agency repeatedly overspends its annual appropriated suppression budget, receiving supplemental appropriations from Congress with little oversight of or accountability for how the money is spent (Dombeck et al., 2004). But as suppression expenditures continue to rise, so have the number of hectares burned, the average size of wildfires, and the numbers of homes destroyed and firefighter lives lost. Equally important, accumulating evidence of the adverse ecological impacts of fighting fires raises serious questions about the effects as well as the effectiveness of wildfire suppression. What are the American people getting in return for their annual expenditures of "blood and treasure" in the seemingly endless "war" against wildfire?

Show Me the Money: Poor Data on Suppression Costs

Rising suppression expenditures have been the subject of dozens of peer-reviewed research articles, internal government reports, inspector general

The Ecological Importance of Mixed-Severity Fires: Nature's Phoenix

audits, blue-ribbon commission analyses, and pundit opinion essays, each of them focusing on various “cost drivers.” By the mid 1990s these reports had cumulatively offered hundreds of recommendations for changes in policies or practices to contain costs or improve fiscal accountability. Analysis of the economic efficiency of wildfire suppression depends on accurate data, but many reports complained that data were inaccessible, lost, or never recorded; were poorly aggregated; and were inconsistently coded or miscoded, and these defects in the data made accurate economic assessments of individual wildfire events or whole wildfire seasons highly problematic (Schuster et al., 1997; Gebert et al., 2008). Also, cost data from state and local agencies are often not obtainable—a growing problem for economic analysts given the increasing number of large-scale, multijurisdictional wildfires (Taylor et al., 2013). Finally, the USFS and Department of Interior land management agencies use different accounting systems that are highly vulnerable to errors in coding various cost items, especially some high-cost items provided by private contractors (Gebert et al., 2008). Efforts to analyze suppression expenditures, gain more fiscal accountability, and attain more economic efficiency in wildfire operations are undermined by a lack of reliable data.

In the context of current fiscal austerity politics where government spending is closely scrutinized and most agency budgets are getting slashed, the fact that so much taxpayer money is being spent on suppression with such little transparency or accountability raises serious questions about how and why this program persists in its current form. Working with the most recent and reliable cost data available, summaries of some our findings on costs are provided (see Tables 12.1–12.4; see the methodological endnote on our data set). The data focus on the 11 western states in the United States because they generate the bulk of wildfire hectares and firefighting expenditures and illustrate the magnitude of the problem. Analysis focuses on the USFS because it is responsible for over 70% of all federal suppression expenditures. What follows are analyses drawn from the peer-reviewed literature discussing various factors that are driving the rising cost of wildfire suppression for the USFS, with extra emphasis on the “human dimensions” of fire management. These human dimensions not only are a major source of the problem but could also offer potential solutions with the most immediate prospects for containing suppression costs.

Size Matters: Larger Fires Mean Larger Costs

Despite the unreliability of suppression cost data, firefighting expenditures clearly have been rising. According to the National Interagency Fire Center (NIFC) (2014), the official keeper of wildfire statistics, suppression costs since 1985 have totaled more than $25.4 billion to fight 2.1 million fires that burned across 83,324,774 ha, with the lion’s share of these expenses ($19.2 billion) spent by the USFS. The 10-year average for annual federal suppression expenditures increased from $620 million in the 1990s to $1.6 billion in the 2000s (inflated

TABLE 12.1 Total Reported Suppression Costs by Year for Fiscal Years 2000 to 2012

Year	Reported Suppression Costs[a]	All Reported Hectares Burned[b]
2000	$795,438,685	2,129,406
2001	$552,558,849	1,053,848
2002	$1,123,052,397	1,626,718
2003	$944,419,924	1,089,209
2004	$509,178,099	558,149
2005	$471,397,270	1,535,598
2006	$1,110,521,349	2,431,536
2007	$1,094,872,834	2,643,259
2008	$1,257,495,618	1,213,015
2009	$895,966,881	617,883
2010	$448,781,350	640,105
2011	$683,773,629	1,441,411
2012	$1,212,528,811	2,840,003

[a]*Costs are not adjusted for inflation.*
[b]*Conversion from acres to hectares (1 hectare = 2.47 acres); numbers may not match exactly because of rounding.*

to constant 2009 dollars) (Gebert and Black, 2012). The years 2002, 2006-2008, and 2012 all surpassed $1 billion in expenditures (Table 12.1). From 2010 to 2013 these expenditures increased further to nearly $2 billion per year (NIFC, 2014).[1]

Suppression costs began increasing when wildfire activity significantly increased in the late 1980s, measured by the growth in the number of hectares burned, the number of large wildfires, and the average size of large wildfires, continuing into the 2000s (Calkin et al., 2005; see Table 12.1). The average annual area burned from 1970 to 1986 was approximately 115,535 ha, but from 1987 to 2002 this increased to over 404,686 ha per year (Calkin et al., 2005). Along with that increase in burned area came greater suppression expenditures. However, the correlation between area burned and suppression costs is not so simple. From 97% to 99% of all wildfires are aggressively suppressed and contained at a small size (<2023 ha). Holmes et al. (2008) reported that from 1980

1. Figures in Tables 12.1 to 12.4 use data provided by the National Wildfire Coordinating Group (NWCG). Cost figures provided by NIFC are higher than those from NWCG. See Methodological Endnote.

TABLE 12.2 Most Expensive Fire by Year from Fiscal Year 2000 to 2012

Year	Fire	State	Suppression Cost[a]	Hectares[b]
2000	Big Bar Complex	CA	$75,790,000	57,040
2001	Clear Creek Complex	ID	$71,500,000	87,801
2002	Biscuit	OR	$152,658,738	202,169
2003	Grindstone Complex	CA	$46,900,000	81
2004	Old	CA	$42,336,057	36,940
2005	Blossom Complex	OR	$28,742,207	6313
2006	Day	CA	$78,000,000	65,843
2007	Zaca Two	CA	$122,553,385	97,208
2008	Klamath Theater	CA	$126,086,065	77,715
2009	BTU Lightning Complex	CA	$95,000,000	26,303
2010	Station	CA	$95,510,000	64,983
2011	Wallow	AZ	$109,000,000	217,741
2012	Chips	CA	$53,300,000	30,526

[a]*Costs are not adjusted for inflation.*
[b]*Conversion from acres to hectares (1 hectare = 2.47 acres); numbers may not match exactly because of rounding.*

TABLE 12.3 Total Costs Ranked by State from Fiscal Year 2000 to 2012

State	Total Costs for 2000–2012[a]	Total Area Burned during 2000-2012 (Hectares)[b]
California	$4,647,895,621.00	3,265,225
Oregon	$1,362,781,801.00	1,964,984
Montana	$1,128,991,325.00	2,153,954
Idaho	$892,061,813.00	3,350,861
Arizona	$685,668,878.00	1,644,756
Washington	$625,514,507.00	808,961
New Mexico	$444,109,025.00	1,654,984
Colorado	$369,132,917.00	542,788
Nevada	$335,074,723.00	2,697,938
Utah	$331,327,216.00	1,089,920
Wyoming	$277,427,870.00	629,582

[a]*Costs are not adjusted for inflation.*
[b]*Conversion from acres to hectares (1 hectare = 2.47 acres); numbers may not match exactly because of rounding.*

TABLE 12.4 Most Expensive Total Cost Years by State from Fiscal Year 2002 to 2012

State	Year	Suppression Costs[a]	Area Burned (Hectares)[b]
Arizona	2011	$211,014,247	416,017
California	2008	$1,093,083,880	712,724
Colorado	2002	$146,181,060	202,056
Idaho	2007	$241,152,064	1,012,635
Montana	2003	$311,375,413	305,330
Nevada	2006	$61,873,505	591,871
New Mexico	2011	$130,180,833	482,011
Oregon	2002	$372,771,201	426,524
Utah	2002	$68,200,851	109,528
Washington	2006	$114,128,342	110,649
Wyoming	2012	$94,221,798	147,094

[a]*Costs are not adjusted for inflation.*
[b]*Conversion from acres to hectares (1 hectare= 2.47acres); numbers may not match exactly because of rounding.*

to 2002 approximately 94% of fire suppression costs used for fires on national forest lands resulted from a mere 1.4% of all wildfires. In 2006 the 20 biggest wildfires accounted for 11.2% of the nearly 4,046,856 ha burned nationwide, but they cost nearly 30% of the $1.5 billion spent by the USFS (ILWCP, 2007). In 1999 the USFS spent over 30% of its national suppression budget fighting two lightning-caused wilderness fires in northern California (SOLFC, 2000). In fact, the largest costs are associated with the largest wildfires. These so-called megafires (see Chapter 2) function as mega budget-busters (Table 12.2).

Total suppression costs paid by the USFS are rising at an annual rate of 12-15%, and those paid by the National Park Service and Bureau of Land Management are rising by 10% (Gebert et al., 2008). This suggests an inflationary rise in costs per hectare, but a 2005 study found that from 1970 to 2002 suppression costs consistently averaged around $308 per hectare nationally (Calkin et al., 2005). Regardless, per-hectare suppression costs can vary substantially among different vegetation types and geographic regions. Wildfires in forested or slash-covered areas are generally two-thirds more expensive to suppress than fires in shrublands or grass-covered areas (Schuster et al., 1997). The California region averaged the highest costs during the period of 1995-2004, at $1039 per hectare (North et al., 2012). In general, smaller wildfires have higher per-hectare costs than larger wildfires because, for large wildfires (defined as ≥2023 ha) (NWCG, 2014), a certain “economy of scale” operates that spreads some fixed suppression costs over a larger landbase (Abt et al., 2008). There can, however, be huge unburned portions of land within the perimeters of large wildfires, and suppression actions mostly occur at the outermost edge of a wildfire. If costs were calculated based on only the specific hectares where suppression actions actually occurred, then the expanse of unburned and unmanaged land inside large wildfires would be irrelevant from a cost standpoint, and per-hectare costs of large wildfires might equal or surpass those of small fires.

Thus suppression expenditures are growing along with the growth in area burned, but the relationship between fire size and suppression costs is not so simple. California (USFS region 5) and Oregon (USFS region 6) have the highest suppression costs compared with all other states and regions but do not necessarily have the most wildfire activity measured either in number of fires or area burned (Table 12.3). Even within the same region, two wildfires with similar conditions of weather, vegetation cover, and terrain can have wide differences in expenditures. In some years the number of fires and burned hectares were below the 10-year averages but suppression costs were above average. So what factors are causing wildfires to grow larger in size in places and suppression costs to increase above the rate of inflation? A host of factors that can be categorized as socioenvironmental, institutional, and operational can affect wildfire activity, firefighting actions, and their subsequent costs. Discussion of these factors follows.

12.2 SOCIOENVIRONMENTAL COST FACTORS

Compounded Interest: Fire Exclusion and Fuel Accumulation

One of the most common assertions in the literature is that large-scale, high-intensity wildfires are burning through unnaturally high fuel loads that have accumulated as a result of fire exclusion from past fire suppression. The ecological impact of past fire suppression varies significantly according to ecosystem and fire regime type. In general, a common view is that many low-elevation dry forest types with a frequent fire regime that have “missed” several fire return cycles (based on average cycles but not variability of cycles) because of past firefighting actions have been the most altered by fire exclusion and may have excess fuels accumulation compared to historical conditions. However, support for this by empirical data is limited or equivocal (see Chapter 1). Both dry and moist mixed-conifer forests and dry ponderosa pine (*Pinus ponderosa*) forests were historically characterized by mixed-severity fire in most, but not all, regions of the western United States, based on dozens of published studies using multiple distinct lines of scientific evidence (Chapter 1). Higher-elevation moist forest types with an infrequent fire regime have been less affected by suppression-caused fire exclusion, and in these systems it is more widely understood that high fuel loads and high-intensity wildfires are natural processes (see Chapter 1). Regardless, large-scale, high-intensity wildfires often are blamed on alleged excess accumulation of fuels.

One of the main claims by proponents of fuels reduction are that "treatments" will compensate for past fire exclusion, reduce the risk of high-severity wildfire, and the costs of fire suppression. Most fuel treatments are designed not to eliminate the need for suppression actions but to facilitate them (Thompson et al., 2013b). When severe fire weather conditions exist, however, high-intensity wildfire can burn over or breach most fuel treatments and make firefighting unsafe or ineffective (Reinhardt et al., 2008; Mercer et al., 2008). Indeed, though there is anecdotal evidence and some modeling exercises that demonstrate that some fuel treatments can reduce fire severity within treated stands, piecemeal fuels reduction projects at the landscape level have not had an appreciable effect on wildfire activity (Williamson, 2007). In recent years the Office of Management and Budget has been cutting funds for fuels reduction, particularly in the National Park Service, because it claims there is no evidence that the millions of dollars invested in treatments to date have actually reduced suppression costs. Thus, although legitimate scientific debate continues over the legacy of past fire suppression and fire exclusion and their effects on current fuels accumulations or wildfire activity (as discussed throughout this book), there is little debate about the fact that recent fuels reduction treatments have not had any real impact on reducing suppression costs (Table 12.4).

Sprawling Suburbs: Wildfire in the Wildland-Urban Interface

Along with excess fuel loads, expansion of the wildland-urban interface (WUI) has been widely blamed for the rapid increase in suppression costs (Gorte, 2013; Liang et al., 2008; Gebert et al., 2007). Public expectations and politics often pressure fire managers to do whatever they can to save homes threatened by wildfire, even if private property protection is not within the scope of federal responsibility, and even if suppression costs surpass the monetary value of the structures being protected (Hesseln, 2001). Despite the billions of dollars spent firefighting, hundreds of homes are burned by wildfires each year. For example, from 1999 to 2010, over $16 billion in federal funds were spent fighting wildfires, yet an annual average of 1179 homes were destroyed from wildfires during this same period (Gude et al., 2013). In the 2010s this more than doubled to an average of 2970 homes burned each year, with over 5000 homes burned in 2007, 2011, and 2012 (Stockmann et al., 2010; Headwaters Economics, 2014).

Several studies have attempted to calculate the influence of WUI protection on suppression costs. A study commissioned by the USFS estimated that approximately one-third of its suppression expenditures went towards WUI protection (SIPFSC, 1994). Based on interviews with USFS managers, estimates from 50% to 95% of total firefighting expenditures on National Forest lands were related to WUI protection (OIG, 2006). However, interviewees based their claims on beliefs that all wildfires can or will eventually threaten private property and homes. Gebert et al. (2007) determined that suppression costs can increase by an average of $1 million for each additional 125 homes near a fire, but their definition of "nearness" included homes up to 32 km away from wildfires, far outside conventional definitions of the WUI.

In some of the most rigorous research assessing the costs of WUI protection, the nonprofit organization Headwaters Economics (2009, 2014) determined that, within the state of Montana alone, the annual cost of WUI protection averaged $28 million, but it estimated that this figure could grow to $40 million by 2025 if rural housing development continues unabated and could explode to between $61 and $113 million under the effects of climate change. Applied nationally, private property protection currently ranges from $630 million to $1.2 billion annually, but a 50% growth in new housing development could increase annual suppression costs up to $2-4 billion.

The post-World War II expansion of logging roads is often an explanation for the alleged increased effectiveness of fire suppression because large amounts of crews and equipment can be transported at relatively low cost. Ironically, however, the presence of roads is associated with higher suppression costs compared with areas with no roads because roads enable managers to order more expensive resources like engines and bulldozers (Gude et al., 2013). Roads also are more often located near high-value assets like homes, private property, or commercial timber stands, so generally more suppression resources are used in areas with roads than in areas without them.

Calkin et al. (2005) pointed out that although the WUI has been expanding since the 1950s, and firefighting agencies are focusing more resources on structure protection, the steady growth of the WUI does not account for the rapid surge in firefighting expenditures that began in the late 1980s. To understand that shift, attention must turn to the effects of weather and climate on the growing frequency of large-scale fires.

The Heat Is On: Global Warming and Wildfires

The increase in suppression costs are highly correlated with the growth in area burned, especially by large wildfires, and large wildfires are primarily driven by weather and climate conditions (Gebert et al., 2008; see also Chapters 1 and 2). Prolonged droughts, high temperatures, and high-wind events all are associated with high wildfire activity in terms of ease of ignition, rapid spread, and high-fireline intensity. Westerling et al. (2006) demonstrated that a significant shift in wildfire activity began in 1987, when wildfire seasons lengthened an extra 2 months because of earlier spring snowmelt and later autumn snowfall at higher elevations, an effect attributed to climate change more than fuels accumulation from past fire exclusion. Contrary to widespread popular belief, the larger size of wildfires since the 1980s has not necessarily resulted in higher severity of wildfires in most forested regions of the western United States (see Chapters 1, 2, 9). The effects of climate change on wildfire activity vary among and within different regions because of complex interactions between temperature, precipitation, relative humidity, and their effects on soil moisture and vegetation cover (see Chapter 9). Some areas experience more frequent severe fire weather conditions and lightning storms, but others have more precipitation throughout the typical fire season. Nationwide, federal agencies predict that climate change will result in wildfires burning 4-5 million hectares annually (USDA-FS et al., 2009) with a predictable rise in suppression costs absent any changes in current policies or practices.

12.3 THE HUMAN DIMENSIONS OF WILDFIRE SUPPRESSION COSTS

The three above-mentioned socioenvironmental factors are the most cited factors fueling increased suppression expenditures (Hand et al., 2014), but Canton-Thompson et al. (2008) argue that the biophysical features that affect fire behavior (e.g., vegetation, weather, and terrain) can explain only half of the variation in suppression costs. Among wildfires that share similar physical characteristics but have significant differences in suppression expenditures, the "human dimensions" of fire management, especially the attitudes and choices of fire managers, play a significant role in cost differences. These human dimensions can be categorized in terms of "external" sociocultural factors and "internal" institutional and operational factors that often pressure fire managers to opt for aggressive suppression actions instead of modified suppression or fire use strategies.

12.4 EXTERNAL SOCIOCULTURAL COST FACTORS

The Smokey Bear Syndrome: Public Pressure for Suppression

"External" influences on suppression costs come from outside federal agencies. A century of anti-wildfire propaganda and pro-suppression policies has created what some fire managers have dubbed the "Smokey Bear syndrome" in American culture. It causes many people to demand that firefighters put out all wildfires. There are strong public expectations that all wildfires should be prevented and/or suppressed, that firefighters will always be effective in their actions, and that no expense should be spared in efforts to protect human life or private property (also see Chapter 13). In the false belief that firefighters actually extinguish large wildfires, extreme public pressure is put on fire managers to use costly and extraordinary suppression methods, even when managers suspect that these efforts will have no meaningful effect on the wildfire and will likely be an economic waste.

Residents of the WUI in particular often demand full suppression of wildfires but do not understand the risks or complexities of firefighting, nor the ecological impacts of fire exclusion (Calkin et al., 2011). Black et al. (2010)

discovered that even citizens who support the use of fire in land management lose their tolerance for wildfires after enduring a few weeks of disruption to their everyday routines while breathing lots of smoke. In addition, there is often pressure on federal managers to hire local private contractors for suppression crews, equipment, or supplies, and fire managers often complain that these private resources are more expensive and require more oversight than public agency resources (Canton-Thompson et al., 2008). Related to this, significant pressure for aggressive suppression can come from state and local cooperators working on multijurisdictional wildfires (i.e., fires burning on federal, state, and private lands). Most state forestry agencies have mandates for full suppression and total fire exclusion and suspect that fire use strategies on federal lands are intended to pass suppression costs onto the states. The legacy of Smokey Bear's anti-wildfire message thus afflicts not only the lay public but also fire management professionals who believe the only good fire is a dead-out fire.

Hot Air: Politicians and the Press

Another form of external pressure pushing fire managers to select costly suppression strategies or tactics comes from local, state, or national politicians. Politicians continually intervene in federal fire management, creating laws or policies that ignore the professional expertise of agency fire scientists and managers and undermine science-based fire management (Fifer and Orr, 2013). Politicians also are prone to public grandstanding in the media during wildfire events, pressuring agencies to aggressively fight fires. Firefighters call their actions "political shows" or "political smokes" to describe situations when, under pressure from politicians or local communities to put out a wildfire, managers select resources, strategies, or tactics that will likely be economically inefficient or ineffective but demonstrate to external audiences that aggressive actions are being undertaken (Cart and Boxall, 2008). Examples include the use of aerial retardant drops that have no chance of success during fire behavior conditions or suppressing interior hotspots that have no chance of escaping the wildfire perimeter just to reduce the public's fear of fire spread. Donovan et al. (2011) were able to quantify the cost of external political or media pressure on fire managers. Increasingly, managers are feeling whipsawed between two opposing political pressures: while wildfires are burning there is intense pressure from local politicians to suppress fires at all costs, but then after fire season is over national politicians criticize agencies for failing to contain costs. In the current politicocultural environment, pressures for managers to aggressively fight wildfires are prevailing over pressures to avoid excessive spending of taxpayer dollars (also see Chapter 13).

12.5 INTERNAL INSTITUTIONAL COST FACTORS

Red Ink: Skewed Budgets and Perverse Incentives

One of the most oft-cited "internal" institutional drivers of rising suppression costs is the system for Congressional appropriations for the USFS that authorizes deficit spending for firefighting. The usual practice is that when the agency's annual appropriation for suppression is exhausted, the agency starts "borrowing" funds from accounts in nonsuppression programs to keep money flowing to firefighting efforts then later asks Congress for supplemental appropriations to replenish the transferred funds. Even when those funds for nonfire programs are fully reimbursed (and sometimes this has not occurred), the disruption caused by the budget transfers causes problems for planning and implementing many research and restoration projects. Relying on emergency funds and supplemental appropriations provides little incentive for cost containment and is a significant human factor in increasing suppression expenditures (Donovan et al., 2008; Donovan and Brown, 2005).

Rising suppression costs and increases in annual and supplemental Congressional appropriations have combined to create an extreme imbalance in the USFS budget that some critics charge is changing the focus of the agency's mission away from managing forests toward fighting fires (the Preface refers to the Forest Service becoming the de facto Fire Service). While funding for fire management has been rapidly growing, the budgets for almost all other nonfire programs in forest management, restoration, research, and recreation have been shrinking. For example, the proportion of the agency's budget devoted to fire management was only 16% in 1995 but swelled to 42% in fiscal year 2014 (USDA-FS, 2014). The numbers of wildfires and acres burned were, surprisingly, below the 10-year average in the 2014 fire season, but regardless, the USFS spent over 50% of its total appropriated budget on firefighting—the first time it passed this threshold, but probably not the last.

Congress almost always delivers on agency requests for supplemental appropriations for firefighting expenditures with almost no questions asked. This has nurtured an "open checkbook" mentality by fire managers, leading them to choose aggressive suppression strategies and order expensive firefighting resources knowing that Congress will eventually pay all of their firefighting bills (Ingalsbee, 2000). This attitude is one of the reasons getting managers to contain costs has been difficult (Snider et al., 2006; Donovan and Brown, 2007; Donovan et al., 2008). Worse, the agency's skewed budget and deficit spending authority has set up a system of "perverse incentives" that encourages the USFS to focus on reactive fire suppression rather than proactive ecosystem restoration or recreation programs (Ingalsbee, 2010). Such programs must be funded by fixed budgets and involve more legal requirements (e.g., environmental analysis and public involvement), but firefighting actions have almost no budgetary limits or legal constraints because of their "emergency" status.

This ongoing practice of Congress failing to appropriate sufficient funds for suppression and then the USFS transferring money from nonfire programs to pay for firefighting expenditures sets up a self-reinforcing system in which the agency keeps reacting to wildfire activity (even in below-average years, as in 2014) while avoiding root problems (e.g., the historic fire deficit, and risks of fire exclusion on rare and imperiled fire-dependent species; see

Chapters 1–5) or the implementation of long-term effective solutions (e.g., wildland fire use to restore mixed-severity fire to ecosystems). In the business management literature this is a classic dilemma known as the "firefighting trap." De Neufville et al. (2013) demonstrated that this emphasis on fire suppression in the short term and neglect of fire restoration over the long term inevitably leads to higher costs over time. Avoiding the vicious cycle of the firefighting trap will be possible only if resources are shifted away from suppression and toward wildland fire use (or "managed wildfire"), ecological restoration projects, and education—but that will be difficult given entrenched mind-sets and vested political and economic interests devoted to fighting fires (Chapter 13). As long as Congress continues allowing deficit spending for firefighting, the USFS will continue to focus on attacking nearly every fire rather than working on root causes or real solutions to rising suppression expenditures.

Tears for Fears: Risk-Averse Managers

External social and political pressures combine with internal agency dynamics to create another suppression cost driver: risk-averse managers. Managing wildfires is inherently risky, hazardous work, and these risks and hazards can be mitigated but never completely eliminated. Current fire management policies are predicated upon "sound risk management," with firefighter safety the highest priority; this involves assessing the exposure of firefighters to safety risks versus the potential effects of wildfire on social and ecological values at risk. New tools for assessing risk are emerging; however, there are several challenges in getting managers to use these new tools in decision-making. First, most managers focus on the immediate or short-term risks of what they perceive as wildfire "damage" rather than the long-term risks of continued fire exclusion to, for example, rare and declining fire-dependent wildlife that benefit from mixed- and high-severity fire effects (see Chapters 2–5). Likewise, managers may try to limit firefighting costs and firefighter exposure by attempting to keep fires small, but this ignores the opportunity costs of failing to get more hectares burned when conditions favor beneficial fire effects. Managers almost always opt to assume short-term risks and costs while externalizing long-term risks and costs to future managers and firefighters.

Second, among managers is a widespread belief that a double standard exists in terms of the risks and consequences of managing wildfires with aggressive suppression versus fire use strategies. Managers believe that if they adhere to approved policies and procedures, then all will be forgiven if the fires they aggressively fight unfortunately result in accidents. If wildfires managed with fire use strategies exceed their desired size, have high-severity effects, or—worse—result in firefighter fatalities or property destruction, then many managers fear that that their agency would not support them, they could lose their careers, and they could be held personally liable for those accidents

(Canton-Thompson et al., 2008). This fear is not entirely unfounded, and it is another source of risk aversion by managers.

There is also a misperception among many managers that increasing firefighter safety and reducing suppression costs are contradictory goals. Consequently, risk-averse managers will order excessive amounts of suppression resources or select more expensive capital-intensive resources like air tankers and engines rather than rely on less expensive labor resources like handcrews to contain and control wildfires quickly and thereby limit firefighter exposure (Calkin et al., 2005). Moreover, managers do not often recognize that foregoing backcountry firefighting reduces unnecessary risks to firefighters. Ironically, much firefighter exposure to hazards occurs during "mop up" (extinguishing all visible smoke and heat sources) that begins after a wildfire has been contained and is no longer spreading. Managers' aversion to the risk of fire escaping containment lines often leads them to order intensive and extensive mop-up activity in which firefighters face increased safety risks from falling snags and health hazards from inhaling large amounts of smoke, ash, and dust. Prolonged mop-up raises total suppression expenditures and can cost several times what other fuels reduction treatments would cost (Gonzalez-Caban, 1984). Allowing more wildfires to burn themselves out over time would raise risks of potential escape but could also result in more fire restoration with less soil disturbance and less risk to firefighters.

Risk aversion extends beyond individual managers' fear of accidental outcomes, but it also includes the agencies' generalized fear of negative publicity. Donovan et al. (2011) demonstrated that managers increase suppression spending in response to media coverage that heightens public fears of wildfire or generates citizen criticisms over alleged government incompetence in fighting fire. The tremendous flexibility in strategies and tactics that federal fire management policy allows is consistently underutilized because much of the public cares only that wildfires are "put out" as quickly as possible. Fire use strategies that achieve management objectives and avoid all accidents can still face public condemnation while smoke plumes stoke people's fear of fire. This causes risk-averse agency officials to avoid potentially negative public reactions by aggressively fighting nearly all fires. Again, this essentially passes on extra risk to future managers, firefighters, and ultimately the ecosystems and species that are negatively affected by fire exclusion.

12.6 OPERATIONAL FACTORS: SUPPRESSION STRATEGIES AND TACTICS

Operational factors driving suppression costs are the least discussed issue in the peer-reviewed literature, but the human factors influencing the objectives, strategies, and tactics that fire managers use to respond to wildfires have huge cost implications. According to incident commanders (the leaders of the teams managing wildfire operations), the number 1 driver of suppression costs is the decisions made by line officers (agency administrators such as regional foresters, forest supervisors, or district rangers) (Canton-Thompson et al., 2008). Incident commanders may recommend a certain strategy and set of tactics, but it is the line officers who make the final decisions authorizing suppression objectives and their estimated costs. Unlike some of the big socioenvironmental problems that are driving up suppression costs and will take many years to solve, the operational decisions guiding wildfire responses have the most potential for immediate cost reductions.

Wildfire responses fall into two basic objectives and five strategies. Protection objectives are intended to limit fire spread and exclude it from burning certain areas. Three strategies include (1) direct or full suppression, which is the most aggressive strategy that attempts to contain and control wildfires at their smallest size feasible; (2) modified suppression, which combines some fireline construction with the use of preexisting natural fire barriers (e.g., bodies of water, rocky areas) to confine wildfires to predefined areas but does not necessarily minimize the area burned; and (3) limited suppression, which is the least aggressive strategy that does not build containment lines along a wildfire's entire perimeter but instead attempts "point protection" tactics to keep fire from burning specific high-value places (e.g., structures), while allowing the fire to spread across a larger area. Cost containment goals may influence the selection of modified and limited suppression strategies, but most often they are chosen because of firefighter safety concerns or a lack of sufficient resources for a more aggressive strategy (Black et al., 2010; Gebert and Black, 2012).

Resource benefit or restoration objectives, on the other hand, aim to promote the benefits of fire on specific natural resources, native species, habitats, or landscapes. The two main strategies for managing wildfires for resource benefits include (1) area monitoring, whereby the fire is permitted to burn freely within a defined area but no management actions are attempted beyond observing and mapping; and (2) area management, which includes monitoring plus other minimum-impact management actions designed to delay, direct, or check fire spread in order to keep the fire within a prescribed area or protect specific features within a wildfire's perimeter (Black et al., 2010; Gebert and Black, 2012).

In 2009 the Obama Administration issued new guidance for implementing the Federal Wildland Fire Management Policy that gives managers tremendous flexibility to manage wildfires for both protection and restoration objectives simultaneously using all the strategies and tactics available in modern fire management. Concretely, there might not be much difference in terms of management techniques and environmental effects between modified suppression versus area management strategies, even though they represent fundamentally different objectives. In terms of costs, managing wildfires for restoration objectives or using less aggressive strategies logically should reduce overall expenditures; however, this assumption has rarely been tested because the USFS continues to apply aggressive suppression strategies on roughly 97% of all wildfire ignitions on National Forest lands and boasts that it is successful in controlling wildfire during the initial attack about 98% of the time (Tidwell, 2014). The few studies that have probed the connection between operational strategies and expenditures have come up with some surprising findings, though.

In a pioneering study of the costs of the full range of fire management strategies, Gebert and Black (2012) found that the average costs per hectare of direct suppression ($730) are higher than modified ($404) and limited ($302) suppression strategies, whereas the average costs per hectare of resource benefit strategies is much lower ($127). Measured on a daily basis, the cost variations are similar although not as great, with direct suppression the most expensive at an average $335,000 per day, but this is only 1.2 times the cost of modified suppression. Considering the big picture, modified suppression can be the most expensive strategy because it allows wildfires to grow larger and especially burn longer—nearly twice as long as direct suppression incidents—and this can considerably increase total expenditures. Indeed, the average total cost for modified suppression incidents is $7.3 million, whereas costs for direct suppression are $4.3 million. Surprisingly, the costs for limited suppression strategies or resource benefit objectives were $3.7 million and $3.6 million, respectively. Less-than-full suppression strategies still cost a considerable amount of money! (but see discussion in "Saving Green in the Black" below).

Focusing on average daily costs and total expenditures per suppression incident seems to justify the rationality of fighting fires aggressively to keep them small or of short duration and thereby limit total expenditures. In fact, in 2012 Deputy Forest Service Chief Jim Hubbard suspended fire use on all National Forests for the duration of that wildfire season, claiming that it was a cost containment strategy. Firefighters were dispatched to wildfires even in remote wilderness areas. In one instance, firefighters spent a whopping $425,000 to keep a lightning fire limited to 0.04 ha in size deep inside the Bob Marshall Wilderness Area, a place where wildfires had most often been monitored rather than fully suppressed. The USFS spent $1.3 billion fighting fires during the 2012 fire season, exceeding its suppression budget by $440 million, and forced Congress to provide emergency supplemental appropriations to cover the agency's budget deficit (FUSEE, 2013). Apparently, going all-out on direct suppression does not necessarily contain costs or keep them within budgetary limits, but the question remains: Is adopting less-than-full suppression strategies a viable means of limiting costs or reducing total expenditures in the near or long-term?

Where's the Beef? Questioning the Efficiency and Effectiveness of Aggressive Suppression

Before answering that question, it is important to note some important research that is raising critical questions about the efficiency and effectiveness of wildfire suppression, especially of large wildfires that are the real budget-busters. Butry et al. (2008) stated, "We find no evidence that large wildfires respond to wildland management .... Instead, large fires appear sensitive only to weather and landscape conditions." Indeed, Finney et al. (2009) and Butry et al. (2008) demonstrated what firefighters have long known: aggressive suppression actions are ineffective in containing the spread of large wildfires unless and until the fire encounters moderate weather or low fuel conditions. However, Calkin (2014) disclosed that, on average, 35% of firefighting crews and resources are ordered after a large wildfire has stopped growing and essentially defined its own boundaries. Aviation resources are typically one of the most expensive resources and comprise a major portion of total suppression expenditures. There is a common misperception by people that aerial retardant drops extinguish flames, but in fact, retardant only slows down the rate of spread. Accordingly, air tankers are best suited for the initial attack, but 75% of fires with air tanker drops escape the initial attack and become large wildfires (Calkin, 2014). Air tankers are least effective on large wildfires because the weather conditions that fuel large fires overwhelm the effect of chemical retardants; nevertheless, the majority of air tankers are used on large wildfires (Thompson et al., 2013a). Moreover, the largest percentage of air drops occur in late afternoons on steep slopes in dense timber stands—the times, places, and conditions in which aerial retardant is least effective (Calkin, 2014). Given the emerging research that questions the effectiveness of some of the more expensive suppression resources and methods, and factoring in the ecological costs of fire exclusion, the case for finding alternatives to aggressive suppression becomes stronger.

Saving Green in the Black: The Eco-Nomics of Fire Use

There is an extensive and growing body of research demonstrating various ecological rationales for managing wildfires with restoration objectives and fire use strategies, but only recently has research raised some economic rationales, too. One of the simplest arguments is that wildfire provides nearly "free" fuel reduction (Houtman et al., 2013). Letting wildfires burn avoids both the cost and damage of fighting the fire and the later cost of fuel reduction treatments in the areas in which suppression had prevented the fire from burning (Dale, 2006). Donovan and Brown (2008) used an "cost method" to model the savings in fuel reduction costs from fire use: what would have cost $39 per hectare for a series of prescribed fires or $121 per hectare for mechanical treatments can be avoided by a single wildfire that is simply monitored. Once a wildfire has burned a stand, reduced fuel loads can be inexpensively maintained over time through periodic prescribed burning or simply monitoring future wildfires. The cost savings are highest when wildfires are allowed to burn within dense stands of small trees without first using expensive mechanical pretreatments. Fire managers often assume that wildfire use is too risky in these stands, but by using the best fire management tools, skills, and experience, it is conceivable that conditions might allow wildfire use where relatively low intensity fires can achieve desired fuel reduction at great cost savings.

Few studies have compared the costs of wildland fire use versus full suppression, mostly because fire use is so rarely authorized by the USFS. Dale et al. (2005) calculated that over a 20-year period the USFS spent an average of $236 per hectare for suppression but only $21 per hectare for fire use. This matches a study by Oppenheimer (2013) where aggressive suppression cost an average $216 per hectare but point protection/area management strategies cost only $20 per hectare. Comparing costs of fire use with those of direct suppression has now become impossible given that fire managers can use both strategies on a single wildfire. Accordingly, fire use is no longer a separate kind of fire, that is, what firefighters used to call a "fire use fire;" instead, it is a strategy or tactic available for every wildfire. Most of the economic arguments supporting fire use are thus based on modeling and commonsense assumptions that less use of suppression resources will equate into less costs.

The same economy of scale that drives down per-hectare costs of suppression as wildfires grow larger works even more dramatically with fire use. With full perimeter-control suppression strategies, more crews and equipment are needed to contain and control a fire as it grows larger, resulting in much higher total expenditures even if per-hectare costs seem to decline. With wildfire use strategies, however, as a fire grows there is not the same necessity to keep adding more resources. A relatively small crew can manage large wildfire events with fire use strategies, and an economy of scale makes fire use an extremely economically efficient means of getting fire on the ground for fuels reduction or ecological restoration objectives.

The real cost driver for wildfire use is not the size as much as it is the duration of a wildfire that can engage crews and a district's management staff for an extended period. Thus low daily costs can accumulate to a large total amount over time. Long-duration wildfire events also normally experience changing weather conditions that can cause fires to make occasional "runs" of rapid fire spread. This might necessitate some temporary scaling up of crews and equipment to apply some limited suppression techniques (e.g., holding or checking actions) as part of an area management strategy, but after rapid fire spread subsides, crew and resource levels can be quickly downsized. On the other hand, fire use strategies typically avoid intensive or prolonged mop-up that is often a major cost of full-suppression strategies.

One of the most prevalent economic rationales for fire use (and fuels treatments in general) is the belief that future wildfires will be easier and cheaper to contain and control, but that argument maintains the assumption that firefighting will continue to be the normal or default response to wildfires and misses the real point of restoration objectives: to allow wildfires to burn to restore and maintain natural fire ecology processes. As Reinhardt et al. (2008) advocated, the primary objective of treating fuels is to make wildfire more acceptable rather than to reduce wildfire extent or make it easier to suppress. The assumption that fire use strategies will make wildfires less costly is most logical and likely if managers choose to safely monitor rather than aggressively fight future fires. Applying fire use strategies spares the land from suppression damage, saves taxpayers money, and therefore is the most ecological and economic—or "eco-nomic"—way to manage wildfires and restore mixed-severity fire to the landscape, particularly in forests where it is in deficit (see Chapter 1).

12.7 BANKING ON CHANGE: RECOMMENDATIONS FOR CONTROLLING COSTS AND EXPANDING BENEFITS OF MANAGING WILDFIRES

There is no question that fire suppression of some sort will be needed as long as there are valued human assets (e.g., structures) at risk of unwanted wildfire damage. But the past century of systematic fire suppression across the landscape, including vast backcountry areas, has been a systemic "policy failure" (Busenberg, 2004) that is simply unsustainable on a social, ecological, and economic level. Political leaders, agency managers, the media, and citizens alike must recognize that complete wildfire exclusion is neither possible nor desirable, and maintaining ecosystem integrity and controlling suppression expenditures require extensive areas of wildlands to be burned by mixed-severity wildfire or prescribed fire each year. This means changing the dominant paradigm of federal fire management from protection to restoration objectives and changing default wildfire responses from aggressive suppression to opportunistic fire use (also see Chapter 13). The following are some ideas for solutions to some of the problematic cost drivers that have been highlighted in this chapter.

Fix the Budget

Congress must end the skewed budgetary structure that authorizes deficit spending, allows budget transfers from nonfire programs, and promotes "emergency" supplemental appropriations exclusively for wildfire suppression. Congress needs to stop signaling to agencies that they will write a "blank check" for wildfire suppression and should consider setting fixed budgets and firm limits for wildfire suppression. A fixed and firm budget for suppression would force managers to be more selective and strategic in their use of suppression resources. Donovan and Brown (2005) proposed that Congress permit budget deficits or surpluses to be carried into the next fiscal year and allow managers who are conservative in suppression spending to use surplus funds for fire planning, fuels reduction, or restoration projects. A fixed budget for suppression might also make managers prioritize aggressive suppression actions near the WUI, where wildfires clearly damage or destroy human assets, while restricting suppression in backcountry wildlands where more often there are net beneficial effects of mixed- and high-severity fire for natural resources and ecosystems

(Calkin et al., 2014). Finally, Congressional appropriations are divided into two programmatic areas called "fire preparedness" and "fire operations." Defining these terms almost exclusively as prevention and suppression has long been a distortion of fire management philosophy. It is time to define fire operations more literally as fire use.

Change the Incentives

Current USFS performance measures focus on fire size rather than fire effects and encourage managers to aggressively fight fires to keep them small rather than wisely manage them to facilitate ecologically appropriate and heterogeneous fire effects across larger areas. The set of "perverse incentives" that reward fire managers for aggressive suppression, fails to recognize their accomplishments in fire use, and harshly penalizes them for any accidents that occur during fire use strategies must be fundamentally changed. The artificial distinction between wildfire and fuels management should be abolished, and managers should be rewarded for accomplishing lower-cost fuels reduction with wildfire use. The agency should fully support managers assuming proper risks for managing wildfires. The current perception of risk should be inverted such that sending firefighters to aggressively attack fires is considered the most risky decision for managers to take, whereas selecting monitoring or fire use strategies with preplanned restoration objectives is the least risky decision from the standpoint of firefighter safety and a manager's professional career.

Convert Costs into Investments

Managing wildfires with less-than-full suppression strategies may actually have higher costs per incident in some cases in the short term than if they had been aggressively attacked and controlled at a small size, though aggressively attacking every small fire creates a cumulatively large expenditure overall. However, the expenditures of fire use strategies should be viewed as investments in forest restoration with long-term payoffs in enhanced biodiversity, ecological integrity, and community security. In addition, fire use strategies involve more labor-intensive than capital-intensive tools and techniques. For the price of one air tanker and its multiple retardant loads, many jobs to manage wildfires for restoration objectives could be funded, providing tangible socioeconomic benefits that better justify the cost to taxpayers. Agencies should thus change their mind-set from viewing large wildfires as costly problems to seeing them as "investment opportunities" yielding multiple social and ecological benefits.

Build a Firewall Against Rural Sprawl

Given that much of the increase in suppression costs has been attributed to the vulnerability of homes in the WUI renovating existing homes with fire-resistant materials, mandating vegetation treatments on private lands adjacent to homes, and preventing new home construction in wildfire-prone wildlands could reduce the need for aggressive suppression and expand fire management options (see Chapter 13). These would require creating local ordinances, land use zoning laws, and other means to regulate or restrict suburbs from sprawling into wildfire-prone rural areas. Another approach would be to charge more of the cost of wildfire protection to those who build homes and the local governments who issue permits for them (Headwaters Economics, 2014). The current system whereby federal and state taxpayers pay most of the costs for firefighting on private lands functions as a de facto subsidy for individual homeowners as well as an incentive for irresponsible new home construction, thereby raising suppression costs for all taxpayers.

Assert the Will to Change

Historically, the USFS has received significant public, political, and fiscal support for its firefighting actions. The current system is not sustainable, however, and it must change given our increased knowledge of fire ecology, understanding of the adverse effects of fire exclusion and the ecological benefits of higher-severity fire effects (e.g., Chapters 2–6), and the evidence of increasing spending but declining effectiveness of suppression on large wildfires. Donovan and Brown (2005) assert that it is not a lack of knowledge that has impeded change in fire management, but a lack of will. The GAO (2007) identified over 300 recommendations for changes in fire management policies and practices to contain or reduce costs, but as Reinhardt et al. (2008) simply stated, the bottom line is that the only sure way to reduce suppression spending is to make a decision to spend less money suppressing fires. In many respects, the USFS has been taking the blame for decisions made by cultural, political, and economic forces outside of its control (Hudson, 2011). Nevertheless, it is time for the USFS to adopt a new philosophy of fire management centered on wildfire use for ecological restoration, or what we call ecological fire use. Working with wildfires rather than fighting against them will ultimately prove to be the safest, surest, most sustainable, most eco-nomical way to control costs while protecting communities and restoring ecosystems in wildfire-adapted areas.

12.8 ENDNOTE ON METHODOLOGY

Suppression costs presented by federal and state agencies reflect a subset of total wildfire costs for which corresponding documentation exists. Here costs are defined as the total expenditures in U.S. dollars spent to suppress a fire, but do not reflect costs incurred due to loss of life and/or property damages that can be much greater than suppression expenditures. The data in Tables 12.1 through 12.4 are based on cost data compiled by the National Wildfire Coordinating Group (NWCG), and from historical incident ICS-209 forms and

SIT-Reports in the FAMWEB database. Each year represents the fire fiscal year (October 1 of the preceding year through September 30). The data reflect all reported wildfires on lands under federal protection that meet federal reporting thresholds (larger than 40 hectares for timber or 121 hectares for shrublands). We acknowledge that these are at best cost estimates that are not complete due to the nature of how these data are tracked and recorded. Wherever possible we have verified much of the information contained in this database via other official sources. Cost figures provided by the NIFC are generally higher because they reflect data from all fires of all sizes on all lands (including state and private lands), but may be subject to more reliability errors from the various reporting sources.

REFERENCES

Abt, K.L., Prestemon, J.P., Gebert, K., 2008. Forecasting wildfire suppression expenditures for the United States Forest Service. In: Holmes, T.P., Prestemon, J.P., Abt, K.L. (Eds.), The Economics of Forest Disturbances: Wildfires, Storms, and Invasive Species. Springer, Dordrecht, The Netherlands, pp. 341–360.

Black, A., Gebert, K., Steelman, T., McCaffrey, S., Canton-Thompson, J., Stalling, C., 2010. The interplay of AMR, suppression costs, agency-community interaction, and organizational performance—a multi-disciplinary approach. Joint Fire Science Program Final Report 08-1-4-01, 23 pp.

Busenberg, G., 2004. Wildfire management in the United States: the evolution of a policy failure. Rev. Policy Res. 21 (2), 145–156.

Butry, D.T., Gumpertz, M., Genton, M.G., 2008. The production of large and small wildfires. In: Holmes, T.P., Prestemon, J.P., Abt, K.L. (Eds.), The Economics of Forest Disturbances: Wildfires, Storms, and Invasive Species. Springer, Dordrecht, The Netherlands, pp. 79–106.

Calkin, D.E., 2014. Effectiveness of suppression resources in large fire management. Webinar for Joint Fire Science Program broadcast on October 8, In: https://www.youtube.com/watch?v=ITE6YKhGwKU&feature=youtu.be.

Calkin, D.E., Gebert, K.M., Jones, J.G., Neilson, R.P., 2005. Forest Service large fire area burned and suppression expenditure trends, 1970-2002. J. For. 103, 179–183.

Calkin, D.C., Finney, M.A., Ager, A.A., Thompson, M.P., Gebert, K.M., 2011. Progress towards and barriers to implementation of a risk framework for U.S. federal wildland fire policy and decision-making. For. Pol. Econ. 13, 378–389.

Calkin, D.E., Cohen, J.D., Finney, M.A., Thompson, M.P., 2014. How risk management can prevent future wildfire disasters in the wildland-urban interface. Proc. Natl. Acad. Sci. U. S. A. 111 (2), 746–751.

Canton-Thompson, J., Gebert, K.M., Thompson, B., Jones, G., Calkin, D., Donovan, G., 2008. External human factors in incident management team decision-making and their effect on large fire suppression expenditures. J. For. 106, 416–424.

Cart, J., Boxall, B., 2008. Air Tanker Drops in Wildfires are Often Just for Show: The Bulky Aircraft are Reassuring Sights to Those in harm's way, but Their use can be a Needless and Expensive Exercise to Appease Politicians. Los Angeles Times, Los Angeles, CA. July 29, http://www.latimes.com/news/local/la-me-wildfires29-2008jul29,0,5666042.story.

Dale, L., 2006. Wildfire policy and fire use on public lands in the United States. Soc. Nat. Resour. 19, 275–284.

Dale, L., Aplet, G., Wilmer, B., 2005. Wildland fire use and cost containment: a Colorado case study. J. For. 103, 314–318.

de Neufville, R., Claro, J., Oliveira, T., Pacheco, A.P., Collins, R.D., 2013. Forest fire management to avoid unintended consequences: a case study of Portugal using system dynamics. J. Environ. Manag. 130, 1–9.

Dombeck, M.P., Williams, J.E., Wood, C.A., 2004. Wildfire policy and public lands: integrating scientific understanding with social concerns across landscapes. Conserv. Biol. 18 (4), 883–889.

Donovan, G.H., Brown, T.C., 2005. An alternative incentive structure for wildfire management on National Forest land. For. Sci. 51 (5), 387–395.

Donovan, G.H., Brown, T.C., 2007. Be careful what you wish for: the legacy of Smokey Bear. Front. Ecol. Environ. 5 (2), 73–79.

Donovan, G.H., Brown, T.C., 2008. Estimating the avoided fuel-treatment costs of wildfire. West. J. Appl. For. 23 (4), 197–201.

Donovan, G.H., Brown, T.C., Dale, L., 2008. Incentives and wildfire management in the United States. In: Holmes, T.P., Prestemon, J.P., Abt, K.L. (Eds.), The Economics of Forest Disturbances: Wildfires, Storms, and Invasive Species. Springer, Dordrecht, The Netherlands, pp. 323–340.

Donovan, G.H., Prestemon, J.P., Gebert, K., 2011. The effect of newspaper coverage and political pressure on wildfire suppression costs. Soc. Nat. Resour. 24 (8), 785–798.

Fifer, N., Orr, S.K., 2013. The influence of problem definitions on environmental policy change: a comparative study of the yellowstone wildfires. Pol. Studies J. 41 (4), 637–654.

Finney, M., Grenfell, I.C., McHugh, C.W., 2009. Modeling containment of large wildfires using generalized linear mixed-model analysis. For. Sci. 55 (3), 249–255.

Firefighters United for Safety Ethics, and Ecology (FUSEE), 2013. Forest Service chief announces return to sensible policy for 2013 wildfire season: ends fiscally and ecologically irresponsible policy against managing wildfires for safety and restoration goals. Press Release on March 5, http://fusee.org/docs/news_releases/FUSEE%20Press%20Release%203-5-13.pdf.

Gebert, K.M., Black, A.E., 2012. Effect of suppression strategies on federal wildland fire expenditures. J. For. 110, 65–73.

Gebert, K.M., Calkin, D.E., Yoder, J., 2007. Estimating suppression expenditures for individual large wildland fires. West. J. Appl. For. 22 (3), 188–196.

Gebert, K.M., Calkin, D.E., Huggett, R.J., Abt, K.L., 2008. Economic analysis of federal wildfire management programs. In: Holmes, T.P., Prestemon, J.P., Abt, K.L. (Eds.), The Economics of Forest Disturbances: Wildfires, Storms, and Invasive Species. Springer, Dordrecht, The Netherlands, pp. 295–322.

Gonzalez-Caban, A., 1984. Costs of firefighting mop-up activities. Research Note PSW-367, USDA Forest Service Pacific Southwest Research Station, 5 pp.

Gorte, R., 2013. The Rising Cost of Wildfire Protection. Headwaters Economics, Bozeman, MT. Accessed December 15, 2014, http://headwaterseconomics.org/wphw/wp-content/uploads/fire-costs-background-report.pdf.

Governmental Accountability Office (GAO), 2007. Wildland fire management: lack of clear goals or a strategy hinders federal agencies' efforts to contain the costs of fighting fires, GAO-07-655. June. 52 pp.

Gude, P.H., Jones, K., Rasker, R., Greenwood, M.C., 2013. Evidence for the effect of homes on wildfire suppression costs. Int. J. Wildland Fire 222, 537–548.

Hand, M.S., Gebert, K.M., Liang, J., Calkin, D.E., Thompson, M.P., Zhou, M., 2014. Economics of Wildfire Management: The Development and Application of Suppression Expenditure Models. Springer, New York, 77 pp.

Headwaters Economics, 2009. Solutions to the rising costs of fighting fires in the wildland-urban interface. December.

Headwaters Economics, 2014. Reducing wildfire risks to communities: solutions for controlling the pace, scale, and pattern of future development in the wildland-urban interface. Accessed on January 10, 2015, http://headwaterseconomics.org/wphw/wp-content/uploads/paper-reducing-wildfire-risk.pdf.

Hesseln, H., 2001. Refinancing and restructuring federal fire management. J. For. 99, 4–8.

Holmes, T.P., Huggett, R.J., Westerling, A.L., 2008. Statistical analysis of large wildfires. In: Holmes, T.P., Prestemon, J.P., Abt, K.L. (Eds.), The Economics of Forest Disturbances: Wildfires, Storms, and Invasive Species. Springer, Dordrecht, The Netherlands, pp. 59–77.

Houtman, R.M., Montgomery, C.A., Gagnon, A.R., Calkin, D.E., Dietterich, T.G., McGregor, S., Crowley, M., 2013. Allowing a wildfire to burn: estimating the effect on future fire suppression costs. Int. J. Wildland Fire 22, 871–882.

Hudson, M., 2011. Fire Management in the American West: Forest Politics and the Rise of Megafires. University of Colorado Press, Boulder, 214 pp.

Independent Large Wildfire Cost Panel (ILWCP), 2007. Towards a Collaborative Cost Management Strategy: 2006 U.S. Forest Service Large Wildfire Cost Review and Recommendations. The Brookings Institution, Washington, DC, 64 pp.

Ingalsbee, T., 2000. Money to Burn: The Economics of Fire and Fuels Management, Part One: Fire Suppression. Western Fire Ecology Center, American Lands Alliance, Washington, DC.http://documents.fusee.org/SuppressionCosts/money_to_burn_copy.html.

Ingalsbee, T., 2010. Getting Burned: A Taxpayer's Guide to Wildfire Suppression Costs. Firefighters United for Safety, Ethics, and Ecology, Eugene, OR, 43 pp.

Liang, J., Calkin, D.E., Gebert, K.M., Venn, T.J., Silverstein, R.P., 2008. Factors influencing large wildland fire suppression expenditures. Int. J. Wildland Fire 17, 650–659.

Mercer, D.E., Haight, R.G., Prestemon, J.P., 2008. Analyzing trade-offs between fuels management, suppression, and damages from wildfire. In: Holmes, T.P., Prestemon, J.P., Abt, K.L. (Eds.), The Economics of Forest Disturbances: Wildfires, Storms, and Invasive Species. Springer, Dordrecht, The Netherlands, pp. 247–272.

National Interagency Fire Center (NIFC), 2014. Federal firefighting costs (suppression only). Accessed January 10, 2015, http://www.nifc.gov/fireInfo/fireInfo_documents/SuppCosts.pdf.

National Wildfire Coordinating Group (NWCG), 2014. Glossary of wildland fire terminology, PMS-205. October.

North, M., Collins, B., Stephens, S., 2012. Using fire to increase the scale, benefits, and future maintenance of treatments. J. For. 110 (7), 392–401.

Office of the Inspector General (OIG), 2006. Audit report: forest service large fire suppression costs. Report Number 08601-44-SF. U.S. Department of Agriculture. November.

Oppenheimer, J., 2013. Fire in Idaho: Lessons for community safety and forest restoration: an analysis of Idaho's 2012 fire season. Idaho Conservation League. 37 pp. http://www.idahoconservation.org/files/fire-in-idaho-2012-report.

Pyne, S.J., 1997. Fire in America: A Cultural History of Wildland and Rural Fire. University of Washington Press, Seattle, WA, 654 pp.

Reinhardt, E.D., Keane, R.E., Calkin, D.E., Cohen, J.D., 2008. Objectives and considerations for wildland fuel treatment in forested ecosystems of the interior western United States. For. Ecol. Manag. 256, 1997–2006.

Schuster, E.G., Cleaves, D.A., Bell, E.F., 1997. Analysis of USDA Forest Service fire-related expenditures 1970-1995. Research Paper PSW-RP-230. USDA Forest Service Pacific Southwest Research Station. March.

Snider, G., Daugherty, P.J., Wood, D., 2006. The irrationality of continued fire suppression: an avoided cost analysis of fire hazard reduction treatments versus no treatment. J. For. 104, 431–437.

Stockmann, K., Burchfield, J., Calkin, D., Venn, T., 2010. Guiding preventative wildland fire mitigation policy and decisions with an economic modeling system. For. Pol. Econ. 12, 147–154.

Strategic Issues Panel on Fire Suppression Costs (SIPFSC), 1994. Large Fire Suppression Costs: Strategies for Cost Management. Wildland Fire Leadership Council and the Brookings Institution, 59 pp.

Strategic Overview of Large Fire Costs Team (SOLFC), 2000. Policy Implications of Large Fire Management: A Strategic Assessment of Factors Influencing Costs. USDA Forest Service State and Private Forestry, 48 pp.

Taylor, M.H., Rollins, K., Kobayashi, M., Tausch, R.J., 2013. The economics of fuel management: wildfire, invasive plants, and the dynamics of sagebrush rangelands in the western United States. J. Environ. Manag. 126, 157–173.

Thompson, M.P., Calkin, D.E., Herynk, J., McHugh, C.W., Short, K.C., 2013a. Airtankers and wildfire management in the US forest Service: examining data availability and exploring usage and cost trends. Int. J. Wildland Fire 22, 223–233.

Thompson, M.P., Vaillant, N.M., Haas, J.R., Gebert, K.M., Stockmann, K.D., 2013b. Quantifying the potential impacts of fuel treatments on wildfire suppression costs. J. For. 111 (1), 49–58.

Tidwell, T., 2014. Keynote address to the conference on large wildland fires: social, political, and ecological effects. Missoula, Montana. May 21.

U.S. Department of Agriculture-Forest Service, 2014. The rising cost of fire operations: effects on the Forest Service's non-fire work. August 20, http://www.fs.fed.us/sites/default/files/media/2014/34/nr-firecostimpact-082014.pdf.

U.S. Department of Agriculture-Forest Service, U.S. Department of Interior, National Association of State Foresters. 2009. Quadrennial Fire Review. 62 pp.

Westerling, A.L., Hidalgo, H.G., Cayan, D.R., Swetnam, T.W., 2006. Warming and earlier spring increases western U.S. forest wildfire activity. Science 313, 940–943.

Williamson, M.A., 2007. Factors in United States Forest Service district rangers' decision to manage a fire for resource benefit. Int. J. Wildland Fire 16, 755–762.

Chapter 13

Flight of the Phoenix: Coexisting with Mixed-Severity Fires

Dominick A. DellaSala[1], Chad T. Hanson[2], William L. Baker[3], Richard L. Hutto[4], Richard W. Halsey[5], Dennis C. Odion[6], Laurence E. Berry[7], Ronald W. Abrams[8], Petr Heneberg[9] and Holly Sitters[10]

[1]Geos Institute, Ashland, OR, USA, [2]John Muir Project of Earth Island Institute, Berkeley, CA, USA, [3]Program in Ecology and Department of Geography, University of Wyoming, Laramie, WY, USA, [4]Division of Biological Sciences, University of Montana, Missoula, MT, USA, [5]Chaparral Institute, Escondido, CA, USA, [6]Department of Environmental Studies, Southern Oregon University, Ashland, OR, USA, [7]Conservation and Landscape Ecology Group, Fenner School of Environment and Society, The Australian National University, Canberra, ACT, Australia, [8]Dru Associates, Inc., Glen Cove, New York, USA, [9]Charles University in Prague, Prague, Czech Republic, [10]Fire Ecology and Biodiversity Group, School of Ecosystem and Forest Sciences, University of Melbourne, Creswick, VIC, Australia

13.1 ECOLOGICAL PERSPECTIVES ON MIXED-SEVERITY FIRE

Throughout this book, we have presented compelling evidence of fire's beneficial ecological role mainly in western North America but with relevant case studies in other regions. Even though most people recognize the importance of maintaining fire on the landscape, few realize the myriad ecosystem benefits associated with large fires of mixed severity. Habitat heterogeneity, which may be maximized by mixed-severity fire that includes large patches of high severity, and the successional mosaic such fire creates, is one of the most dependable predictors of species diversity (Odion and Sarr 2007, Sitters et al., 2014). This ecological tenet has yet to be fully realized in management circles. If such fires are operating within historical bounds, then ecosystems will remain resilient to them; indeed, deficits of these fires relative to the natural range of variability, in places such as montane forests of western North America, are degrading to fire-dependent biodiversity (Odion et al., 2014a; Sherriff et al., 2014). This is particularly the case when reductions in fire extent and/or severity occur in combination with forest management practices, such as post-fire logging, that undermine development of complex early seral forests (Chapter 11).

The Ecological Importance of Mixed-Severity Fires: Nature's Phoenix

Natural heterogeneity in vegetation types, stand structures, and successional age classes at all spatial scales and environmental settings is emerging as a strategy for enhancing forest ecosystem resilience to climate change, at least in North America (Moritz et al., 2014). This will help ensure that there will be enough habitat for species with varying postfire habitat requirements. The fire dynamic is changing in places, however, with climate change now poised in some systems to recalibrate fire behavior (Chapter 9). With the addition of ongoing pre- and postfire logging in forests and other development pressures, particularly in shrublands, this is having a combined negative impact on native biodiversity associated with both complex early seral and old-growth forest and chaparral ecosystems (e.g., Chapters 2–5).

Beneficial Fire Effects Often Take Time to Become Fully Realized

In general, for ecological acceptance of postfire landscapes to translate into improved management practices, as a prerequisite fire ecologists, land managers, and the general public all must recognize both pre- and postfire landscapes as irreplaceable habitat for fire-associated biodiversity. To a large extent, this depends on how one views the postfire landscape.

When considering the effects of fire, patience is clearly a virtue; postfire processes may take years, decades, or longer to unfold. However, land managers often rely on quick indices to assess fire effects, and this can have negative consequences. For instance, in the western United States, the US Forest Service's "burn area emergency response" (http://www.fs.fed.us/eng/rsac/baer/; accessed February 22, 2015) uses satellite images and other geospatial data in real time to classify soil "damages" immediately after fire. Similarly, the US Forest Services' Rapid Assessment of Vegetation Condition (RAVG) after Wildfire (http://www.fs.fed.us/postfirevegcondition/whatis.shtml; accessed February 22, 2015) provides estimates of "basal area losses" in forests 30-45 days following fires >400 ha. We saw in Chapter 11 that these types of rapid assessments can overestimate tree mortality given their immediate timeline compared with the delayed response of fire-affected trees. In forests, particularly pine and mixed conifer, this can lead to premature conclusions about fire "damages" and fire "catastrophes," as well as erroneous notions about high-severity fire patch size, along with a rush to "take action" at any cost and to advance "restoration" or "recovery" approaches that do far more harm than good (Box 13.1; see also DellaSala et al., 2014; Hanson, 2014).

Notably, differences in whether postfire vegetation is viewed as fuel or habitat (Haslem et al., 2011) most often are at the heart of heated conflicts between natural resource managers and conservationists. Witness these polar opposites: fire suppression (including both mechanical thinning and actions to halt active fires) versus let-burn approaches for wildlife habitat (Chapter 12); postfire logging versus a pulse of biological legacies produced by higher-severity fires (Chapter 11); thinning versus habitat for closed-canopy species; and

BOX 13.1 Rapid Assessment of Vegetation Condition after Wildfire "Treatments" as Defined by the US Forest Service

According to the US Forest Service RAVG assessments, the term *treatment* "describes any of a set of management activities that can assist the prompt recovery of forestlands. Management actions include any combination of live, dead, and dying wood removal, or disposal (with or without commercial value) by any feasible method, including but not limited to logging, piling, masticating, and burning, for site preparation. In addition, planting, seeding, and monitoring for natural regeneration without site preparation are appropriate management activities designed to foster the prompt recovery following wildfire. Treatments also include follow up activities to control vegetation that is believed to compete with desired trees during the early establishment period, usually 1 to 5 years after establishment, using any viable method that meets Land and Resource Management Plan direction."

reseeding/replanting and shrub removal versus the montane chaparral component of complex early seral forest (Chapters 3, 4, and 7). Where one stands on this debate can be a matter of principle and perspective, but can also stem from a lack of a comprehensive understanding of the effects of mixed-severity fire and successional processes after fire (see, e.g., Chapters 2–5). Further, while the public may consider fire to be a necessary change agent (see "Understanding the Public's Reaction to Fire," below), this seems to be tempered by whether fire is operating within "safe limits," constrained by prescribed (or "controlled") fire or reduced in intensity by tree thinning and shrub mastication. While prescribed fire is most appropriate for low-severity, high-frequency fire systems, it is not a replacement for the ecosystem benefits produced by large and higher-severity fire because prescribed fire does not mimic the patch mosaics or pulses of biological activity that higher-severity fires produce (Moritz and Odion 2004, DellaSala et al., 2014). Thus, understanding one's perspective is a starting point for potentially settling differences and developing ways to coexist safely and beneficially with fire. Being willing to respond competently to the cognitive dissonance created when perspectives do not align with new scientific information is also vital to the development of successful and ecologically sound fire management strategies (e.g., Chapter 7).

13.2 UNDERSTANDING THE PUBLIC'S REACTION TO FIRE

If ecologists and conservationists want a new discourse on fire that improves ecological understanding and fire management practices, then informed and sustained communications with the public, land managers, the media, and decision makers are vital. A common understanding is needed to move the public and land management agencies from a view of fire as the harbinger of death (Kauffman, 2004) to fire as nature's phoenix. Here we provide some insights from a public poll on fire attitudes in the United States that reaffirms our personal experiences about the prevailing attitudes of the public and of land managers when it comes to fire.

Attitudes Toward Fire

In 2008 The Wilderness Society and The Nature Conservancy got together to construct a 10-year fire communications framework that was informed by a large national sample of public attitudes ($n=2000$ respondents), focus groups in six regions of the United States where fire was a concern, and communications experts (Metz and Weigel, 2008). The task was to develop ecological messaging on fires that would "complement Smokey Bear's message" about being careful with fire.

Based on a summary of the survey findings, important messages on fire can be gleaned from survey data, some of which are remarkably aligned with fire ecology, whereas others are at odds with basic ecological principles. Most notably, the poll demonstrated the public's sophistication regarding the role of fire in ecosystems, but it was clearly tempered by safety concerns (Smokey Bear), notions regarding the importance of "controlled" burns, and a desire to let "some" fires burn in "natural areas." Education (higher levels) was associated with positive attitudes toward fires, and gender was a factor, with men being more risk tolerant and women more risk averse. Some of the poll's most relevant findings are displayed in Box 13.2. We

BOX 13.2 Key Findings on Public Fire Attitudes from the Study by Metz and Weigel (2008)

- Some fires can be beneficial, and a history of fire suppression has led to more large and destructive fires. (*Note that dramatic changes in fire behavior actually are associated with very few forest types in western North America* (Odion et al., 2014a)).
- Strong negative emotional reactions to fire persist based on safety issues (most view fire as "scary").
- Public understanding of fire's ecological role has increased over time.
- Public concerns about wildfire rank very low compared with other conservation issues.
- The most significant fire concerns pertain to effects on people and firefighters rather than ecosystem benefits.
- Allow fire teams to use "controlled burns" when and where doing so will safely reduce the amount of fuel for fires (*controlled burns are most relevant in low-severity rather than mixed-severity systems*).
- Cut and remove overgrown brush and trees in natural areas that act as fuel for fires (*this is largely true for low-severity systems, not higher-severity fires that are largely controlled by extreme weather*).
- Allow naturally started fires that do not threaten homes, people, or the health of natural areas to take their natural course, rather than putting them out.
- Shift some government funds from putting out practically all fires to proactively cutting and removing overgrown brush and trees and using controlled burns to reduce the amount of fuel for fires (*removing brush/trees and controlled burns are mostly ways to reduce fire severity in low-severity systems*).

also highlight in parentheses those beliefs that seem to be at odds with the ecological literature on mixed-severity fires.

Communication experts then advised the conservation groups that successful fire messaging should have the following five fundamental communication themes:

1. Protect people, property, and communities
2. Safeguard the health and regeneration of natural areas
3. Safely manage controlled burns to clear fuels (*this management is appropriate in low-severity systems only during the natural fire season*)
4. Save taxpayer money through controlled burns
5. Protect air and water by protecting the health of forests and natural areas and giving plants and wildlife the exposure to fire they need to survive

From focus groups and polling results, according to communication experts the following cogent messages are likely to reach the public:

- Safety is always the number one priority when it comes to fire. By putting out every single fire, however, we are actually creating more dangerous conditions (*in western North America, higher-severity fires are operating at an historical deficit*). Using controlled burns to thin out overgrowth and carefully managing natural fires help ensure the safety of neighborhoods in outlying areas.
- Forests and natural areas are important to our health; they act as natural filters to give us clean air and are the source of clean drinking water. We must ensure the health of forests and natural areas by allowing some fires to take their natural course.
- Taxpayer money is being wasted putting out fires that are far from people and their property. A far more cost-effective approach is to use controlled burns to prevent large, severe fires from spreading into areas where people live and to allow some fires to take their natural course (*and they are ecologically inappropriate when applied outside the natural fire season*).

For higher-severity fires, a good portion of this messaging may work to bridge the divide between science and public attitudes, whereas some of the recommendations of the communications experts in 2008 (refer to the italicized text in the parentheses above) do not incorporate the ecological importance of maintaining, and managing for, complex early seral forest created by mixed-severity fire. In particular, the poll's findings that fire safety matters most is still very much relevant; thus putting out fires that are dangerous to human communities is clearly of primary importance. From a safety standpoint, Smokey Bear's cautionary fire safety tale needs to be updated so that the focus of fire management is on creating "defensible space" around homes, the home ignition zone (HIZ), and introducing land use zoning to allow fire to run its course unimpeded in natural areas under safe conditions ("Making Homes Fire Safe", see below). And, while the poll found the public generally agreed that fire is necessary in natural areas, how far this tolerance would go in relation to large or higher-severity fires is unclear given that the poll's questions were geared toward low-severity fires that can be either "controlled" or suppressed (through thinning or the use of fire retardants). Notably, in Chapter 12 we discussed how runaway expenditures in fire suppression have been ecologically damaging and fiscally irresponsible, and the public seems to agree with these fiscal concerns. In combination with economics, whether public attitudes will change, or are changing, regarding large or higher-severity fires is still unknown; this will require polling that is more specific to these kinds of fires along with enhanced public education (e.g., the videos referenced in the preface) regarding ways to coexist with large fires.

A core message—and one that will most certainly be difficult for much of the public to accept despite being fact based—is that large fires in any given location each year, at least in western North America, cannot be stopped no matter what we do. We at least need to be honest about that and clearly state the damages that can ensue from large-scale pre- and postfire management that attempts to control large, mainly climate-driven fires that are uncontrollable. We also need to clearly communicate to the public the current state of scientific knowledge regarding the ecological benefits and values of the habitats created by mixed-severity fire. This is especially so given the still all-too-common notions that such areas have been categorically damaged by fire, which in turn leads to misguided assumptions that they are in need of "restoration" or "recovery" management actions.

13.3 SAFE LIVING IN FIRESHEDS

Based on public attitudes toward fire there are important challenges to coexistence with fire. These can be overcome, however, if we not only increase public education about current fire ecology but also act responsibly in reducing risks where they matter most. We note that by far the biggest challenge to coexistence with fire is the explosion of exurban sprawl in many rural communities triggered by those moving out of congested cities.

A case in point is Kalispell, Montana, the gateway to Glacier National Park. A November 17, 2014, article in *Greenwire*, the online source of information on the environment ("Where property rights are king, development continues despite growing wildfire threat"), reported that during the 1990s the county's population grew at twice the state's average as more and more people seeking a rural quality of life purchased 16-ha "ranchettes" scattered across Big Sky fire country. They were able to do so as a result of lax and often resisted land use zoning standards. Based on data provided by Headwaters Economics (2014), 11,000 houses in this Montana county lie within the wildland-urban interface (where towns, homes, and other built structures abut fire-prone wildlands)—more than any other county in Montana—and this number is growing at a phenomenal rate. As reported in the online article, public attitudes included the notion that fire will not directly affect them and strong views about private property rights (i.e., "don't tell me what to do on my land"). Some of the same people vocally oppose government actions in general then demand that public money be spent to remove "fuels" from wildlands. In essence, the lack of homeowner fire risk reductions and inappropriate fuel treatments is setting in motion the perfect storm of land use and fire conflicts.

To minimize these kinds of conflicts, landowners need to practice fire-safe (also known as "fire-wise" in the United States) planning to protect home structures. We suggest that landowners first declare a common "fireshed" boundary, as they do for watersheds. Firesheds are multidimensional spaces. They begin at the scale of a watershed and encompass the residential community with similar fire risks (Figure 13.1a). Within a fireshed, homeowners can take fire risk reduction measures together (preferably) or on their own (Figure 13.1b).

Making Homes Fire-Safe

Probably no research results are as relevant to fire safety science than those of Dr. Jack Cohen (e.g., Cohen 2000, 2004), whose seminal fire safe research recommendations are now standard risk reduction measures taken by many homeowners[1] and have caught on with risk-averse insurance companies[2]. The work of Syphard et al. (2012, 2014) on home loss in chaparral systems of southern California is strikingly similar.

According to Dr. Cohen, fire planning within an HIZ begins with defensible space nearest the home. Notably, research on HIZ risks shows that homes whose owners reduced vegetation and flammables within 10-18 m of the structure and built with nonflammable roof materials had an 86% (Foote, 1996) to 95% (Howard et al., 1973) "survival" rate when fires swept through an area (cf. Syphard et al. (2014) for more recent and similar home structure protection distances). Combined with home fire simulations by the insurance industry (http://www.extension.org/pages/63495/vulnerabilities-of-buildings-to-wildfire-exposures#.VHUr00snRNs; accessed February 15, 2015), Box 13.3 provides measures that are most critical for living safely in firesheds.

An example from a town in Idaho during an intense 2007 fire is instructive regarding the importance of the HIZ and fireshed management. As the *Idaho Statesman* newspaper reported (Druzin and Barker, 2008):

> *We spend billions attacking almost every wildfire, but scientists say that's bad for the forest, can put firefighters in unnecessary danger and doesn't protect communities as well—or as cheaply—as we now know how to do. A wall of fire barreled through the forest with a jet-engine roar near Secesh Meadows last August, and local fire chief Chris Bent knew his work was about to be tested.*

1. http://www.firewise.org/wildfire-preparedness/firewise-toolkit.aspx?sso=0; accessed November 25, 2014.
2. http://www.extension.org/pages/63495/vulnerabilities-of-buildings-to-wildfire-exposures#.VHUr00snRNs; active November 26, 2014.

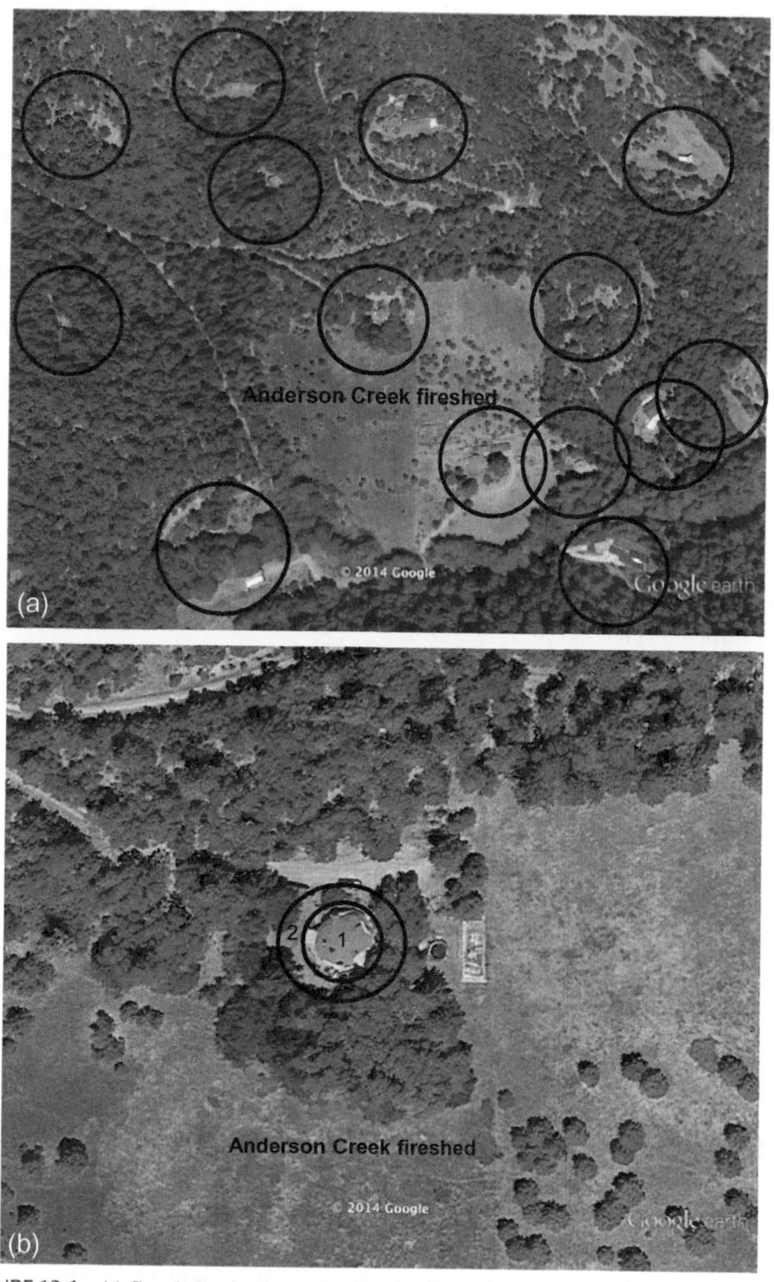

FIGURE 13.1 (a) Google Earth image of the Anderson Creek watershed and community fireshed in Talent, Oregon, showing a housing development (circled; the center house is depicted in b). Most members of this community reduced lower-strata fuels via thinning small trees in the surroundings, although tree densities are beginning to fill in and require repeat treatments. (b) Two fire-safety zones where the landowner built with fire-resistant material in the inner most zone (home ignition zone 1) and cleared most vegetation within a 10 m radius around the structure (zone 2). Tree crowns are touching in zone 2; however, lower branches were pruned to 3 m, and there are few ladder fuels to carry fire from the ground into tree crowns. Downslope grasses may pose a fire hazard but may not crown out given the precautions taken in zones 1 and 2.

BOX 13.3 Prudent Fire Risk Reduction Measures for Homeowners

- Build homes with noncombustible roof covering and siding; keep roof and gutters clear of leaves/needles; keep firewood away; keep vegetation adjacent to homes to a minimum; cut overhanging limbs of trees closest to the home; and install ember-resistant attic vents.
- Clearing vegetation within 5-20 m of a home is the most effective treatment: Carefully space plants, reduce wood plant cover to <40% around the structure, and use varieties that grow low and are free of resins, oils, and waxes that burn easily; mow the lawn regularly and prune trees up to 3 m from ground; space conifer crowns ~3 m apart and remove lower limbs; trim back trees overhanging the house; create a "fire-free" area within 1.5 m of the house using noncombustible landscaping; remove dead vegetation; use fire-resistant furniture; remove firewood and propane tanks; and water plants or use xeriscaping.
- Additional measures include low-growing, well-irrigated, and relatively noncombustible vegetation in low planting densities; a mix of deciduous and conifer trees; fuel breaks like driveways and gravel walkways and lawns.
- Treatments >30 m from the home structures offer no additional protection (Syphard et al., 2014).

Flames danced atop lodgepole pines, smoke darkened the sky, and residents of the tiny mountain hamlet north of McCall prepared for the worst. Just a month earlier, a forest fire had burned 254 homes near Lake Tahoe and the 2007 fire season appeared ready to claim its next community. But as the raging East Zone Complex fire reached the cluster of loosely-spaced homes, the flames dropped to the ground, crackling and smoldering. The fire crept right up to doorsteps. But without the intense flames that spurred the fire just moments before, no homes burned—a feat fire managers attributed largely to Bent's push to clear flammable brush from around houses in the community. "It just blew through the area," Bent said. "We were well prepared." The town's ability to withstand a frontal assault by a major wildfire demonstrates what fire behavior experts have been saying for more than a decade. Clearing brush and other flammables and requiring fireproof roofs will protect houses even in an intense wildfire—without risking firefighters' lives. More provocatively, the research suggests that fighting fires on public lands to protect homes is ineffective and, in the long run, counter-productive. It is also far more expensive.

Importantly, clearing vegetation nearest a home is not enough, as fire risk reduction also needs to include the home structure itself (Figure 13.2). This is often missed in discussions about homeowner fire safety, and it is a crucial step in responsible fire risk reduction, as we illustrate in the following examples.

In a recent research paper concerning why homes burn in wildfires, Syphard et al. (2014) concluded that geography is key: where the house is located and where houses are placed on the landscape. Syphard and her coauthors gathered data on 700,000 addresses in the Santa Monica Mountains and part of San Diego

FIGURE 13.2 Homes burn because they are flammable. Many homes with adequate defensible space still burn in wildland fires because embers land on flammable materials around the home or enter through openings such as attic vents. These two homes burned during the 2014 Poinsettia Fire in Carlsbad, California, despite fire-safe landscaping, a firewall, and thinned wildland vegetation. Focusing exclusively on wildland vegetation clearing ignores the main reasons homes burn: they are flammable. *(Photo credit: Richard W. Halsey.)*

County. They then mapped the structures that had burned in those areas from 2001 to 2010, a time of significant wildfire activity in the region. Buildings on steep slopes, in Santa Ana wind corridors, and in low-density developments intermingled with wildlands were the most likely to have burned. Nearby vegetation was not a major factor in home destruction.

Looking at vegetation growing within roughly 800 m of structures, Syphard et al. (2014) concluded that the exotic grasses that often sprout in areas cleared of native habitat like chaparral could be more of a fire hazard than shrubs. Interestingly, they found that homes that were surrounded mostly by grass actually ended up burning more than homes with higher fuel volumes such as shrubs.

Similarly, during the 2007 Witch Creek Fire (San Diego County, CA), houses in Rancho Bernardo started burning by ember contact when the fire front was nearly 6 km away. Two-thirds of the burning homes were set on fire by embers (Maranghides and Mell, 2009).

During the 2007 Grass Valley Fire near Lake Arrowhead in California's San Bernardino Mountains, approximately 199 homes were destroyed or damaged. This happened despite the fact that the US Forest Service had thinned the surrounding forest. The main cause of the losses was that individual homeowners failed to understand that vegetation management is only one part of the fire risk reduction equation. Fire will exploit the weakest link—and it did so in Grass Valley. In the detailed report of the fire, Forest Service researchers (Rogers et al., 2008) concluded: "Post-fire visual examination indicated a lack of substantial fire effects on the vegetation and surface fuels between burned homes. Lack of surface fire evidence in surrounding vegetation provides strong evidence that house-to-house ignitions by airborne firebrands were responsible for many of the destroyed homes."

Investments in making homes and communities fire safe are clearly fiscally prudent and responsible homeownership that can save lives and homes by reducing risks to all, especially firefighters. Moreover, proper land use zoning that reduces housing development in firesheds is key to the survival of home structures over the larger area (Syphard et al., 2014).

In sum, these recent studies show that overcoming misperceptions about homeowner losses is urgently needed because those misconceptions are a driving factor in many inappropriate fuel reduction projects in wild areas. We hypothesize that with stepped-up planning directed at proper homeowner safety (as demonstrated in the above studies), public attitudes about large and intense fires may begin to shift from fear-based primal responses to more of a neocortex-like awareness of fire as nature's phoenix. This could be tested using before-and-after polling about large, higher-severity fires with and without proper public safety measures in places.

13.4 TO THIN OR NOT TO THIN?

One of the most significant challenges involved in changing the way land managers think about fire in the forests is how the US Forest Service views forest fires. The agency is deeply invested in continuing the fire management trajectory of the past—a situation compounded by the budgetary issues associated with the agency's direction of much, and often most, of their tax-based support to selling timber from public lands, and the agency's retention of most of the revenue from such timber sales to fund staff salaries and operations. Though in recent years we have learned much about the ecological benefits of higher-severity fire and the risks to fire-dependent wildlife species from further suppressing these fires, which are deficient in most western US conifer forests (Chapters 1–5), the Forest Service continues to aggressively promote landscape-level mechanical thinning (North, 2012; Stine et al., 2014) and postfire logging (Collins and Roller, 2013) ostensibly to reduce fuels and prevent and mitigate future fire. These forest management policies are promoted based on the assumption that decades of fire suppression have created forests "overloaded with fuel, priming them for unusually severe and extensive wildfires" (Stine et al., 2014; see also North, 2012). The basic concept being articulated by the Forest Service is that, because of decades of fire suppression and "fuel accumulations," we cannot simply allow wildland fires to burn because long-unburned forests will "uncharacteristically" burn almost exclusively at higher severities (North, 2012; Stine et al., 2014). Under this premise, recommendations focus on how to manage forests through logging and fire suppression to further reduce and prevent the significant occurrence of mixed-severity fire (North et al., 2009; North, 2012; Stine et al., 2014). Yet these sources do not include a discussion of the current deficit of these fires in most forests of western North America (Odion et al. 2014a; see also Chapters 1, 2, and 9) or meaningful content on the ecological importance of mixed-severity fire for many rare and imperiled wildlife species (Chapters 2–5). Nor do they explore the validity of the basic premise that long-unburned forests will burn much more severely.

Studies that empirically investigated the "time-since-fire" issue in the Sierra region of northern California and the Klamath Mountains of Oregon and California tended to find that, contrary to popular assumptions, the most long-unburned forests experience mostly low- and moderate-severity fire and do not have significantly higher levels of higher-severity fire than more recently burned forests (Odion et al., 2004, 2010; Odion and Hanson, 2006, 2008; Miller et al., 2012; van Wagtendonk et al., 2012). One modeling study predicted a modest increase in fire severity with increasing time since fire, but the strength of inference was limited by a lack of data for all but long-unburned stands, especially in the largest forest types, such as mixed-conifer forest. Even the most long-unburned forests were predicted to have ~70-80% low/moderate-severity effects (Steel et al., 2015), well within the range of natural variability (see Chapter 1). In fact, long-unburned forests sometimes have the lowest levels of higher-severity fire; understory vegetation and the lower limbs of conifers self-thin as canopy cover increases and available sunlight in the understory decreases with increasing time since fire (Odion et al., 2010). Therefore the argument that we cannot allow more wildland fires to burn without suppression in natural areas is not valid for many dry montane forests in western North America (Odion et al., 2010).

Problems with Fuel Models and Fire Liabilities

Government programs that aim to make forests safe places for people to live are based on theory rather than actual evidence about historical forests. As discussed above, the common argument has been that fuels have unnaturally accumulated from fire exclusion and land uses, and if fuels are restored to low levels, fires will burn primarily at low intensity rather than as high-intensity crown fires (e.g., Agee and Skinner, 2005). Thus forests can be restored while also making them safe places to live—a win-win solution that is appealing to the public. Little evidence about actual historical fuel amounts in forests to support this argument was available, however; instead, evidence is mostly based on the idea that frequent fires would have kept fuels at low levels. When records from land surveys before fire exclusion were examined (Baker, 2012, 2014; Baker and Williams, 2015; Hanson and Odion, in press), understory fuels (shrubs, small trees) that would naturally have promoted intense fires were found to have been common and often abundant in many areas, and small trees were dominant, not rare. This direct evidence suggests that fuel treatments would typically have to artificially remove natural shrubs and small trees and adversely alter habitat for native species in a quest to make forests safer places for people to live.

Fuel reduction also has been overpromised to be effective, using questionable logic and unvalidated models. First, fire intensity in most forest types is much more strongly affected by wind than by fuel. High fire-line intensity, the primary fire characteristic that promotes crown fires, is the product of the energy released by burning fuel and the rate of spread of fire (Alexander, 1982). Energy release by fuel varies over perhaps a 10-fold range, however, whereas rate of spread can vary over more than a 100-fold range; thus a high rate of spread caused by strong winds can easily overcome the limited reductions in fuel that are feasible (Baker, 2009). This was confirmed by a recent analysis of the 2013 Rim Fire in California, which concludes: "Our results suggest that even in forests with a restored fire regime, wildfires can produce large-scale, high-severity fire effects under the type of weather conditions that often prevail when wildfire escapes initial suppression efforts.... During the period when the Rim fire had heightened plume activity... no low severity was observed [in thinned areas], regardless of fuel load, forest type, or topographic position" (Lydersen et al., 2014, p. 333). Second, common fire models used to show that forests would be fire-safe after fuel reductions have an underprediction bias and are not validated. These flawed models include NEXUS, FlamMap, FARSITE, FFE-FVS, FMAPlus, and BehavePlus (Cruz and Alexander, 2010; Alexander and Cruz, 2013; Cruz et al., 2014). The underprediction bias means that these models often predict that fuel reductions would reduce or eliminate the potential for crown fires in forests, when in fact fuel reductions do not achieve this effect. Fixing these models would be difficult and has not yet occurred (Alexander and Cruz, 2013). Also, these models have not been sufficiently tested and validated using a suite of actual fires, in which case they would likely be shown to fail (Cruz and Alexander, 2010). Alternative validated models are available and could be further developed, but they are not being used (Cruz and Alexander, 2010). Further, studies of tree mortality in thinned areas following fire do not typically take into account the mortality caused by the logging itself before the fire, leading to further biased results.

These concerns should raise red flags about the effectiveness of fuel treatments, as well as issues regarding liability and responsibility. Imagine if a company sold airplanes with identified flawed designs and without adequate test flights, which then crashed. There are thus sound scientific reasons to closely scrutinize government wildland fuel-reduction programs. Meanwhile, we need to be honest and warn the public that living within or adjacent to natural forests prone to burn is inherently hazardous. Only treating fuels in the immediate vicinity of the homes themselves can reduce risk to homes, not backcountry fuel reduction projects that divert scarce resources away from true home protection (Cohen, 2000; Gibbons et al., 2012; Calkin et al., 2013; Syphard et al., 2014).

Finally, another land management liability that is frequently overlooked when assessing fire-related economic losses is the role of silviculture. For instance, before the 2013 Rim Fire, a significant portion of the Stanislaus National Forest in central California's Sierra Nevada Mountains consisted of even-aged monoculture tree plantations (following past clearcuts) distributed across large landscapes (Figure 13.3). Land managers often claim that clearcutting over large landscapes

FIGURE 13.3 Google image of the Stanislaus National Forest, central Sierra Nevada, taken on July 8, 2012, before the August 25, 2013, Rim Fire. The red boundary is where the Rim fire burned. Note numerous clearcuts within the burn area, where the fire later burned intensely. Figure provided by J. Keeley.

like this reduces fire spread, yet based on preliminary findings from the Rim Fire, clearcutting did nothing to stop the fire. In fact, the area with the most clearcutting had the largest contiguous area of high-severity fire of any portion of the Rim Fire (see Figure 13.3 and compare with Figure 11.11). In other areas with large portions of the landscape in tree plantations from past clearcutting, fires have a tendency to burn uncharacteristically severe, presumably because of homogenized fuel loads (e.g., Odion et al., 2004). Despite these observations, in postfire assessments land managers rarely discuss this effect or the liabilities it creates for economic losses related to intense burns.

13.5 FIRE SAFETY AND ECOLOGICAL USE OF WILDLAND FIRE RECOMMENDATIONS

Based on the ecological importance of higher-severity forest fires (e.g., Reinhardt et al., 2008; DellaSala et al., 2014; Hanson, 2014; Moritz et al., 2014) and home safety concerns (e.g., Cohen, 2000; Headwaters Economics, 2014), there are ways for people to live safely in firesheds and still allow fire to perform its vital ecosystem service. Below we provide some summary recommendations that, if widely implemented, would allow fire to take its natural course (i.e., ecological use of wildland fire) while reducing risks to people.

Fire Safety Recommendations (mainly summarized from Headwaters Economics, 2014)

- Prepare to live safely with fire so that it can perform its ecologically beneficial functions. (The bulk of fire risk reduction should occur immediately adjacent to homes.)

- Develop negative financial consequences for landowners who increase fire risk within firesheds by not taking precautionary measures versus providing financial incentives for those who reduce risks (e.g., cost sharing for fire safety). As an example, mortgage and/or insurance rates could be increased for high risks from lack of fire safety and discounted for those who practice fire risk management principles. In this manner, planning for home fire safety would become as routine as taking out a mortgage to buy a home.
- Include HIZ and fire-safe principles in rural land use planning, including zoning restrictions that limit housing densities in firesheds deemed too risky for development.
- Require mandatory disclosure of fire risks to homebuyers.
- Have local and state governments contribute to firefighting costs to create a powerful incentive for improved land use planning, including zoning restrictions, which reduce fire suppression needs.
- Offer technology transfer to local governments and financial assistance to plan communities that are fire safe.
- Map high-risk areas where fire-safe standards are most prudent within a local county or other land use unit.
- Discourage rebuilding in the same high-risk place or require that building occurs with risk management conditions.
- Redirect funding away from backcountry fire suppression and fuel reduction programs and toward aiding willing homeowners in creating defensible space and reducing the ability of homes to ignite.
- Initiate strategies to reduce human-caused fire ignitions, especially along roadsides. Many wildland fires start along highways and streets.

Wildland Fire Recommendations

- Postfire "salvage" logging is especially damaging to complex early seral forests. If such forests were ecologically valuable or protected before fire, then they should also be recognized as uniquely valuable and protected after fire.
- Wildlands cannot be fireproofed by suppression (mechanical thinning or aerial retardants) or clearcutting; fuel treatments (thinning) are more likely to work in low-severity frequent fire systems and much less so in mixed- and higher-severity fire systems that tend to burn under extreme conditions, when suppression is least effective.
- Large fires, including high-severity patches, are the most efficient means of restoring fire-dependent ecosystems and natural heterogeneity where fire has been excluded for decades. When a fire burns under these conditions, fire-dependent communities are therefore restored. This should be encouraged, with public safety assured.
- The best way to buffer fire-dependent ecosystems from climate change is to increase ecological resilience, particularly in areas where a fire deficit

exists, by allowing fires to burn naturally under safe conditions. This will require relatively large protected landscapes with proper land use zoning and logging restrictions.

- Implement strategies to reduce human-caused fires in ecosystems with excessive fire frequencies, such as the chaparral in southern California.

13.6 LESSONS FROM AROUND THE GLOBE

Africa

Of the five communication themes that arose from the polling in North America, the one most applicable to attitudes in sub-Saharan Africa is number 5 (as mentioned in the above "Attitudes Toward Fire"), a broad statement to protect natural resources for the ecosystems services they provide (see Chapter 8). The public in South Africa, for example, assumes number 3, safety in controlled burns, because the public is already attuned to the widespread use of fire for habitat management, and when accessible, fuel wood is collected for heat and cooking. Of course, the South African public is not deluged by media reports of catastrophic losses caused by wildfire, so items 1, 2 and 4 are not part of a daily discourse in countries where wildfires in large forests are rare and most of the managed habitat is the much thinner type of woodland associated with savanna (see Chapter 8).

In terms of such issues as woodland thinning (directed silviculture or ad hoc management), in African savanna the public and policy makers are more concerned with maintaining herbivore populations as part of ecotourism and for the love of Africa's "big five" megafauna wildlife species. South Africa practices extensive silviculture, and it often is blended into wilderness areas (Tsitsikama National Forest lies adjacent to extensive tracts of forest plantation, where fire suppression is practiced because of economics of the wood industry). It seems the "fear" of fire so prevalent in North America is absent from rural areas of Africa for multiple reasons, but this results in a more sane approach to fire ecology. In Kruger Park managers learned over time that allowing wildfire is acceptable, and it is now a tool (although not frequent) integrated with controlled burns. They even seek to achieve as hot a fire as they can in certain habitat conditions to clear the invasive vegetation or just to suppress woody growth. The lesson learned in South Africa over 50 years of "experimenting," and from many decades of following the Serengeti system, is that monitoring is critical, and adapting to those results (adaptive management) is imperative.

Australia

In Australia prescribed burning is considered a staple part of the land management tool kit and is routinely applied with the aim of reducing the risk of large, unplanned wildfires to property and infrastructure (Clarke, 2008). In some cases fire is applied to the landscape in efforts to "restore" ecosystems or to create fine-scaled fire mosaics of mixed successional stages to encourage greater faunal and floral diversity (Bradstock et al., 2005). In response to the perceived need to apply fuel-reduction burns, the Victorian state government implemented a policy that mandated that 5% of the total land area under state jurisdiction be burned each year. This policy did not discriminate fire prescriptions between ecosystems and has been subject to widespread criticism from fire ecologists in Australia; it is currently under review (DELWP, 2015a). Although appropriate fire regimes have positive ecological outcomes in many systems, application of prescribed burning can lead to species declines and in some cases can cause irreversible changes in ecosystem state (Pardon et al., 2003, Pennman et al. 2011, Pastro et al., 2011).

Recent large wildfires in Australia have spurred new policies to address the growing public concern over the dangers presented by these fires (McLennan and Handmer, 2012; Whittaker et al., 2013). The royal commission that followed the 2009 "Black Saturday" fires suggested the implementation of new policies to encourage clearing around homes and to shift public perceptions toward recognition of bushfires as defensible events (i.e., homes can be effectively protected) that require early planning and avoidance actions (Teague et al., 2010). Residents in areas of high fire risk are now able to clear all vegetation within 50 m of their homes. These new measures, coupled with the 5% burn target, aim to reduce the potential of a repeat of the 2009 fires. This home protection approach is partially supported by science. Gibbons et al. (2012) highlighted that houses with vegetation cleared within 50 m were 70% more likely to survive a fire than those with no clearing. They revealed, however, that there was no effect of fuel reduction burning in nearby state forest or ecological reserves on house preservation following the 2009 fires in Victoria, Australia. Furthermore, in some of the most potentially pyrogenic systems, such as mountain ash (*Eucalyptus regnans*) forests, fuel reduction burns are rarely applied because moisture levels are normally high, and risk of fire spread is considered unacceptable when conditions are dry (DELWP, 2015b). A growing body of literature indicates that inappropriate fire regimes are contributing to species declines globally (Driscoll et al., 2010). In response to the increased fire risk caused by climate change, policy makers should seek to implement strategies with a proven ability to protect homes, while avoiding ineffective actions that detrimentally impact biodiversity.

Central Europe

In central Europe forest fires are relatively infrequent and mainly limited to regions with pine forest plantations growing on sands, gravel-sands and sandstone rocks. Any burned areas are mandatorily reclaimed within just 2 years of their formation; exceptions are possible in forests protected as national parks or nature reserves. The option to request avoidance of logging and replanting is used only rarely, however, and nearly all forests affected by fires are quickly logged and replanted.

Available evidence suggests that fire-induced bare soil patches, charred trunks, and dead wood resulting from the postfire dieback represent unique nesting resources for numerous species. The areas subject to mixed- and high-severity fires are associated with dynamic assemblages of plant and animal species, many of which are rare or even absent in the surrounding landscape. The burned forests serve as key habitats, particularly for aculeate Hymenoptera associated with cavities in dead wood (such as *Dipogon vechti*). Such cavities are considered limiting nesting resources, and their absence (and targeted removal of any newly emerging snags, which is mandatory by law) causes numerous specialized cavity adopters to be red-listed or extinct. Mounting evidence suggests that specific groups of organisms are strictly dependent on the occurrence of repeated fires. As long as sites of natural disturbances become extremely rare in the intensively cultivated landscape of central Europe, bare soil specialists and species that specialize in cavities of decaying wood will be completely absent where forests are subject to intense cultivation and rigorous dead wood removal. Dead wood thus should be considered an important habitat resource deserving conservation measures. Mosaic management of burned forest sites and retaining charred trunks are suggested as management measures supporting biodiversity at the sites of recent forest fires (Bogusch et al., 2015).

Canadian Boreal

There is emerging a new paradigm about the role of fire in the Canadian boreal forest. Historically, it was perceived as a simple system where "catastrophic" fire created landscapes of young, even-aged stands and where species diversity was poor. The reality is much more complex. There is an impressive range of fire cycle estimates—some as long as several centuries—suggesting that for at least part of the boreal forest region the abundance of old-growth forests in preindustrial times was much greater than expected (see Chapter 8). Associated with these old-growth forests is high understory diversity in black spruce (*Picea mariana*) stands and a number of rare species of nonvascular plants associated with balsam fir (*Abies balsamea*) stands. Similar findings have been made in boreal forests of Europe and Asia.

At the other end of the disturbance spectrum, there is now compelling evidence showing the importance of early seral burned habitats for the pyrocommunity, led by saproxylic insects (dependent on dead or decaying wood) and followed by primary cavity nesting birds (see Chapter 8). The retention of a wide range of burn conditions enhances saproxylic insect diversity. A link between this saproxylic community and nutrient cycling has been found, indicating a connection between biodiversity and ecosystem function in Canadian boreal forests. Large fires produce significant pulses of dead wood, which drive biodiversity and ecosystem processes through natural succession over time. Fire skips, or remnants left after large burns, also are critically important for biodiversity, species persistence, and recolonization and ecosystem recovery.

For a long time, forest management was driven with a strong focus on timber extraction and developed a jargon that infiltrated the dialect of forestry, with words like "decadent" for old-growth forests, "waste wood" for trees that had been killed by natural disturbances, and "salvage" as the practice used to recover that "wasted" timber. Today, management in the boreal forest is increasingly driven by themes like ecosystem-based management and sustainable development. The new era will require conservation of boreal forests at different ends of the disturbance spectrum from newly created, postfire habitat to multicentury, old-growth forests.

13.7 ADDRESSING UNCERTAINTIES

Even though most people recognize the importance of maintaining fire on the landscape, there remain important questions about what might be the optimal postfire conditions for the broad suite of species with varying fire tolerances. For instance, we do not know whether there is a certain amount of burned forest or spatial distribution of burned forest patches, patch sizes, and fire frequencies necessary to maintain species at polar ends of the successional gradient. However, we hypothesize that in large, intact forested landscapes where fire is allowed to burn and logging is restricted (e.g., wilderness areas, large national parks, and other protected ecosystems) there should be ample habitat for all seral species over the long term and the best opportunities for coexistence with fire as a process (see Chapters 3–5). By contrast, in highly degraded landscapes, particularly those close to towns and homes, an optimal condition of recently burned and long-unburned patches is more difficult to ascertain because it may involve tradeoffs for public safety reasons (DellaSala et al. 2004).

Currently, megafires in western North American forested landscapes burn in mixed-severity patterns and seem to provide the necessary patch mosaics for a broad array of species (Chapters 2–6). Fire-related change of late seral habitat to complex early seral forest (Swanson et al., 2011; DellaSala et al., 2014; Hanson, 2014) has not been a threat to species dependent on such mature forest habitat, particularly given that there is generally much less high-severity fire in mixed-conifer and pine forests of western North America than there was historically (Odion et al., 2014a). Rates of old forest recruitment, as a result of growth, also outpace rates of high-severity fire in old forest by several times (Hanson et al., 2009; Odion and Hanson, 2013; Odion et al., 2014b). The situation is less clear in portions of Australia, however, where fewer vertebrate species have thus far been found to be fire dependent (see Chapters 3 and 4) and there are more species associated with late seral conditions that are especially at risk (Kelly et al., 2015). By contrast, other Australian research found bird species richness to be highest where there is the most successional diversity from higher-severity fire (Sitters et al., 2014) (see Chapter 8). Human-caused fires in North American chaparral, the Great Basin, and many desert ecosystems, which mostly replace stands, have exceeded historical bounds, adversely affecting this diverse shrubland community (Chapter 7). Thus, whether or not fire mosaics are correlated with high levels of biodiversity (cf. Martin and Sapsis, 1991 versus Parr and Andersen, 2006; Taylor et al., 2012; Kelly et al., 2015) depends on differences in biogeography, fire histories, land use histories, and life history requirements (including fire tolerances and dependencies) of species over long time lines and large landscapes (e.g., Scott et al., 2014; see Chapters 3–5).

In addition, climate change introduces uncertainty in how forests will respond to changes in fire extent, longer fire seasons, and higher severities in places, how soon the current fire deficit in places will remain that way before exceeding historical bounds, and whether existing deficits will be exacerbated in some forests with increasing precipitation driven by climate change (see Chapter 9). Nonetheless, at least for mixed-severity fire systems there is no magic thinning or suppression bullet to forestall climate-mediated fire changes. Changes in fire behavior are a consequence of human-caused climate change. It is best to treat the cause—climate change—rather than the symptom (fire behavior) if we are truly concerned about climate effects on ecosystems and people.

13.8 CLOSING REMARKS

When viewing the natural world, as a matter of perspective, we are reminded of discussions we have often had with foresters regarding how we each see the value of postfire landscapes. Clearly, we see the world differently depending on our professional judgment and value system.

A professional forester views the fruits of his or her labor, imagining what the future "production" forest will look like after decades of growing wood fiber, and then being frustrated by nature run amuck when the forest goes up in flames.

For the fire-trained ecologist, the initiating fire is but a glimpse into a vibrant community that begins with a pulse of biological activity and ensures successional events, just one of the many important links to follow in a long chain of ecosystem changes. Even the most charred forest is transformed by fire on one of nature's grandest stages. Among the first actors to arrive on the postfire stage are the biological legacies that provide the supporting foundation for other postfire actors to enter with the passage of time. If we imagine what the stage will look like years after a severe burn (often only 1 year), we see a floral phoenix arising from the ashes, we hear a cacophony of songbirds and drumming woodpeckers, and the rhythmic buzzing of bees and other insects as they go about their business of pollinating the next explosion of flowering plants. Up close and personal, we see tiny native beetle larvae tucked neatly into galleries beneath the outer charred tree bark, wood-boring scorpion wasps recoiling long abdomens after depositing eggs into open crevices in tree bark, centipedes and millipedes working charred humus, and ravenous insect-loving bats and fly-catching birds feasting on all the buzz.

The postfire landscape is indeed a transformative place if we humans are willing to have the patience to look beyond the brief snapshot in time right after the initiating event. Only then will the postfire esthetic become apparent. Our human world of instant gratification pales in comparison to nature's seemingly infinite horizon. Meticulous observations by trained ecologists too often are drowned out by the noise of a fast-paced society preoccupied with one-size-fits-all solutions, impulses to do something at any cost, myopic economic benefits, and a fear-based media blitz of fire catastrophe reporting. But if we wait for the ecosystem actors to emerge in synchronicity, the postfire habitat unveiled is remarkably resilient, brilliant like the mythical phoenix, and even musical if we know how to listen. We hope that we have sufficiently portrayed an ecological awareness for this postfire symphony in the chapters of this book.

In this closing chapter we also have discussed the importance of education and outreach for a communications framework and improved ecological understanding of fire that follows fundamental ecological and safety principles.

From a communications standpoint, fire operates very much like an apex predator, thinning out and culling its prey, sometimes in large numbers, sometimes not. Apex predators are indeed vital to fully functioning ecosystems, yet they are either loved or hated based on one's perspective, which simply boils down to either an appreciation for wild things or a fear of being attacked or of losing a commodity. People view fire in much the same way. Decades of public outreach and campaigns in many places (most notably Europe and North America) have shifted public opinion to be more accepting of predators, and even to relish them in national parks and other protected landscapes where predators roam free and tourists flock to witness nature primeval. Clearly, fires, like apex predators, cannot be restricted to inside national parks, as the parks are not big enough to sustain them.

There is a lesson to be learned regarding the message of fear in both instances: As with predators, the risks of losses to people and property can be successfully mitigated by taking precautionary measures (e.g., just don't feed the bears, and remember to make loud noises while hiking in grizzly bear country!). In the case of fire, public safety of those living in firesheds is based on prudent fire risk reduction that with stepped-up outreach one day may become common knowledge. With a shift in this direction, we envision a move toward fire tolerance, and eventually coexistence, so that fire, in all its severities and forms, can continue to shape ecosystems into the next millennium. This will take a concerted effort of sophisticated and sustained message framing, an infusion of funds for stepped-up education that at least rivals predator-friendly campaigns, a commitment from land management agencies and the media to become more ecologically literate (including replacing Smokey Bear with nature's phoenix), conservation groups to see the value in mixed-severity and not just low-severity fire, and politicians to see the big picture that the postfire landscape has irreplaceable ecological value and is not just a money tree to be ravaged for short-term profit. Then nature's phoenix will truly take flight, reborn out of the ashes of a postfire landscape mosaic that is alive and well!

REFERENCES

Agee, J.K., Skinner, C.N., 2005. Basic principles of forest fuel reduction treatments. For. Ecol. Manag. 211, 83–96.

Alexander, M.E., 1982. Calculating and interpreting forest fire intensities. Can. J. Bot. 60, 349–357.

Alexander, M.E., Cruz, M.G., 2013. Are the applications of wildland fire behaviour models getting ahead of their evaluation again? Environ. Model. Softw. 41, 65–71.

Baker, W.L., 2009. Fire Ecology in Rocky Mountain Landscapes. Island Press, Washington, D.C.

Baker, W.L., 2012. Implications of spatially extensive historical data from surveys for restoring dry forests of Oregon's eastern Cascades. Ecosphere 3, 23.

Baker, W.L., 2014. Historical forest structure and fire in Sierran mixed-conifer forests reconstructed from General Land Office survey data. Ecosphere 5, 79.

Baker, W.L., Williams, M.A., 2015. Bet-hedging dry-forest resilience to climate-change threats in the western USA based on historical forest structure. Frontiers in Ecology and Evolution 2 (88), 1–7.

Bogusch, P., Blažej, L., Trýzna, M., Heneberg, P., 2015. Forgotten role of fires in Central European forests: critical importance of early post-fire successional stages for bees and wasps (Hymenoptera: Aculeata). Eur. J. For. Res. 134, 153–166.

Bradstock, R.A., Bedward, M., Gill, A.M., Cohn, J.S., 2005. Which mosaic? A landscape ecological approach for evaluating interactions between fire regimes, habitat and animals. Wildl. Res. 32, 409–423.

Calkin, D.E., Cohen, J.D., Finney, M.A., Thompson, M.P., 2013. How fire risk management can prevent future wildfire disasters in the wildland-urban interface. Proc. Natl. Acad. Sci. U. S. A. 111, 746–751.

Clarke, M.F., 2008. Catering for the needs of fauna in fire management: science or just wishful thinking? Wildl. Res. 35, 385–394.

Cohen, J.D., 2000. Preventing disaster: home ignitability in the wildland-urban interface. J. For. 98, 15–21.

Cohen, J.D., 2004. Relating flame radiation to home ignition using modeling and experimental crown fires. Can. J. For. Resour. 34, 1616–1626.

Collins, B.M., Roller, G.B., 2013. Early forest dynamics in stand-replacing fire patches in the northern Sierra Nevada, California, USA. Landsc. Ecol. 28, 1801–1813.

Cruz, M.G., Alexander, M.E., 2010. Assessing crown fire potential in coniferous forests of western North America: a critique of current approaches and recent simulation studies. Int. J. Wildland Fire 19, 377–398.

Cruz, M.G., Alexander, M.E., Dam, J.E., 2014. Using modeled surface and crown fire behavior characteristics to evaluate fuel treatment effectiveness: a caution. For. Sci. 60, 1000–1004.

DellaSala, D.A., Williams, J., Deacon-Williams, C., Franklin, J.R., 2004. Beyond smoke and mirrors: a synthesis of forest science and policy. Cons. Bio. 18, 976–986.

DellaSala, D.A., Bond, M.L., Hanson, C.T., Hutto, R.L., Odion, D.C., 2014. Complex early seral forests of the Sierra Nevada: what are they and how can they be managed for ecological integrity? Nat. Area J. 34, 310–324.

DELWP, 2015a. Monitoring, Evaluation and Reporting Framework for Bushfire Management on Public Land. The State of Victoria Department of Environment, Land, Water and Planning, Melbourne, Australia.

DELWP, 2015b. Strategic Bushfire Management Plan East Central Bushfire Risk Landscape. The State of Victoria Department of Environment, Land, Water and Planning, Melbourne, Australia.

Driscoll, D.A., Lindenmayer, D.B., Bennett, A.F., Bode, M., Bradstock, R.A., Cary, G.J., Clarke, M.F., Dexter, N., Fensham, R., Friend, G., Gill, M., James, S., Kay, G., Keith, D.A., Macgregor, C., Russell-Smith, J., Salt, D., Watson, J.E.M., Williams, R.J., York, A., 2010. Fire management for biodiversity conservation: key research questions and our capacity to answer them. Biol. Conserv. 143, 1928–1939.

Druzin, H., Barker, R., 2008. Fire Wise Series Part One: Are We Wasting Billions Fighting Wildfires? Idaho Statesman, Boise, Idaho (July 20, 2008).

Foote, E.I.D., 1996. Structural survival on the 1990 Santa Barbara "Paint" fire: a retrospective study of urban-wildland interface fire hazard mitigation factors. Master's thesis, University of California at Berkeley.

Gibbons, P., van Bommel, L., Gill, A.M., Cary, G.J., Driscoll, D.A., Bradstock, R.A., Knight, E., Moritz, M.A., Stephens, S.L., Lindenmayer, D.B., 2012. Land management practices associated with house loss in wildfires. PLoS One 7, e29212.

Hanson, C.T., 2014. Conservation concerns for Sierra Nevada birds associated with high-severity fire. Western Birds 45, 204–212.

Hanson, C.T., Odion, D.C., DellaSala, D.A., Baker, W.L., 2009. Overestimation of fire risk in the Northern spotted owl recovery plan. Conserv. Biol. 23, 1314–1319.

Hanson, C.T., Odion, D.C., Historical forest conditions within the range of the Pacific Fisher and Spotted Owl in the central and southern Sierra Nevada, California, USA. Natural Areas Journal (in press).

Haslem, A., Kelly, L.T., Nimmo, D.G., Watson, S.J., Kenny, S.A., Taylor, R.S., Avitabile, S.C., Callister, K.E., Spence-Bailey, L.M., Clarke, M.F., Bennett, A.F., 2011. Habitat or fuel? Implications of long-term, post-fire dynamics for the development of key resources for fauna and fire. J. Appl. Ecol. 48, 247–256.

Headwaters Economics, 2014. Reducing wildfire risks to communities. Solutions for controlling the pace, scale, and pattern of future development in the wildland-urban interface. http://headwaterseconomics.org/wphw/wp-content/uploads/paper-reducing-wildfire-risk.pdf; (Accessed 22.02.2015.).

Howard, R.A., Warner, D.W., Offensend, F.L., Smart, C.N., 1973. Decision Analysis of Fire Protection Strategy for the Santa Monica Mountains: An Initial Assessment. Stanford Research Institute, Menlo Park, CA, 159 pp.

Kauffman, J.B., 2004. Death rides the forest: perceptions of fire, land use, and ecological restoration of western forests. Conserv. Biol. 18, 878–882.

Kelly, L.T., A.F. Bennett, M.F. Clarke, and M.A. McCarthy. 2015. Optimal fire histories for biodiversity conservation. Conservation Biology 29(2), 473–481.

Lydersen, J.M., North, M.P., Collins, B.M., 2014. Severity of an uncharacteristically large wildfire, the Rim Fire, in forests with relatively restored fire regimes. For. Ecol. Manag. 328, 326–334.

Maranghides, A., Mell, W., 2009. A Case Study of a Community Affected by the Witch and Guejito Fires. National Institute of Standards and Technology Technical Note 1635. U.S. Department of Commerce.

Martin, R.E., Sapsis, D.B., 1991. Fires as agents of biodiversity: pyrodiversity promotes biodiversity. In: Proceedings of the Symposium on Biodiversity of Northwest California, October 28-30, 1991. Santa Rosa, CA.

Mclennan, B.J., Handmer, J., 2012. Reframing responsibility-sharing for bushfire risk management in Australia after Black Saturday. Environ. Haz. 11, 1–15.

Metz, D., Weigel, L., 2008. Key public research findings on the ecological role of fire. Prepared by FMM&A and POS for Partners in Fire Education. http://www.wildlandfire.com/docs/2008/fed/08SummaryFireAccptnc.pdf (accessed 26.11.14).

Miller, J.D., Skinner, C.N., Safford, H.D., Knapp, E.E., Ramirez, C.M., 2012. Trends and causes of severity, size, and number of fires in northwestern California, USA. Ecol. Appl. 22, 184–203.

Moritz, M.A., Odion, D.C., 2004. Prescribed fire and natural disturbance. Science 306, 1680.

Moritz, M.A., Batllori, E., Bradstock, R.A., Gill, A.M., Handmer, J., Hessburg, P.F., Leonard, J., McCaffrey, S., Odion, D.C., Schoennagel, T., Syphard, A.D., 2014. Learning to coexist with wildfire. Nature 515, 58–66.

North, M.P. (Ed.), 2012. Managing Sierra Nevada forests. U.S. Forest Service, General Technical Report PSW-GTR-237. Pacific Southwest Research Station, Albany, California.

North, M., Stine, P., O'Hara, K., Zielinski, W., Stephens, S., North, M., Stine, P., Hara, K., Zielinski, W., Stephens, S., 2009. An ecosystem management strategy for Sierran mixed-conifer forests. U.S. Forest Service General Technical Report PSW-GTR-220, Pacific Southwest Research Station, Albany, California.

Odion, D.C., Hanson, C.T., 2006. Fire severity in conifer forests of the Sierra Nevada, California. Ecosystems 9, 1177–1189.

Odion, D.C., Hanson, C.T., 2008. Fire severity in the Sierra Nevada revisited: conclusions robust to further analysis. Ecosystems 11, 12–15.

Odion, D.C., Hanson, C.T., 2013. Projecting impacts of fire management on a biodiversity indicator in the Sierra Nevada and Cascades, USA: the Black-backed Woodpecker. Open For. Sci. J. 6, 14–23.

Odion, D.C., Sarr, D.A., 2007. Managing disturbance regimes to maintain biological diversity in forested ecosystems of the Pacific Northwest. For. Ecol. Manage. 246, 57–65.

Odion, D.C., Strittholt, J.R., Jiang, H., Frost, E., DellaSala, D.A., Moritz, M., 2004. Fire severity patterns and forest management in the Klamath National Forest, northwest California, USA. Conserv. Biol. 18, 927–936.

Odion, D.C., Moritz, M.A., DellaSala, D.A., 2010. Alternative community states maintained by fire in the Klamath Mountains, USA. J. Ecol. 98, 96–105.

Odion, D.C., Hanson, C.T., Arsenault, A., Baker, W.L., DellaSala, D.A., Hutto, R.L., Klenner, W., Moritz, M.A., Sherriff, R.L., Veblen, T.T., Williams, M.A., 2014a. Examining historical and current mixed-severity fire regimes in ponderosa pine and mixed-conifer forests of western North America. PLoS One 9, e87852.

Odion, D.C., Hanson, C.T., DellaSala, D.A., Baker, W.L., Bond, M.L., 2014b. Effects of fire and commercial thinning on future habitat of the Northern Spotted Owl. Open Ecol. J. 7, 37–51.

Pardon, L.G., Brook, B.W., Griffiths, A.D., Braithwaite, R.W., 2003. Determinants of survival for the northern brown bandicoot under a landscape-scale fire experiment. J. Anim. Ecol. 72, 106–115.

Parr, C.L., Andersen, A.N., 2006. Patch mosaic burning for biodiversity conservation: a critique of the pyrodiversity paradigm. Conserv. Biol. 20, 1610–1619.

Pastro, L.A., Dickman, C.R., Letnic, M., 2011. Burning for biodiversity or burning biodiversity? Prescribed burn vs. wildfire impacts on plants, lizards, and mammals. Ecol. Appl. 21, 3238–3253.

Peterson, D.W., Dodson, E.K., Harrod, R.J., 2015. Post-fire logging reduces surface woody fuels up to four decades following wildfire. For. Ecol. Manage. 338, 84–91.

Reinhardt, E.D., Keane, R.E., Calkin, D.E., Cohen, J.D., 2008. Objectives and considerations for wildland fuel treatment in forested ecosystems of the interior western United States. For. Ecol. Manag. 256, 1997–2006.

Rogers, G., Hann, W., Martin, C., Nicolet, T., Pence, M., 2008. Fuel Treatment Effects on Fire Behavior, Suppression Effectiveness, and Structure Ignition. Grass Valley Fire. San Bernardino National Forest. USDA R5-TP-026a.

Scott, A.C., Bowman, D.M.J.S., Bond, W.J., Pyne, S.J., Alexander, M.E., 2014. Fire on Earth: An Introduction. John Wiley and Sons Ltd, Hoboken, New Jersey.

Sherriff, R.L., Platt, R.V., Veblen, T.T., Schoennagel, T.L., Gartner, M.H., 2014. Historical, observed, and modeled wildfire severity in montane forests of the Colorado Front Range. PLoS One 9, e106971.

Sitters, H., Christie, F.J., Di Stefano, J., Swan, M., Penman, T., Collins, P.C., York, A., 2014. Avian responses to the diversity and configuration of fire age classes and vegetation types across a rainfall gradient. For. Ecol. Manag. 318, 13–20.

Steel, Z.L., Safford, H.D., Viers, J.H., 2015. The fire frequency-severity relationship and the legacy of fire suppression in California's forests. Ecosphere 6, 8.

Stine, P., Hessburg, P., Spies, T., Kramer, M., Fettig, C., Hansen, A., Lehmkuhl, J., O'Hara, K., Polivka, K., Singleton, P., Charnley, S., Merschel, A., White, R., 2014. The ecology and management of moist mixed-conifer forests in eastern Oregon and Washington: a synthesis of the relevant biophysical science and implications for future land management. U.S. Forest Service, General Technical Report PNW-GTR-897, Pacific Northwest Research Station, Portland, Oregon.

Swanson, M.E., Franklin, J.F., Beschta, R.L., Crisafulli, C.M., DellaSala, D.A., Hutto, R.L., Lindenmayer, D.B., Swanson, F.J., 2011. The forgotten stage of forest succession: early-successional ecosystems on forested sites. Front. Ecol. Environ. 9, 117–125.

Syphard, A.D., Keeley, J.E., Massada, A.B., Brennan, T.J., Radeloff, V.C., 2012. Housing arrangement and location determine the likelihood of housing loss due to wildfire. PLoS One 7, e33954.

Syphard, A.D., Brennan, T.J., Keeley, J.E., 2014. The role of defensible space for residential structure protection during wildfires. Int. J. Wildland Fire 23, 1165–1175.

Taylor, R.S., Watson, S.J., Nimmo, D.G., Kelly, L.T., Bennett, A.F., Clarke, M.F., 2012. Landscape-scale effects of fire on bird assemblages: does pyrodiversity beget biodiversity? Divers. Distrib. 18, 519–529.

Teague, B., Mcleod, R., Pascoe, S., 2010. Final Report, 2009 Victorian Bushfires Royal Commission. Parliament of Victoria, Melbourne Victoria, Australia.

van Wagtendonk, J.W., van Wagtendonk, K.A., Thode, A.E., 2012. Factors associated with the severity of intersecting fires in Yosemite National Park, California, USA. Fire Ecol. 8, 11–32.

Whittaker, J., Haynes, K., Handmer, J., Mclennan, B.J., 2013. Community safety during the 2009 Australian 'Black Saturday' bushfires: an analysis of household preparedness and response. Int. J. Wildland Fire 22, 841–849.

Index

Note: Page numbers followed by *f* indicate figures, *b* indicate boxes and *t* indicate tables.

C

D

E

I

J

K

L

M

N

O

P

T

U

V

W

Y

Z